FJORD OCEANOGRAPHY

NATO CONFERENCE SERIES

I   Ecology
II   Systems Science
III   Human Factors
IV   Marine Sciences
V   Air—Sea Interactions
VI   Materials Science

IV MARINE SCIENCES

FJORD OCEANOGRAPHY

Edited by
Howard J. Freeland
David M. Farmer
Institute of Ocean Sciences
Sidney, British Columbia, Canada

and
Colin D. Levings
Fisheries and Oceans Canada
West Vancouver Laboratory
West Vancouver, British Columbia, Canada

Published in coordination with NATO Scientific Affairs Division by
PLENUM PRESS · NEW YORK AND LONDON

Library of Congress Cataloging in Publication Data

Nato Conference on Fjord Oceanography, Victoria, B. C., 1979.
    Fjord oceanography.

    (NATO conference series: IV, Marine sciences; v. 4)
    Includes index.
    1. Fjords—Congresses. 2. Oceanography—Congresses. 3. Marine ecology—Congresses.
I. Freeland, Howard J. II. Farmer, David M. III. Levings, Colin D. IV. North Atlantic
Treaty Organization. Division of Scientific Affairs. V. Nato Special Program Panel on
Marine Science. VI. Title. VII. Series.
GC28.N37 1979   551.46'09   80-12273
ISBN 0-306-40439-7

Proceedings of the NATO Conference on Fjord Oceanography,
held in Victoria, British Columbia, Canada, June 4–8, 1979, and
sponsored by the NATO Special Program Panel on Marine Sciences.

© 1980 Plenum Press, New York
A Division of Plenum Publishing Corporation
227 West 17th Street, New York, N.Y. 10011

Printed in the United States of America

PREFACE

Fjords are deep, glacially carved estuaries that are peculiar to certain coastlines, and have several characteristics that distinguish them from shallower embayments. At higher latitudes they indent the western coastlines of Scandinavia, North and South America, and New Zealand. They are also a common feature of much of the arctic coastline.

The papers contained in this volume were presented at a workshop funded by the NATO Advanced Studies Institute in Victoria, British Columbia. It may seem curious to the reader that this special class of estuaries should have attracted an international gathering of oceanographers from several different disciplines. The reason for this interest stems from both practical and scientific considerations. On the one hand, fjords are a feature common to the coastlines of several countries that depend heavily on the oceans for communication, fisheries and other resources. The impact of man's activities on these coasts has created a demand for new knowledge of the physical, biological and chemical aspects of fjords. Sometimes man's influence on the ocean is intentional as, for example, in the artificial control of ice cover; often it is the more insidious build-up of toxic wastes that is of concern. These problems are particularly acute where the conflicting demands of fisheries, industrial development and recreation meet in a single fjord; and indeed, this is a common occurence along several of the fjords in Scandinavia and Canada. The need here is for a deeper understanding of nature's resilience, of the long term effects of pollution and of the flushing ability of the estuary.

On the other hand, there are also scientific aspects having an intrinsic interest that command attention. As a peculiar oceanographic feature fjords have been studied both from a descriptive, or synoptic, point of view and also as a convenient place to study specific processes. Important contributions to this volume include the description of a wide range of geographically and biologically distinct fjords. Thus we identify particular T-S distributions characteristic of high latitude fjords containing icebergs, or particular biological and chemical features of fjords which are

infequently flushed below sill depth. The broad range of scientific
approaches to the study of fjords makes classification difficult,
and different oceanographers would probably each seek different ways
of grouping them. But this very diversity of fjord types emphasizes
the other aspect that makes them peculiarly attractive to oceanographers: a place in which to study specific processes.

Fjords exhibit many oceanographic processes occuring over a
wide range of space and time scales. At the high frequency end of
the spectrum, for example, tidal flows can generate turbulence,
internal waves and eddies, while processes in deep enclosed basins
may have time scales of tens of years. Being semi-enclosed bodies
of water they can be studied with less trouble than the deep ocean;
moreover the boundary conditions, for example the river discharge
and tidal fluxes and nutrient input, can often be determined with
accuracy. This makes them very convenient for scientific experiments
as will be evident from many of the papers in this volume. Fjords
that have deep, infrequently ventilated basins offer biologists
and chemists an opportunity to study the progressive shift towards
anoxia. Likewise, the progressive change in sedimentation, and in
the interaction between sediment and adjacent waters, occur over a
time scale that is long enough to study the biological and chemical
aspects in detail. At intermediate depths, problems associated with
internal tides and the exchanges over the sill, and similar processes, together with corresponding biological phenomena can be
studied, and there are many papers in this volume to indicate the
rich diversity of topics of interest. In strongly stratified fjords
with high runoff near surface zones, the top 5 to 20 metres can
itself represent a whole class of fascinating problems ranging from
wind mixing to frontogenesis.

The study of fjords has a long history, most of the early work
having been carried out by Scandinavian scientists. More recently
oceanographers in other parts of the world, notably the west coast
of Canada and the United States and New Zealand, have started to
study these features and it seemed appropriate that this workshop
be held in Canada. An intensive study of one inlet has recently
been undertaken by several scientists and a number of papers in
this volume are devoted to this work. The workshop has also
provided us with an opportunity to discuss new techniques of measurement; techniques of particular interest include the use of
radioactive tracers that are valuable in looking at long term
movements of water over considerable distances. A second technique,
that has proven very revealing in the Knight Inlet studies, has
been the use of high frequency acoustic sounding. It is perhaps
appropriate to note here that biologists have used echo sounding
to study zooplankton for many years, but it is only recently that
physical oceanographers have taken to using the same equipment to
look at the physical processes with the biological scatterers acting
as the acoustic tracer.

The papers are arranged under general subject headings; however, it is inevitable that many of the papers might equally well have been placed in alternative sections. Indeed, in organising the workshop it was our intention to encourage contributions of an interdisciplinary nature. Many of the reports are only brief summaries, but it was felt of greater value to present a number of short accounts together with some longer invited papers rather than to curtail the number of contributions too severely. This approach has the advantage of providing a rich source of information on current research in many areas, while still retaining the more detailed reviews presented by our invited speakers.

The meeting in Victoria provided a valuable opportunity for exchanges of information and ideas both from within and between the various disciplines represented, and for a detailed discussion of new ideas and questions. We hope that this volume reflects the proceedings of the workshop. We are particularly grateful to the NATO Science Committee for their generous support of the workshop.

Howard J. Freeland

David M. Farmer

and Colin D. Levings

CONTENTS

CIRCULATION STUDIES

PELAGIC ECOLOGY

FJORD CHEMISTRY

PACIFIC FJORDS - A REVIEW OF THEIR WATER CHARACTERISTICS

G.L. Pickard and B.R. Stanton*

Department of Oceanography
University of British Columbia
2075 Wesbrook Mall
Vancouver, B.C. Canada V6T 1W5

INTRODUCTION

Fjord Regions

The regions with which this review is concerned are the deep
fjord regions of Alaska and British Columbia in the northern
hemisphere and Chile and New Zealand in the southern. The relative
locations in latitude are shown in Fig. 1. The latitude zones for
the Pacific fjords are: North America from 50^0 to 60^0N, South
America from 40^0 to 55^0S and New Zealand from 45^0 to 47^0S. (The
Atlantic fjord regions are generally at higher latitudes, i.e.
approximately 50^0 to 75^0N for eastern Canada, 64^0 to 75^0N for
Greenland and 60^0 to 72^0N for Norway.)

I have limited myself to the regions with which I have personal
experience and therefore left to others the description of the
Puget Sound region which forms the southern extremity of the western
North American fjord region. Also, many detailed studies of the
Alaskan fjords have been made by the University of Alaska and the
present review covers only the broad characteristics.

Figs. 2 to 5 show the regions in more detail than in Fig. 1.
Only the major fjords etc and those referred to in this text are
named in these figures.

*D.S.I.R., New Zealand Oceanographic Institute, P.O. Box 12-346,
Wellington North, New Zealand.

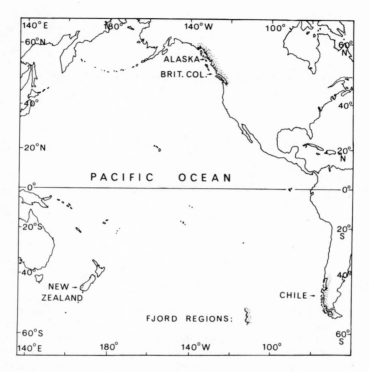

Figure 1. Fjord regions of the Pacific.

Although the western North American fjords are in one continuous geographic region this is an extensive one, nearly 1500 km long, and it is convenient for description to use the political divisions of Alaska and British Columbia. Also, in this review the Canadian fjords will be divided into the British Columbia (mainland) fjords and the Vancouver Island ones. In this case there are some physical differences. For Chile, the terms north, middle and south zones will be used when there is significant oceanographic difference between the groups of fjords.

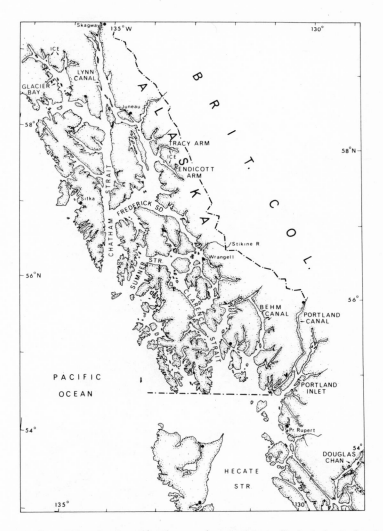

Figure 2. Fjords in Alaska and northern British Columbia.

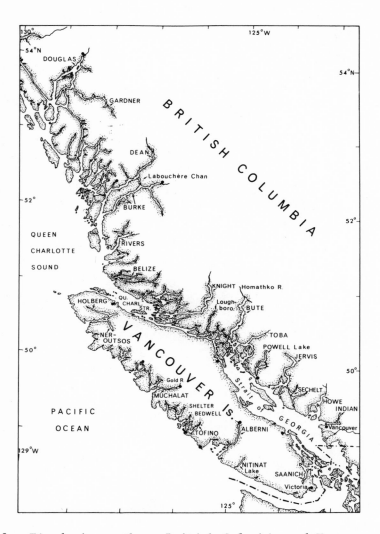

Figure 3.   Fjords in southern British Columbia and Vancouver Island.

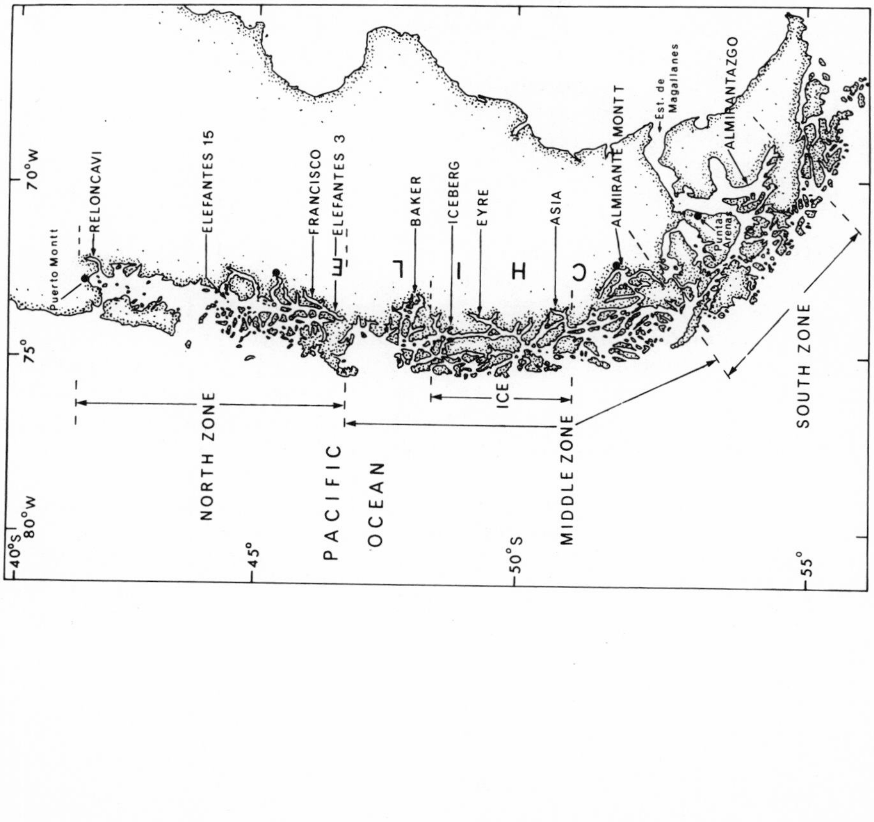

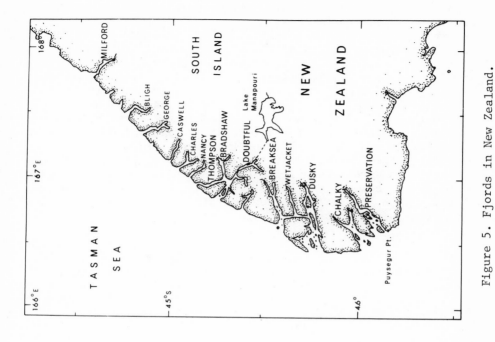

Figure 5. Fjords in New Zealand.

Figure 4. Fjords in Chile.

Many of the fjords have large rivers discharging into them, usually at the head, and all have some fresh water inflow. They are estuaries according to the definition of Cameron and Pritchard (1963) and an 'estuarine circulation' is presumed to occur. In this circulation, the fresh water from rivers ('runoff') flows seaward in the upper layer, entraining salt water from below and carrying it out of the fjord. In consequence, to maintain the average salt content of the fjord there must be an inflow of saline water from the sea and this more dense water must enter below the outflowing low salinity layer. Direct measurements in a few of the North American fjords have confirmed this circulation pattern, although the periodic effects of tidal flow and aperiodic modification due to wind stress complicate the pattern over short time-periods. Also, the mean outflow of the upper layer is evident in many fjords by the drift of flotsam and silt, while changes which are consistent with an estuarine circulation occur in the deeper waters.

Some of the fjords in Alaska and Chile have glaciers coming down to the sea and icebergs from them drift along the fjords. These are referred to as 'iceberg' fjords as some of the water characteristics in them are different from those in neighbouring fjords without icebergs.

References to earlier papers

This paper is intended as a review and a comparison of the regions and item by item references are not given. A list of the earlier papers in which the regions are described individually is given in the Bibliography and it will be evident in which of these papers more detail will be found. The description of the New Zealand fjords is being prepared, the original data being on file with the New Zealand Oceanographic Institute.

Data available

The data used are mainly those gathered by the Institute (now Department) of Oceanography of the University of British Columbia (UBC) for the American fjords, with some data for British Columbia from Fisheries Research Board of Canada establishments, and those gathered by the New Zealand Oceanographic Institute for New Zealand. The quantity of data available is very much biassed toward British Columbia. The UBC data for the British Columbia mainland (41 fjords and many neighbouring passages) have been collected for 28 years since the first survey cruise in 1951. A few fjords have been visited only once, many have been visited several times and some fairly extensive time-series of observations have been made in some of the southern fjords. For Vancouver Island (17 fjords) UBC data are available for several years and for Alaska (15 fjords) for

three years. For Chile (32 fjords) the UBC observations made in
March 1970 as part of the 'Hudson 70' cruise around the Americas
are believed to be the only ones available for that region.   The
New Zealand Oceanographic Institute visited all the South Island
fjords (14) in 1977 and earlier data are available for some of them.
Statements in this review about the South Pacific fjords are made
on the assumption that the limited data are typical.

     Because the data for Alaska, Chile and New Zealand are for the
summer period, data for the same season have been used for British
Columbia when comparing regions.   Comments on seasonal variations
of characteristics are limited to the data available for British
Columbia, for some of the southern fjords.

General topography of the regions

     The Pacific fjord regions are in the Coast Mountains of Alaska
and British Columbia, the southern Andes of Chile and the Fiordland
mountains of the South Island of New Zealand, all relatively young
mountains.   The fjords extend along coastline lengths of about 550
km for Alaska, 850 km for British Columbia, 300 km for Vancouver
Island, 1500 km for Chile and 230 km for New Zealand.   Elevations
of the mountains surrounding the fjords are from 2000 to 4000 m
along Alaska/British Columbia, to 2200 m in Vancouver Island, to
3500 m in Chile and to 2760 m in New Zealand.   The fjords themselves
are largely the result of earlier glacial gouging and they extend
from the coastline to as much as 150 km inland.

     The prevailing westerly air flow to all the coasts brings
copious precipitation, generally as rain to lower elevations and as
snow to higher ones.   Extensive snowfields cover the higher mountain
areas from north of 60^0N to the southern limits of the North American
fjords, south of 49^0N, and there are glaciers from north of the
Alaskan fjords to about 55.5^0N and in British Columbia from about
52^0N to 49.5^0N.   There are permanent snowfields on the mountains
of Vancouver Island but no glaciers.   In Chile, the major glaciers
on the mainland are from 46.5^0 to 51.5^0S with smaller ones from
42^0 to 44.5^0S, 46^0 to 46.5^0S and 51.5^0 to 53.3^0S and there are
fairly extensive glaciers from 53.7^0 to 55.2^0S on the islands south
of Magellan Strait.   In New Zealand there is a small glacier contributing to one river into Milford Sound but only snowfields into
the remaining fjords.

     One other topographic feature has a significant influence on
the water characteristics of the fjords - the degree of access to
the open ocean outside, the source of their saline water.   In
Alaska, most of the shorter fjords open into larger ones which then
open to the ocean.   A major exception is the Lynn Canal/Chatham
Strait system which is long (380 km) but generally deep and with

a deep sill before the ocean. In the northern half of British
Columbia, the fjords open on to the 100-150 km wide Hecate Strait and
Queen Charlotte Sound, much of which is only 20 to 100 m deep,
before connecting to the ocean. In the southern half of British
Columbia, the fjords open on to Queen Charlotte Strait or the Strait
of Georgia, with mean depths of 50 to 100 m. In Chile, all of the
mainland fjords open on to passages between the mainland and outer
archipelagoes for the middle and south zones and on to a relatively
shallow inland sea or strait in the north. In contrast, for
Vancouver Island and for New Zealand the fjords all open directly
to the ocean over sills of only tens of metres deep but over a
continental shelf of a few kilometres width.

    The fjord areas are largely uninhabited. In Alaska only two
of the fjords have established communities, (with three in neighbouring passages), in British Columbia five, in Vancouver Island four,
in Chile two, and in New Zealand two have small tourist resorts.

Morphology of the fjords

    The fjords are generally long, narrow and deep, with one or
more sills, although these are usually deeper than the halocline and
rarely prevent sub-surface exchange with outside waters. Sharp
bends (e.g. 90^0) are common, and in the North American and New
Zealand fjords, interconnections between fjords occur in many cases.
An example is the case of Dean and Burke Channels (Fig.3) running
roughly perpendicular to the coastline with Labouchere Channel
connecting them at about mid-length and lying roughly parallel to
the coast.

    In referring to the fjords, the term 'head' is used for the
inland end, where there is usually a major river bringing in fresh
water, and the term 'mouth' for the seaward end. The position of
the head is usually clearly defined but the mouth is often less so,
particularly if the fjord leads into an archipelago, and sometimes
an arbitrary choice has to be made for the location of the mouth.
Some of the smaller fjords are tributary to larger ones, so that
their mouth does not open either to the ocean or to an inland sea.

    In length, the fjords range from a few kilometres to nearly
400 km and in mean width from 0.6 to 15 km. The dot diagrams of
Figs. 6a, b show the lengths and widths of individual fjords in
the five regions. 90% of the lengths lie in the range 10 to 200 km
and 90% of the widths from 1 to 6 km. Length to mean width ratios
range from 3 to 70 with 90% being in the 5 to 30 range as shown in
Fig. 6c.

    In Fig. 7 are shown the individual sill, mean and maximum
depths at lowest normal tides. The effective depths will be greater

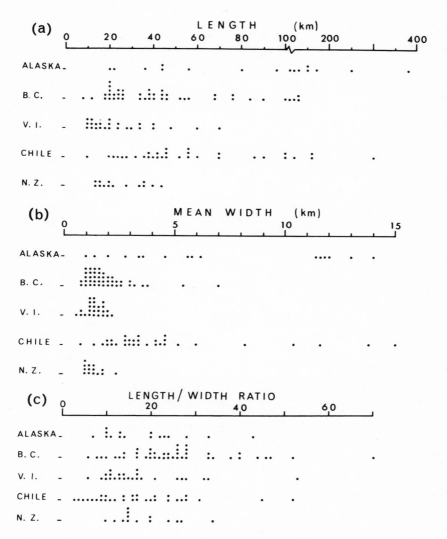

Figure 6.   Length, mean width and length/width ratios of Pacific
             fjords.

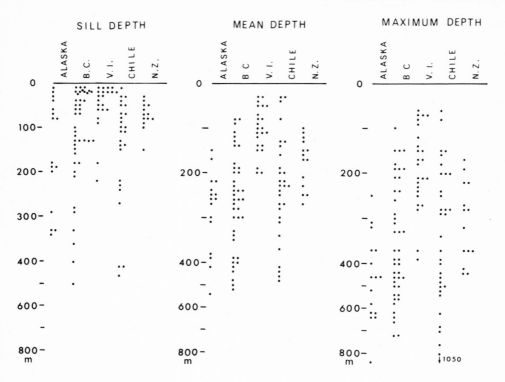

Figure 7.   Sill, mean and maximum depths of Pacific fjords.

at higher stages of the tide; typical values for the tidal range
are given in a later section on the tides. The mainland American
fjords generally have greater sill depths than the Vancouver Island
and New Zealand ones, and it is noted that very shallow entrance
sills are uncommon. (The definition adopted here for a 'very
shallow' sill is that it should be shallow enough to hinder inflow
of saline water from outside, i.e. that its depth should be equal
to or less than the thickness of the low salinity upper layer which
is usually less than 10 m).

Climate

The fjord regions being very sparsely occupied, the number of
climate recording stations is small. In particular, almost all of
the stations are close to sea-level and there is very little information for higher elevations which are important in connection with
freshwater runoff.

Air temperature

In all regions there is a seasonal cycle with a maximum in the local summer and a minimum in the local winter; sample annual temperature cycle surves are shown in Fig. 8 (full lines). The annual mean value for Alaska/British Columbia decreases from about 10^0C at 49^0N to 5^0C at 58^0N, for Chile it decreases from 10^0C at 41^0S to 5^0C at 55^0S and for New Zealand it is about 10^0C. The annual ranges for Alaska/British Columbia are from 10^0 to 20^0C, for Chile from 3^0 to 5^0C and for New Zealand 9^0C.

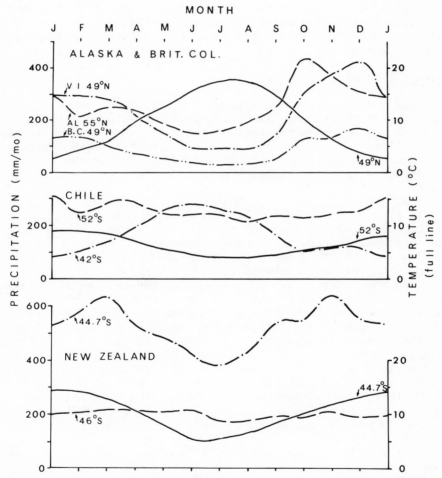

Figure 8.   Typical mean annual precipitation (dashed line) and air temperature (full line) patterns for Pacific fjord regions.

Precipitation Patterns

        Precipitation also shows an annual cycle and sample curves
are given in Fig. 8 (dashed lines). For Alaska/British Columbia
there is a summer minimum in all cases and a winter maximum which
sometimes divides into two maxima, fall and late winter. In
Chile, a winter maximum is most common but in the south there
is little variation through the year with only a slight summer
maximum. For New Zealand, the pattern at the head of Milford is
of a spring and fall maximum with a total annual precipitation of
6200 mm, more than twice that for Alaska/British Columbia and
Chile regions. The pattern at two other New Zealand stations in
mid-coast is almost identical but at Puysegur Point at the southern
extremity of the fjord coast the annual variation is small. This
station is on a headland at the coast and presumably orographic
effects are minimal.

Freshwater input

        This is probably the most important factor affecting water
properties in the fjords. Contributions are as precipitation on
the water surface, a minor factor, and river runoff which is the
major factor. This has two components, direct rainfall drainage
into the rivers and fjord, and stored runoff in the form of winter
snow in the mountains which melts in the summer and contributes
to the rivers.

        In Alaska and British Columbia, some of the rivers are or
have been metered, although not always near their mouth so that the
data often underestimate the runoff into the fjords. The large
rivers which come from the higher mountains show a peak runoff from
snow melt in June or July with values as high as 3400 m^3/s. Typical
long-term mean monthly values are smaller and samples are shown in
Fig. 9. The first two (Stikine and Homathko Rivers) show the
summer maximum while the Vancouver Island one (Gold River) shows a
double peak, spring and fall, corresponding to the seasonal precipitation cycle and suggesting that direct drainage to the rivers is
the major factor. The only river metered in New Zealand is the
Cleddau River into Milford and a 10-year mean flow is shown in
Fig. 9. The annual mean runoff is about 24 m^3/s. In addition to
rivers, there is the freshwater discharge from Lake Manapouri
through a power station into Doubtful. The measured flow for 1977
is shown in Fig. 9 the mean value being 368 m^3/s.

        Expressed as the thickness of a layer of water spread over the
whole fjord area, the annual runoff total for moderate to high
runoff fjords varies from 2 to 10 m for Alaska and 10 to 40 m for
British Columbia except for Portland Inlet for which it is about
100 m. For New Zealand, the Cleddau River annual discharge

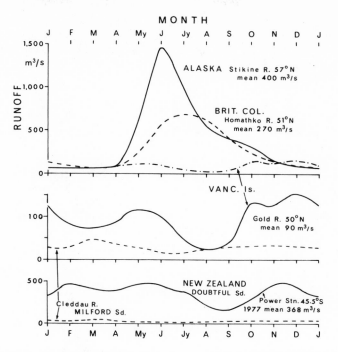

Figure 9.   Typical annual river runoff patterns into some Pacific fjords.

corresponds to a layer of 30 m thickness while the Lake Manapouri input into Doubtful corresponds to a thickness of 90 m of fresh water over the whole of the interconnected Doubtful/Thomson system or 200 m if limited to Doubtful only, large values compared with those for the North American fjords.

To make some bulk comparisons between the fjord regions (lacking any runoff data for Chile) the measured watershed areas were multiplied by the means of such annual precipitation figures as were available and divided by the coastline length to obtain the data in Table 1, a 'specific freshwater input' in cubic metres per kilometre of coastline. It seems that Alaska/British Columbia and Chile have the same specific freshwater input while Vancouver Island and New Zealand have smaller values by factors of four and three respectively. This is only a very crude estimate of the possible freshwater input because of the very limited precipitation data and the fact that values are only available for sealevel stations.

Precipitation values are likely to be significantly higher at higher
elevations; this would probably increase the ratios of the Alaska/
British Columbia and Chile values to the Vancouver Island and
New Zealand values.

Table 1

Specific freshwater input to coasts.

Region	Coast length	Watershed mean width	Precipitation annual	Specific annual freshwater input
Alaska & B.C.	1500 km	220 km	2000 mm	4.4×10^8 m^3/km
Vanc. I.	280	60	2000	1.2×10^8
Chile	1500	180	2500	4.5×10^8
New Zeal.	230	25	6000	1.5×10^8

Glacier meltwater runoff

     In Alaska, all of the large fjords and most of the small ones
visited receive glacier runoff, made evident by the 'rock flour'
in suspension. In British Columbia, about 40% of the fjords receive
it but none in Vancouver Island. In Chile about 40% have rock flour
in suspension in the upper layers while in New Zealand none of them
do.

Glacier ice

     Four of the sixteen fjords visited in Alaska had ice from
glaciers in the water in 1964 and 1965, and in Chile 12 of the 32
fjords had glacier ice toward the heads of fjords in 1970.
According to the U.S.N.O.O. Sailing Directions for South America,
Vol. 11, glacier ice often extends further down the fjords in
winter and may also be encountered in the inside passage between
the mainland and the outer islands.

Tides

     In Alaska and British Columbia, the tides are semi-diurnal with
marked diurnal inequality. In Alaska, the diurnal range (mean
higher high water to mean lower low water) is 5 m, in British

Columbia it decreases from 5 m in the north to 3.5 m in the south. In Chile and New Zealand, the tides are semi-diurnal with some diurnal inequality, with a diurnal range in Chile from 5.5 m in the north to 1.5 m in the south, while in New Zealand the range is about 2 m for all the fjord coast.

In Alaska and British Columbia, the times of high and low water and the range of tide are substantially the same at the head as at the mouth of each fjord (and presumably the same is true in Chile and New Zealand) as their natural periods of oscillation are substantially less than the semi-diurnal tidal component periods.

WATER CHARACTERISTICS

General ranges of water properties

Tables 2 and 3 present a general overview of the ranges of temperature, salinity and dissolved oxygen values found during spring and summer in the fjords and in the oceans outside, while Figs. 10 and 11 present relationships between properties in the form of mean characteristic diagrams and property variations.

Distributions of surface values of temperature, salinity, dissolved oxygen and turbidity

Table 2 gives values for the surface (upper metre) for stations near the head and mouth of the fjords and in the oceans outside. The numbers in the columns show the percentage of cases for which the values lie between the column limits. The ranges of values in the fjords are seen to be greater than in the ocean. Lower fjord water temperatures than in the ocean are generally due to cool run-off, higher temperatures than in the ocean possibly due to less vertical wind-mixing of solar heated upper water in low-runoff fjords than in the ocean. The range of salinity values is obviously a consequence of freshwater runoff diluting saline water from the ocean and having its maximum effect near the heads of the fjords because most large rivers enter there.

Dissolved oxygen values for the surface are of doubtful value as a water characteristic or tracer because these waters are often supersaturated, due to photosynthetic processes, though some values are given later in the descriptions of the individual regions.

One characteristic of the surface waters which does present some systematic features is the water turbidity as represented by Secchi disc depth measurements, and a major factor in causing high turbidity (low transparency), especially near the fjord head, is the silt or glacial rock flour brought down by the major river. Table 3 shows ranges of values and mean values. For the North American fjords there are sufficient data to present values for

Table 2

Percentage frequency of occurrence of temperature and salinity values in the surface layer.

TEMPERATURE ^{0}C — Head

Region	0	5	10	15	20
Alaska		62	38		
Brit.Col.		12	62	26	
Vanc. Is.		6	70	24	
Chile	13	38	44	7	
New Zealand			62	38	

SALINITY $^0/_{00}$ — Head

Region	0	5	10	20	30	35
Alaska	50	10	20	20		
Brit.Col.	53	13	17	17		
Vanc. Is.	24	19	19	38		
Chile	21	18	11	46	4	
New Zealand	15	15	8	15	47	

TEMPERATURE — Mouth

Region	0	5	10	15	20
Alaska		8	84	8	
Brit. Col.		3	63	34	
Vanc. Is.			88	12	
Chile	6	32	59	3	
New Zealand			100		

SALINITY — Mouth

Region	0	5	10	20	30	35
Alaska			8	61	31	
Brit. Col.		8	39	47	6	
Vanc. Is.			7	73	20	
Chile	4	7	19	48	22	
New Zealand			15		85	

TEMPERATURE — Ocean

Region	0	5	10	15	20
Alaska			100		
Brit. Col.			100		
Vanc. Is.		10	80	10	
Chile			75	25	
New Zealand			100		

SALINITY — Ocean

Region	0	5	10	20	30	35
Alaska					100	
Brit. Col.					100	
Vanc. Is.					100	
Chile					100	
New Zealand					100	

the head and mouth separately. For Alaska, all the fjords are lumped together, the majority having significant runoff, generally silt or rock flour laden. For British Columbia, it is practical to separate into high and low runoff fjords. The very low Secchi disc values (<0.5 m) are found in the fjords receiving glacial runoff and are due to the rock flour brought down by the river. These very low values are found within a few kilometres of the head as the rock flour tends to aggregate and sink as the upper layer picks up salt water. The Secchi disc depth then increases along the fjord. The other factor reducing transparency sometimes in the outer half of the fjords is the occurrence of phytoplankton blooms in spring and summer. No systematic data are available on the blooms in these fjords.

Table 3

Secchi disc depths, in metres, by regions.

Regions	MOUTH			HEAD		
	Range of values		Mean	Range of values		Mean
Alaska	3	to 11	6	0.1 to 4		1.5
Brit. Col.-high runoff	2	to 10	4.5	0.1 to 2		0.7
" " - low runoff	3	to 11	6	2.5 to 10		6
Vanc. Is.	3.5 to 8		6	1.5 to 15		6

	Range		Mean
Chile - north	1	to 11	6
- middle	0.1 to 5		2
- south	1	to 9	5.5

For Vancouver Island there is no glacial runoff and only a small increase in turbidity toward the head (generally due to phytoplankton blooms, judging from the colour of the water). The anomalously high maximum value (15 m in Holberg Inlet) at the head than at the mouth is not an error. (It no longer occurs as there has been a massive dumping of mine tailings into the Rupert/Holberg system in recent years which has radically increased the turbidity.)

In Chile, about one-half of the stations were taken at night, with no systematic distribution between mouth and head of fjords, and so separation in terms of position along the fjord is not practicable. However, in terms of Secchi disc depth (as well as other characteristics) the coast can be divided into north, middle and south zones. The low Secchi disc values in the middle zone result from the fact that glacial water inflow with its rock flour is limited to the middle zone; the north and south zones are not significantly different.

Secchi disc depths were not measured in the New Zealand fjords but estimates from observations of the reversing water bottles suggest values of the order of 5 m or more, with no notable variation along the fjords. There is certainly no significant silt input from rivers.

Distributions of deep water values for temperature, salinity and
dissolved oxygen

Table 4 gives values at 200 m to represent the deep water in
the fjords and in the oceans outside. The 200 m depth was chosen
as a standard depth as this permitted most of the fjords to be
included in the tabulation. There are generally only small changes
beyond this depth.

The relative uniformity for temperature and salinity along the
lengths of fjords makes it unnecessary to present separate mouth
and head values for the deep water. Again somewhat greater ranges
of values are found in the fjords than in the ocean, partly because
of the river runoff effect on salinity but also because many of the
fjords have outer sills shallower than 200 m, so that ocean water
inflow will have properties somewhere between the ocean values in
Table 2 (surface) and in Table 4 (200 m depth).

For dissolved oxygen at 200 m, the range of values is much
larger in the fjords than in the ocean. There are several reasons
for this. The majority of the fjords have sills which are deeper
than the upper, low salinity layer but nevertheless do restrict
exchange with the ocean, and in some cases there are both outer
and inner sills. When there is little river runoff and therefore
only a weak estuarine circulation, the basin water is not exchanged
rapidly and oxygen depletion occurs. This is the case in many of
the Vancouver Island fjords where inner basins often have lower
dissolved oxygen values than the outer basins. In a few British
Columbia mainland fjords with low runoff, a mid-depth oxygen minimum
occurs toward the head of the fjord and some of the low values in
Table 4 are due to this phenomenon. Very low values of dissolved
oxygen are uncommon in the fjords.

The higher values in the fjords than in the ocean are, to some
extent, an artifact of the choice of 200 m to represent deep values.
In both the north and south Pacific waters outside the American
fjords, the dissolved oxygen decreases rapidly below 50 to 100 m.
Off Alaska/British Columbia it falls from about 7 mL/L in the upper
50 to 100 m to 3 mL/L at 200 m and to a minimum of less than 0.5
mL/L at 800 m; off the Chile fjords it falls from 6 mL/L for the
upper layer to a minimum of 2 mL/L at 200 m. Therefore the fjord
values greater than 4 mL/L probably indicate that significant
inflow takes place from the upper 50/100 m of the water column
outside.

The very small range of dissolved oxygen values in the main
basins of the New Zealand fjords suggest relatively free and continued exchange with the Tasman Sea which, in the south, has only a
small range of values compared with the east Pacific ranges.

Table 4

Percentage frequency of occurrence of temperature and salinity values at 200 m depth.

TEMPERATURE °C

Fjord

Region	2	4	6	8	10	12
Alaska	11	68	21			
Brit.Col.		14	66	20		
Vanc. Is.			33	67		
Chile			11	29		
New Zealand					100	

Ocean

Region	2	4	6	8	10	12
Alaska	50	50				
Brit. Col.		25	75			
Vanc. Is.		10	90			
Chile			25	75		
New Zealand					100	

SALINITY °/₀₀

Fjord

Region	28	30	32	34	36
Alaska		36	57	7	
Brit.Col.	9	36	55		
Vanc. Is.	19	31	50		
Chile	15	12	61	12	
New Zealand				100	

Ocean

Region	28	30	32	34	36
Alaska			100		
Brit. Col.			100		
Vanc. Is.			100		
Chile				100	
New Zealand				100	

DISSOLVED OXYGEN mL/L

Fjord

Region	0	1	2	3	4	5	6	7
Alaska		8	40	19	13	12	8	
Brit.Col.	3	10	39	30	14	3	1	
Vanc. Is.	22	13	15	18	22	5	5	
Chile		2	13	22	39	20	4	
New Zealand						100		

Ocean

Region	0	1	2	3	4	5	6	7
Alaska			60	40				
Brit. Col.			60	40				
Vanc. Is.				50	50			
Chile				50	50			
New Zealand						100		

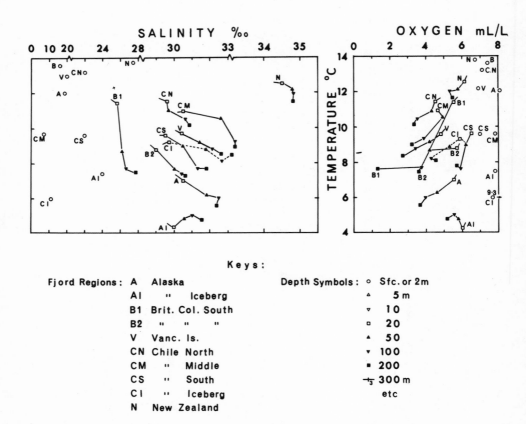

Figure 10.   Mean T,S and T,O₂ diagrams for Pacific fjord regions.

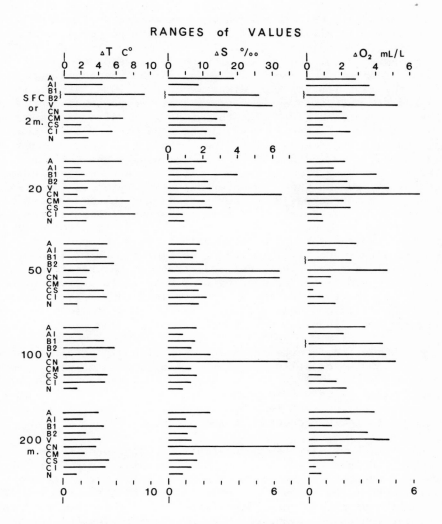

Figure 11.   Ranges of temperature, salinity and dissolved oxygen
values for Pacific fjord regions.

Mean characteristic diagrams

T,S and T,O_2 scatter diagrams were first prepared for all the
fjords in a region using mean values along each fjord at depths of 0–
2 m, 20, 50, 100 and 200 m and envelopes were drawn for each region.
The envelopes were quite large in many cases and overlapped
considerably when all the envelopes were drawn on one diagram,
so that it was confusing. To simplify it, mean values for the
three properties in each region at each depth were determined
and plotted on two characteristic diagrams and connected with lines
to form the mean T,S and T,O_2 diagrams of Fig. 10. This figure
then presents and compares mean aspects of the regional characteristics. The surface (or 2 m depth for oxygen) values are shown
but not connected to the 20 m values because many of the lines
cross and make the diagram confusing. (Note the scale changes
for salinity at 20, 28 and 33 $^0/_{00}$.) For Alaska and Chile the
iceberg fjords have significantly lower temperatures than those
without icebergs and they have been shown separately. For British
Columbia, the northern fjords (north of Knight Inlet at about 51^0N)
have properties very similar to the Alaska non-iceberg fjords and
have not been shown separately. For southern British Columbia
(Knight and south) there are two distinct T,S envelopes. The
lower salinity group (B.1) including smaller fjords with moderate
runoff, sills and relatively narrow entrances; the remainder (B.2)
include high runoff fjords opening over deep sills on to inside
passages or straits where the deep water is 2 to 3 $^0/_{00}$ less saline
than in the Pacific Ocean outside. The B.1 group is seen to have
the lowest mean dissolved oxygen values for all the fjords.

For Chile, the scatter diagrams show significant differences
along the 1500 km coastline and three main zones are identified
and shown separately in Fig. 10. The New Zealand fjords show the
smallest range of mean values on both diagrams, probably because
they occupy the shortest length of coastline of the four regions
and are most freely connected to the ocean.

Fig. 10 shows only the mean values of water properties
obtained from the T,S and T,O_2 scatter diagrams. In Fig. 11 are
shown bars representing the ranges of values for each regional
group. There is no overall systematic trend to these ranges. The
Chile North salinities show larger ranges than the other regions,
except for Vancouver Island at 50 m for which region the low values
which extend the range are for two fjords only, both having inner
sills.

Density profiles

Density, represented by σ_t = (density – 1000) kg m^{-3}, always
increases with depth (with two exceptions described in the following

section on salinity profiles). There may be a mixed layer of a
few metres depth from the surface, then a strong pycnocline to
10 to 20 m, followed by a reduced rate of increase to about 100 m
and a still slower icrease from there to the bottom. The density
usually reaches 90% of the deep water value by 10 - 15 m depth. In
Fig. 12 (top) are shown typical density, depth (σ_t,z) profiles
for the head, middle and mouth of a high runoff fjord in spring
(Bute Inlet, B.C.) with a well-defined mixed layer of 3 to 6 m
depth from head to middle of the fjord becoming less defined
toward the mouth as the stability in the pycnocline decreases and
permits increased vertical mixing. In low runoff fjords, the mixed
layer is generally absent and the pycnocline starts at the surface,
as for Loughborough Inlet in Fig. 12 (bottom).

     The ranges of density values are markedly different for the
different fjord regions. Fig 13 shows envelopes of σ_t,z profiles
for the five regions. The New Zealand fjords appear to be the most
freely connected with the outside ocean (Tasman Sea) and have the
smallest range of values. In fact, the curves for most of the fjords
lie in the high density half of the envelope area with the northern
fjords having the higher density and the main decrease occurring
south of Breaksea. Many of the Alaskan fjords have fairly direct
deep connection to the North Pacific and the range of σ_t values is
the next smallest of the regions, with the high values for the
envelope area (for Lynn Canal) being only slightly smaller than for
the Pacific outside. The British Columbia fjords show a larger range
of σ_t values because the southern ones have the enclosed Strait of
Georgia between them and the Pacific. While the northern ones are
less restricted in access to the Pacific, the connection is still
across a shallow coastal sea (Hecate Strait) and is therefore a less
direct connection than for the New Zealand and Alaska fjords. In
Chile, the south and middle zone fjords open on to an inside passage,
and thence through an archipelago to the Pacific. The low σ_t values
from 20 to 24 kg m^{-3} are from several inner fjords which open on
to Elefantes (fjord) which acts as an inside passage and partially
isolates these inner fjords from the ocean. (See also later section,
Characteristic Diagrams, and Figure 16.)

     The Vancouver Island fjords show a surprisingly large range
of σ_t values for relatively short fjords which open directly on
to the ocean. The range from 20 to 23 kg m^{-3} is due to a group
of three fjords at the south (Shelter, Bedwell and Tofino) but
otherwise the fjord σ_t,z profiles are scattered within the remainder
of the envelope area with no geographic zonation. The group of
three referred to above have relatively shallow sills and, being
short, probably do not have much tidal inflow to promote exchange
of their basin waters.

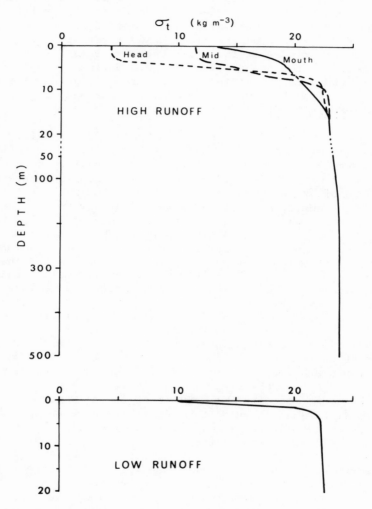

Figure 12. Typical density (σ_t), depth profiles for high and low
runoff fjords.

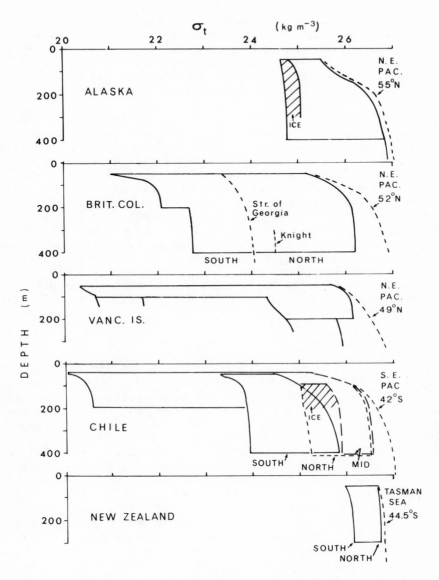

Figure 13. Envelopes of density, depth profiles for Pacific fjord regions.

Salinity profiles

     As density is determined much more by salinity than by
temperature in the fjords, particularly in the upper 100 to 200 m,
the salinity, depth (S,z) profiles are similar to the density
profiles whose general characteristics were described in the previous
section. As salinity is the measured quantity, whereas density is
derived from it, the details of the structure will be described in
terms of the salinity rather than density.

     Salinity-depth profiles in the fjords fall into one of two
types (see Fig. 14a), type 1 in which there is a surface mixed or
near-mixed layer below which there is a marked halocline, and type
2 in which the halocline starts at the surface, as nearly as can
be determined. (It should be noted that in routine surveys the
sampling bottles are rarely set closer than two metres so that a
mixed layer of less than this thickness will not be detected.)
However, in a number of locations in early studies in the British
Columbia fjords, salinity and temperature profiles were determined
in the upper 5 m or so using an Industrial Instruments (now Beckman)
salinometer whose conductivity sensor is about 10 cm in diameter and
which effectively measures conductivity, and hence salinity, with
a depth sensitivity of ±0.2 m or better. The surface mixed layer
was found to be less than 0.5 m thick in most of the type 2 salinity
profiles.

     The frequency of occurrence of the S,z types is given in Table 5
for the heads and mouths of fjords. The well-defined type 1.a
profile is most common at the head of high runoff fjords, and
mixing during passage along the fjord weakens the halocline toward
types 1.b and 1.c. Type 2 profiles occur generally in low runoff
fjords. In the same table are given some vertical dimensions for
the profiles. For type 1 profiles, the thickness of the surface
layer is given as the mean and the range of values observed. Also
the mean and range of depths at which the salinity reaches 50% and
90% of the deep water value in the same fjord are given. Small
values for these depths, for example, indicate a small content of
freshwater resident in the upper zone, while larger values indicate
a larger freshwater content. The Alaska and Chile fjords have the
greatest values while the New Zealand fjords have the smallest values
and little salinity variation below the top few metres.

     Almost invariably, the salinity increases with increase of
depth in all the American fjords, rapidly below the surface mixed
layer for 10 to 20 m and then more slowly. The two exceptions
(in 26 years of observations in British Columbia) occurred during
one cruise when salinities at 50 m in Bute of up to $0.05^0/_{00}$
greater than in the deep water and in Belize of up to $0.13^0/_{00}$
greater than in the deep water were observed. In the second case,

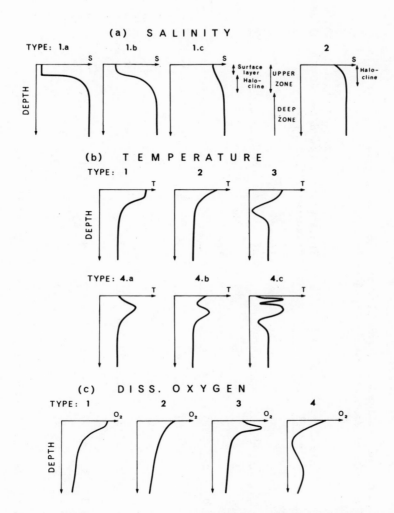

Figure 14.   Types of vertical profiles for temperature, salinity and dissolved oxygen in Pacific fjords.

Table 5

Percentage frequency of occurrence of salinity profile types (Fig.14), depth of mixed layer and depths for 50% and 90% of deep water salinity (depths in metres).

Region	Percentage of each profile type — Mouth 1a	1b	1c	Mouth 2	Head 1a	1b	1c	Head 2	Thickness of surface mixed layer at head † Mean	Min / Max	Depth at which salinity reaches stated percentage of deep water salinity 50% † Mouth	50% † Head	90% † Mouth	90% † Head
Alaska		38	62		30	40	10	20	5	2 / 12	0 (0–2)	2.5 (0–12)	N* 4 (0–8)	12 (2–22)
Brit. Col.	6	24	45		32	26	6	35	4	1 / 9	1.5 (0–8)	4 (0–10)	S*16 (14–17)	8 (0–20)
Vanc. Is.		69	31		13	50	6	31	2	1 / 5	0 (0–0)	1 (0–5)	2 (0–8)	9 (0–7)
Chile	3	30	37		4	42	12	42	8	1 / 45	<1 (0–6)	<2 (0–6)	14 (0–40)	14 (0–40)
New Zealand	31	38	31		8			92	3	2 / 4	<0.5 (0–3)	0.5 (0–2)	1 (0–4)	2 (0–4)

* N = northern fjords, S = southern fjords.

† Left number is mean value, right numbers are minimum and maximum values.

the salinities were quite irregular with depth at several stations
on the first day of observations and also on the second day when
several stations were repeated because the irregular salinity
profiles were considered so unusual. There was no reason to suspect
errors in technique. Corresponding temperature irregularities
compensated in part for the salinity when density was calculated.
It was suspected that large inflows of water from outside might
have taken place shortly before the observations and the water
bodies had not yet settled down to a stable density stratification.

In the New Zealand fjords, the northern group from Milford to
Doubtful show a salinity maximum at 50 to 100 m of 0.01 to $0.07^0/_{00}$
greater than in the deep water but the fjords south of Doubtful
do not show this.

Temperature profiles

It is less easy to categorize the temperature, depth (T,z)
profiles than those for density and salinity, probably because in
the upper 200 m or so of the water column temperature is a minor
factor in determining the stability which is controlled mainly by
the salinity distribution.

In many cases the T,z profile is, in shape, a mirror image of
the S,z profile, with a thermocline starting below a surface mixed
layer or from the surface itself. These T,z profiles are referred
to as types 1 and 2 (Fig. 14b), without subdividing the type 1
profiles as was done for salinity. However, a feature of many
T,z profiles is the occurrence of one or more minima or maxima
below the surface. These profiles are divided into types 3 and 4
with subdivisions for the last type (Fig. 14b).

In Table 6, the frequency of occurrence of the different
types is given. Types 1 and 2 occur in all the fjord regions,
with type 1 being most common for high runoff and type 2 for low
runoff fjords. The type 3 profile occurs in the North American
fjords, with the temperature minimum being most marked in the
larger mainland fjords, chiefly those in British Columbia. Here
the type 3 form is most conspicuous in spring or early summer and
it has been attributed to effects of winter cooling occasioned
by strong down-fjord flows of cold, dry continental arctic air
which mix and cool the upper tens of metres until they are colder
than the deep water (although less dense because of the lower
salinity). Subsequent warming, through the surface, of the upper
layer in spring and summer results in the temperature minimum which
is therefore a relic of the winter cooling. This process has been
documented during several time-series studies in southern fjords;
the temperature depression, in the minimum, below the deep water
temperature and the thickness of the cool layer have been shown

Table 6

Percentage frequency to occurrence of temperature profile types
as sketched in Fig. 14.

Temperature profile type

Region	1	2	3	4		
				a	b	c
Alaska	18	43	30	5	3	
Brit. Col.	35	40	16	5		3
Vanc. Is.	14	66	17	3		
Chile	11	30		14	14	32
New Zealand	27	60		13*		

*Thin cool surface layer - see text comment.

to be at least semi-quantitatively related to the magnitude of the
cooling process during the previous winter. It will be noted that
this T,z profile does not occur in the Chile or New Zealand fjords
where the cold, polar air outbreaks would not be expected.

    The complex T,z profiles of type 4 are most common in Chile.
They occur, in all but two cases, in the middle zone fjords into
which there is glacial melt-water inflow and often glacier ice in
the water. In this group of fjords there were three to seven minima
plus maxima (not counting the surface or bottom extrema) with an
average of four. The T,z profiles were obtained both with a mechanical bathythermograph and with Bisset-Berman Model 9060 STD,
verified by reversing thermometer measurements. (The latter
would have been quite inadequate to delineate these complex T,z
profiles as the maxima and minima were all located in the upper 50
to 75 m.) The most complex profiles occurred in fjords which had
glaciers coming down to sea-level and small icebergs in the water
at the time of observation. A few T,z profiles of type 4 have been
observed in the North American fjords but they are not common.

    The profiles in the New Zealand fjords are chiefly of types 1
and 2. A few which are technically of type 4a have only a very thin
(2 to 4 m) surface layer of cool water and could be basically of
type 2 but in which the surface layer has been cooled locally and
the type 4.a occurrence is only transitory.

    It must be remarked that the above categorization of T,z
profiles is a simplification and cannot be regarded as complete.
For instance, in some fjords the type 1 characteristic continues
from head to mouth (as for the S,z profile in high runoff fjords)

but in others a type 1 at the head changes to a type 2 toward the
mouth. For the larger fjords, where this progressive change is
most common, the frequency of occurrence of profile types was
calculated for Table 6 by giving equal weight to the inner and to
the outer half of the fjord, whereas the shorter fjords, in which
one type prevails, have been counted once only. For some of the
Vancouver Island fjords the T,z profile type changes from type 2
for most of the fjord to type 1 at the mouth but this change has
been disregarded as it appears to be simply due to the intrusion
of the upper mixed layer/thermocline structure characteristic
of the ocean outside.

Also, for the type 3 (mid-depth minimum) fjords, most have a
type 1 upper layer T,z profile but a few have type 2. For the type
4 profiles, the details of the upper 5 to 10 m T,z shape have been
disregarded.

Dissolved oxygen profiles

Again, the dissolved oxygen content profiles show a greater
range of shapes than do the density and salinity profiles.
Characteristic features are a decrease from values at or above
saturation at the surface to lower values in deep water, with a
mid-depth minimum in certain fjords.

The most common shapes for oxygen, depth (O_2,z) profiles are
similar to those shown (in smoothed form) in Fig. 14c. The profiles
of types 1 and 2 commonly occur in conjunction with S,z profiles
of mirror image form (Fig. 14a, types 1 and 2). However, in some
cases, O_2,z profiles of type 2 with the oxycline reaching to the
surface or 2 m (whichever was the shallowest sample depth for
oxygen) occur with type 1.a or 1.b S,z profiles with a surface
mixed layer (for salinity) of more than 2 m depth. Type 3 O_2,z
profiles are characterized by a strong oxygen maximum, often to
150% saturation, coincident with or slightly below the halocline,
i.e. at 5 to 20 m depth. These maxima occur with both type 1 and
type 2 salinity profiles. ('Saturation value' for dissolved oxygen
means the value appropriate to the temperature and salinity of the
water sample and at one atmosphere pressure.)

The type 4 O_2,z profile is most characteristic of a few fjords
with small runoff but it is observed occasionally in high runoff
fjords at times when runoff has been judged to be low. The minimum
occurs at 75 to 100 m depth and is most marked near the head of the
fjord.

Some indication of the frequency of occurrence of these profiles
is given in Table 7. It will be seen that the type 2 profile is the
most common overall. However, it must be noted that sometimes two

Table 7

Percentage frequency of occurrence of oxygen profile types as sketched in Fig. 14.

Oxygen profile type

Region	1	2	3	4
Alaska	6	91	3	(15)
Brit. Col.	20	53	27	(7)
Vanc. Is.	23	62	15	(8)
Chile	20	80		(17)
New Zealand		100		

or more types of profile are observed in a single fjord during the same cruise, or different types may be observed during different cruises. Also, the mid-depth minimum (type 4) is generally a sub-type of types 1 or 2, the frequencies of occurrence being calculated relative to the total numbers of type 1 + 2 + 3. (Minor mid-depth minima, i.e. less than 0.2 mL/L below that of the deep water, have been ignored.)

In the Alaska fjords, the 2 m depth values (the shallowest used for oxygen in routine sampling) are generally at or above saturation value (i.e. 8 to 6.6 mL/L for salinities from 0 to $30^0/_{00}$ and at 10^0C), decreasing to 2 to 4 mL/L in the deep water, most of the decrease taking place by 50 to 100 m depth. Exceptions are in the fjords with glacier ice where deep values are higher, i.e. to 6 mL/L in Glacier and 4 to 6 mL/L in Endicott and Tracy. At the head of Lynn Canal, where ice forms in some winters and deep mixing also occurs due to strong winds, values of 4 to 6 mL/L are observed. The lowest values are generally in the deep water toward the mouth of Lynn Canal/Chatham Strait where the values are close to North Pacific values of 2 mL/L at sill depth (350 m). (The North Pacific values off Alaska typically decrease to a minimum of less than 0.5 mL/L at about 800 m depth, see Fig. 17.)

In the British Columbia fjords, surface values at or above saturation are usual. Type 3 profiles with a strong maximum in the halocline depth zone are more common than in the other regions, although this observation may be an artifact of the much more extensive data base for British Columbia than for the other regions. The halocline maximum is most common in high runoff fjords but it is intermittent in occurrence; in some cruises in a particular fjord it is the most conspicuous feature of the profile while in another cruise there is no sign of it. It has been suggested that the maximum (always above 100% saturation) is related to photosynthesis by phytoplankton trapped at or below the halocline by the strong density gradient inhibiting upward movement. This feature remains to be investigated fully and explained.

The mid-depth minimum is again an intermittent feature. It occurs in low runoff fjords where the estuarine circulation may be expected to be weak and the density and density gradient therefore to be a little lower than in strong estuarine circulation fjords. The suggested reason is that biological oxygen demand takes place throughout the water column but a steady, though slow, inflow of water dense enough to sink to the bottom occurs and refreshes the bottom water, leaving an oxygen minimum at mid-depth. A time-series study in Sechelt Inlet demonstrated clearly the development of a minimum oxygen layer at the head of one arm of the fjord and its elimination when a major flushing of most of the fjord water occurred (revealed by major temperature and salinity distribution changes).

Oxygen values of less than 1 mL/L have been observed in 27% of British Columbia fjords and 38% of Vancouver Island ones, and of less than 0.5 mL/L in 10% of fjords in British Columbia and in 31% of Vancouver Island ones. These low values are always found near the bottom in inner basins, i.e. inside an inner sill, or at the fjord head in the three fjords which show the mid-depth minimum. The volume of water with these low values is estimated at only about 1% of the total fjord volume. Values of less than 1 mL/L have not been observed in Alaska or Chile and only in one small basin at the head of a fjord in New Zealand.

There are four special locations having low oxygen waters in British Columbia. Three of these are 'lakes' in name, although of fjord basin form. Nitinat Lake at the south end of Vancouver Island is probably a fjord and has depths of 200 m but has a very shallow entrance bar giving a sill depth of 2 m at low water, with a tidal rise of 4 m at higher high water. There is a strong halocline to about 20 m depth with a deep salinity of over $30^0/_{00}$ and the oxygen content is zero below 30 m. On the east side of the Strait of Georgia, just west of Sechelt, Saginaw Lake, which is of fjord shape and has a maximum depth of 140 m, has its surface level maintained at 3 to 4 m above mean sea level outside by a dam but it has a salinity of up to $11^0/_{00}$ in the deep water and the oxygen content becomes zero at about 30 m with H_2S present below this. A rise in temperature of 2 to 3 C occurs from below the halocline to the bottom. A third similar case is Powell Lake at 49.9^0 N on the east side of the Strait of Georgia. A dam here raises the lake level to 55 m above sea level and the basins have depths to 350 m. From the surface to 100 m the water is fresh but from there the salinity rises to about $17^0/_{00}$ at the bottom (by

silver nitrate titration but uncorrected for possible ion ratio
differences from sea water or for the H_2S present). The temperature
falls from the surface to about 5^0C at about 100 m and then rises
to 9^0C at the bottom. Dissolved oxygen decreases from the surface
to zero at about 180 m, below which H_2S is very evident in the water
and voluminous degassing occurs (mostly methane) when samples
are drawn from the reversing water bottles.

The fourth case is Saanich Inlet in southeastern Vancouver
Island, opening on to the Strait of Georgia, with depths to 235 m
and a sill depth of 55 m. In this basin, the oxygen content is
usually close to zero in the bottom 50 m or so but at intervals
this water is partially refreshed by inflow from the Strait of
Georgia. The occurrence of stagnation with such a deep sill is
attributed to the very small runoff now occurring into the basin
and consequent absence of significant estuarine circulation.

In the Vancouver Island fjords, the surface values are at or
above saturation with two exceptions. Near the head of Neroutsos
Inlet, values below 1 mL/L have been observed at 2 m depth, a
consequence of the oxygen demand of the low density effluent from
the sulphite pulp mill near the fjord head at the time. Values
at 2 m depth of 90% of saturation have also been observed at the
head of Alberni Inlet, again probably due to effluent demand.
Deep water values are often at 3 to 4 mL/L but inside inner sills
of some fjords values fall to less than 0.5 mL/L. Two of the
Vancouver Island fjords (low runoff types) show the mid-depth oxygen
minimum (type 4). It is possible that the reason for this is the
same as for the mainland fjords, such as Sechelt described above.

In Chile, the majority of the fjords show a sharp decrease
in oxygen content from the surface to about 50 m and then a slower
decrease with further increase of depth. For the north and middle
zones, values are from 6 to 9 mL/L at the surface, 3.5 to 5.5 mL/L
at 50 m and 2.5 to 4.5 mL/L in the deep water. For the south zone,
the respective values are 6 to 11 mL/L, 6 mL/L and 5 to 6.5 mL/L.
The higher values in the deep water in the south zone than in the
north and middle ones are to be noted. In the north and south zones
there is usually a small minimum at 75 to 200 m depth (0.2 to
1 mL/L less than in the deep water).

In the New Zealand fjords the salient characteristic is that
although the oxygen content decreases with depth, the decrease is
much less than in the other fjord regions, i.e. from a mean of 6.6
mL/L at the surface to 5.3 mL/L near the bottom. Most of the
decrease is in the upper 50 to 100 m and the profiles are not always
smooth but there are no obvious common structural features which
would take the profiles out of the type 2 category.

Characteristic diagrams

The T,S and T,O_2 diagrams of Fig. 10 give an idea of the general character of the characteristic diagram shapes for the various fjord regions. However, as these are for mean values, most of the detail is smoothed out. Some of the characteristic features of individual diagrams are now described, using actual curves for specific stations to illustrate the features.

Some of the distinguishing features of the Alaska and British Columbia fjords are shown in Fig. 15. Curves (a) in this figure are for a moderate to low runoff fjord in which temperature and oxygen decrease and salinity increases monotonically from surface to bottom (type 2 profile). Curves (b) show the low salinity surface layer and the characteristic mid-depth temperature minimum which occurs in many mainland high runoff fjords (type 1 profiles for salinity, type 3 for temperature and type 4 for oxygen). The temperature minimum is most conspicuous after a cold winter and it is sometimes accompanied by the mid-depth oxygen minimum and maximum shown in this figure. Both the temperature minimum and the oxygen maximum are attributed to vertical mixing to the surface in winter. Curves (c) show the marked mid-depth oxygen minimum which occurs in a few low runoff fjords; curve (c.1) is for a station near the fjord head and (c.2) for one nearer the mouth. The long-dash curve on the upper T,O_2 diagram shows the oxygen saturation values as a function of temperature at one atmosphere pressure in sea water (salinity $30^0/_{00}$) and in fresh water ($0^0/_{00}$).

In Fig. 16, part of the reason for the large range of salinity values for the north zone of Chile (Fig. 11) is shown with T,S curves (d.1 to d.4) for a series of stations from the mouth of Elefantes (d.4) where the water is quite saline, to the head (d.1) where the values are quite low, a distance of 300 km. The curves (e) for the nearest ocean station are also shown.

Curves (f) show two examples of T,S curves for fjords which have several temperature extrema. Curve (f.1) is for a station near the fjord head and curve (f.2) for one toward the mouth, showing that the complex temperature structure is most marked at the fjord head where there is most glacier ice in the water. The relatively high values for dissolved oxygen compared to the examples above for non-iceberg fjords are to be noted.

Finally Fig. 17 (curve g) shows a feature which occurs regularly only in the New Zealand fjords, a slight salinity maximum in mid-depth, rather than at the bottom as in all the other regions. It is presumably related to the salinity maximum observed in the Tasman Sea outside (Fig. 17, curve j.1). The southern fjords do not show this feature. The small range of oxygen values is also characteristic of the New Zealand fjords.

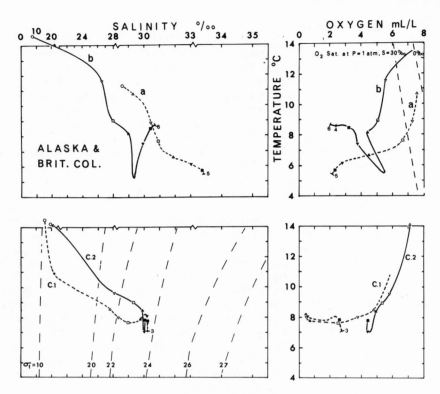

Figure 15.   Sample T,S and T,O$_2$ diagrams for Alaska and British
             Columbia.   Curve (a) low runoff fjord (Clarence, Alaska),
             (b) high runoff fjord (Bute, B.C.), (c) mid-depth oxygen
             minimum in low runoff fjord (Belize, B.C.); c.1 - near
             head, c.2 - near mouth. (Note that the salinity scale
             changes at 20, 28 and 33^0/$_{00}$.   Selected isopycnals and
             dissolved oxygen saturation curves at one atmosphere
             pressure are shown.   See Fig. 10 for depth symbol key.)

        It will be appreciated that the figures described only show some
of the salient features of the fjord characteristic diagrams.   Many
variations on these themes, together with fine structure features,
occur in different fjords, occur along individual fjords and occur
from year to year in the same fjord (at least in the British Columbia
ones for which most data are available).

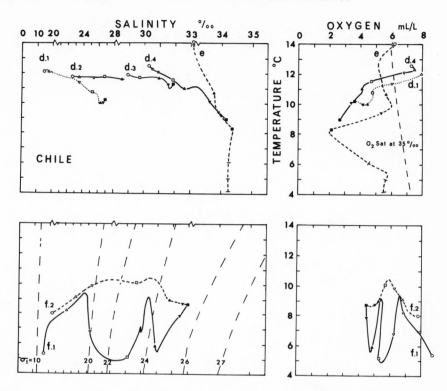

Figure 16.   Sample T,S and T,O₂ diagrams for Chile. Curves (d) for stations in Elefantes from head (d.1) to near mouth (d.4) (e) Pacific Ocean off mouth of Elefantes, (f) iceberg fjord (Asia), (f.1) near head, (f.2) near mouth. See note to Fig. 15 legend.)

The oceans outside the three main regions each have their own characteristics which are shown in Fig. 17 (lower). The features which distinguish these regions are the small temperature minimum in the north Pacific (h), the salinity maximum in the south Pacific, more marked in the Tasman Sea (j.1) than in the eastern South Pacific (i), the marked oxygen minima in the eastern Pacific at about 800 m in the north (h) and at 200 m in the south Pacific (i) off the fjord regions and the small range of oxygen values in the Tasman Sea. The T,S curve j.1 for the Tasman Sea was obtained during the 1977 study of the fjords but the T,O₂ curve j.2 is for January 1961 as oxygen values were not observed offshore in 1977. The oxygen minimum values are lower off Chile north of the fjord region.

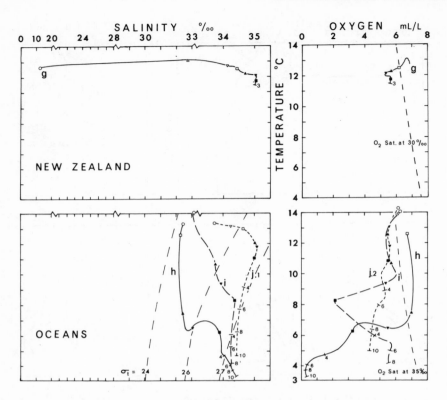

Figure 17.   Sample T,S and T,O$_2$ diagrams for New Zealand and for
             ocean stations off the fjord regions. Curve (g) middle
             of Doubtful, (h) north-east Pacific Ocean off Alaska/
             British Columbia, 49.7^{0}N, (i) southeast Pacific Ocean
             off Chile, 53.75^{0}S, (j.1) Tasman Sea off south-east
             New Zealand, 45^{0}S, 1977, (j.2) Tasman Sea off south-
             east New Zealand, 45^{0}S, 1961. (See note to Fig. 15
             legend.)

Longitudinal vertical sections

     These have not been presented systematically because in most
cases they show little that is not seen as well or better in profiles
and characteristic diagrams.

Longitudinal sections are only useful to display properties whose values change significantly along the fjord, and a characteristic of most of the fjords studies is that although there may be marked longitudinal changes in the upper layer (of 20 m or so) there is generally little longitudinal variation below this at any particular depth in each basin. As density always, and salinity almost always, increases monotonically with depth, the longitudinal profiles of these properties consist of substantially horizontal isopleths whose only feature is increased spacing with depth, and the related vertical gradients are shown more clearly on profiles than on sections.

For temperature, a longitudinal profile in some cases is useful in displaying features. For the mid-depth temperature minimum which occurs in the North American fjords, the series of longitudinal sections for Bute Inlet for August 1972 to August 1973 in Fig. 18 shows that final decline in August to October 1972 of the 1971-72 winter formed minimum, the cooling through the surface in December 1972 to February 1973 and the subsequent warming through the surface which leaves the temperature minimum at about 60 m in March 1973 and causes its gradual attrition through the following summer.

For oxygen, the section of Belize Inlet in Fig. 19 shows an example of the mid-depth oxygen minimum which develops from time to time in some British Columbia fjords.

TEMPORAL VARIATIONS OF WATER PROPERTIES

Although time-series information in the U.B.C. data set is only available for British Columbia, and therefore no comparison between Pacific regions can be made, it is worthwhile to describe briefly what is known of the temporal variations of water properties in British Columbia. (The time sequence giving rise to the mid-depth temperature minimum in British Columbia fjords has already been described in the section on temperature profiles.

Temperature, salinity and dissolved oxygen

In Fig. 20 are shown temperature, salinity and dissolved oxygen plots against time for three years for the surface (2 or 5 m for oxygen) and 10 m depth to represent the surface layer for Knight, a high runoff fjord, and for Jervis, a low runoff fjord. Curves are presented for a station near the fjord head (dashed line) and for one near the mouth (full line) in each case. At the surface and 10 m the annual cycles are of basically the same form for both fjords with temperature maxima in summer and minima in winter, following the head input cycle, and salinity minima in summer and maxima in winter, following the runoff cycle. The annual range of salinity at the surface is greatest at the fjord head near the

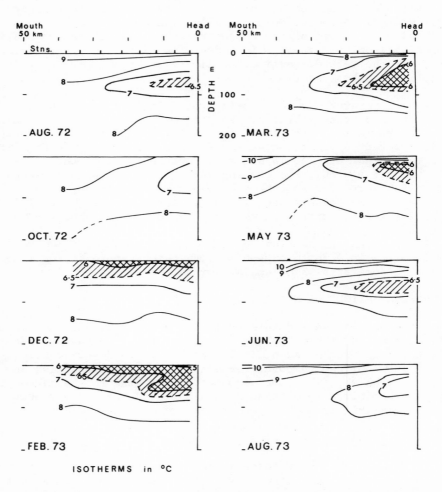

ISOTHERMS in °C

Figure 18.   Longitudinal T,z sections for Bute, British Columbia
             to show the sequence of development and decay of the mid-
             depth temperature minimum.

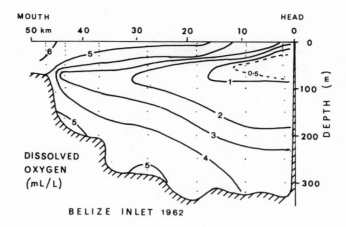

Figure 19. Longitudinal O_2, z section for Belize, British Columbia
to show the mid-depth oxygen minimum.

main river input and is much greater for Knight than for Jervis,
but at 10 m depth there is little difference between the two fjords
and little difference between mouth and head. For temperature,
both the ranges of values and the maxima are smaller for Knight
than for Jervis because of the large volumes of cold river water
flowing into Knight in Summer.

    The density curves follow the salinity curves because salinity
changes predominate in changing density. (A typical seasonal
change of salinity of $+20^0/_{00}$ changes sigma-t by about $+15$ kg m^{-3}
whereas a typical temperature change of $+10\,^0C$ changes sigma-t by
only -1.3 kg m^{-3}.)

    The rapid decrease in seasonal variations with increasing
depth, particularly for salinity, is a feature of the temporal
variations. For Bute, a high runoff fjord situated between Knight
and Jervis, the variations of temperature and salinity for 1972–
74 are similar to those for Knight. At 50 m (not shown) the
temperature traces for the head of Bute are 1 to 2^0C lower than at
the mouth because of the mid-depth temperature minimum (near the

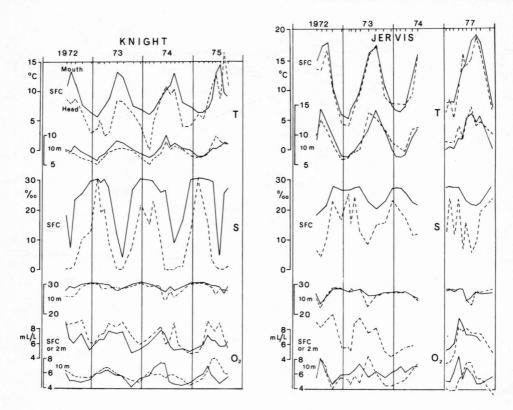

Figure 20.   Time series of values of temperature, salinity and
             dissolved oxygen in the upper layer for Knight (high
             runoff) and Jervis (low runoff), British Columbia, for
             surface or near surface and 10 m depth for head (dashed
             line) and mouth (full line).

head), which is usually more apparent in Bute than in Knight.

    Dissolved oxygen also shows an annual cycle, more clearly in
Knight (and Bute) than in Jervis, with maxima in the spring.

    In the deep water there are essentially two types of annual
variation. In one, the properties vary with time in a quasi-
sinusoidal fashion with some indication of a phase lag with increase
of depth for temperature. The other type is of saw-tooth form with
a gradual change of property values for most of the year and then
a sudden change, usually in the winter, corresponding to a major
flushing event.

The annual variations of properties decrease in magnitude with increase of depth but are generally evident to 100 m depth and usually deeper. Fig. 21 shows property variations of the first type for three years for Knight, a high runoff fjord, in which it will be seen than an annual cycle is evident to 300 m and possibly to 500 m. Series for Bute (high runoff) and for Jervis (low runoff) show less clear evidence of annual cycles to 300 m for temperature and dissolved oxygen; for salinity, while marked variations take place, a regular cycle is not clear. A study of Bute indicates that inflows at mid-depth probably take place several times a year without any apparent regular annual cycle.

The longest series of observations in the U.B.C. data set are for Indian Arm, a small, low runoff fjord which connects to the Strait of Georgia through Vancouver Harbour across two shallow sills. The two time-series are for 1956-1963 with an average of 9 observations per year and for 1968-1975 with an average of 13 observations per year. This fjord shows the second type of variation and Fig.22 presents property variations for the second series. The large changes in temperature and salinity in 1968-69 and 1970-71 and for oxygen in 1974-75 are conspicuous, while smaller jumps in properties occurred in all winters except possibly 1973-74. The earlier series shows very similar patterns of change.

Turbidity and attenuation coefficient

For three southern British Columbia fjords there are more Secchi disc data than for the others and some winter measurements (December to March) are available to compare with the spring and summer ones (May to July) described earlier. Table 8 compares these values. Bute and Knight have large runoff from glaciers and snowfields while Jervis has no glacial runoff. The difference between the two types of fjord is marked during the summer (high runoff) period but less so in the winter when the glacial contribution is reduced to zero.

The water colour in the fjords is generally white or dirty white ('silty') from the silt and rock flour for Secchi disc depths of less than 1 m, is 'silty-green' from 1 to 5 m and green or blue-green for depths over 5 m.

Some measurements as a function of depth have been made by a light scattering method on samples obtained by water sampling bottles. Table 9 shows values for the attenuation coefficient at a wavelength of 5460 nm for Bute (high runoff) and for Jervis (low runoff). The winter values are not significantly different between the fjords but Bute shows larger values in the upper layers in summer during the high river runoff period. For comparison, the attenuation coefficient at the same wavelength for pure water is

Table 8

Seasonal variations of Secchi disc depths for high runoff fjords
(Bute/Knight, glacial component) and for low runoff fjord (Jervis,
no glacial component).

	MOUTH		HEAD	
Season/Region				
Spring/summer	Range of values (m)	Mean (m)	Range of values (m)	Mean (m)
Bute/Knight	2 to 10	5	0.1 to 0.7	0.4
Jervis	3 to 9	6	4 to 10	5
Winter				
Bute/Knight	4 to 17	11	1 to 7	6.5
Jervis	7 to 17	12	7 to 13	11.5

Table 9

Attenuation coefficient (m^{-1}) at 5460 nm.

	Winter				Summer			
Fjord	Upper		Deep	Bottom	Upper		Deep	Bottom
	Mouth	Head			Mouth	Head		
Bute	0.2	0.7	0.12	0.3	6	36	0.6	1.5
Jervis	0.15	0.9	0.1	0.2	2	12	0.8	1.5

(cf. Sargasso Sea ~ 0.08 m^{-1}, Caribbean ~ 0.2 m^{-1},
Baltic Sea ~ 0.8 m^{-1})

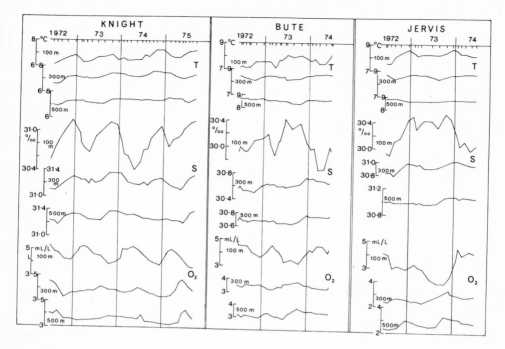

Figure 21.   Time-series of values of temperature, salinity and
             dissolved oxygen in the deep water for Knight and Bute
             (high runoff) and Jervis (low runoff), British Columbia.

about 0.06 m^{-1}, for the Sargasso Sea about 0.08 m^{-1}, for the
Caribbean Sea about 0.2 m^{-1} and for the Baltic Sea about 0.8 m^{-1}.

WAVES

Surface Waves

Only visual estimates have been made. These indicate that swell
does not occur in the fjords except near the mouth of Chatham
Sound in Alaska and at the mouths of the New Zealand fjords. Table
10 shows the frequency of occurrence of wind speeds (anemometer
measurement) and similtaneous wave heights (visual estimate) for
Alaska and British Columbia combined (20 years of spring and summer
data) and for Chile (one late summer cruise).

The waves are locally generated by the wind which generally
blows along the fjord, usually toward the head during the summer,
and is stronger near the mouth than near the head. (In the winter,
in British Columbia, strong down-fjord (outflow) winds occur
frequently when the arctic front moves south over the coast.)

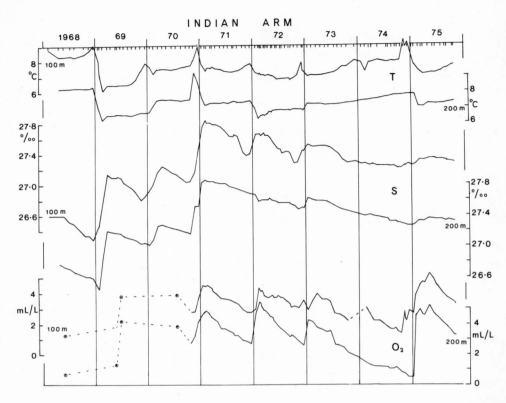

Figure 22.   Time series of values of temperature, salinity and dissolved oxygen in the deep water in Indian, B.C.

Internal waves

        Internal waves of both short period (minutes) and longer
period (hours) are common features of the fjords.

        The shallow, short-period waves are most conspicuous in the
vicinity of the halocline, i.e. at depths of 5 to 20 m.   Intervals
between crests of 1 to 5 minutes and amplitudes of 5 to 8 m were
measured from an anchored vessel in some of the early cruises in
British Columbia.   Wavelengths of 30 to 100 m have been estimated
from the separation of the 'ruffled' bands on the surface which
usually accompany these waves (resulting from the convergence which
occurs behind the crest of the waves).   Still-water wave speeds
of about 1 m/s are estimated, although the speed over the ground
may be different because the upper layer and deeper water are
usually moving at different speeds and often in opposite directions
when the shallow waves occur.   These shallow waves are often

Table 10

Percent frequency of occurence of wind speeds and wave heights in Alaska and British Columbia (left columns) and Chile (right columns)

Wave Heights		Wind Speed (knots)							Wave Height Frequency Totals	
m		0 to 3		4 to 12		13 to 24		25		
	0.1	26	33	5	10	0	1	0	0	31	44
0.1 to	0.3	13	7	30	17	8	11	0	0	51	35
0.35	1.0	1	0	5	5	9	8	1	2	16	15
1.1	2.0	0	0	0	0	1	2	1	4	2	6
Wind Speed Frequency Totals		40	40	40	32	18	22	2	6	100%	100%

Alaska/British Columbia - 20 years (1950-1969) - 850 observations
Chile - one cruise in 1970 (85 observations).

observed in the vicinity of sills at a change of the tide; they are also generated as a ship slows down to take an oceanographic station. Hughes and Grant (1978) have described a technique for generating waves with a ship in order to study them.

The longer period waves (of tidal period) are observed with amplitudes of 20 - 25 m, at depths of 20 to 150 m, by the vertical motion of isotherms in the presence of the mid depth temperature minimum.

More detailed measurements of the internal waves and associated currents have been made recently by the Institute of Ocean Sciences, Patricia Bay, with the University of Washington, Seattle.†

BUTE INLET WAX - A CURIOSITY

A number of times during winter samples of a white waxy substance have been found in Bute Inlet in kilogram to tens of kilogram amounts. Studies of the material show it to be essentially a wax, i.e. fatty alcohols combined with fatty acids but no glycerides (characteristic of petroleums). Carbon-14 determinations suggest an age of less than 300 years. Above 11^0C it is a light

†Ed. Note: see section on Knight Inlet Observations, pages 251-298.

brown oil, below this a soft white solid. It is found on beaches
in the upper half of Bute (and occasionally near the mouth of the
neighbouring Toba) but nowhere else along the North American coast
and no references have been found to such a substance occurring
in other fjords.

ACKNOWLEDGEMENTS

We wish to acknowledge the efforts of many members of our
institutions over the years in collecting the data used in this
review and in assembling Data Reports. For the American fjords,
the scientific work was supported by grants from the Defence
Research Board of Canada and from the National Research Council
(especially N.R.C. Grant A.5091 to G.L.P.) and measurements were
made from research vessels made available by many Canadian
Government agencies and the generous help given by the officers
and crews of all the research vessels is gratefully acknowledged.

REFERENCES

Adams, K.R. and R. Bonnett. 1969. Bute Inlet was. Nature 223:
     943–944.

Anderson, J.J., and A.H. Devol. 1973. Deep water renewal in
     Saanich Inlet, an intermittently anoxic basin. Estuarine &
     Coastal Mar.Sci. 1: 1–10.

Anon., 1979. Pacific Coast Tide and Current Tables, Vols. 5
     and 6. Canadian Hydrographic service, Fisheries and Environ-
     ment Canada, Ottawa.

Batham, E.J. 1965. Rocky shore ecology of a southern New Zealand
     fjord. Trans. Roy. Soc. N.Z., Zoology, 6: 215–227.

Bell, W.H. 1973. The exchange of deep water in Howe Sound basin.
     Pacific Marine Science Report 73–13, Marine Sciences Direct-
     orate, Pacific Region, Environment Canada.

Cameron, W.M. and D.W. Pritchard. 1963. Estuaries in "The Sea;
     ideas and observations on progress in the study of the sea".
     M.N. Hill (ed.), Interscience.

Carter, N.M. 1932. The oceanography of the fjords of southern
     British Columbia, Fish. Res. Bd. Canada, Prog. Rept. Pac. Coast
     Sta., No. 12, 7–11.
     1934. Physiography and oceanography of some British Columbia
     fjords. Proc. 5th Pac. Sci. Cong., 1933, Vol. 1, pp. 721–733.

Dawson, W.B.   1920.   The tides and tidal streams with illustrative
    examples from Canadian waters.   King's Printer, Ottawa, 43 pp.

Garner D.M.   1969.   The hydrology of Milford Sound.   pp. 25-33 in
    Skerman, T.M. (ed.) "Studies of a Southern Fiord", Memoir N.Z.
    Oceanogaraphic Inst. No. 17.

Gilmartin, M.   1962.   Annual cyclic changes in the physical ocean-
    ography of a British Columbia fjord.   J. Fish. Res. Board Can.
    19:   921-974.

Herlinveaux, R.H.   1962.   Oceanography of Saanich Inlet in Vancouver
    Island, British Columbia.   J. Fish. Res.Board Canada. 19:   1-37.

Hughes, B.A. and H.L. Grant.   1978.   The effect of internal waves
    on surface wind waves.   1.   Experimental measurements.   J.
    Geophys. Res. 83:   443-454.

Jillet, J.B. and S.F. Mitchell.   1973.   Hydrological and biological
    observations in Dusky Sound, south-western New Zealand.   pp.
    419-427 in Fraser, R. (comp.), "Oceanography of the South
    Pacific 1972".   N.Z. Nat. Commis. for UNESCO, Wellington.

Kendrew, W.G. and D. Kerr.   1955.   The climate of British Columbia
    and the Yukon Territory.   Queen's Printer, Ottawa, 222 pp.

Lafond, C. and G.L. Pickard.   1975.   Deepwater exchanges in Bute
    Inlet, British Columbia.   J. Fish. Res. Board Canada.   32:
    2075-2089.

Northcote, T.G., Mildred S. Wilson, and D.R. Hurn.   1964.   Some
    characteristics of Nitinat Lake, an inlet on Vancouver Island,
    British Columbia.   J. Fish. Res. Bd. Canada, 21(5):   1069-1081.

Peacock, M.A.   1935.   Fiord-Land of British Columbia.   Bull. Geol.
    Soc. Amer. 46:   633-696.

Pickard, G.L.   1961.   Oceanographic features of inlets in the British
    Columbia mainland coast.   J. Fish. Res. Bd. Canada 18:   907-999.
         1963.   Oceanographic characteristics of inlets of
    Vancouver Island, British Columbia.   J. Fish. Res. Bd. Canada
    20:   1109-1144.
         1967.   Some oceanographic characteristics of the
    larger inlets of southeast Alaska.   J. Fish. Res. Bd. Canada.
    24:   1475-1506.
         1973.   Water structure in Chilean fjords.   Ocean-
    ography of the South Pacific 1972, comp. R. Fraser, New Zealand
    Nat. Comm. for UNESCO, 95-104.

Pickard, G.L. 1975. Annual and longer term variations of deepwater
    properties of the coastal waters of southern British Columbia
    J. Fish. Res. Bd. Canada, 32: 1561-1587.

Pickard, G.L. and L.F. Giovando. 1960. Some observations of tur-
    bidity in British Columbia inlets. Limn. and Oceanog., 5(2):
    162-170.

Pickard, G.L. and Keith Rodgers. 1959. Current measurements in
    Knight Inlet, British Columbia. J. Fish. Res. Bd. Canada,
    16(5): 635-678.

Pickard, G.L. and R.W. Trites. 1957. Fresh water transport deter-
    mination from the heat budget with applications to British
    Columbia inlets. J. Fish Res. Bd. Canada, 14(4): 605-616.

Richards, F.A., J.D. Cline, W.W. Broenkow and L.P. Atkinson. 1965.
    Some consequences of the decomposition of organic matter in
    Lake Nitinat, an anoxic fjord. Limnol. Oceanog., 10(Suppl.):
    R185-R201.

Stanton, B.R. 1978. Hydrology of Caswell and Nancy Sounds. pp.
    73-82 in Glasby, T.P. (Ed.) "Fiord studies: Caswell and Nancy
    Sounds, New Zealand", Memoir N.Z. Oceanographic Institute No.79.

Stanton, B.R. and G.L. Pickard. Physical oceanography of the New
    Zealand fjords. To be published as a Memoir N.Z. Oceanographic
    Institute.

Tabata, S. and G.L. Pickard. 1957. The physical oceanography of
    Bute Inlet, British Columbia. J. Fish. Res. Bd. Canada, 14(4):
    487-520.

Thomas M.K. 1953. Climatological Atlas of Canada. Nat. Res. Coun-
    cil, Ottawa, 256 pp.

Thompson, T.G. and K.T. Barkey. 1938. Observations of fjord-waters.
    Trans. Amer. Geophys. Union, 19: 254-260.

Tully, J.P. 1937. Oceanography of Nootka Sound. J. Biol. Bd.
    Canada, 3(1): 43-69.
            1949. Oceanography and prediction of pulp mill
    pollution in Alberni Inlet. Fish. Res. Bd. Canada Bull. No.
    83, 169 pp.

Waldichuk, M. 1958. Some oceanographic characteristics of a
    polluted inlet in British Columbia. J. Mar. Res., 17: 536-
    551.

Williams, M.Y. 1957. Bute Inlet wax. Trans. Roy. Soc. Canada, Ser.
    111, 51: 13-17.

Theses

Coote, A.R. MS, 1964. A physical and chemical study of Tofino
    Inlet, British Columbia. M.Sc. Thesis, Univ. British Columbia,
    Vancouver, B.C., 74 p.

Elliott, J.A. MS, 1965. A study of the heat budget components for
    the British Columbia and southeast Alaska coast. M.Sc. Thesis,
    Univ. British Columbia, Vancouver B.C. and MS Rept. No. 18,
    Inst. Oceanog., Univ. British Columbia, Vancouver, B.C., 101 p.

Lazier, J.R.N. 1963. Some aspects of the oceanographic structure
    of the Jervis Inlet system. M.Sc. Thesis. Institute of
    Oceanography, University of British Columbia, Vancouver, B.C.,
    54 p.

McNeill, M.R. 1974. The mid-depth temperature minimum in B.C.
    inlets. M.Sc. Thesis University of British Columbia, Vancouver,
    B.C., 91 p.

Trites, R.W. 1955. A study of the oceanographic structure in
    British Columbia inlets and some of the determining factors.
    Ph.D. thesis, University of British Columbia, Vancouver, B.C.,
    125 p.

Data Reports from British Columbia Inlets are contained in "Data
    Reports of the Institute of Oceanography", University of
    British Columbia, Vancouver, British Columbia; numbers 1-19,
    21, 23-28, 30-35, 37, 41-43, 45 covering cruises from 1953
    to 1979.

Anon. Ms. 1957 Physical and chemical data record, Alberni Inlet
    and Harbour, 1939 and 1941. Fisheries Research Board of Canada,
    Nanaimo, B.C. 89 pp., 7 figures.
            1957-59. Data records. Fish. Res. Bd. Canada, MS Repts.
    Oceanogr. and Limnol., Nos. 16, 17, 29, 36, 43, 47, 52.

WHY BIOLOGISTS ARE INTERESTED IN FJORDS

Torleiv Brattegard

Institute of Marine Biology
University of Bergen
Bergen, Norway

All kinds of biologists who use marine organisms in their research can find objects for their studies in a fjord. In the context of fjord oceanography a fjord should be regarded as a biological entity, the domain of ecologists who study the interactions that determine the distribution and abundance of organisms.

The answers to the question "Why are biologists interested in fjords?" will depend on who is asked. To avoid being too biassed in my answer I sought help and put the question to a number of my Norwegian colleagues. Their answers could be classified under fairly well-defined headings:
- the existence of fjords,
- special species composition,
- gradients in environmental factors,
- mini-ocean,
- manageable ecosystem,
- acute problems
- fjord management.

The common denominators of these answers are three basic types of questions: Where? How many? and Why? Problems of these types are approached from two angles and on three fronts - the descriptive point of view and the functional point of view, and by field methods, laboratory methods and mathematical methods. Usually there are two main pitfalls to be aware of: those who stress the descriptive point of view have a tendency to get lost in details because the research goals are often too loosely defined, and those who stress the functional approach have a tendency to get removed from reality in the absence of detailed biological knowledge. Whatever point of view and methods applied, the distribution and abundance of organisms

in fjords will be the ultimate criterion of comparison and the
basic standard against which the results must be judged.

THE EXISTENCE OF FJORDS

Fjords as we know them today have existed for more than
10,000 years. Shores along fjords have been populated by man for
at least 7,000 years. People's activities have been, and are still,
intimately connected with the use of fjords (food, transportation,
etc.) and through time they have encountered many practical problems
in their use of fjords. Some people have been curious about the
contents, processes and function of fjords, and in recent time have
even made such curiosity their career. Today the man depending on
fjords is confronted with new problems and new questions almost
every day and his methods and means of solving them are better
today than yesterday. It is an inherent part of being man that he
explores his environment and there is no difference in principle
between the urge to explore and investigate fjords than to study
the marvels of the Himalayas, the secrets of the deep sea, or the
strange moons of the planet Jupiter.

As biologists we want to know what kind of organisms live and
have lived in fjords; to know when and how organisms populated
fjords or disappeared from them; to find out how many different
kinds there are; and to be able to understand why the community
structures are as they are. These questions are simple but basic.

SPECIAL SPECIES COMPOSITION

Scientific studies on the flora and fauna of fjords have been
carried out for more than 150 years in northern Europe. The
accumulated floristic and faunistic literature and the more recent
more sophisticated literature that deals with distribution and
abundance of algae and animals in fjords has revealed:
- that the various habitats within a fjord together support
  a highly diverse flora and fauna,
- that the multitude of environmental factors in fjords
  makes it possible to investigate many types of problems
  concerning distribution and abundance,
- that species composition and relative abundance of dominating species may be drastically different even in
  neighbouring fjords,
- and that the deeper parts of fjords act as conservative
  environments due to the inertia of the resident basin water
  that reduces the effects of short-time (season-to-season
  or even year-to-year) meteorological changes.

Due to the variety of fjords, which will be commented upon later,

the possibilities of studying effects of specific environmental
factors, or sets of environmental factors, on species or assemblages
of species, seem almost unlimited. Some fjord organisms may be
very suitable objects for biologists concerned with problems
within the domain of **genetics and evolution.**

The Fauna of Deep Fjord Basins: A Special Case of Special Species Composition

The deepest fjord that is known is known is Sognefjord in
Norway, and has a long, narrow basin with a maximum depth of about
1,300 m. The maximum depth of fjords in Alaska, British Columbia,
Chile and New Zealand is about 840 m, 670 m, 1,050 m and 420 m,
respectively (Pickard and Stanton 1979). Do such deep fjords have
deep water faunas mainly composed of shallow water species, or do
we find faunas similar to those found at corresponding depths in the
ocean, or do we find elevated deep-sea faunas?

The answer is that at least some of the species belong to
deep-sea fauna. Carpine (1970) compared bathyal fauna of the
northwestern Mediterranean with data from 1,200-1,300 m in Sognefjord (Brattegard 1967) and found a remarkable similarity between
the two faunas.
Pogonophora is essentially a deep-sea group. Four species
have been found in Norwegian fjords. None of the species has been
recorded from the Norwegian continental shelf.

The entire isopod fauna of deep fjords, with the exception of
one genus (*Pleurogonium*), consists of the same kinds of isopods
as at abyssal depths (Dr. R.R. Hessler, pers. comm.).

Examples of typical deep-sea types living in deep fjords can
be found in many other groups, for example among the polychaetes,
bivalves and fishes.

By studying the deep water fauna of fjords there might be a
chance of understanding the deep-sea fauna better. The deep-sea
is characterized as being a highly predictable environment. Many
of the operating factors in deep fjords are also remarkably stable
over long periods of time, but in contrast to the deep-sea environment fjord basin waters are, sooner or later, subject to overturn
or some sort of periodicity.

Studies on living deep-water animals and communities are wanted.
Modern technology allows us to do *in situ* studies for example on
growth (Grassle 1977) and respiration (Smith 1977, Smith *et al.*
1979) at great depths. Problems related to colonization or rate
of recovery after a disturbance are extremely difficult or costly

to study in the deep-sea but might be reasonably possible to do in
a fjord environment.

Even if modern technology allows us to do the most astonishing
studies in deep water we must not forget that for a number of types
of studies animals must be brought live to a laboratory. This is
possible during winter time when the upper water masses of fjords
have the same or a lower temperature than the basin water.

Personally I believe that fjords are ideally suited for studies
of the possible dynamic interactions between plankton and benthos.
The results obtained by Beyer (1958), **Isaacs & Schwartzlose (1965)**,
Lagardère (1977), Hopkins *et al.* (1978) and Smith *et al.* (1979)
indicate that such studies would be highly rewarding. At the
Institute of Marine Biology, University of Bergen, a number of
students are contributing to the knowledge of dominant bentho-
pelagic and epibenthic animals.

Since fjords are different, a special deep-water problem may
be best studied in a particular fjord.

GRADIENTS

A property of fjords that attracts the attention of many biologists is the presence of gradients in a number of abiotic factors.
The factors may affect organisms, and their functional responses
can be measured in terms of tolerance, rates of metabolism and
activity, reproduction and distribution. The gradients may exist
in the horizontal or the vertical or both and may be permanent or
changing.

Gradients in abiotic factors induce gradients in the assemblages
of organisms, or in other words the biotic factors. A species being
able to tolerate variations in abiotic factors may be forced by
interspecific competition to a limited zone in the biological
gradient.

By carefully selecting locations studies of effects of specific
environmental factors on organisms may be performed *in situ* with
a fruitful combination of descriptive and functional approached to
the problems in question.

Examples of thorough studies of intertidal and subtidal algae
are the works of Sundene (1953), Klavestad (1967) and Bokn & Lein
(in prep.) from the Oslofjord and Jorde & Klavestad (1963) from the
Hardangerfjord. Brattegard (1966) gives an account of the horizontal
distribution of animals on rocky shores in the Hardangerfjord.
Pearson & Rosenberg (1978) give a review of changes in benthic

communities brought about by organic enrichment and refer to a number of studies in smaller fjords.

MINI-OCEAN

Biological oceanographic processes are complicated. Fjords as ecosystems have the same basic properties as the much more open ecosystem of the ocean. Biological processes in a fjord can be treated as a model of processes in much larger oceanic systems. Hence, some biological processes of the ocean, may, with advantage, be studied in fjords. Protected fjords are particularly well suited for experimental field-studies within the field of biological oceanography.   An example is grazing - the effects of herbivores on the phytoplankton community - a key problem in biological oceanography. Some of the advantages of looking at and using fjords as mini-oceans are:

practical - smaller scale
        - easier positioning
        - better working conditions resulting in higher quality of sampled data
        - rates of processes may be higher
        - seldom seasick
economical - easy access
        - do not need large research vessels which are costly to run
        - logistics far less complicated

A series of reports on work done on the deep-water pelagic community in Korsfjorden, western Norway can illustrate this. Matthews & Bakke (1977) studied the monthly biomass over 3 years of 23 macroplanktonic and microplanktonic species. The composition of this set of the pelagic community changed from month to month and from year to year. Despite this variation the biomass estimates showed a consistent group pattern when the species were grouped as herbivores, omnivores and carnivores. Their conclusion is that there can be a considerable flexibility in the species composition of a planktonic community without corresponding marked effects on its trophic structure.

Studies of population dynamics of oceanic species seldom give results of sufficiently high quality due to the difficulties of sampling from the same population over time. Deep fjords often contain oceanic species whose populations may be isolated from other populations for long periods of time. Such populations are close to the ideal population for studies of growth, mortality, life span, and succession of generations. Examples of results that appear to be very difficult to obtain in the open sea are given in Matthews, Hestad & Bakke (1978):

Calanus finmarchicus has, at least from copepodid III onwards, a high instant mortality rate (= 0.02-0.4). Virtually the whole population in Korsfjorden appears to fall prey to carnivores. On the other hand, *Calanus hyperboreus* has a consistently low instant mortality rate (=0.003-0.007) except at the end of its life span. A large proportion of its biomass seems free to pass to the decomposers.

MANAGEABLE ECOSYSTEM

A fjord is geomorphologically a well-defined entity. In biological terms it is an easily defined almost closed ecosystem showing less variability than the open coast or ocean. In practical research the system can be conveniently divided into three parts according to the hydrodynamics: the upper part where the estuarine circulation takes place, the part influenced by the intermediate water masses, and the deep part below the level of the sill. It is possible, at least in theory, to estimate water exchange and inputs and outputs of energy and matter. Thus a fjord can be regarded as an ecosystem of manageable size. The prospects of being able to find out what is going on within the ecosystem are good.

An illustrative paper is that of Rosenberg, Olsson & Ölundh (1977) in which they suggest an energy flow model for a well-defined ecosystem in the inner part of a Swedish fjord. As usual the model is based on measurements and estimates of the phytoplankton production and the faunal production. In the manageable ecosystem context the importance of their paper is that they demonstrate the possibility of measuring and giving good estimates of in- and outflux of organic matter across the boundaries of their ecosystem.

A long-term project to study the marine ecosystem and the self-sustaining herring stock in a small, land-locked fjord (Lindaspollene) in western Norway was initiated in 1970 (Dahl, Østvedt & Lie 1973). Its ultimate aim is to build a dynamic mathematical model of the energy flow through, and the cycling of matter in, the ecosystem. Some of the important papers published up to now are Dale (1978), Lie, Bjørstad & Oug (1978), Lie, Dahl & Østvedt (1978), Oug (1977) and Skjoldal & Lännergren (1978).

ACUTE PROBLEMS

This particular point concerns man's misuse of fjords. Waste products put or dumped into a fjord may be of different kinds:
    - heat

- fresh-water discharge
- litter, mud, inert materials
- inorganic chemicals
- sewage, organic wastes, oil
- pesticides, petrochemicals, other organic chemicals
- radioactive materials, etc.

The sources are also of different kinds:
- industry
- urban areas
- rural areas
- ships, fishing vessels, small boats
- military installations and activities.

Problems usually surface when concerned people point to loss of amenities or obvious damage to fisheries – that is biological damage. The usual way of tackling the problem is:
- **to search for the responsible factor(s),**
- to locate the source(s)
- to guess at the tolerance limits of the ecosystem before it was disturbed
- to recommend some sort of action to be implemented at the source(s) of the pollution, or somewhere else in the fjord
- to hope that the system will change and return to or develop into an acceptable system.

In some cases the damage may be repaired by appropriate actions but it is likely that the recovery in many cases will take a long time, from some years to perhaps tens or hundreds or years depending on the nature of the pollutant, flush time of the fjord, etc.

FJORD MANAGEMENT

As in estuaries, biological production and productivity in fjords is generally high. Some of the products (species of fish, crustaceans, bivalves, algae) are harvested. In some fjords some species have been harvested beyond optimum yield, but in general the natural biological resources are underutilized. The magnitude of the harvesting of living resources is closely connected with available harvesting techniques and demands for the products. It seems quite possible to increase the harvest of demanded species with the advance of technology. In northern Norway a fishery for echinoids has started. Japanese importers pay up to $180.00 per kilo roe w.w.

By simple means it is possible to increase production of commercially valuable species. In Norway, e.g. oyster spat was for some time produced in large quantities in large quantities in

special polls for export to oyster farms in Denmark. New techniques
allow us to produce *Mytilus edulis* in enormous quantities at low
cost with little manipulating of the environment. Only the market
is lacking.

Many biological oceanographers are investigating problems
concerned with biological production. In countries with fjords
efforts should be directed towards studies of the processes
governing biological production in fjords with the specific aim
of being able to manipulate the ecosystem to increase harvestable
production and improve the potential for aquaculture.

The main objective of the project in Lindaspollene, which was
mentioned in connection with manageable ecosystems, is to gain
sufficient insight to be able to manipulate the ecosystem for the
benefit of man. Optimal management of the system would involve
attempts to increase the yield of the local herring fishery, to
cultivate marine molluscs and crustaceans and to guarantee continued
recreational use of the area (Dahl, Østvedt & Lie 1973).

The fjord environment is also used for other purposes than
providing food. It is used for transport, leisure, source of non-
living raw materials, and as a recipient for wastes that are un-
reflected or deliberately dumped into it. The different uses of
fjords leads to conflicts among the users. I believe we can foresee
the time coming when fjord areas important for human society must
be managed according to a specified policy.

FJORDS AND COMMON SENSE IN BIOLOGY

The occurrence of fjords is restricted to coasts in higher
latitudes: in the northern hemisphere in Canada, Alaska, Greenland,
Iceland, Scotland, Norway, Sweden, and USSR, and in the southern
hemisphere in Chile and New Zealand. All these areas have at some
times been covered by ice and the opinion among geologists is that
fjords are mainly the result of glacial erosion. Thus they are
relatively young geomorphological features and all of them have
got their flora and fauna from adjoining areas outside.

The last glaciation period in western Norway reached its
maximum about 18,000 YBP. The ice-sheet reached out over the
shelf and the fjords were certainly filled with ice. As the ice
retracted warm Atlantic water reached the inner shelf about **13,000**
YBP. Four thousand years later, about 9,000 YBP, the warm Atlantic
water could for the first time after the last glaciation penetrate
into the deep basins of the fjords.

Aarseth et al. (1975) studied a 550 cm long core taken from

the top of the 80-90 m thick sediment carpet in the outer basin
of the Korsfjord, western Norway, at 603 m depth. Analysis of the
sub-fossils show that before 9,000 YBP the fauna was dominated
by cold-water species (e.g. the benthic foraminiferans *Elphidium
clavatum* and *Nonion labradoricum*, the pelagic radiolarian *Amphi-
melissa setosa* and the bivalve *Yoldiella lenticula*). The
ecological conditions seem to have been stable for some time. Then
the fauna changed leading to a new stable period that began about
2,500-3,000 YBP dominated by species now having a lusitanian and
boreal distribution (e.g. the benthic foraminiferans *Hyalina
balthica* and *Uvigerina peregrina*, the pelagic radiolarian
Cromyechinus borealis and the bivalve *Kelliella miliaris*).

In terms of biogeographical classification all fjords in the
northern hemisphere lie within the cold-temperate and the arctic
regions. The flora and the fauna of fjords in these areas is
composed of species belonging to different types of geographical
distribution - the cosmopolitan, the warm-temperate, the cold-
temperate, the sub-arctic and the arctic. The relative numbers of
these biogeographical components vary within wide limits within a
fjord, between fjords and with time.

The composition of the intertidal fauna along the Hardangerfjord shows that the boreal species dominate the fauna, that the
southern (lusitanian-boreal) species are most common in the outer
part and rare in the inner part, and that northern (arctic-boreal)
species are distributed all along the fjord but are, relatively,
most conspicuous in the inner part (Brattegard 1966).

The deep parts of fjords may act as biogeographical enclaves,
e.g. the inner part of the Oslofjord in southern Norway (Beyer 1958),
some small fjords(polls) with shallow sills in western Norway
(Greve & Samuelsen 1970, Dahl, Østvedt & Lie 1973), and fjords in
northern Norway (Soot-Ryen 1924). The Saguenay fjord in southeastern Canada has been described as being formed of two strikingly
different and superimposed ecological milieus. Its deeper part
constitutes a biogeographical arctic enclave within the boreal
region (Drainville 1970).

This does not surprise the ecologist and the biogeographer.
Current ecological theory tells us that different organisms live
under different sets of optimal conditions. Organisms are distributed in different ways according to their modes of dispersal,
tolerance limits concerning habitat, interactions with other
organisms and the regime of abiotic factors. Even populations of
a given species vary in their ecological properties from place to
place or in a given area through time. Slight differences in
temperature will for example affect metabolism, growth rate,
breeding and longevity as has been demonstrated for the boreo-

arctic deep-water *Pandalus borealis* by e.g. Rasmussen (1953).

Former and recent members of the fjord community can at a given time be classified into one of four groups:

1. Species absent, but a population was present earlier. Immigration from an outside population was possible for some time but later on at least one factor exceeded the tolerance limit for the population.
2. Species occasionally present. Immigration from outside populations occur occasionally, but immigrants are unable to found lasting populations. Factors related to habitat selection, species interactions and abiotic factors exceed tolerance limits.
3. Species always present, but population not self-sustaining over long periods of time. Immigration from outside populations necessary. One or more factors related to species interactions and/or abiotic factors varying about tolerance limits.
4. Species always present, population self-sustaining. Immigration from outside population only necessary during the period the fjord population was established.

A natural question is how individuals from a population outside a fjord can disperse into a fjord in sufficient numbers. The possibilities are: 1) passively - by drifting as spores, eggs, larvae or adults, or being transported by moving living or non-living objects, and 2) actively - by swimming, walking, crawling. Algae and most animals disperse by drifting with moving water masses. The exchange of water between a fjord and the areas outside is thus the agent responsible for the structure and perhaps the dynamics of the total community within a fjord at any time.

What determines the processes of water exchange between fjords and areas outside fjords? Following Pedersen (1978) the observed differences in hydrodynamics of fjords originate mainly from fjord variation in:

1) The geometry of fjords (e.g. sill depth, basin depth, length, width, volume).
2) The hydrology of the adjacent watershed (e.g. low to high level watershed, yearly precipitation).
3) The oceanographic conditions outside the fjord (e.g. weak or strong stratification, tides, upwelling).
4) Meteorological conditions (e.g. wind fields in the fjord area, distribution of high and low pressure areas, extreme weather conditions).

Within each group the parameters can vary within wide limits. We can hardly expect two geographically separated fjords to be identical with respect to their geomorphological, physical and chemical characteristics nor their hydrodynamics.

If we choose a specific fjord, the parameters can be grouped in another way, namely as being

1) Constant (sill depth, basin depth, extent of watershed, etc.)
2) Variable but highly predictable (tides).
3) Variable but relatively predictable (yearly freshwater discharge, etc.).
4) Variable and stochastic (windfields, upwelling, extreme weather conditions, etc.).

I think it is safe to say that we can hardly expect a fjord to be the same from one year to another with respect to its physical and chemical characteristics or its hydrodynamics.

The biologist is of course very much interested in having an understanding of the hydrodynamics of his study fjord. But the biologist is also dependent on facts, good estimates, because the planktonic organisms follow the water masses as they move. He wants to know the kind of water masses that leave and enter the fjord, the volume of water transported, the depth at which the water masses move, and the time or time periods when water transport takes place. Information of this kind is necessary for the biologist if he should need to estimate types and numbers of organisms imported to or exported from a fjord.

In bringing these biological and abiological considerations together I feel we are compelled to conclude that neither structures nor dynamics of fjords allow us to expect one fjord to mirror another with respect to the composition and dynamics of their flora and fauna. This implies that those who try to construct a theory of fjord biology have to be content with a theory of fairly low resolution power. Even so, such a theory is badly needed but we must be aware that it will only be one of several tools needed to solve biological problems in fjords.

The conclusions or perhaps *recommendations* is a much better word are

- that studies on organisms and communities in fjords should continue and hopefully increase in quality and quantity just because fjords are part of our environment,
- that structures and processes unique to the biology of fjords should be identified and studied,
- that experimental use of fjords should be encouraged and suitable fjords or polls set aside for field-experiments,
- that we should increase the use of fjords as suitable study areas for biological studies that are too complicated or too expensive to do elsewhere,
- that research should be specifically designed to make contributions towards the goal of getting a sound scientific basis for policies of fjord mangement.

REFERENCES

Aarseth, I., K. Bjerkli, K.R. Björklund, D. Böe,J.P. Holm,
    T.J. Lorentzen-Styr, L.A. Myhre, E.S. Ugland, and J.Thiede.
    1975. Late quaternary sediments from Korsfjorden, western
    Norway. Sarsia 58: 43-66.

Bokn, T. & T.E. Lein, in prep. Long term changes in the fucoid
    associations in the inner Oslofjord.

Brattegard, T. 1966. The natural history of the Hardangerfjord.
    7. Horizontal distribution of the fauna of rocky shores.
    Sarsia 22: 1-54.

Brattegard, T. 1967. Pogonophora and associated fauna in the deep
    basin of Sognefjorden. Sarsia 29: 299-306.

Carpine, C. 1970. Écologie de l'étage bathyal dans la Méditerranée
    occidentale. Mémoires de l'Institut océanographique, Monaco.
    2: 1-146.

Dahl, O., O.J. Østvedt, U.Lie. 1973. An introduction to a study
    of the marine ecosystem and the local herring stock in Linda-
    spollene. FiskDir. Skr. Ser. HavUnders. 16: 148-158.

Dale, T. 1978. Total chemical and biological oxygen consumption
    of the sediments in Lindaspollene, western Norway. Marine
    Biology 49: 333-341.

Drainville, G. 1970. Le fjord du Saguenay. II. La faune ichty-
    ologique et les conditions écologiques. Le Naturaliste
    Canadien. 97: 623-666.

Grassle, J.F. 1977. Slow recolonisation of deep-sea sediment.
    Nature 265: 618-619.

Greve, L. & T.J. Samuelsen. 1970. A population of *Chlamys islandica*
    (O.F. Müller) found in western Norway. Sarsia 45: 17-23.

Hopkins, C.C.E., S. Falk-Petersen, K. Tande & H.C. Eilertsen. 1978.
    A preliminary study of zooplankton sound scattering layers in
    Balsfjorden: structure, energetics, and migrations. Sarsia
    63: 255-264.

Isaacs, J.D. & R.A. Schwartzlose. 1965. Migrant sound scatterers:
    Interaction with the sea floor. Science 150: 1810-1813.

Jorde, I. & N. Klavestad. 1963. The natural history of the
    Hardangerfjord. 4. The benthonic algal vegetation. Sarsia
    9: 1-99.

Klavestad, N. 1967. Undersøkelser over benthosalgevegetasjonen i
    indre Oslofjord 1962-1965. 119 pp. + 44 maps. Part 9 of:
    Oslofjorden og dens forurensningsproblemer. I. Undersøkelsen
    1962-1964. NIVA, Oslo.

Lagardere, J.P. 1977. Recherches sur le régime alimentaire et le
    comportement prédateur dés decapodes benthiques de la pente
    continentale de l'Atlantique Nord Oriental (Golfe de Gas-
    cogne et Maroc). In: Keegan, B.F., P.O. Ceidigh & P.J.S.
    Boaden. Biology of benthic organisms. Pergamon Press, New
    York. pp. 397-408..

Lie, U., D. Bjørstad & E. Oug. 1978. Eggs and larvae of fish from
    Lindaspollene. Sarsia 63: 163-167.

Lie, U., O. Dahl & O.J. Østvedt. 1978. Aspects of the life history
    of the local herring stock in Lindaspollene, western Norway.
    Fisk.Dir.Skr.Ser.HavUnders. 16: 369-404.

Matthews, J.B.L., J.L.W. Bakke. 1977. Ecological studies on the
    deep-water pelagic community of Korsfjorden (western Norway).
    The search for a trophic pattern. Helgoländer wiss. Meeres-
    unters. 30: 47-61.

Matthews, J.B.L., L. Hestad & J.L.W. Bakke. 1978. Ecological
    studies in Korsfjorden, western NOrway. The generations and
    stocks of Calanus hyperboreus and C. finmarchicus in 1971-1974.
    Oceanologia Acta 1: 277-284.

Oug, E. 1977. Faunal distribution close to the sediment of a
    shallow marine environment. Sarsia 63: 115-121.

Pearson, T.H. & R. Rosenberg. 1978. Macrobenthic succession in
    relation to organic enrichment and pollution of the marine
    environment. Oceanogr. Mar. Biol. Ann. Rev. 16: 229-311.

Pedersen, B. 1978. A brief review of present theories of fjord
    dynamics. In: Nihoul, J.J. (ed.). Hydrodynamics of estuaries
    and fjords. Elsevier Oceanography Series 23. pp. 407-422.

Pickard, G.L. & B.R. Stanton. 1979. Pacific fjords - a review of
    their water characteristics. (In this volume).

Rasmussen, B. 1953. On the geographical variation in growth and
    sexual development of the deep sea prawn. Fisk. Dir. Skr.
    Ser. HavUnders. 10 (3): 1-60.

Rosenberg, R., I. Ölsson & E. Olundh. 1977. Energy flow model of
    an oxygen-deficient estuary on the Swedish coast. Marine
    Biology 42: 99-107.

Skjoldal, H.R. & C. Lännergren. 1978. The spring phytoplankton
    bloom in Lindaspollene, a land-locked Norwegian fjord. II.
    Biomass and activity of net- and nanoplankton. Marine Biology
    47: 313–323.

Smith, K.L. 1978. Benthic community respiration in the N.W.
    Atlantic Ocean: *in situ* measurements from 40 to 5200 m.
    Marine Biology 47: 337–347.

Smith, K.L. & G.A. White & M.B. Laver. 1979. Oxygen uptake and
    nutrient exchange of sediments measured *in situ* using a free
    vehicle grab respirometer. Deep-Sea Research 26A: 337–346.

Smith, K.L., G.A. White, M.B. Laver, R.R. McConnaughey & J.P. Meador.
    1979. Free vehicle capture of abyssopelagic animals. Deep-
    Sea Research. 26A: 57–64.

Soot-Ryen, T. 1924. Faunistische Untersuchungen im Ramfjorde.
    Tromsø Mus. Arsh. 45 (6): 1–106.

Sundene, O. 1953. The alagal vegetation of Oslofjord. Skr. N.
    Vidensk.-Akad. Oslo. I. Mat. -Naturv. Kl. 1953 No.2: 1–244.

THE FLUID MECHANICAL PROBLEM OF FJORD CIRCULATIONS

Robert R. Long

Department of Earth & Planetary Sciences
The Johns Hopkins University
Baltimore, Maryland 21218

INTRODUCTION

There are too many theoretical aspects of fluid mechanical research on fjords and estuaries to treat all of them in a single paper even if, contrary to present circumstances, the author could claim overall competence. So, we will treat in some detail the single topic of classical estuarine circulations in which, on the average, low-salinity water, originating from run-off, flows out to sea over salty water moving slowly into the estuary. There are many associated physical considerations but the basic mathematical problem is to solve the set of equations of conservation of momentum, salt and volume together with boundary conditions to yield the mean velocity and density distributions.

The problem is difficult in part because of the diversity of estuarine geometry, of the amount and location of the fresh-water inflow, of conditions outside the mouth, and of external conditions of wind and tide and, in part, because of the basic non-linearity of the equations. Doubtless, these difficulties will be overcome eventually with the help of numerical integration. Our basic ignorance of the phenomenon of turbulence is more serious. The upper levels of a fjord are probably always turbulent and important turbulence must exist below the halocline, associated with the breaking of internal gravity waves and the rubbing of tidal flows against the bottom and sides. The importance of turbulence in fjord research is basic. It, rather than molecular processes, is the agency for the redistribution of momentum, salt, heat and pollutants.

We want to be helpful to those who are interested in fjord

problems (pollution, for example) but who are not, perhaps, specialists in basic fluid mechanics. It may not be amiss, therefore, to
begin by discussing the importance and difficulties of turbulence
and the current state of knowledge based on major research over the
past ten or fifteen years.

THE TURBULENCE PROBLEM

     Experience of the past century indicates that the equations of
fluid mechanics and heat and mass transfer (Schlichting, 1955, Ch. 3)
are a remarkably accurate set of laws governing most fluid bodies
including oceans and estuaries. The equations are, to be sure,
highly non-linear but in *laminar* motions, in which accelerations
are small or only moderately large compared to the important forces,
the problem is usually tractable or, at least, will become so in
the foreseeable future through the use of electronic computers. But,
in typical geophysical cases the mean fluid speeds, for example,
in a current of water flowing out to sea in an estuary, are too
large for laminar flow to exist and a chaotic type of *turbulent*
instantaneous flow field develops which presents enormous mathematical and physical difficulties. Classically, the criterion for
the onset of turbulent flow is usually given as a critical Reynolds
number $R = Uh/\nu$ where U is a characteristic speed, h is a characteristic length and ν is molecular viscosity. However, this single
criterion is not adequate in geophysics. In the oceans and atmosphere the dimensions are so great that any reasonable choice of
U and h yields values of R which should certainly be large enough
for turbulence; yet much of the deeper portions of the oceans and
the upper portions of the atmosphere are apparently in laminar
motion because other effects, in addition to molecular friction,
notably density stratification, are stabilizing and keep the flow
laminar. But even this conclusion that flow is frequently laminar
must be used carefully because of the complexity of geophysical
problems. In a deep fjord the motion may be laminar nearly
everywhere except in a few sporadic regions of internal wave-breaking
and yet, because of the weakness of molecular processes, *this small
amount of turbulence may transport nearly all of the heat and salt.*
The importance of internal wave-breaking in this connection has
been discussed theoretically by Csanady (1967, 1968a, b), Stigebrandt
(1976, 1979) and Walin (1972a, b) who direct attention to breaking
of internal waves (and Kelvin waves) against the sides and bottoms
of fjords and by Smith and others (Smith, 1974, Partch and Smith,
1977, Gardner and Smith, 1978, Farmer and Smith, 1978) who connect
the breaking of the internal waves to shear-flow instability.
Observations of internal waves in fjords were made long ago by
Pickard and Rodgers (1959) and more recently by Perkins and Lewis
(1978). Additionally, turbulence in the upper layer of an estuary
may excite internal waves in the halocline and below, and the

breaking of these may have the important significance it apparently
has in laboratory experiments (Long, 1978a, b, Long and Kantha,
1978).   Some reserve is required, however, because, with the
exception of direct observations by Woods (1968), evidence for this
kind of mixing in the field is indirect.   Furthermore, we have
no indication of the quantitative importance of these effects.

The central problem in turbulent flows is that the governing
equations for the mean field quantities are indeterminate and it is
impossible, except for a very few special cases to determine useful
solutions without *ad hoc* assumptions of various kinds.   The simplest
problem in turbulence is the decay of homogeneous, isotropic turbulence filling infinite space (Hinze, 1976, Chp. 3), but the
basic behaviors for the time-dependence of the rms velocity and
integral scale have not been solved for except in the case of very
special initial conditions (Long, 1978c).   Similarly the energy
spectrum of isotropic turbulence is of basic importance but controversy still exists regarding the form of the spectrum even in the
one range - the inertial subrange - in which theoretical arguments
have progressed the most (Kraichnan, 1974).   The problem of the
spectrum introduces the fundamental difficulty that turbulent
motions exist on a wide range of scales (eddies or Fourier components)
from very small to very large wave numbers, but the viscous terms
in the equations are contributed to only by the eddies of very
large wave numbers in Fourier space or by regions of small size
(boundary layers) in physical space.   Yet these small regions in
wave number or physical space are of the greatest importance despite
their small extent.   In the problem of the spectrum (Hinze, 1976
p. 336), energy is transformed from large energy-containing eddies
by nonlinear transfer effects to the small eddies in which the viscous dissipation of energy is accomplished entirely by the large
wave-number portion of the spectrum.   This transfer process is
poorly understood and, in addition, the small dissipative eddies
seem to be confined to very limited regions of space according to
recent experiments (Kuo and Corrsin, 1971, 1972), rather than
generally distributed as originally thought.   This is referred to
as *intermittency of fine-structure* and recent work by the author
(Long, 1979a) connects this to the existence of a new region in
wave-number space, called the *mesoregion*, lying between the viscous
range and the inertial range.   We will discuss mesolayers and meso-
regions in more detail below.

Isotropic decaying turbulence has no energy source and no
energy sink except friction and this makes such turbulent motions
remote from geophysical problems in which there are several energy
sources and sinks.   A most important energy source in geophysics
is the conversion of mean-flow energy to turbulent energy when, as
is usual, velocity shear exists.   Shear is found in simpler situations, for example, in pipe flow and much research has been conducted
on the problem.   Indeed, until recently, it was thought that the

problem of turbulent shear flow near a surface was essentially
solved. It is simple and useful to reconstruct the basic argument:
In Fig. 1 we portray turbulent flow over a smooth surface. The
mean velocity profile, as observed, consists of a lower thin inner
layer near the plate called the viscous sublayer. According to
classical ideas, in the outer region viscosity is unimportant but
conditions are influenced by the outer length H, e.g., the radius
of the pipe. In the inner region ν is important, and H is unimportant. It is further assumed that the two regions overlap at large
enough R and, therefore, in the overlap region (inertial sublayer)
neither H or ν is important. This argument leads directly to the
logarithmic boundary layer (Monin and Yaglom, 1971, p 274). The
existence of the overlap is inferred from arguments like those of
Tennekes and Lumley (1972) and portrayed graphically in Fig. 2. The
argument depends utterly on the assumption regarding the limits of
the two regions. If the viscous sublayer is $0 \leq z \ll H$ and the
outer region is $\nu/u_* \ll z \leq H$ where u_* is the friction velocity, the
overlap will exist at large enough Reynolds numbers $u_* H/\nu = R$ as
we see in Fig. 2. Nondimensionally, these limits are $0 \leq z_+ \ll R$,
$1 \ll z_+ \leq R$ for inner and outer regions, respectively, where
$z_+ = z u_*/\nu$. However, this choice of limits is purely arbitrary. If
we choose for the limits of the two regions, say,

$$0 \leq z_+ \ll R^{\frac{1}{2}}, \quad R^{\frac{1}{2}} \ll z_+ \leq R \tag{1}$$

respectively, there is obviously no overlap as we see in Fig. 3.
This difficulty with classical theory has been mostly overlooked
although others have challenged the overlap assumption recently
(Malkus, 1979). Similar remarks may be made about the inertial
subrange in spectrum arguments. The classical approach (Tennekes
and Lumley, 1972) is to assume an inner region of wave-number space
$k \gg \ell^{-1}$ in which the outer length ℓ (integral scale) is unimportant
and an outer region $k \sim \ell^{-1}$ in which ν is unimportant. An overlap
of these regions is assumed in which the spectrum function E depends
only on the energy dissipation ε and k. The famous $-5/3$ law follows.
Again, the argument stands of falls on the assumption of the overlap.

Objections to classical arguments may be advanced also by citing
observations contrary to predictions. For example:

1) In flow near smooth surfaces, classical arguments lead to
$\ell \simeq z$ where ℓ is the integral scale, but observations indicate that
horizontal integral scales are far larger than z, in fact as large
as the outer length H, as close to the wall as can be measured, and
measurements have been made down to $z_+ < 1$ *deep within the sublayer*
(Mitchell and Hanratty, 1966, Tritton, 1967). Similarly in the
atmospheric surface layer (Kaimal, 1978) the integral scale was
measured to be 2 km. at levels of 4 m above the ground!

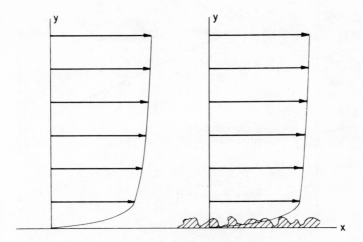

Figure 1.   Schematic flow over smooth and rough surfaces.

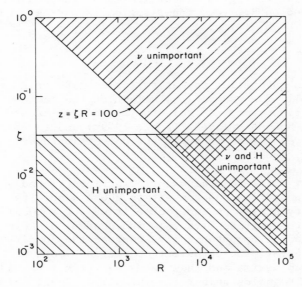

Figure 2.   Picture of classical theory for the overlap of inner and
             outer regions in shear flow $\zeta = z/H$ where z is distance
             from wall and H is outer length. R is the Reynolds number.

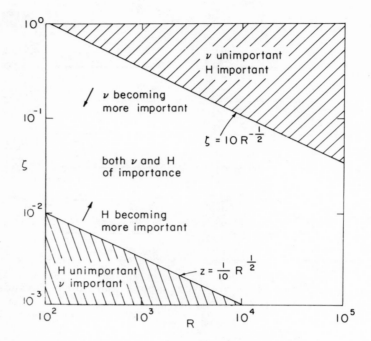

Figure 3.   Similar to Fig. 2 except limits of inner and outer region
            depend on R, i.e., inner region is $z \ll R^{\frac{1}{2}}$, outer region
            is $\zeta \gg R^{-\frac{1}{2}}$.

        2)   Certain quantities appear to scale on $R^{\frac{1}{2}}$ or $R^{\frac{1}{4}}$ in a region
close to the wall, for example the height of the maximum of momentum
stress in pipe flow (Long, 1979b), the maximum of transverse momentum
stress in Ekman boundary layers (Long, 1979c) and the maxima of the
horizontal rms velocities in turbulent convection over a hot plate
(Chern and Long, 1980).

        3)   The very different behavior of horizontal and vertical rms
velocities in shear flow and convection above a surface.   The
vertical velocity scales on inner quantities for a considerable
distance above the surface (i.e., is independent of Reynolds number)
increases with height monotonically, whereas the longitudinal rms
velocity profile (or, for turbulent convection, both horizontal
velocities) reach maxima and then decrease further up and, furthermore, vary with Reynolds number (Laufer, 1954, Long, 1979b, Chern
and Long, 1980).   Similar observations have been made recently in
the atmospheric boundary layer (Kaimal, 1978).

        4)   Observations indicate that the classical ideas of continuous
turbulence in wave-number and physical space are quite wrong.   Instead

the fine structure which does the dissipating is highly intermittent
in fully turbulent regions (Kuo and Corrsin, 1971, 1972) and near
the plate in shear flow, *bursts* appear intermittently which contain
much of the *turbulence* and transmit most of the momentum (Ueda and
Hinze, 1975, Rao, *et al.*, 1977). In short, the basic concept of
eddies filling the space near a wall and increasing in size with
distance from the wall is inconsistent with observations.

The author is in the process of constructing new theories incor-
porating, basically, a new boundary layer called a *mesolayer* near
a wall of thickness $R^{\frac{1}{2}}$ when scaled on inner variables ($R^{\frac{1}{4}}$ for thermal
convection) and a new mesoregion in wave number space of dimensions
$k_m^{-1} \sim R^{\frac{1}{2}}$ (i.e., proportional to Taylor's microscale for the outer
flow) which intrudes between the classical inner and outer regions
and prevents the classical overlap in the way discussed in connection
with Figs. 2 and 3. The turbulent motion is divided into two
terms $\underset{\sim}{u}_1$ and $\underset{\sim}{u}_2$ such that u_1 alone survives as the outer length
tends to infinity. Then the u_1 motion is similar to what would be
expected from classical theory, because, of course, a mesolayer
does not exist if H or R is infinite. For example u_1 is logarithmic
above the sublayer. The $\underset{\sim}{u}_2$ motion, however, is influenced basically
by the outer dimensions of the system. For example, u_2 is dominated
by large eddies very near the wall of size H. These have a major
effect on the rms σ_u profile (causing Reynolds-number dependence,
for example). On the other hand, the rms vertical velocity σ_w is
dominated by the first component of motion and is Reynolds-number
independent near the wall. The mesolayer is essentially a boundary
layer for the large flattened eddies "sloshing" against the plate.
There are many other interpretations. For example in the sublayer
(now embedded in the mesolayer) viscous stresses and Reynolds
stresses are, of course, of the same order. In the rest of the
mesolayer (which is nearly a constant stress layer) the *forces* of
viscosity and the *forces* associated with the Reynolds stress are
of the same order. These two effects are connected because the
decrease of Reynolds stress in a pipe comes from the contribution
of the $\underset{\sim}{u}_2$-motion. The mesolayer is similar to the viscous boundary
layer in "shear-free" turbulence above a plate in a wind tunnel
moving along at the speed of the free stream. (Uzkan and Reynolds,
1967, Thomas and Hancock, 1977, Hunt and Graham, 1978).

The new theories are complete for the spectrum (Long, 1979a),
for the neutral planetary boundary layer (Long, 1979c), for shear
flow near a surface (Long, 1979b), and nearly so for thermal
convection above a plate (Chern & Long, 1980). In all of these there
are major differences with classical theory. In estuarine-type
problems the turbulence is mainly in the upper layers in the
presence of a free surface rather than a no-slip rigid surface,
so that a mesolayer associated with the free surface may not exist.
One problem of importance to oceanographers is the entrainment

velocity at the interfacial region below the upper mixed layer and
the deep layers, e.g., the upper mixed layer in a fjord. In this
case the importance of flattened eddies appears to be fundamental
as discussed by the author (Long, 1978a).

The reader may be justifiably skeptical of the existence of
the mesolayer, since it changes so profoundly the classical concepts.
We will, therefore, present an item of evidence related to the Ekman
boundary layer (Long, 1979c). For a homogeneous fluid above a
rotating plate, the velocities far above are geostrophic with
components $(u_g, v_g, 0)$. The Ekman layer thickness for a homogeneous
fluid is $H = au_*/f$ where u_* is the friction velocity, f is the
Coriolis parameter and a is a constant. The theory makes a number
of predictions. One is that the mean velocity near the ground along
the direction of the surface-stress force increases logarithmically
to begin with and then drops off sharply at a height z_m of order
$(\nu/u_*)R^{\frac{1}{2}}$ for a smooth plate or $h_0 R_0^{\frac{1}{2}}$ for a rough plate where R is
a Reynolds number $u_* H/\nu$ and $R_0 = H/h_0$, and where h_0 is the height
of the roughness elements of a rough plate. Additionally, the transverse Reynolds stress $-\overline{v'w'}$ is predicted to be a maximum at a height
of the same order (mesolayer thickness). In Fig. 4 we see the
results of two quite different experiments (Caldwell, *et al*. 1972,
Kreider, 1973) designed to reproduce the Ekman layer. Of the 20
pieces of data, 19 may be said to lie along the line corresponding
to $R^{\frac{1}{2}}$ or $R_0^{\frac{1}{2}}$ and, remarkably, to the same proportionality constant.
We submit this as strong evidence for the existence of the mesolayer.

MIXED LAYERS AND ENTRAINMENT

Rouse and Dodu (1955) ran an experiment in which an oscillating
grid created turbulence in a stratified fluid as in Fig. 5. They
found that the turbulence remained confined to a layer near the grid
in which there are weak mean buoyancy differences. Below, the fluid
was non-turbulent and there was an interfacial region over which
the turbulence died out. The mixed layer increases in depth with
an "entrainment" velocity w_e. The situation is not dissimilar to
that in an estuary in which there is an upper layer containing a
mixture of fresh and salt-water moving over a nearly inert layer
below. In the estuary the level of the interface remains constant
over periods of time of the order of days and longer yet the entrainment goes on as the downward motion of the interface is just matched
by the upward mean velocity of the lower water to maintain a quasi-
steady state (Long, 1975 a, 1976a).

Much attention has been given to theories and experiments related to this experiment. Experiments with oscillating grids have
been run by Cromwell (1960), Bouvard and Dumas (1967), Kraus and
Turner (1967), Turner (1968), Crapper (1973), Linden (1973, 1975),
Crapper and Linden (1974), Thompson and Turner (1975) and similar

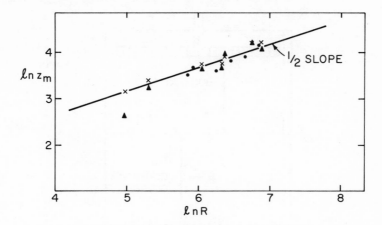

Figure 4. Experiments on the Ekman layer. The triangles correspond
to heights z_m above the rotating plate at which the
transverse Reynolds stress $-\overline{u'v'}$ reaches a maximum. The
crosses and dots are levels at which the logarithmic
behavior of $\bar{u}$ stops rather abruptly (crosses are from
Caldwell, *et al.*, 1972 and the dots are from Kreider
(1973). z_m is scaled on the height of the roughness
elements h_0 over the rough plate of Caldwell, *et al.* and
on ν/u_* for the smooth plate of Kreider. The line
corresponds to $z_m = 1.92\ R^{\frac{1}{2}}$ where $R = u_*/fh_0$ or $u_*^2/f\nu$,
respectively. The points z_m should lie in the mesolayer
of Long (1979c), which is predicted to have a thickness
of order $R^{\frac{1}{2}}$.

experiments with other sources of turbulence by Kato and Phillips
(1969), Brush (1970), Moore and Long (1971), Wolanski (1972),
Baines (1975), Wolanski and Brush (1975), Hopfinger and Toly (1976),
Kantha, Phillips and Azad (1977), Kantha and Long (1979), Long and
Kantha (1979a,b). More recently, experiments have been run in which
a flux of buoyancy q_0 is imposed at the upper surface and we will
discuss these below, but if q_0 is zero, experiments indicate that
entrainment continues indefinitely with an entrainment velocity

$$(w_e/\sigma) = f(h\Delta b/\sigma^2), \quad Ri = h\Delta b/\sigma^2 \qquad (2)$$

where σ is a representative turbulence velocity, Ri is a Richardson
number, h is the depth of the layer and Δb is the buoyancy jump
across the interface. For an oscillating grid in which turbulence
decreases rapidly with depth, w_e is proportional to $Ri^{-3/2}$ according
to most measurements (Turner, 1968), Wolanski and Brush, 1975) and
$Ri^{-7/4}$ according to a theory of the author (Long, 1978a). In
experiments in which a screen is drawn along the water surface to

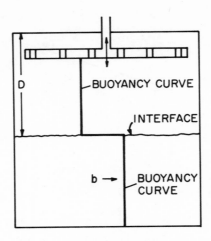

Figure 5.   Rouse and Dodu experiment

create a stress and a mean current, the turbulence at distances
away from the surface is supplied in part by the generation of
turbulence energy from the shear term and this leads to a faster
entrainment velocity in which w_e is apparently proportional to Ri^{-1}
at least for moderate Richardson numbers where σ is now u_*, the
friction velocity (ρu_*^2 is the stress at the surface).

    The entrainment takes place, probably, because the turbulent
eddies cause breaking waves at the interface, if there is a discontinuity of density, or within an interfacial layer. Most workers
in this area assume a discontinuity which, of course, cannot exist
strictly and for mean conditions certainly does not exist even approximately. Thus, an *instantaneous* mathematical discontinuity nevertheless leads to an interfacial layer of the order of thickness of
the interfacial wave-amplitude when averaged over a long time. It
is controversial whether there is always an interfacial layer in
other than a statistical sense, e.g., a layer much thicker than the
interfacial waves. In fact, however, for experiments with oscillating
grids the interfacial layer of thickness h_i is quite thick, of order
of 20% of h and indeed proportional to h according to measurements
by Linden (1973). For high Richardson numbers, at least, the
thickness does, in fact, exceed the amplitude of the disturbances
(Wolanski, 1972), and one is led to a picture of entrainment in
which the turbulent eddies of the mixed layer induce waves within the
interfacial layer some of which must be breaking to yield the
observed salt flux. This idea has been used to construct a theory
by the author leading to a $Ri^{-7/4}$ law which seems close enough to
the experimental $Ri^{-3/2}$ relation (Long, 1978a). Physically, the

theory suggests that the entrainment law is of an Ri^{-1} form if one uses for σ the velocity of the eddies near the interface to form the Richardson number Ri. For a grid, of course, this value of σ is small compared to rms velocities near the grid and the result is a weaker entrainment than Ri^{-1} in terms of the characteristic σ. When a surface stress exists, it is likely that the rms velocity near the surface is proportional to that near the interface suggesting a higher entrainment of the order of Ri^{-1} as apparently observed.

Recent experiments with a non-zero buoyancy flux q_o at the surface are interesting. If q_o is positive (destabilizing) turbulence is created by the convection which is set up in the absence of a grid. The situation that results is similar to that in the upper layers of the sea with cooling imposed as at night or in the fall of the year or due to evaporational cooling, or in the lower layer of the atmosphere during the day due to heating. This "penetrative convection" is discussed in a paper by Long and Kantha (1978) and experiments are reported on in a paper in this volume (Kantha, 1979a).

If q_o is negative (stabilizing) as in heating at a water-air surface and a grid is oscillating, the situation is quite different. If q_o is imposed suddenly on a fluid of uniform density, experiments indicate that a layer forms quickly, probably in one eddy time and remains at this depth although the density difference continues to increase. The depth h is proportional to the Monin-Obukhov length σ^3/q_o where σ is a representative velocity of the turbulence. According to theory and experiment by Long (1978d) and Dickinson and Long (1978) the grid acts like a plane energy source at, say, z = 0 characterized by a single dimensional quantity K called the "action" of the grid. K has the dimensions of eddy viscosity which, in fact, may be constant near the grid, although observations are not conclusive (Thompson and Turner, 1975, Hopfinger and Toly, 1976). In the theory dimensional analysis yields $\sigma(z) \propto K/z$ and $\ell \propto z$ so that eddy viscosity $\sigma\ell$ is constant. Thus the representative rms velocity σ may be represented by K/h. This yields $h \propto K^3/h^3 q_o$ and this is substantiated by experiments also reported on in this volume (Kantha, 1979b).

Mention may be made of some recent experiments which may relate to the existence or non-existence of mesolayers near free surfaces. In the experiment of Dickinson and Long (1978), in Fig. 6 a screen is started oscillating with a stroke s and with frequency ω in a resting, homogeneous fluid. A mixed layer of depth h(t) develops and propagates downward. If the screen is similar to the plane energy source of the author, h should be a function of the action K and t only, i.e., $h \propto \sqrt{Kt}$. Indeed, in the experiment of Dickinson and Long, this is confirmed when there was a free surface and K is taken as $K \propto \omega s^2$. However, we have recently discovered (Kantha and Long, 1980) that when this was replaced by a rigid surface, the exponent of t was reduced to values as low as 0.25. It is possible

to explain this in part by supposing that a mesolayer develops below
the rigid surface in the second case but that a free surface, which
does not impose a no-slip condition, does not yield a mesolayer. If
so, it is unlikely that mesolayers in the mixed layer will develop
below the water-air surface or above the interface, The large eddies
in the layer will still be flattened near both these surfaces, however, and this must be considered in any theory of the entrainment
of the interface.

MOMENTUM TRANSFER IN TWO-LAYER SYSTEMS

     In the previous section, we considered salt-transfer problems
between two layers when one is turbulent and came to a rough understanding of the processes involved including the dependence of the
entrainment velocity on the Richardson number. As we will see,
this will be useful in formulating one of the basic equations of
estuarine circulations.

     In the present section, we consider a problem of equal importance
to that of salt transfer between layers, namely momentum transfer
between layers. A similar estuarine problem is the drag of the lower
fluid on the upper as the latter flows toward the sea.

     An experiment of interest in this connection is one mentioned
earlier by Kato and Phillips (1969) and more recently by Kantha,
et al. (1977). The newer experiments involve a screen at the surface
of a two-fluid system in an annular channel rotating under the influence of a force per unit area ρu_*^2 applied to the screen and hence
to the upper surface of the fluid. The moving screen creates
turbulence and mixing and entrainment as does the grid in the Rouse
& Dodu and the Turner experiments. In addition, however, there is
momentum and momentum transfer perhaps with a velocity profile as
in Figure 7.

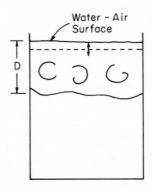

Figure 6.   Propagating turbulent front experiment.

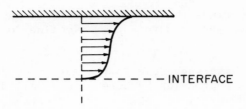

Figure 7. Possible velocity profile in Kato and Phillips experiment.

If we assume, for the moment, that the experiment accurately reproduces flow in an infinite fluid, the mean equation of motion in the direction of the flow is

$$\frac{\partial \bar{u}}{\partial t} = \frac{\partial \tau}{\partial z} \tag{3}$$

where $\tau = \nu \bar{u}_z - \overline{u'w'}$ is the stress. Using a quite accurate theory of Long (1975a) for the relation between screen speed U_s and friction velocity u_*,

$$(U_s/u_*) = (2/\kappa) \ln(h/z_0) + const \tag{4}$$

where κ is Kármáns's constant and z_0 is the roughness length of the screen, we find, for the middle of the layer,

$$\frac{\partial \tau}{\partial z} \simeq w_e u_* / (\kappa h) \tag{5}$$

where we estimate $\bar{u}$ in the middle of the layer to be approximately half the speed of the screen (Kantha, et $al.$, 1977). Then, for the layer as a whole,

$$(\Delta \tau / u_*^2) \simeq w_e / \kappa u_*^2 \tag{6}$$

This is very small and we conclude that there is little change of the stress within the layer. If the interfacial region has a thickness h_i over which the velocity drops from say, $U_s/2$ to zero, the change of τ in this region may be estimated to be

$$(\Delta \tau / u_*^2) \simeq (w_e / u_*) \ln(h/z_0) / 2\kappa \tag{7}$$

From an order-of-magnitude viewpoint, this is small compared to one although the numerical magnitude may be close to one for smaller Richardson numbers. Thus, for larger Ri, at least, we are at a loss to explain the apparent reduction of τ to zero in the lower layer if we continue to assume horizontal homogeneity. This problem was first brought to my attention in a private conversation with

Bo Pedersen and he finds the same difficulty to explain the apparent
decreasing momentum stress across the halocline in an estuary
(Pedersen, 1978).

A possibility, also suggested to the author by Pedersen, is
that the stress is decreasing because of sidewall drag in the experiment. This, however, has been shown by Kantha, *et al*. (1977)
to be a small effect. Another possibility is that the fluid is
actually in motion below the interface[1] and that the momentum stress
is not zero below but decreases downward by the action of friction
or of internal gravity waves. The transfer of momentum by internal
gravity waves will be discussed later with respect to estuaries but
may be dismissed now because the lower fluid has a homogeneous
density. In the experiment, molecular friction is certainly too
small to account for the large decrease of τ and since the lower
layer is not turbulent there seems little likelihood that there are
appreciable stresses below the interface. The only other possibility
seems to be that the curvature of the apparatus is causing a secondary
circulation in the upper layer. If so, the momentum equation includes
other terms:

$$\frac{\partial \overline{u}}{\partial t} + \overline{v}\,\frac{\partial \overline{u}}{\partial r} + \overline{w}\,\frac{\partial \overline{u}}{\partial z} + \frac{\overline{uv}}{r} = \frac{\partial \tau}{\partial z} \tag{8}$$

In Kato and Phillips (1969) it was argued that the stability in the
upper layer was, perhaps, large enough to reduce the secondary
circulation substantially but it seems likely that potential and
kinetic energies are of the same order in the mixed layer (Long,
1975a) so that a parcel can make the journey from top to bottom
and bottom to top of the layer. (In fact, visual observations reveal
a strong downward motion at the outer wall indicating part of secondary circulation). If a secondary circulation exists which substantially lowers the momentum stress and therefore, possibly, the
turbulence with distance downward toward the interface, it may be
a dominant effect in the erosion mechanism and, likely, substantially
reduces the applicability of the results to geophysical problems
in which such circulations are absent.

We may seek to find the effect of the secondary circulation by
considering first the case of a *full* screen of radius R_S rotating
above a two-fluid system. In a coordinate system moving with the
screen, the fluid below, say, near the interface where the velocities
are weak, is rotating nearly as a solid with angular velocity ω
and there is a balance of the Coriolis force and the "pressure gradient" force where the pressure includes the centrifugal force term
$\omega^2 r^2$. In the "Ekman" layer, the shear should be of order u_*/h and
there will be an unbalanced radial force in the layer of order u_*^2/R_S
where R_S is the radius of curvature. In an infinite fluid this leads
to $\overline{v}$ accelerations

[1] We are being very conservative here. Visual observations indicate
very weak mean motions indeed below the instantaneous interface.

such that

$$\bar{v}\,\frac{\partial \bar{v}}{\partial r} \sim u_*^2/R_S \qquad (9)$$

and because $\partial/\partial r \sim R_S^{-1}$, we get $\bar{v} \sim u_*$. However, in the annular channel, $\partial/\partial r \sim W^{-1}$, where W is the width, and we then get

$$\bar{v} \sim u_*\sqrt{W/R_S} \qquad (10)$$

Then

$$\frac{\partial \tau}{\partial z} \sim \bar{v}\,\frac{\partial \bar{u}}{\partial r} \sim \frac{\Delta\tau}{h} \sim u_*^2/\sqrt{WR_S} \qquad (11)$$

In the experiments $W \sim R_S/3$ and $h \sim W$, so

$$\Delta\tau/u_*^2 \sim 0.6 \qquad (12)$$

We may infer then that τ decreases substantially over the layer because of the secondary circulation. We will see that the Coriolis force may play a similar role in the oceans and in fjords but, even so, I think we must be skeptical of the application of these experiments to geophysics.

THE PROBLEM OF ESTUARINE CIRCULATIONS

Let us consider a model of a fjord-type estuary similar, hopefully, to Knight Inlet (Long, 1975c). We will allow wind effects on upper-level mixing but we will assume that the mean wind is zero so that there is no mean surface stress. In addition, we will neglect the rotation of the earth and the curvature of the channel, although we will discuss these two approximations later.

The lower fluid is deep compared to the mixed layer (100's of metres compared to 3 - 20 metres over most of the length and the sill depth is well below the pycnocline so that we may, perhaps, model this with a lower layer of infinite depth. Notice that the mean velocity of the lower layer tends to zero as the depth tends to infinity but, contrary to a remark of Pedersen (1978), the lower layer is not inert because we place no restrictions on the velocity distribution. If this were uniform, it would, of course, be zero and the layer at rest and inert but this is not assumed and, indeed, we expect higher mean velocities landward in the lower layer up near the interface or in the interfacial layer if there is appreciable drag on the upper fluid (sizeable τ_i). Thus we do not assume the velocity is discontinuous. The density gradient is assumed discontinuous for convenience in handling the pressure gradient force but this is not essential and for purposes of physical reasoning we recognize the existence of an inter-facial layer of thickness of order h_i. Below in the deep water the density gradient is small to be sure but so are velocities and the layer must be regarded as stratified. Indeed, Pickard and Rodgers (1959) found internal gravity waves as deep as 270 m.

If we use the Boussinesq approximation, the longitudinal momentum equation averaged over a long time may be written

$$\nabla \cdot \underset{\sim}{y}u = - \frac{1}{\rho_f} \frac{\partial p}{\partial x} + \nabla \cdot \underset{\sim}{\tau} \tag{13}$$

where $\underset{\sim}{\tau} = (\tau_x, \tau_y, \tau_z)$ are the Reynolds stresses associated with the turbulence and ρ_f is the density of fresh water. Since the inward flux of salt water is finite, it follows from the integral of this equation over a cross-section of the whole two-fluid system that

$$gH_x = \Delta b \frac{dh}{dx} - h \frac{d\bar{b}}{dx} \tag{14}$$

where H_x is the slope of the free surface, h is the upper-layer depth, $\Delta b = b_0 - \bar{b}$ is the buoyancy jump across the interface, $\bar{b}(x)$ is the mean buoyancy of the upper fluid, and b_0 is the uniform buoyancy in the lower layer. We repeat, however, that this does not imply anything about the velocity distribution in the upper portions of the lower layer.

An integral of the momentum equation over the upper fluid leads to

$$\frac{\gamma}{W} \frac{d}{dx} (\bar{u}^2 Wh) = \frac{h^2}{2} \frac{d\bar{b}}{dx} - h\Delta b \frac{dh}{dx} - K_d \bar{u}^2 + w_e \bar{u}_i \tag{15}$$

where γ is 1 or 2 or so, depending on the velocity distribution in the upper fluid, W(x) is the width of the estuary, and we have put the Reynolds stress at the interface $\tau_i = K_d \bar{u}^2$ where K_d is a drag coefficient, assumed constant. Finally w_e is the mean upward (entrainment) velocity at the interface and $\bar{u}_i$ is the mean velocity there. In Long (1975c) it was argued that K_d should be large compared to typical drag coefficients, perhaps $K_d = 1$, and on this basis $w_e \bar{u}_i$ is comparatively small. However, Gade and Svendsen (1978) in application of this model to the Sognefjord in Norway and, very recently, Freeland (1979) in application to Knight Inlet, found that K_d is far smaller than this. In such cases, $w_e \bar{u}_i$ may be important and we retain it in this paper.

We also need the two Knudsen relationships

$$\overline{ub}h + \bar{u}_0 b_0 h_0 = 0, \quad \bar{u}_0 h_0 W + \overline{uh}W = q_f W_h \tag{16}$$

for conservation of buoyancy and volume, where $h_0(x)$ is the depth of the lower layer (h_0 very large), $q_f W_h$ is the fresh-water influx and W_h is the width at the head of the estuary. Eq. (16) neglects longitudinal diffusion which, in fact, may dominate in cases of low fresh-water influx. This may be true in Oslofjord (Long, 1975c) and A. Stigebrandt has evidence of a similar character in other Norwegian fjords (personal communication).

A final equation is derived from the considerations above in the mixing experiments. We use for the entrainment velocity[1] (Long, 1975c)

$$w_e = K_n \sigma^3 / h \Delta b \tag{17}$$

where K_n is a constant estimated to be 0.1 and σ is a velocity, which we associate with the rms turbulent velocity near the interface in the mixed layer, perhaps within an eddy length ℓ of the interface. According to arguments of the author (Long, 1975a) $q\ell/\sigma^3$ is a constant equal approximately to $1/150$ and $qh/\sigma^3 \simeq 1/10$. The continuity equation for the upper layer now becomes

$$\frac{d}{dx} (Wh\bar{u}) = K_n \frac{\sigma^3 W}{h \Delta b} \tag{18}$$

We may now nondimensionalize using

$$Q = \bar{u} h R_w / q_f, \quad \xi = x\sigma / q_f, \quad \eta = h\sigma / q_f \tag{19}$$

$$D = \Delta b / b_o, \quad m = \gamma \sigma^3 / b_o q_f, \quad R_w = W / W_h \tag{20}$$

If we adopt the convenient assumption that $\bar{u}_i = -A\bar{u}$ where A is a constant, Eq. (15), despite the addition of the $w_e \bar{u}_i$ term, for a channel of constant width $R_w = 1$, is identical to the problem in Long (1975c) and we obtain the author's solution of that paper

$$(\eta/\eta_h) = (\zeta/\eta_h^3)^{s/3} \{(1 + c^3 \eta_h^3)/(1 + c^3 \zeta)\}^{(s+1)/3} \tag{21}$$

$$\xi = \frac{\gamma}{m K_n} \int_{Q_c}^{Q} (\eta/Q) \, dQ \tag{22}$$

---

[1] This law was inspired, of course, by mixing experiments but, in fact, Eq. (17) can be interpreted simply as a definition $K_n \sigma^3 \equiv w_e h \Delta b$ of a quantity with the dimensions of (velocity)^3. Since $w_e \Delta b$ is the buoyancy flux q_e near the interface, σ^3 may also be regarded as a quantity proportional to buoyancy flux. Whether the use of the law is justifiable or not depends utterly on the assumptions made about the size and the variations of σ^3 which, in turn, should depend on the nature of the mixing processes in a given fjord, wind, tides, breaking of internal waves coming from the sill, etc. Here, we assume σ^3 is constant in the fjord. This seems to be a poor assumption in Knight Inlet (Freeland 1979) but at this stage of our search for understanding, it is, perhaps, excusable.

where

$$\zeta = (\eta/Q)^3, \quad s = 1 + 1/(1 + K/mK_n), \quad C^3 = (s-1)/2m \quad (23)$$

except that K is no longer the drag coefficient K_d but, rather

$$K = K_d + \frac{AK_n m}{\gamma} \quad (24)$$

In Knight Inlet, according to Freeland (1979) and in the Sognefjord, according to Gade and Svendsen (1978), $K/K_n m$ is probably 3 or less. Freeland believes that the above model with small K is in reasonable agreement with what is measured in Knight Inlet. We will not make detailed comparisons, in deference to the careful discussion in this volume but one computation may be presented. If we use

$$\frac{K}{K_n m} = 3, \quad K = 3 \times 10^{-4}, \quad K_n = 0.1, \quad q_f = 2000 \text{ cm}^2 \text{sec}^{-1} \quad (25)$$

$$b_o = 25 \text{ cm sec}^{-2}, \quad \gamma = 1.3, \quad L = 10^7 \text{ cm} \quad (26)$$

where L is the length of the inlet, we compute

$$m = 10^{-3}, \quad \sigma = 3.38 \text{ cm sec}^{-1}, \quad C = 5.0 \quad (27)$$

Using Figs. 4 and 5 in Long (1975c), we get

$$Q_c = 4.9, \quad h_h = 3.07 \text{ m} \quad (28)$$

where Q_c is the ratio of the flux in the upper layer at the mouth to the river runoff, and h_h is the depth of the mixed layer at the head. All values including the ones derived in (28), are in reasonable agreement with observations and with similar calculations by Freeland. Other choices are possible but values of K much smaller than 1 are needed to get the small values of h_h observed. If, as seems likely, A in Eq. (24) is considerably less than 1, the $w_e u_i$ -contribution to the drag is small, i.e., the upward flux of momentum due to the entrainment velocity is negligible.

The small values of K_d and the large rms velocities σ should be considered together. $K_d \bar{u}^2$ is of order $-\overline{u'w'}$ near the interface and perhaps in the bulk of the mixed layer. In typical shear flows, say in pipes, this is of order σ^2 but here, apparently, it is orders of magnitude smaller. To the extent that the tidal oscillations contribute to σ, there will be no corresponding contribution to $-\overline{u'w'}$ because no net momentum is involved. Similarly, a wind varying from up-estuary to down-estuary creates turbulence but no net contribution to $-\overline{u'w'}$. According to the above calculations for Knight Inlet and using $\bar{u} \sim 10$ cm/sec, we find $-\overline{u'w'} = u_*^2 \sim 3 \times 10^{-4} \times 10^2 \text{ cm}^2 \text{sec}^{-2}$ or $u_* \sim 0.17$. Notice, however, that these small values of $-\overline{u'w'}$ and K_d do not justify the conclusion that friction is unimportant. Thus, for $K_d \sim 3 \times 10^{-4}$, friction forces in the layer, of order $K_d \bar{u}^2/h$, are, in fact, considerably larger than the longitudinal acceleration $\bar{u}\bar{u}_x$ if we use 100 km as the longitudinal length scale.

The small values of $-\overline{u'w'}$ relate to the objection of Pedersen (1978) to the author's model, that the stress must be zero below the interface and that there is no mechanism to reduce it to zero from its finite values in the mixed layer. The objection is removed, however, by noting that, without turbulence in the lower layer, salt flux is zero but *momentum flux can be effected by internal waves* (Booker and Bretherton, 1967). Thus, the stress need not, in fact, be zero in the lower layer and it seems likely that the small $-\overline{u'w'}$ can be transmitted. Of course, if the upper layer depth is close to the Ekman-layer depth, the stress will be close to zero at the interface anyhow. Our rough calculations of K_d suggest that the stress at the interface is appreciable with respect to inertial forces, at least, but uncertainties are great.

The data of Freeland (1979) show that σ^3 in the entrainment law is about 8 times as large in the lower straight reach of Knight Inlet than in the upper sinuous portion and suggest that this is because of the breaking of gravity waves on the halocline propagating up-inlet from the region of the sill where they are generated by tidal motions. They are observed in this region, to be sure, but the extent to which they break and contribute to the mixing is unknown. An alternative is that the mixing is not strongly affected by the waves but is caused by the turbulence in the upper layer associated with the observed large shears there. The extra mixing in the lower reaches of Knight Inlet is easily accounted for by assuming that this section, which lies east-west along the prevailing wind, has winds which are $(8)^{1/3} = 2$ times larger (in peak periods when the main mixing is accomplished) than in the upper section which is nearly normal to the prevailing wind and which may be protected from the wind to this extent. Additionally, the summer sea-breeze may be restricted to the lower section of the inlet. Careful, long-period wind observations over the entire year are needed as well, perhaps, as model experiments to determine if the waves propagating up-inlet from the sill after the ebb-tide do, in fact, tend to break or lose energy in some other way, say, to upper-layer turbulence.

We have neglected the rotation of the earth and the curvature of the channel. We know that rotation has some effects in estuaries, leading, for example, to reduced salinities to the right of the direction of flow and to effects on dispersion (Pritchard, 1952, Smith, 1976, 1978). In a channel, the Coriolis force may be estimated first by comparing accelerations $\bar{u}^2/R_s$ to $f\bar{u}$, i.e., by computing the Rossby number $\bar{u}/fR_s$ where $\bar{u}$ is a typical longitudinal velocity and R_s is a typical radius of curvature. If we use

L = 20 km, $\bar{u}$ = 6 cm sec^{-1} as typical, say, of Knight Inlet, we get extremely small values for $\bar{u}/fR_s$. We conclude that the flow is nearly geostrophic and that the pressure-gradient force toward the left of the flow nearly cancels out the Coriolis force toward the right by a piling up of water on the right bank. Any imbalance, however, will lead to secondary circulations. If the unbalanced radial force is $C_1 f \bar{u}$, where C_1 is some factor, we may estimate $\bar{v} \partial \bar{v}/\partial y \sim C_1 f \bar{u}$ or $\bar{v} \sim \sqrt{C_1 f \bar{u} W}$ if the transverse length scale is the width of the estuary. Then

$$\frac{\partial \tau}{\partial z} \sim \bar{v} \frac{\partial \bar{u}}{\partial y} \sim \overline{vu}/W$$

so that if we denote by $\Delta \tau$ the increment of τ over the mixed layer, we get

$$\Delta\tau/u_*^2 \sim (C_1 f/W)^{\frac{1}{2}} \bar{u}^{3/2} h/u_*^2 \tag{29}$$

This yields $\Delta\tau/u_*^2 \sim 12 C_1^{\frac{1}{2}}$ so that rotation is of strong importance if, as seems likely, C_1 is as large as 0.01 or so. Indeed the Ekman-layer thickness, $0.4 u_*/f$, is of order 6 metres which is the order of the thickness of the mixed layer.

Let us now investigate curvature effects. The same arguments as those leading to Eq. (10) now yield for the secondary circulation $\bar{v} \sim \bar{u}(W/R_s)^{\frac{1}{2}}$ so that $\partial\tau/\partial z \sim \bar{v}\partial\bar{u}/\partial y \sim \bar{u}^2/(WR_s)^{\frac{1}{2}}$ or

$$\Delta\tau/u_*^2 \sim \bar{u}^2 h/u_*^2 (WR_s)^{\frac{1}{2}} \tag{30}$$

This yields $\Delta\tau/u_*^2 \sim 2.6$ so that the secondary circulation due to curvature seems also to be important. We conclude that our neglect of rotation and curvature is not justifiable on a quantitative basis but our qualitative conclusions are perhaps useful.

In the theoretical model, the Froude number becomes one at the mouth and this is considerably larger than any values found in Knight Inlet (Freeland, 1979). Of course, the width of the inlet does not vary in the same way as in the model near the mouth but, in addition, even though the sill is well below the pycnocline, observations reveal large internal waves near the sill (Smith, 1979) with much accompanying turbulence far below. This must have a major effect not contemplated by the theory. One possibility is that conditions near the sill may favor, instead of a smooth transition from subcritical to supercritical flow, a discontinuous jump as in Fig. 8. This is impossible in single-fluid hydraulics and we must make a special investigation in the present two-layer problem. We do this in the next section.

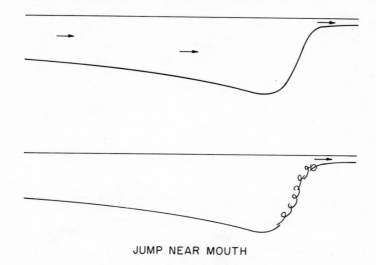

JUMP NEAR MOUTH

Figure 8.   Smooth transition to supercritical flow contrasted with hydraulic jump near mouth.

HYDRAULIC JUMP IN A WIDENING ESTUARY

Let us consider a situation pictured in Fig. 9 in which we may neglect mixing entirely. This is appropriate for the short distance along the fjord involved in the present discussion. The lower layer is at rest and the longitudinal pressure gradient along the estuary is zero in the lower water. From hydrostatics (and using the Boussinesq approximation) the pressure in the lower layer is

$$p/\rho = g(H-z) + \Delta b (H-h-z) \qquad (31)$$

where ρ is the density of the upper fluid which we now take to be the reference density, and Δb is the buoyancy difference. The equation of the free surface is $z = H$, and h is the thickness of the upper layer. Thus in the lower layer, to within the Boussinesq approximation,

$$gH_x = h_x \Delta b \qquad (32)$$

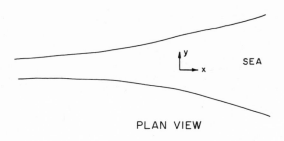

PLAN VIEW

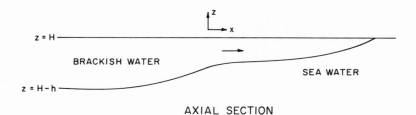

AXIAL SECTION

Figure 9.   Plan view and axial section in a fjord.

or

$$gH = h\Delta b + const \qquad (33)$$

We assume irrotational motion in the upper layer so that the Bernoulli equation is

$$p/\rho + u^2/2 + gz = const \qquad (34)$$

where u is the longitudinal velocity which we assume to be uniform across a section. We neglect the other two components of the kinetic energy. They will be small if the length scale along the estuary is large compared with the depth and if the change of width is gradual. Hydrostatics leads to

$$gH + u^2/2 = const \qquad (35)$$

or, using (33) and the continuity relation, $uhW = R_f$,

$$h(1 + \tfrac{1}{2}F^2) = E \qquad (36)$$

where E is a constant proportional to the sum of kinetic and potential energies, and F is the densimetric Froude number

$$F^2 = R_f^2/(W^2 h^3 \Delta b) \tag{37}$$

where R_f is the upper-layer flux and W is the width of the channel. Let us non-dimensionalize using the depth of the upper layer h_r and the channel width W_r at some up-inlet section. We have

$$2h' + F_r^2/(h'W')^2 = E' \tag{38}$$

where

$$h' = h/h_r, \quad W' = W/W_r \tag{39}$$

and E' is non-dimensional constant proportional to the energy. If energy is conserved along the flow $E' = 2 + F_r^2$. Differentiating (38), we get

$$\frac{dh'}{dW'} = h'F^2/(1-F^2)W' \tag{40}$$

This shows that the depth of the upper level decreases with distance downstream (assuming the width increases downstream) if and only if conditions are supercritical, i.e., $F^2 > 1$. Eq. (36) shows that F^2 then increases along the estuary. Thus, if we require energy conservation and that h decreases everywhere toward the mouth, $F_r^2 \geq 1$, i.e., conditions must be supercritical at the reference section and everywhere down-inlet. In fact, it is physically obvious that h must decrease ultimately as we proceed seaward so that conditions must ultimately be supercritical in any case. Let us consider, however, the typical case in which $F_r^2 < 1$. We see from (36) and (40) that h increases and that F^2 decreases seaward. These behaviors continue toward the mouth and h can never decrease as required ultimately by physical considerations. We conclude that if energy is conserved along the estuary, conditions must be supercritical everywhere and h must decrease monotonically.

Another possibility exists, however, namely that the flow at the reference section is supercritical with h increasing downstream as in the theory of the author, but then changes discontinuously at some section (with loss of energy in an internal hydraulic jump) to a smaller h and supercritical conditions, with h then decreasing with further distance downstream as required by physical arguments. To see if this can be, we must make two investigations (Long, 1976b). In the first, we calculate the momentum balance in a jump as in Fig. 10. We consider the rate of change of (longitudinal) x-momentum of the fluid contained between sections A and B at time t. At time t + dt, this fluid is between sections A' and B'. Its rate of change of momentum, M, given by,

$$\frac{dM}{dt} = u_b^2 W h_b \rho - u_a^2 W h_a \rho \tag{41}$$

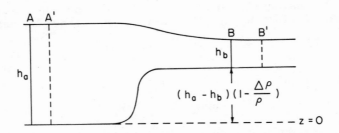

Figure 10.   Internal hydraulic jump

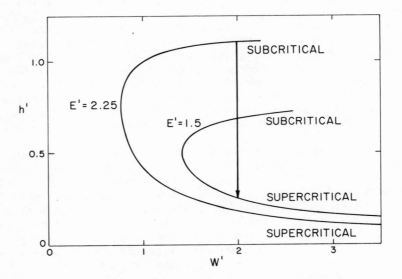

Figure 11.   Energy curves.

equals the pressure forces on the fluid, i.e.,

$$W\int_0^{h_a} \rho g(h_a - z)\,dz - W\int_{h_1}^{h_2} \rho g\{h_a - \frac{\Delta\rho}{\rho}(h_a - h_b) - z\}\,dz$$
$$-W\int_0^{h_1} \rho g\{h_a - (1 + \frac{\Delta\rho}{\rho})z\}\,dz \tag{42}$$

where $h_1 = (h_a - h_b)(1 - \Delta\rho/\rho)$ and $h_2 = h_a - (\Delta\rho/\rho)(h_a - h_b)$ and we have neglected mixing and non-hydrostatic pressures in the jump. Integration yields, to within the Boussinesq approximation,

$$\frac{u_b^2 h_b}{g} - \frac{u_a^2 h_a}{g} = \frac{\Delta\rho}{2\rho}(h_a^2 - h_b^2) \tag{43}$$

Using continuity, $u_a h_a = u_b h_b$, we may write this in either of two ways:

$$F_a^2 = \tfrac{1}{2}(h_b/h_a)(1 + h_b/h_a) \tag{44}$$
$$F_b^2 = \tfrac{1}{2}(h_a/h_b)^2(1 + h_b/h_a) \tag{45}$$

If $h_b \leq h_a$, we get $F_a^2 \leq 1$, $F_b^2 \geq 1$ so that a jump downstream from subcritical to supercritical is possible from a momentum viewpoint.

Let us now consider energy. For illustrative purposes, we may choose a representative subcritical initial flow $F_r^2 = 0.25$. Then, if energy is conserved, $E' = 2.25$ in Eq. (38). Fig. 11 contains this curve plus one corresponding to an arbitrarily chosen lower energy level $E' = 1.50$. We see that there is a possible jump to a lower energy from the subcritical branch as shown by the arrow. To within the approximations used in the argument, we conclude that supercritical flow in the main portions of the fjord is possible with h increasing gradually but that there must then be a hydraulic jump to a smaller h followed by a further monotonic reduction of h as the upper fluid spreads out at sea. This may be what is occurring in Knight Inlet with the jump precipitated by the widening at Hoeya Head or by the presence of the sill at the same location. The control of up-inlet conditions by such a "statistical" jump may be similar to that by the smooth transition to $F^2 > 1$ in the original theory of the author or, on the other hand, the smooth transition may be impossible.

FURTHER REMARKS ON AN ESTUARY OF VARIABLE WIDTH

The theory of the author (Long, 1975c) led to two equations for η and $F^2 = mQ^3/\eta^3 R_w^2$ along a channel of variable width. They are

$$\frac{d\eta}{d\xi} = \frac{1}{1-F^2}\left\{\frac{2K_n mR_w}{\gamma}\left(\frac{1}{4} - \frac{KF^2}{2mK_n R_w} - F^2\right) + \frac{\eta F^2}{R_w}\frac{dR_w}{d\xi}\right\} \qquad (46)$$

$$\frac{dF^2}{d\xi} = \frac{3F^2 K_n mR_w}{\gamma\eta(1-F^2)}\left\{F^2 + \frac{1}{2} + \frac{KF^2}{mK_n R_w}\right\} - \frac{F^2(F^2+2)}{R_w(1-F^2)}\frac{dR_w}{d\xi} \qquad (47)$$

where Eq. (46) reduces to Eq. (40) if mixing and friction are neglected.

Eqs. (46) and (47) have not been solved except for $R_w = 1$ and it is possible in view of the discussion of the previous section that for decreasing or increasing widths no continuous solution exists, e.g., hydraulic jumps must occur and solutions must be obtained by imposing jump conditions across them.

We may outline the procedure for finding the solution for an estuary of arbitrary shape, however, if one does exist over the whole length. We first notice that Eqs. (46) and (47) are invariant for transformations $\eta = \alpha\eta^*$, $\xi = \alpha\xi^*$. Using $R_w = f(\xi^*)$, we may write

$$f(1-F^2)\frac{d\eta^*}{d\xi^*} = \frac{2K_n m}{\gamma}f^2(\tfrac{1}{4}-F^2) - \frac{K}{\gamma}fF^2 + \eta^*F^2\frac{df}{d\xi^*} \qquad (48)$$

$$\eta^*(1-F^2)f\frac{dF^2}{d\xi^*} = \frac{3K_n m}{\gamma}F^2 f^2(F^2+\tfrac{1}{2}) + \frac{3F^4 Kf}{\gamma} - F^2(F^2+2)\eta^*\frac{df}{d\xi^*} \qquad (49)$$

Let us investigate the solution in the region just past the mouth ($\xi^* = 0$) where, by assumption, the width of the estuary increases linearly. At $\xi^* = a$, the flow becomes critical. Then, if $\zeta^* = \xi^*-a$ and R_{wm} is the value of R_w at $\xi^* = 0$,

$$\eta^* = \eta_c^* + b_1\zeta^* +\ldots, \quad F^2 = 1 + a_1\zeta^* +\ldots, \quad f = R_{wm} + c_1 a + c_1\zeta^* \qquad (50)$$

Substituting (50) into (48) and (49) and equating coefficients of ζ^*, we obtain to zero order two identical equations which may be used to solve for $R_{wm} + c_{1_a}$:

$$R_{wm} + c_1 a = \{-K/\gamma + (K^2/\gamma^2 + 6c_1 K_n m\eta_c^*/\gamma)^{\frac{1}{2}}\}/(3K_n m/\gamma) \qquad (51)$$

The other two equations obtained from equating coefficients of ζ^* may be written

$$A_1 b_1 + B_1\eta_c^* = E_1 \qquad (52)$$

$$A_2 b_1 + B_2\eta_c^* = E_2 \qquad (53)$$

where

$$A_1 = a_1(R_{wm} + c_1a) + c_1, \quad B_1 = c_1a_1, \quad A_2 = -3c_1,$$

$$B_2 = a_1^2(R_{wm} + c_1a) - 4c_1a_1 \tag{54}$$

$$E_1 = + \frac{2K_n m}{\gamma} \{1.5 c_1(R_{wm}+c_1a)+a_1(R_{wm}+c_1a)^2\} + \frac{K}{\gamma}\{a_1(R_{wm}+c_1a)+c_1\} \tag{55}$$

$$E_2 = - \frac{3K_n m}{\gamma} \{2.5 a_1(R_{wm}+c_1a)^2+3c_1(R_{wm}+c_1a)\} - \frac{3K}{\gamma}\{2a_1(R_{wm}+c_1a)+c_1\} \tag{56}$$

Thus

$$b_1 = (E_1B_2-E_2B_1)/(A_1B_2-A_2B_1) \tag{57}$$

$$\eta_c^* = (A_1E_2-A_2E_1)/(A_1B_2-A_2B_1) \tag{58}$$

The integration may then proceed as follows. We specify R_{wm} and c_1 and the constants K_n, m, γ. We also specify a_1. This permits calculation of b_1 and η_c^* from (57) and (58), and $R_{wm}+c_1a$ from (51). We may then compute η^* and F^2 from (50) for, say, $\xi^* = 0$, and $\xi^* = -\Delta\xi^*$ and then use the finite difference forms of (48) and (49) to compute F^2 and η^* at subsequent grid points until $R_w = 1$. This corresponds to the head of the estuary, where $F^2 = F_h^{2w}$, $Q = 1$, η_h, $= \eta_h^*\alpha$. Then we have

$$\alpha^3 = \frac{m}{F_h^2\eta_h^{*3}} \tag{59}$$

and we have a solution for an estuary of width

$$R_w = f(\frac{\xi}{\alpha}) \tag{60}$$

Also possible is a perturbation approach to (46) and (47) for slowly varying widths.

DISCUSSION OF TURBULENCE IN THE MIXED LAYER

The origin of the turbulence in the mixed layer is of great interest. The possible sources are turbulence associated directly with tidal motions or with the breaking of internal waves generated by tides, turbulence driven by the mean-shear production, turbulent thermal convection and turbulence produced by the wind.

Various fjords will have different major sources of turbulence and all of the above may be important in one fjord or another. In

the Baltic Sea, for example, (Fonselius, 1969, Long, 1976a) there
is only a weak tide and most of the turbulence probably arises from
the conversion of energy of the mean current, set up by winds blowing
over a period of days, to turbulent energy through the shear-gener-
ation term in the energy equation (Long, 1977) and by thermal
convection over short periods at night and over longer periods from
late summer to winter with seasonal cooling. The diurnal heating
and cooling probably contributes negligibly to the entrainment.
Thus, if wind is absent, the stability created by day-time heating
will suppress all turbulence. Then, for equal cooling at night,
this thin heated layer must be cooled down again before any
convection starts and the net mixing effect is zero. Of course,
winds do occur and distribute the warm water over the whole mixed
layer during the day but the rms velocities should be reduced by
this stabilizing effect as much as they are increased during the
subsequent night cooling. The longer-period cooling ultimately
drives downward and weakens the summer thermocline in the Baltic
as well as in fresh-water bodies. When it is destroyed in the fall
in the Baltic, the turbulence then penetrates to the permanent
pycnocline and entrainment processes take place there. If we
assume a temperature drop in the upper layer of the Baltic of say
10^0C over five months, $\partial \bar{T}/\partial t \sim 7 \times 10^{-7}$ deg/sec or $\partial \bar{b}/\partial t \sim 7 \times 10^{-9}$
cm/sec^3.

Since

$$\frac{\partial \bar{b}}{\partial t} = \frac{\partial q}{\partial z} \qquad\qquad (61)$$

where q is the buoyancy flux, we get for the top of the mixed layer
above the halocline, which is at depth of about 60 m,

$$q_0 \sim 4.6 \times 10^{-5} cm^2/sec^3 \qquad\qquad (62)$$

The convection velocity near the interface w_* (Long, 1977) is of
order $(qD)^{1/3} \sim 0.65$ cm/sec. The quantity of importance for the
entrainment velocity is σ^3 near the interface and estimates by the
author (1976a) indicate the need for $\sigma \sim 1.4$ cm/sec or $\sigma^3 \sim 2.74$
cm^3/sec^3 over the entire year to accomplish the observed salt flux
from the bottom layer. Since the entrainment occurs only over a
five-month period at most, we need $(12/5) \times 2.74 = 6.6$ cm^3/sec$^3 \sim \sigma^3$
compared to $w_*^3 \sim 0.3$. We conclude that the contribution of convection to the entrainment of the halocline is negligible, that seasonal cooling simply acts to wipe out the thermocline built up by the
seasonal heating and that the entrainment across the halocline in
the Baltic is caused by turbulence from wind action at the surface.
The conclusion that convection is negligible may also be drawn from
similar reasoning for the shallower mixed layers in typical fjords.

Let us now consider wind effects. We have adopted an entrainment law in which w_e (which is proportional to salt flux at the halocline) is proportional to σ^3. Probably σ is proportional to the wind speed so that the contribution to w_e from a two-day storm with velocities of 20 m sec^{-1}, say, is the same as a steady wind of 4.7 m sec^{-1} acting over a period of five months! Indeed, 5 m sec^{-1} or so is a typical wind over the Baltic and such storms probably occur once each winter. Such considerations apparently overcome the difficulty of explaining the rather large value of $\sigma \sim 1.4$ cm sec^{-1} cited above needed to explain entrainment in the Baltic, in comparison with u_*-values of $\frac{1}{2}$cm sec^{-1} produced by typical average winds over large bodies of water. The average winds experienced apparently have little impact on the entrainment which, in the Baltic, is accomplished by one or two severe storms each winter. The importance of rare events is acknowledged for many estuarine phenomena, for example, deep-water renewal (Gade, 1970).

Another influence of importance, in addition to density stratification, for a body of water of the size of the Baltic (or the oceans) is the earth's rotation. If the water is homogeneous and the wind blows at a speed W, the stress exerted may be taken to be equal to $\rho_a K_d W^2$ where ρ_a is the density of air and K_d is the drag coefficient estimated by Kullenberg (1971) for the Baltic to be $K_d = 0.0012$. The drag of the air equals $\rho_w u_*^2$ where u_* is the friction velocity in the water and ρ_w is the density of water so that $u_* \simeq 0.0012W$. The Ekman layer depth is $H \simeq 0.4u_*/f$, according to observations in a laboratory experiment (Caldwell, *et al.*, 1972). If we take W to be an average of 5 m sec^{-1}, we get $H \simeq 21$ m which is much less than the halocline depth. The situation in this respect is not very clear but we may conjecture as follows: In winter there is certainly little density variation in the upper 50-60 m in the Baltic so that the momentum stress should vanish at 20 m or so under typical wind speeds. A first impression is that turbulence should also vanish below H in which case no entrainment processes should be acting at the halocline under typical winds. This may be true in fact because, as we have seen, there is little contribution to entrainment in typical wind conditions anyhow. But we should mention that in some preliminary experiments by my student, Stuart Dickinson, in a rotating container with grid-produced turbulence in a homogeneous fluid, there appears to be a layer of some thickness below the grid in which the turbulence looks more or less isotropic. Below this, there are strongly anistropic motions consisting of vortices extending all the way to the bottom of the container. If the upper layer is an "Ekman layer", i.e. one of depth proportional to σ/Ω where σ is the characteristic velocity, we have $H \sim (K/H)\Omega$ or $H \sim (K/\Omega)^{\frac{1}{2}}$ where K is the action and Ω is the angular velocity of the vessel. The layer depth is difficult to define, let alone measure, but there are some preliminary indications that this prediction regarding the depth is correct. The existence of vortices below the "Ekman Layer" suggest that motion does not cease below the Ekman Layer in

field conditions in which the layer may exist, e.g., when the
Monin-Obukhov length exceeds the Ekman depth. Such vortices would,
seemingly, permit non-zero, but weak, entrainment at an interface
below the Ekman depth. But, in the case of the Baltic, such weak
entrainment need not be invoked because the depth of the interface,
which changes very little from day-to-day under normal circumstances
will, in fact, be less than the Ekman depth during the storms which
are capable of causing the major entrainment and the turbulence can
then reach the halocline.

Walin (1972,a,b) believes that the main effect of wind in
creating turbulence in the upper layers of the Baltic is through
the creation of combined rotational-internal gravity waves which
break along the sides of the basin, and he has measured up-and-
down motions on the east coast of Sweden which appear much greater
near shore than further out, suggesting such behavior. This may
be true and yet the entrainment ideas advanced in this paper may
still be valid approximations if we regard σ^3 as an average over
the whole basin. Nevertheless, there is no evidence that such
localized effects are dominant.

The effects of wind and tide in typical fjords is an uncertain
matter. In the measurements of Pickard and Rodgers (1959), the
wind strongly affects the motions more than half of the time
frequently reversing the seaward flow in the upper layer, for
example! It is simpler, however, to combine the wind and tidal
effects in explanation of the typical velocities and velocity
shears $\bar{u}_z$ in the upper layer. Pickard and Rodgers' data show typical
velocity shears in the upper level which we may crudely estimate
to be 30 cm sec^{-1}/5 m. If we assume a vertical eddy size ℓ of 50 cm,
using again $\ell \sim h/14$, we get $\sigma \sim \bar{u}_z \ell \sim 3$ cm sec^{-1} which we have
calculated above as needed to account for the observed entrainment
velocities. These shears and lengths are conservatively small and,
although the calculations are uncertain, it may be a reasonable
conclusion that turbulence from turbulent shear production appears
to be sufficient to explain the level of turbulence needed for
entrainment.

This conclusion that shear-production from shears in the upper
levels associated with wind and tide, are sufficient to explain the
entrainment in fjords such as Knight Inlet, conflicts with the
efforts of a number of workers (Stigebrandt, Smith and others)
mentioned above to explain much or most of the mixing by the breaking
of internal gravity waves[1], either against the sides of the fjord

[1] Internal wave breaking, used here, includes situations in which
either shear instability (with local Richardson numbers less than ¼,
perhaps) or gravitational instability (local reversals of density
gradient) occurs.

or within the basin itself in the manner of hydraulic jumps or
bores or by breaking locally in regions of high shear. These are
important studies but a conservative approach should note the
following: (1) It is not known that internal waves break against
the bottom and sides of a fjord although laboratory experiments are
suggestive. (2) There is no direct evidence that internal waves
break for other reasons in fjords to produce mixing. (3) Even if
such breaking occurs, it is not known to what <u>extent</u> this contributes
to the mixing, i.e., whether the produced turbulence is sufficient
to explain the needed entrainment. (4) Gravity waves, such as
those propagating up Knight Inlet, may lose their energy before
breaking by interactions with turbulence in the upper layers so
that it is not true, as frequently stated, that the ultimate sink
for internal wave energy is mixing and entrainment. If the gain
of energy is in turbulence well removed from the interface, little
entrainment will result.

The gravity-wave phenomenon, indeed all wave phenomena, are
favorite subjects for investigation because they can be investigated
mathematically as well-determined problems once linearization and
other approximations are applied, in contrast to the messy problems
of turbulence, but all wave theories break down long before conditions approach those even close to turbulence (with the exception
of the hydraulic jump phenomenon). I advise caution regarding the
importance of gravity waves (and, in the Baltic, Kelvin waves), in
the mixing process, not because I am sure that other mixing effects
are dominant, but because I have seen indications that hardened
attitudes are already developing in favor of such sources of mixing
and I think we are far from a state of understanding that would
justify this.

OTHER THEORIES OF FJORD CIRCULATIONS

This paper emphasizes the author's opinion that future solutions
of problems of fjord dynamics will be judged primarily on the basis
of the adequacy of the treatment of turbulence. Much work in the
past has used eddy viscosity and eddy diffusion techniques to handle
the problem of turbulence. These techniques have proved useful for
obtaining order-of-magnitude estimates of certain effects but, in
fact, the introduction of eddy coefficients is self-defeating
because one of the few simplifications which work well in turbulence
theory is the neglect of molecular coefficients and to bring an
analogous quantity like eddy viscosity (which is actually a new
unknown) back into the analysis and then to adjust its overall
magnitude and its variation in space and time to bring theory and
observation into agreement is to assume, in effect, the answer one
is seeking. But, worse than this, the eddy viscosity concept often
obscures the problem. For example, if a wind blows over a large
body of water, exerting a kinematic stress u_*^2 in the water, an

Ekman layer is produced, as we have discussed. Its depth for a
homogeneous fluid is often estimated by oceanographers, assuming,
by analogy to Ekman's theory for the laminar case,

$$H \sim (K/f)^{\frac{1}{2}}$$

where K is eddy viscosity. But to estimate K, one may then assume
$K \sim u_* H$, i.e. that the eddy velocity is of order u_* and the eddy
size of order H. Then

$$H \sim u_*/f$$

But, of course, the latter is obtainable directly by dimensional
analysis and the introduction of eddy viscosity to get the same
result is decidedly inferior because the right answer requires two
unnecessary assumptions.

If velocity and density distributions are continuous, for example, as in so-called type 2 fjords (Pickard, 1961), we may hope
to obtain information using the K-approach. Some recent papers are
by Perrels and Karelse (1978), Smith (1976, 1978), Connor and Wang
(1974). Similarity solutions with various assumptions about the
variation of eddy coefficients have been found by Rattray, Winter
and others (Rattray, 1967, Rattray and Hansen, 1965, Winter, 1972,
1973) and realistic flow patterns obtained. However, for the
typical, strongly stratified fjords, near-discontinuities of density
occur and, as pointed out in Moore and Long (1971), eddy coefficients may change by three orders of magnitude or more from one
layer to the next so that any assumption of constant or even
continuous variation is quite useless.

A number of recent theories of fjord circulations, including the
classical theory of Stommel (1951) as well as more recent ones by
Pedersen (1972, a,b), Ottesen-Hansen (1975), Long (1975c), have
been described by Pedersen (1978) and we need not discuss them
further. We will, however, complete this section by referring to
a recent theory of Pearson and Winter (1978) of circulation in a
fjord. The theory is indeterminate† because it requires a knowledge
of density distributions from field observations as input. But the
paper raises an interesting question regarding the possibility of
a two-way flux of salt and momentum and we will briefly describe
these fluxes and the difficulties that an idea of two-way flux
entails. The authors use a two-fluid system which we portray in
Fig. 12. In this figure the density and felocity of the two fluids
are ρ_1, u_1 and ρ_2, u_2 where these densities and velocities are
uniform vertically but variable along the direction of the estuary.
Pearson and Winter assume discontinuities of ρ and u across the
interface but this introduces indeterminacies, and we retain a
thin interfacial layer of thickness $h_i = h_2 - h_1$ to permit the

---

† *Editor's note: Pearson and Winter's model is strictly indeterminate, but they prefer to use it as, and call it, a diagnostic model.*

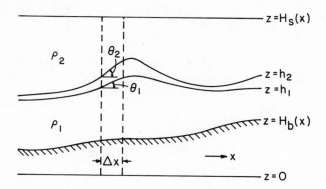

Figure 12.   Schematic picture of a two-layer fjord. The upper layer
has a thickness H_s-h_2 and density ρ_2; the lower layer has
a thickness h_1-H_b and density ρ_1. There is a thin interfacial layer.

density and velocity to vary continuously. We may take mean values
$\rho_i = (\rho_1 + \rho_2)/2$ and $u_i = (u_1 + u_2)/2$. In the limit as $h_i \to 0$, this
yields the model of Pearson and Winter.

Let us integrate the equations $\nabla \cdot y = 0$, $d\rho/dt = 0$ for volume
and mass conservation over the region of dimensions W and Δx in the
transverse and longitudinal directions, where breadth W is considered
constant for simplicity. We get, exactly,

$$\frac{d}{dx}\{u_1(h_1-H_b)\} = -\frac{w_{e1}}{\cos\theta_1}, \quad \frac{d}{dx}\{(u_1+u_2)(h_1-h_2)/2\} = \frac{w_{e1}}{\cos\theta_1} - \frac{w_{e2}}{\cos\theta_2} \quad (63)$$

$$\frac{d}{dx}\{u_2(H_s-h_2)\} = w_{e2}/\cos\theta_2 - R' \tag{64}$$

$$\frac{d}{dx}\{\rho_1 u_1(h_1-H_b)\} = -w_{e1}\rho_1/\cos\theta_1 \,, \quad \frac{d}{dx}\big(\tfrac{1}{4}(\rho_1+\rho_2)(u_1+u_2)(h_1-h_2)\big) =$$

$$w_{e1}\rho_1/\cos\theta_1 - w_{e2}\rho_2/\cos\theta_2; \frac{d}{dx}\big(\rho_2 u_2(H_s-h_2)\big) = w_{e2}\rho_2/\cos\theta_2 - \rho_f R' \tag{65}$$

where $R'(x)$ is river runoff and w_{e1} and w_{e2} are normal (nearly vertical) entrainment velocities at the lower and upper boundaries of
the interfacial layer. Pearson and Winter write down the first and
third equations in (64) and (65) but with the following notation:

$$-w_{e1}/\cos\theta_1 = F_d - F_u \,, \quad w_{e2}/\cos\theta_2 = F_u - F_d \,, \tag{66}$$

$$- \frac{w_{e1} \rho_1}{\cos\theta_1} = \rho_2 F_d - \rho_1 F_u, \frac{w_{e2} \rho_2}{\cos\theta_2} = \rho_1 F_u - \rho_2 F_d \qquad (67)$$

where F_u is said to "represent the upward volume rate of flow of fluid from the deep layer to the near-surface layer per square metre of interfacial area. Likewise, F_d denotes the downward volumetric flux rate from the upper to the lower layer". However, multiplying (66) by ρ_1 and ρ_2, respectively, and comparing with (67), we get

$$\rho_1 F_d = \rho_2 F_d , \rho_2 F_u = \rho_1 F_u \qquad (68)$$

so that either $F_d = F_u = 0$ or $\rho_1 = \rho_2$, both of which are trivial. Pearson and Winter also use F_u and F_d to compute momentum fluxes but, if the above arguments are correct, the theory is in obvious difficulty. Eqs. (64) and (65) may be used to derive the two Knudsen relations but it is not apparent that they are otherwise useful.

STRONGLY MIXED ESTUARIES

The ideas of Stommel on "overmixing" advanced in the early 1950's (Stommel, 1951, Stommel and Farmer, 1952, 1953) have been much used in estuarine research but confusion exists about them and we will attempt to explore them further. Fig. 13 shows conditions in the vicinity of the mouth of an estuary (on the left) connecting the estuary to the open sea (on the right). The estuarine circulation near the mouth has fluid of mean buoyancy b_1 flowing out with mass flux q_1 over the incoming fluid with the buoyancy b_0 of sea water and mass flux q_0. In this portion of the analysis we do not need to assume uniform buoyancies in the two layers although we will do this later. The mean depth of the water averaged across the channel is $\bar{h}$, the mean width is $\bar{W}$. Other lengths required to specify the geometry of the estuary and its mouth may be denoted by L_1, L_2,... . There is fresh-water flux R into the estuary and some kind of mixing. In the present discussion the precise nature of the mixing processes is not important (as we will see) but to be definite and simple we will assume the mixing is caused by a rotating propellor[1] with angular speed ω. Its geometry and location are specified by lengths a_1, a_2,... .

A quantity of importance in the discussion is the buoyancy difference $b_0 - b_1 = \Delta b$. From dimensional analysis we may write

$$\Delta b / b_0 = f_1(Q_f, \Omega, ...), \quad Q_f = R/(b_0 \bar{h}^3)^{\frac{1}{2}} \, \bar{W}, \quad \Omega = \omega a_1/(\bar{h} b_0)^{\frac{1}{2}} \quad (69)$$

where a number of non-dimensional ratios of lengths of the form

---

[1] Other possibilities are a wind exerting a stress u_*^2 and a tide of amplitude h_t and frequency ω, yielding $\Omega = u_*/(\bar{h} b_0)^{\frac{1}{2}}$, $\omega h_t/(\bar{h} b_0)^{\frac{1}{2}}$, respectively.

$L_1/\bar{h}$, $L_2/\bar{h}$, ..., $a_1/\bar{h}$, $a_2/\bar{h}$, ... have been omitted. Certainly $\Delta b/b_o$ will vary with the quantity Ω; indeed $\Delta b/b_o$ will decrease as Ω increases from the maximum value of one when the mixing is zero. Ultimately, if the estuary has finite dimensions, a sufficiently large Ω, say Ω_m, will just cause the estuary to be thoroughly mixed and any further increase of Ω will not further decrease Δb and it will assume its minimum value Δb_m. Since $\partial(\Delta b/b_o)\partial\Omega = 0$ when $\Delta b = \Delta b_m$, we get

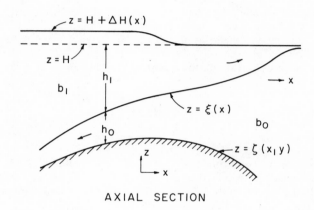

AXIAL SECTION

Figure 13. Conditions near the mouth of an estuary.

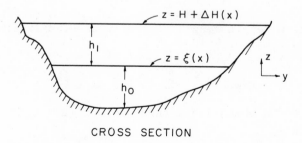

CROSS SECTION

Figure 14. Cross-section in the constriction of an estuary.

$$\Delta b_m/b_o = f_2(Q_f, \ldots) \tag{70}$$

The ratio η of the mean depth of the lower layer to $\bar{h}$ is

$$\eta = f_3(Q_f, \Omega, \ldots) \tag{71}$$

When the estuary is just thoroughly mixed and $\Delta b = \Delta b_m$, any further increase in the mixing Ω can only affect conditions at the mouth by the tendency to further increase the turbulence in the outflowing fluid. This effect is difficult to analyse and we will neglect it. Subsequently, we will assume, in fact, perfect-fluid flow in the vicinity of the mouth so that the turbulence is assumed to be negligible compared to the mean motions. Then for $\Omega \geq \Omega_m$, the ratio η assumes a fixed value η_m such that

$$\eta_m = f_4(Q_f, \ldots) \tag{72}$$

In general we notice that a combination of (69) and (71) permits us to write

$$\Delta b/b_o = f_5(Q_f, \eta, \ldots) \tag{73}$$

for arbitrary mixing.

Let us attempt to find the form of the function in Eq. (73) subject to several simplifying assumptions. With reference to Fig. 13 and Fig. 14, the Bernoulli equations in the two layers are

$$\tfrac{1}{2}u_1^2 + g\Delta H = \text{const} \tag{74}$$

$$\tfrac{1}{2}u_o^2 + g\Delta H - h_1\Delta b = \text{const} \tag{75}$$

where h_1 is the distance from the free surface to the interface, and where we assume irrotational motion. Let us justify the neglect of the other two components of the kinetic energy in (74) and (75). The equation of continuity leads to

$$u/L \sim v/\bar{W} \sim w/\bar{h} \tag{76}$$

where L, $\bar{W}$ and $\bar{h}$ denote length scales in the longitudinal, transverse and vertical directions. In each layer v^2 and w^2 will be negligible compared to u^2 if

$$(\bar{W}/L)^2 \ll 1, \quad (\bar{h}/L)^2 \ll 1 \tag{77}$$

We assume that the channel has these properties.

Eliminating ΔH between Eqs. (74) and (75) and introducing the constant fluxes q_1 and q_o, we have

$$\frac{q_o^2}{2\Delta b B_o^2\{H-h_1-\bar{\zeta}(x)\}^2} - \frac{q_1^2}{2\Delta b \bar{B}_1^2 h_1^2} - h_1 = \text{const} \tag{78}$$

where B_o is the width of the channel at the level of the interface, $\bar{B}_1$ is the average width in the upper layer and $\bar{\zeta}$ is the average height of the bottom of the channel above $z = 0$. Let us now differentiate Eq. (78) with respect to x. We get

$$\frac{dh_1}{dx}(F_o^2+F_1^2-1) = -\frac{d\bar{\zeta}}{dx}F_o^2 + \frac{1}{\bar{B}_o}\frac{dB_o}{dx}(H-h_1-\bar{\zeta})F_o^2 - \frac{1}{\bar{B}_1}\frac{d\bar{B}_1}{dx}h_1 F_1^2 \tag{79}$$

where F_o and F_1 are the densimetric Froude numbers defined by

$$F_o^2 = q_o^2/(\Delta b B_o^2\{H-h_1-\bar{\zeta}\}^3) \quad , \quad F_1^2 = q_1^2/\Delta b \bar{B}_1^2 h_1^3 \tag{80}$$

Suppose we have a channel with a minimum of width at a certain section (i.e., $dB_o/dx = 0$, $d\bar{B}_1/dx = 0$, presumably at the same section). If the bottom is level there (or if the mean depth is also a minimum there), the left-hand side of Eq. (79) is zero at that section. Since we would expect a strong slope of the interface as the upper level moves out of the estuary and spreads laterally and thins vertically in the widening channel, we infer that conditions will be critical at the section, i.e.

$$F_o^2 + F_1^2 = 1 \tag{81}$$

Such a section is referred to as a control section. Experiments by Stommel and Farmer (1953), Assaf and Hecht (1974) and Anati, Assaf, and Thompson (1977) have shown that critical conditions do tend to occur in a variety of straits in laboratory models. If the strait is long, friction becomes important and critical conditions occur at the ends of the strait.

At the control section we let W_o and $\bar{W}_1$ be the values of B_o and $\bar{B}_1$ and η be the ratio of the mean thickness of the lower fluid to the mean depth of the water $(H-\bar{\zeta}) = \bar{h}$ both evaluated at the control section. We also use the equations for conservation of mass and buoyancy (salt)

$$q_o + R = q_1 \tag{82}$$

$$q_o b_o = q_1 b_1 \tag{83}$$

where R is the fresh water discharge. The critical condition becomes

$$\eta^2 Q_f^2(1-\frac{\Delta b}{b_o})^2/\eta^3 + Q_f^2/(1-\eta)^3 = (\frac{\Delta b}{b_o})^3 \tag{84}$$

where

$$Q_f = R/\bar{W}_1(b_o\bar{h}^3)^{\frac{1}{2}} \ , \ n = \bar{W}_1/W_o \tag{85}$$

For a given value of Q_f, Eq. (84) represents a curve like that shown in Fig. 15. It is of the form of Eq. (73) yielding $\Delta b/b_o$ as a function of η instead of the mixing parameter Ω. Physically, if we increase Ω from a small value, we expect $\Delta b/b_o$ to decrease monotonically from the value 1 corresponding to zero mixing. η will increase or decrease with increase of Ω, and we move on the curve of Fig. 15 toward the minimum point. At a certain value Ω_m the estuary will be thoroughly mixed, Δb will no longer decrease and $\partial(\Delta b/b_o)/\partial\Omega = 0$. We have

$$\frac{\partial(\Delta b/b_o)}{\partial\Omega} = \frac{\partial(\Delta b/b_o)}{\partial\eta} \frac{\partial\eta}{\partial\Omega} \tag{86}$$

Obviously if $\partial(\Delta b/b_o)/\partial\eta = 0$, corresponding to the minimum point of the curve in Fig. 15, $\partial(\Delta b/b_o)/\partial\Omega = 0$ and $\Delta b/b_o$ will have the value appropriate to the thoroughly mixed estuary. The condition

$$\frac{\partial(\Delta b/b_o)}{\partial\eta} = 0 \tag{87}$$

together with Eq. (84) determines the problem of a thoroughly mixed, or in the terminology of Stommel and Farmer, an "overmixed" estuary (Long, 1977b). Notice that the conservation equations (82), (83) yield

$$q_1\Delta b/b_o = R \tag{88}$$

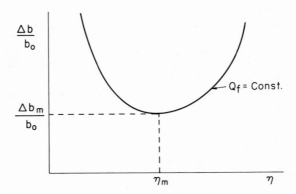

Figure 15.   Curve of density difference as a function of non-dimensional thickness of lower layer.

so that a minimum of $\Delta b/b_o$ corresponds to a maximum of the discharge q_1.

The variation of n^2 with η depends on the particular geometry of the estuary, and although this variation may have some quantitative importance we will consider instead the idealized channel with vertical sides so that $n^2 = 1$. Eq. (87) then yields

$$(1-\Delta b/b_o) = n^2/(1-\eta)^2 \tag{89}$$

In terms of the densimetric Froude numbers, we may write

$$F_o^2 = \eta, \quad F_1^2 = 1-\eta, \quad Q_f^2 = (1-2\eta)^3/(1-\eta)^2 \tag{90}$$

One special case involves the condition in which the densities of the two fluids are nearly equal. Then from (89), $\eta \simeq \frac{1}{2}$, $R \simeq 0$. Experiments by Stommel and Farmer (1953) and Anati, Assaf and Thompson (1977) confirm the prediction that the two fluids have equal thicknesses in thoroughly mixed estuaries with small fresh water influx and I have derived this result theoretically (Long, 1976a). A second special case is one in which $\eta = 0$ so that no salt water enters and fresh water fills the estuary. In this case the solution yields $\Delta b/b_o = 1$, $F_o^2 = 0$ as also required by physical considerations. In the general case (90) shows that $\eta \geq \frac{1}{2}$ so that the interface is always at or below mid-depth.

Also of interest is the height ΔH_o of the water surface in the estuary above the level of the ocean surface. If we evaluate the constants in Eqs. (74) and (75) in the estuary and in the ocean respectively, we get

$$\tfrac{1}{2}u_1^2 + g\Delta H = g\Delta H_o \tag{91}$$

$$\tfrac{1}{2}u_o^2 + g\Delta H - \Delta b h_1 = 0 \tag{92}$$

where we now assume that the widths in the ocean and estuary are very large compared to the width of the channel. The constant in Eq. (78) is $-g\Delta H_o/\Delta b$ and this equation may then be written

$$\tfrac{1}{2}F_o^2\eta - \tfrac{1}{2}F_1^2(1-\eta) - (1-\eta) + \beta = 0 \tag{93}$$

where

$$\beta = \Delta H_o/(\bar{h}\Delta\rho/\rho_f) \tag{94}$$

Using (90), we get

$$\beta = 3(1-4\eta/3)/2 \tag{95}$$

In the overmixed state β ranges between 0.5 and 1.5.

We may contrast this strongly mixed case with the case in which the mixing is zero. Then $F_o^2 = 0$, $F_1^2 = 1$ and Eq. (93) yields $\beta = 3(1-\eta)/2$.

THREE-LAYER CIRCULATIONS IN ESTUARIES AND HARBORS

A number of writers beginning, apparently, with Hachey (1934) have remarked that if a density variation with depth exists in the water outside of an embayment and if there is a source of mixing in the embayment from tides and wind, there will be a tendency for a density distribution as shown in Fig. 16 and a resulting three-layer circulation from the density-pressure effect even in the absence of a fresh-water influx. Stroup, Pritchard and Carpenter (1961) have calculated that this type of circulation is dominant as a flushing mechanism for Baltimore Harbor in the Chesapeake Bay. Here the less salty water in the upper layer outside of the harbor originates from the fresh water discharge of the Susquehanna River at the head of the Chesapeake Bay. It has also been suggested that the outside density distribution is important for the circulations in the Oslofjord in Norway (Gade, 1970) and in the Gullmarfjord on the west coast of Sweden (Rydberg, 1975). Rydberg observed inflow in the layer below the halocline at the sill and inferred from considerations of conservation of salt that there must be an influx of lower salinity water somewhere in the upper level. Conservation of mass requires an efflux of water at some intermediate level. The lower salinity water outside of the Oslofjord and the Gullmarfjord originates as the brackish water flowing out of the Danish sounds from the Baltic Sea. The rise and fall of this halocline with the storm cycle over the Baltic may send gravity waves into Gullmarfjord which break to cause mixing. The three-layer type of circulation has also been discussed by Hansen and Rattray (1972).

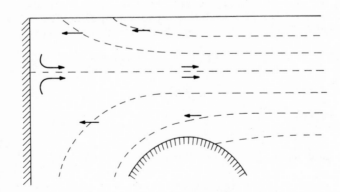

Figure 16.   Schematic density distribution in an embayment producing
               three-layer circulation.

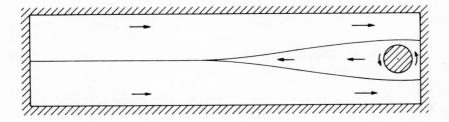

Figure 17.   Hachey's Experiment

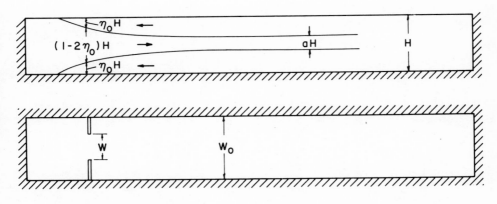

Figure 18.   Axial and plan view of three-layer experiment.

A simple laboratory model of a three-layer circulation was constructed by Hachey (1934). It is shown schematically in Figure 17. A long channel contains, initially, two layers of water of different density. At one end of the channel a cylinder at the interface is rotated as shown to provide a source of mixing. The resulting pressure-density distribution produces the three-layer circulation in the figure.

An investigation by the author involved an experiment portrayed in Figure 18 in which there are initially two fluids of different density and equal depth in a long channel. In the harbor to the left intense mixing suddenly produces a thoroughly mixed fluid of intermediate density which flows out at the interface. Its thickness at the mouth was half that of the total thickness of the two-fluid system in accordance with a theory based on the assumption of an "overmixed" state.

ACKNOWLEDGEMENTS

This research was supported by the Office of Naval Research Fluid Dynamics Branch, under Contract N00014-75-C-0805 and by the National Science Foundation, Oceanographic Section, under Grant No. OCE 76-18887.

REFERENCES

Assaf, G., and A. Hecht. 1974. Sea straits: a dynamical model.
    Deep Sea Res., 21, 947-958.

Anati, D.A., G. Assaf, and R. Thompson. 1977. Laboratory models
    of sea straits. J. Fluid Mech., 81, 341-351.

Baines, W.D. 1975. Entrainment by a plume or jet at a density
    interface. J. Fluid Mech., 68, 309-320.

Booker, J.R., and F.P. Bretherton. 1967. The critical layer for
    internal gravity waves in a shear flow. J. Fluid Mech., 27,
    513-539.

Bouvard, M. and H. Dumas. 1967. Application de la methode de
    fil chaud a la mesure de la turbulence dans l'eau. Houille
    Blanche, 22, 257-723.

Brush, L.M. 1970. Artificial mixing of stratified fluids formed
    by salt and heat in a laboratory reservoir. N.J. Water
    Resources Res. Inst. Res. Project B-024.

Caldwell, D.R., C.W. Van Atta, and K.N. Helland. 1972. A labor-
    atory study of the turbulent Ekman layer. Geophy. Fluid Dyn.,3,
    125-160.

Chern, C.-S., R.R. Long. 1980. A new theory of turbulent convection.
    In preparation.

Connor, J.J. and J.D. Wang. 1974. Finite element model of two-
    layer coastal circulation. 14th Coastal Engineering Conference,
    Copenhagen.

Crapper, P.F. 1973. An experimental study of mixing across density
    interfaces. Ph.D. thesis, University of Cambridge.

Crapper, P.F., and P.F. Linden. 1974. The structure of turbulent
    density interfaces. J. Fluid Mech., 65, 45-63.

Cromwell, T. 1960. Pycnoclines created by mixing in an aquarium
    tank. J. Mar. Res., 18, 73-82.

Csanady, G.T. 1967. Large scale motion in the Great Lakes. J.
    Geophys. Res., 72, 4151-4162.

Csanady, G.T. 1968a. Wind-driven summer circulation in the Great
    Lakes. J. Geophys. Res., 73, 2579-2589.

Csanady, G.T.   1968b. Motions in a model Great Lake due to a
    suddenly imposed wind. J. Geophys. Res., 73, 6435-6447.

Dickinson, S.C., and R.R. Long. 1978. Laboratory study of the
    growth of a turbulent layer of fluid. Physics of Fluids 21,
    10, 1698-1701.

Farmer, D. and J.D. Smith. 1978. Nonlinear internal waves in a
    fjord. Hydrodynamics of Estuaries and Fjords (Ed. J. Nihoul)
    Elsevier Oceanography Series No. 23, 465-494.

Fonselius, S.H. 1969. Hydrography of the Baltic deep basins III.
    Fishery Bd. of Sweden, Series Hydrography, No. 23, 97 pp.

Freeland, H.J. 1979. The hydrography of Knight Inlet, B.C. in the
    light of Long's model of fjord circulation. Proceedings of the
    Fjord Oceanographic Workshop, NATO, Sidney, B.C. Canada.

Gade, H. 1970. Hydrographic investigations in the Oslofjord. A
    study of water circulations and exchange processes, Vols.
    I, II, III. Geophys. Inst., Univ. of Bergen. Norway.

Gade, H., and E. Svendsen. 1978. Properties of the Robert R. Long
    model of estuarine circulation in fjords. Hydrodynamics of
    Estuaries and Fjords. (Ed. J. Nihoul). Elsevier Oceanography
    Series, No. 23.

Gardner, G.B. and J. D. Smith. 1978. Turbulent mixing in a salt
    wedge estuary. Hydrodynamics of Estuaries and Fjords (Ed.
    J. Nihoul) Elsevier Oceanography Series No. 23, 79-106.

Hachey, H.B. 1934. Movements resulting from mixing of stratified
    waters. J. Biol. Bd. Canada, 1, 133-143.

Hansen, D.V., and M. Rattray, Jr. 1972. Estuarine circulation
    induced by diffusion. J. Mar. Res., 30, 281-294.

Heidrick, T.R., S. Banerjee, and R.S. Azad. 1977. Experiments on the
    structure of turbulence in fully developed pipe flow, Part 2.
    A statistical procedure for identifying "bursts" in the wall
    layers and some characteristics of flow during bursting periods.
    J. Fluid Mech., 82, 705-723.

Hinze, J.O. 1976. Turbulence, McGraw-Hill, N.Y.

Hopfinger, E.J., and M.-A. Toly. 1976. Spatially decaying turbulence
    and its relation to mixing across density interfaces. J.
    Fluid Mech., 78, 155-175.

Hunt, J.C.R., and J.M.R. Graham. 1978. Free-stream turbulence near plane boundaries. J. Fluid Mech., 84, 209-235.

Kaimal, J.C. 1978. Horizontal velocity spectra in an unstable surface layer. J. Atmos. Sci., 35, 18-23.

Kantha, L. H.,O.M. Phillips, and R.S. Azad. 1977. On turbulent entrainment at a stable density interface. J. Fluid Mech. 79, 753-768.

Kantha, L.H. 1979a. Turbulent entrainment at a buoyancy interface due to convective turbulence. Proceedings of NATO Fjord Oceanographic Workshop, Sidney, B.C. Canada.

Kantha, L.H. 1979b. Experimental simulation of the retreat of the thermocline by surface heating. Proceedings of NATO Fjord Oceanographic Workshop, Sidney, B.C., Canada.

Kantha, L.H. and R.R. Long. 1979. Turbulent mixing with stabilizing surface buoyancy flux. Submitted to J. Fluid Mech.

Kantha, L.H. and R.R. Long. 1980. The depth of a mixed layer in an oscillating grid experiment. In Preparation.

Kato, H., and O.M. Phillips. 1969. On the penetration of a turbulent layer into a stratified fluid. J. Fluid Mech., 37, 643-655.

Kraichnan, R.H. 1974. On Kolmogorov's inertial-range theories. J. Fluid Mech. 62, 305-330.

Kraus, E.B., and J.S. Turner. 1967. A one-dimensional model of the seasonal thermocline. II. The general theory and its consequences. Tellus 19, 98-106.

Kreider, J.F., 1973. A laboratory study of the turbulent Ekman layer. Ph.D. dissertation. Univ. of Colorado.

Kullenberg, G. 1971. Vertical diffusion in shallow waters. Tellus 23, 129-135.

Kuo, A., and S. Corrsin. 1971. Experiments on the geometry of the fine-structure distribution functions in fully turbulent fluid. J. Fluid Mech., 50, 285-319.

Kuo, A.Y. and S. Corrsin. 1972. Experiment on the geometry of the fine-structure regions in fully turbulent flow. J. Fluid Mech., 56, 447-479.

Laufer, J. 1954. The structure of turbulence in fully developed
    pipe flow. Natl. Advis. Comm. Aeronaut., Rep. 1174.

Linden, P.F. 1973. The interaction of a vortex ring with a sharp
    density interface: A model for turbulent entrainment. J.
    Fluid Mech. 60, 467-480.

Linden, P.F. 1975. The deepening of a mixed layer in a stratified
    fluid. J. Fluid Mech. 71, 385-405.

Long, R.R. 1975a. The influence of shear on mixing across density
    interface. J. Fluid Mech. 70, 305-320.

Long, R.R. 1975b. On the depth of the halocline in an estuary.
    J. Phys. Oceanogr. 5, 551-554.

Long, R.R. 1975c. Circulations and density distributions in a
    deep, strongly stratified, two-layer estuary. J. Fluid
    Mech., 71, 529-40.

Long, R.R. 1976a. Mass and salt transfers and halocline depths
    in an estuary. Tellus, 28, 460-472.

Long, R.R. 1976b. Lectures on Estuarine Circulations and Mass
    Distributions. Tech. Rep. No. 9 (series C), Department of
    Earth & Planetary Sciences, The Johns Hopkins University.

Long, R.R. 1977a. Some aspects of turbulence in geophysical systems.
    Adv. Appl. Mech., Volume 17, Academic Press, N.Y.

Long, R.R. 1977b. Three-layer circulations in estuaries and harbors.
    J. Phys. Oceanogr., 7, 415-421.

Long, R.R. 1978a. A theory of mixing in a stably stratified fluid.
    J. Fluid Mech. 84, 113-124.

Long, R.R. 1978b. The growth of the mixed layer in a turbulent
    stably stratified fluid. GAFD, 11, 1-11.

Long, R.R. 1978c. The decay of turbulence. Tech. Rep. No. 13
    (series C), Dept. of Earth & Planetary Sciences, The Johns
    Hopkins Univ.

Long, R.R. 1978d. Theory of turbulence in a homogeneous fluid
    induced by an oscillating grid. Physics of Fluids, 21, 1887-
    1888.

Long, R.R. 1979a. A theory for portions of the energy spectrum
    and for intermittency of fine-scale turbulence. Proceedings
    2nd Symposium on Turbulent Shear Flows. Also available as

Tech. Rep. No. 15 (series C), Department of Earth & Planetary
Sciences, The Johns Hopkins University.

Long, R.R. 1979b. A new theory of turbulent shear flow. Submitted
to J. Fluid Mech.

Long, R.R. 1979c. A new theory of the neutral planetary boundary
layer. Submitted to J. Fluid Mech.

Long, R.R., and L.H. Kantha. 1978. The rise of a strong inversion
caused by heating at the ground. Proceedings of Twelfth
Symposium on Naval Hydrodynamics, Vol. VII-VIII, Washington, D.C.

Long, R.R., and L.H. Kantha. 1979a. On the depth of a mixed layer
under a surface stress and stabilizing surface buoyancy flux.
Submitted to J. Fluid Mech.

Long, R.R., and L. H. Kantha. 1979b. Experiments on penetrative
convection. In Preparation.

Malkus, W.V.R. 1979. Turbulent velocity profiles from stability
criteria. J. Fluid Mech., 90, 401-414.

Mitchell, J.E., and T.J. Hanratty. 1966. A study of turbulence
at a wall using an electrochemical wall shear-stress meter.
J. Fluid Mech., 26, 199-221.

Monin, A.S., and A. M. Yaglom. 1971. Statistical Fluid Mechanics:
Mechanics of Turbulence, Vol. 1. MIT Press, Cambridge, Mass.

Moore, M.J., and R.R. Long. 1971. An experimental investigation
of turbulent stratified shearing flow. J. Fluid Mech., 49,
635-655.

Ottesen-Hansen, N.-F., 1975. Entrainment in two-layer flow.
Series Paper 7, Institute of Hydrodynamics and Hydraulic
Engineering, Tech. Univ. Denmark. 97 pp.

Partch, E.N., and J.D. Smith. 1978. Time-dependent mixing in a salt
wedge estuary. Estuarine and Coastal Marine Science, 6, 3-19.

Pearson, C.E. and D.F. Winter. Two-layer analysis of steady
circulation in stratified fjords. Hydrodynamics of Estuaries
and Fjords. (Ed. J. Nihoul) Elsevier Oceanographic Series, 23.

Pedersen, Fl. B., 1972a. Gradually varying two-layer stratified
flow in fjords. Intern. Sympos. on Stratified Flows., Paper
No. 19, Novosibirsk, 413-429.

Pedersen, Fl. B.   1972b.   Gradually varying two-layer stratified
    flow. ASCE J. Hydraulic Div. 98, No. HY1.

Pedersen, Fl. B.   1978.   A brief review of present theories of
    fjord dynamics. Hydrodynamics of Estuaries and Fjords. (Ed.
    J. Nihoul) Elsevier Oceanographic Series. 23.

Perels, P.A.J., and M. Karelse.   1978.   A two-dimensional numerical
    model for salt intrusion. Hydrodynamics of Estuaries and
    Fjords. (E. J. Nihoul). Elsevier Oceanographic Series 23.

Perkins, R.C., and E.L. Lewis.   1978.   Mixing in an arctic fjord.
    J. Phys. Oceanogr. 8, 873-880.

Pickard, G.L. and K. Rodgers. 1959.   Current measurements in Knight
    Inlet, British Columbia, J. Fish. Res. Bd of Canada, 18,
    635-678.

Pickard, G.L.   1961.   Oceanographic features of inlets in the
    British Columbia mainland coast. J. Fish. Res. Bd. of Canada,
    18 (6), 907-999.

Pritchard, D.W.   1952.   Salinity distribution and circulation in
    the Chesapeake Bay estuarine system. J. Mar. Res. 11,
    106-123.

Rao, K.N., R. Narasimha, and M.A. Badri Narayanan.   1971.   The
    "bursting" phenomenon in a turbulent boundary layer. J. Fluid
    Mech., 48, 339-352.

Rattray, M., Jr. and D.V. Hansen.   1965.   Gravitational circulation
    in straits and estuaries. J. Mar. Res., 23, (2), 104-122.

Rattray, M., Jr.   1967.   Some aspects of the dynamics of circulations
    in fjords. Estuaries (Ed. G.H. Lauff) 52-62, AAAC, Washington,
    D.C.

Rousse, H., and J. Dodu.   1955.   Turbulent diffusion across a
    density discontinuity. Houille Blanche 10, 405-410.

Rydberg, L.   1975.   Hydrographic observations in the Gullmarfjord
    during April 1973. Report No. 10, Göteborgs Universitet
    Oceanografiska Institutionen, Sweden.

Schlichting, H.   1955.   Boundary Layer Theory, McGraw-Hill, N.Y.

Smith, J.D.   1974.   Turbulent structure of the surface boundary
    layer in an ice-covered ocean. Rapp. P.-v. Reun. const. int.
    Mer., 53-65.

Smith, J.D. 1979. Mixing induced by internal hydraulic disturbances in the vicinity of sills. Proceedings of the NATO Fjord Oceanographic Workshop, Sidney, B.C., Canada.

Smith, R. 1976. Longitudinal dispersion of a buoyant contaminant in a shallow channel. J. Fluid Mech., 79, 677-688.

Smith, R. 1978. Coriolis curvature and buoyancy effects upon dispersion in a narrow channel. Hydrodynamics of Estuaries and Fjords. (Ed. J. Nihoul) Elsevier Oceanographic Series 23.

Stigebrandt, A. 1976. Vertical diffusion driven by internal waves in a sill fjord. J. Phys. Oceanogr. 6, 486-495.

Stigebrandt, A. 1979. Observational evidence for vertical diffusion driven by internal waves of tidal origin in the Oslofjord. J. Phys. Oceanogr. 9, 438-441.

Stommel, H. 1951. Recent development in the study of tidal estuaries. WHOI Tech. Rep. Ref. No. 51-33, Woods Hole Oceanographic Inst., Woods Hole, Mass.

Stommel, H. and H. G. Farmer. 1952. Abrupt change in width in a two-layer open channel flow. J. Mar. Res., 11, 205-214.

Stommel, H. and H.G. Farmer. 1953. Control of salinity in an estuary by a transition. J. Mar. Res., 12, 13-20.

Stroup, E.D., D.W. Pritchard, and J.H. Carpenter. 1961. Final Report Baltimore Harbor Study, Tech. Rep. Rep. 26, Chesapeake Bay Institute, The Johns Hopkins University.

Tennekes, H., and J.L. Lumley. 1972. A First Course in Turbulence. MIT Press, Cambridge, Mass.

Thomas, N.H., and D.E. Hancock. 1977. Grid turbulence near a moving wall. J. Fluid Mech., 82, 481-496.

Thompson, S.M., and J.S. Turner. 1975. Mixing across an interface due to turbulence generated by an oscillating grid. J. Fluid Mech., 67, 349-368.

Tritton, D.J. 1967. Some new correlation measurements in a turbulent boundary layer. J. Fluid Mech., 28, 803-821.

Turner, J.S. 1968. The influence of molecular diffusivity on turbulent entrainment across a density interface. J. Fluid Mech. 33, 639-656.

Ueda, H., and J.O. Hinze. 1975. Fine-structure turbulence in the
    wall region of a turbulent boundary layer. J. Fluid Mech.,
    67, 125-143.

Uzkan, T., and W.C. Reynolds. 1967. A shear-free turbulent
    boundary layer. J. Fluid mech., 28, 803-821.

Walin, G. 1972a. On the hydrographic response to transient
    meteorological disturbances. Tellus, 24, 169-186.

Walin, G. 1972b. Some observations of temperature fluctuations in
    the coastal region of the Baltic. Tellus, 24, 187-198.

Winter, D.F. 1972. A similarity solution for circulation in
    stratified fjords. Intern. Symp. on Stratified Flows. Novo-
    sibirsk. 715-724.

Winter, D.F. 1973. A similarity solution for steady state gravi-
    tational circulation in fjords. Estuarine and Coastal
    Marine Science, 1, 387-400..

Wolanski, E.J. 1972. Turbulent entrainment across stable density-
    stratified liquids and suspensions. Ph.D. thesis, The Johns
    Hopkins University.

Wolanski, E.J. and L.M. Brush. 1975. Turbulent entrainment across
    stable density step structures. Tellus, 27, 259-268.

Woods, J.D. 1968. Wave-induced shear instability in the summer
    thermocline. J. Fluid Mech., 32, 792-800.

TOPOGRAPHIC EFFECTS IN STRATIFIED FLUIDS

Herbert E. Huppert

Department of Applied Mathematics and
Theoretical Physics
University of Cambridge
Silver Street
Cambridge CB3 9EW, England

1. Introduction

As will be clear from reading other papers in this volume, the study of fjords represents a scientific challenge in a large number of different areas. There are practical problems, such as those connected with pollution, the construction of hydroelectric schemes and the maintenance of an ice-free surface during the winter. There are also more fundamental problems, such as those concerned with the determination of the circulation in a fjord, the temperature and salinity structure, and the occurrence of internal waves, mixing and deep-water renewal. The consideration of these problems is of interest in itself, as well as possibly pointing the way to understanding similar events in the open ocean.

Because of the presence of sills and the varying bottom bathymetry of fjords, topographic effects play a large role in many fjord problems. This review surveys some of those topographic interactions with the flow of stratified fluids which may be important in fjords. The review was prepared for presentation at the conference and was aimed at explaining fundamental principles in a manner understandable to scientists who were not necessarily specialists in fluid mechanics.

The hydrostatic flow of thin layers of fluid are considered in the second and third sections. Some effects of rotation are discussed in the fourth. Non-hydrostatic effects are reviewed in the fifth section. A brief discussion of some time-dependent aspects completes the presentation.

2. Steady flow of a Single Layer

Consider the flow of a thin inviscid layer of fluid of density ρ_1 and depth $d(x)$ in a channel with a lower floor at $h(x)$ and a width of $w(x)$, as depicted in Figure 1a. The discussion to be given below applies also to the case when the flow is under a deep layer moving at constant horizontal velocity if the gravitational acceleration g is replaced by $g' = g(\rho_1 - \rho_2)/\rho_1$, where ρ_2 is the density of the upper layer. The concepts to be introduced can also be applied to the flow <u>over</u> a deep quiescent layer as depicted in Figure 1b, which more closely resembles that encountered in fjords. The only modification is that the change in depth of the lower interface must be taken into account. A thin upper layer such as this can be readily swept along by the force of the wind and be dynamically decoupled from the lower denser layer. Such a condition occurred during the 1968 Olympic Sailing Contest and played an important role in the tactics of the winners (Houghton and Woods 1969).

The assumptions that the channel width and depth are slowly varying functions and that the length scales of such variations are large compared to the depth allow the hydrostatic approximation to be made. Thus vertical accelerations can be neglected and the pressure at any depth is entirely due to the weight of the overlying fluid. Application of Bernoulli's theorem just above and below the interface combined with the equation of continuity

$$dwu = Q \qquad\qquad (2.1)$$

leads to the governing equation

$$Q^2/(2gw^2d^2) + d = C - h(x) \equiv e(x), \qquad\qquad (2.2)$$

where $C = d_\infty(1 + \tfrac{1}{2}F_\infty^2)$, the Froude number

$$F_\infty = Q/(gw^2d^3)^{\frac{1}{2}} \qquad\qquad (2.3)$$

is the ratio of the velocity of the flow to that of very long waves of infinitesimal amplitude and the subscript ∞ indicates that the variable is to be evaluated far upstream (at $x = -\infty$). The relationship (2.2), often called the specific energy curve, is sketched in Figure 2. It is easy to determine that $e(x)$ has a minimum of 3/2 $(Q^2/gw^2)^{1/3}$ at $d_{min} = (Q^2/gw^2)^{1/3}$, that is at $F = 1$. For given conditions, that is for given $e(x)$, there is in general either two values of d – one for which $d < d_{min}$ and $F>1$, called supercritical, and one for which $d > d_{min}$ and $F<1$, called subcritical – or no values of d which satisfy (2.2). The special intermediate case when there is just one value of $d = d_{min}$ and $F = 1$ occurs when the flow velocity is exactly equal to the velocity of long waves of infinitesimal amplitude. When the flow is everywhere supercritical no small

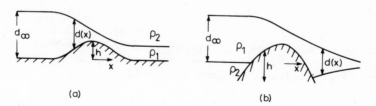

(a)   (b)

Figure 1.   The geometry and parameters for flow in a single fluid
             layer.

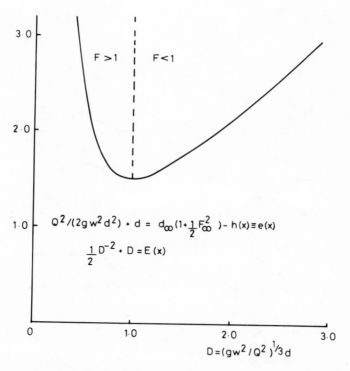

$$Q^2/(2gw^2d^2) + d = d_\infty(1 + \frac{1}{2}F_\infty^2\ ) - h(x) \equiv e(x)$$

$$\frac{1}{2}D^{-2} + D = E(x)$$

$$D = (gw^2/Q^2)^{1/3}d$$

Figure 2.   The specific energy, e, as a function of depth, d.

disturbances are able to propagate upstream, while if the flow is
subcritical everywhere disturbances of sufficiently long wavelength
can propagate both upstream and downstream.

The possible flow patterns have been analysed by Long (1954,
1970, 1972) and Houghton and Kasahara (1968) for channels of
constant width using conservation of mass, momentum and the Bernoulli
equation as necessary. They find that at least five possible flow
patterns are possible, as mapped out in Figure 3.

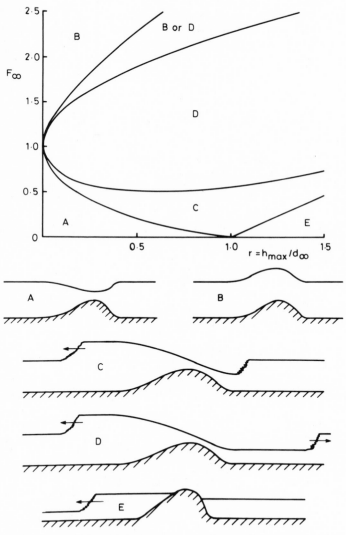

Figure 3.   The possible flow patterns (A - E) as a function of
             F_∞ and h_{max}/d_∞.

A) The flow is subcritical upstream and the channel is such that the minimum value of $e(x)$, $d_\infty(1 + \frac{1}{2}F_\infty^2) - h_{max} > e_{min} \equiv 3/2(Q^2/gw^2)^{1/3}$. The flow then remains on the right-hand branch in Figure 2, is subcritical everywhere and the interface bends down over the topography. The condition for this to occur is that $F_\infty < 1$ and $r < 1 - 3/2\ F_\infty^{2/3} + \frac{1}{2}F_\infty^2$, where

$$r = h_{max}/d_\infty. \qquad (2.4)$$

B) The flow is supercritical upstream and $e(x) > e_{min}$ everywhere. The flow is then everywhere supercritical and the interface bends up over the topography. This form of flow can occur if $F_\infty > 1$ and $r < 1 - 3/2\ F_\infty^{2/3} + \frac{1}{2}\ F_\infty^2$.

C) The flow is subcritical upstream, but $e(x) < e_{min}$. Differentiating (2.2) we see that

$$d'\ (1-F) - Fdw'/w = -h'. \qquad (2.5)$$

where the variation in w, currently being neglected is included in (2.5) for future use. At the minimum point, $F = 1$ and so $h' = 0$. The flow at the highest point of the topography (whether the topography is symmetric or not) is thus just critical, $F = 1$. In order for this to be possible there is a bore which propagates upstream. Differentiating (2.5) indicates that $d' \neq 0$ at $F = 1$ and the flow changes from subcritical upstream of the crest to supercritical downstream. In order to return to the far downstream conditions there is a stationary hydraulic jump in the lee of the topography. Both the bore and the hydraulic jump can be quite intense and generate a lot of mixing across the interface.

D) For larger values of F_∞ the flow pattern is similar to that described in C) except that the hydraulic jump in the lee of the topography is swept downstream. If the oncoming flow is supercritical, this form of motion can occur if $r > \{\{1 + (8F_\infty^2+1)^{3/2}\}/16F_\infty^2\} - 1/4 - 3F_\infty^{2/3}/2$. This indicates that there is a region of the (r, F_∞) plane where either a flow of this form or one without any jumps, as described in B), occurs, depending on the details of how the flow is initiated.

E) The topography is so high that it completely prevents the oncoming flow surmounting it. A travelling hydraulic jump is then necessary in order to balance the prescribed volume flux. Such a completely blocked flow occurs when $F_\infty^2 < (r^2-1)(r-1)/2r$.

A similar pattern must exist for the flow in a channel of varying width as well as depth. However, as seen from (2.5), the position of the critical point is dependent on the flow itself, and not specified by the geometry of the channel alone. To pursue the analysis, it would be convenient to introduce nondimensional variables

$$D = (gw^2/Q^2)^{1/3}d \quad E = (gw^2/Q^2)^{1/3}e \qquad (2.6,7)$$

into (2.2), which becomes

$$\tfrac{1}{2} D^{-2} + D = E(x). \tag{2.8}$$

The details have not yet been evaluated.

The discussion has so far tacitly assumed that the flow is unseparated. Conditions for flow separation are usually difficult to determine, but in this case a simple condition can be obtained. We put forward the criterion that separation occurs when the flow is subject to an adverse pressure gradient. Since the pressure is linearly related to the depth, the criterion becomes that flow separation occurs whenever the layer thickness in the lee of the topography increases in the downwards direction, or beneath a hydraulic jump. Examination of Figure 3 indicates that only in region A does the layer depth increase as required; and thus only in this region should topographic separation occur (as distinct from separation beneath a hydraulic jump). Experiments conducted in an open channel flume have confirmed the prediction (Huppert & Britter, 1980).

We conclude this section by mentioning that as the length scale of horizontal variations of the topography decreases, so does the validity of the hydrostatic relationship. For sufficiently short topographic features, lee waves can be generated, as shown in figure 4. Such waves are the main subject of study in section 5.

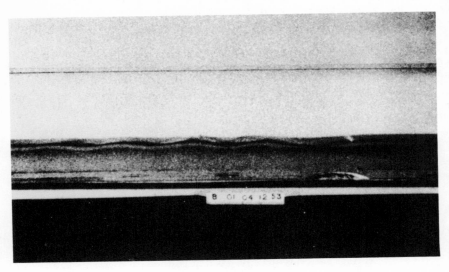

Figure 4.   Interfacial lee waves behind an obstacle being towed
            from left to right. Taken from Long (1954), Figure 4.

3. The Steady Flow of Two Fluid Layers

We briefly consider here, the extension of the ideas of the previous section to the hydrostatic flow of two inviscid layers, as depicted in Figure 5. For almost all practical situations, the velocity in and thickness of the upper layer is such that the Froude number of the upper interface is substantially less than unity. Also the density difference across the interface between the two layers is considerably smaller than that across the upper interface and thus deformations in the interface between the layers greatly exceed those at the free surface. As a consequence, the simplification that the upper interface is rigid is usually made. With this assumption, the specific energy equation, obtained by the use of mass continuity and of Bernoulli's equation at the interface between the two layers, becomes

$$Q^2/(2g'w^2d^2) - Q_2^2/\{2g'w^2(H-d-h)^2\} + d \tag{3.1}$$

$$= d_\infty(1 + \tfrac{1}{2}F_{1,\infty}^2) - \tfrac{1}{2}S(H - d_\infty)\,F_{2,\infty}^2 - h(x)$$

(cf. (2.2)), where $S = \rho_2/\rho_1 < 1$, H is the (assumed constant) total depth of the two layers

$$F_i = U_i/(g'd_i)^{\tfrac{1}{2}} = Q_i/(g'w^2d_i^3)^{\tfrac{1}{2}} \ (i = 1,2) \tag{3.2}$$

and the subscript 1 is dropped where this can be done without confusion. Differentiation of (3.1) indicates that critical condtions occur when

$$F_1^2 + SF_2^2 = 1. \tag{3.3}$$

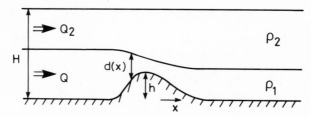

Figure 5.   The geometry and parameters for flow in a two-layer fluid.

An interesting application of some aspects of these results
was put forward by Stommel and Farmer (1952). They suggested that
the transition in width that frequently occurs at the mouth of an
estuary, or fjord, might control the upstream flow. To explore this
idea further, they conducted a large number of experiments in a
flume with an abrupt increase in width. Above an almost quiescent
lower layer, $F_2 \simeq 0$, a flow was forced in the upper layer which was
supercritical far upstream ($F_{1,\infty} > 1$). Stommel and Farmer report
measurements of d at the transition which were in good agreement
with those predicted by (3.3). They also conducted a series of
experiments in a flume with an abrupt decrease in width. With flow
parameters chosen such that there was subcritical flow in the upper
layer, and that nowhere along the flume was the minimum of the right-
hand side (3.1) reached, they confirmed that the flow was described
by (3.1). It should be noted that the experimental geometry, and the
title of the paper, incorporated an abrupt transition, which led to
results in agreement with theoretical ideas applied to a slowly
varying transition. This is frequently observed in such hydraulic
problems, and will be commented upon again in the next section. It
is probably due to the fluid arranging by itself for the flow to be
slowly varying; nearly stagnant eddies close to the transition are
set up which result in a quite smooth flow field.

In an accompanying paper, Stommel and Farmer (1953) noted that
the control condition may exert a surprisingly strong influence on
the mixing that can take place between the two layers. Consider
the fresh water outflow of a river over the salty water in an estuary
connected at a transition in width to the sea as an example. Suppose
that far upstream the lower layer is quiescent and the upper layer
has a fixed discharge. Suppose further that beyond the constriction
the width is very large and that the flow is subcritical there.
Allow the two layers to mix vertically in the estuary, thereby
transferring salt to the upper layer and momentum to the lower
layer and altering the relative depths of the two layers. The
control condition (3.3), which is applicable to the flow in the
transition provided that it is supercritical upstream, limits the
amount of mixing and places a lower limit on the density difference
that can be achieved. Stommel and Farmer conclude with a list of
five estuaries on the east coast of the United States in which the
mixing in the estuary may be constrained by this mechanism.

Detailed investigations of flows in two-layer systems, rather
than just examinations of the consequence of the existence of critical flow at a control, have been undertaken by Yih and Guha (1955)
and Long (1954, 1970, 1972, 1974).

Yih and Guha present a theory for internal hydraulic jumps in
a two-layer system which conserves mass and momentum in each layer
and thereby neglects any transfer of momentum across the interface.
Three series of experiments - one with a lower layer at rest, one

with the upper layer at rest and one with equal downstream velocities
in the two layers - result in the measured ratio of the layer depths
upstream and downstream of the jump as being in good agreement with
the theory.

Long has extended his analysis of single-layer flow over
topography in a channel of constant width to two-layer flow. The
analysis, and corresponding experiments, assume the same far upstream
velocities in the layers, Thus no effects of velocity shear (to be
discussed in section 5) are included. Except for one difference,
the theory is conceptually similar to that for a single layer of
fluid, though the algebraic complexity is considerably greater. The
one different concept involved is that the theory predicts that if
either:
    (a)   the maximum height of the topography exceeds the depth
           of the lower layer; or
    (b)   the lower layer is deeper than the upper layer
a hydraulic drop may occur.

In case (a) the predicted hydraulic drop occurs in the lee of
the topography, while in case (b) the prediction may place the drop
either upstream or downstream of the topography, depending upon the
flow parameters. The accompanying experimental investigations are
not yet complete, but so far indicate that downstream hydraulic drops
have not yet been observed. Details of the theoretical predictions
and experiments so far performed are contained in the various papers
by Long.

4.   Some Effects of Rotation

In this section we discuss some influences of the earth's rotation. The application of such effects to fjords seems not to have
been considered, though they are likely to play a role in fjord circulation. Rotation effects will be more influential for larger
fjords.

Aside from the possible wave motions induced, the most important
influence of rotation of flows discussed in the previous sections is
that it introduces a cross-stream variation. The length scale of
this cross-stream variation is known as the Rossby radius of deformation. For a one-layer flow, as depicted in Figure 1a, the Rossby
radius $(g'd)^{\frac{1}{2}}/|f| = a_0$ say, where f is twice the vertical component
of the earth's rotation. The Rossby radius is thus the ratio of the
long-wave speed to the rotation. This relationship can be usefully
extended to continuously stratified fluids by introducing the nth
baroclinic Rossby radius, $a_n = c_n/|f|$ where c_n is the wave speed of
the nth mode in a non-rotating system. The quantity c_n is obtained
by solving

$$h" + (N^2/c_n^2) \ h = 0 \qquad\qquad (4.1)$$

$$h(0) = 0 \qquad\qquad (4.2)$$

$$h(d) = 0 \qquad\qquad (4.3)$$

(c.f. (5.1)) where $N(z)$ is the buoyancy frequency. For a layer
of depth d with uniform buoyancy frequency N, $a_o = Nd/|f|$.

When the geometrically imposed scale, the width of the fjord,
for example, is much less than the Rossby radius, rotational effects
are small compared to buoyancy effects. However, if the geometric
scale becomes comparable to or larger than the Rossby radius,
rotational effects will be important and completely change the flow
field.

The sort of change that can occur will be illustrated by a few
particular examples. The first is due to Whitehead, Leetmaa & Knox
(1974), who considered the forced flow of a layer of fluid from a
deep, wide basin through a narrow channel and into another deep,
wide basin. They developed a theory which basically assumed that the
flow was subcritical upstream and was slowly varying there and became
critical in the channel (just like the similar calculation of Stommel
& Farmer described in the previous section). Their results indicated
that, as expected, cross-stream variations of the height of the fluid
layer and the downstream velocity scaled with the Rossby radius, a_o.
The depth of the fluid layer decreased to the left (looking downstream) so as to apply a downstream pressure gradient while the
velocity increases to the left because of the anticyclonic potential
vorticity effect as the fluid flows up to the channel. Whitehead
et al. also predicted that if $w > \sqrt{2}a_o$, where w is the width of the
channel, the cross-stream variation would be so strong that the
layer would no longer adhere to the left-hand wall. The experiments
yielded results which agreed well with the predicted upstream height,
but separation of the layer from the wall was never observed.

The second example is due to Sambuco & Whitehead (1976) who
considered the flow over a very long crested weir. Again the flow
was subcritical upstream, critical at the crest of the weir and then
supercritical. Sambuco & Whitehead predict the volume flux Q over
the weir as a function of the upstream height and also the strength
of the accompanying cross-stream velocity. Their experiments, for
moderate rotation rates, yielded results which agreed well with the
theoretical predictions. For higher rotation rates, the effects of
the walls became important and an additional analysis would be required.

Gill (1977) considered a fairly wide class of problems of the
flow in rotating channels with slowly varying downstream geometry.

He finds that the effects of the walls can be very significant and
the determination of the flow requires the prescription of the far-
upstream flow in the main body of the fluid as well as that in each
of the boundary layers which can be set up at either of the walls.

5.   Lee Waves Due to Steady Currents

     In this section we begin the consideration of non-hydrostatic
effects. The most important new concept, which occurs when stratified fluid flows steadily over a topographic feature, is the existence of lee waves downstream of the topography. The fundamental
physical idea involved in these waves is straightforward. Fluid
particles approaching the (two-dimensional) topography move upwards
in order to clear the topography and being then heavier than their
surroudings are acted upon by a downward buoyancy force which causes
them to move downwards once they are beyond the topography. The
potential energy gained from going over the topography is converted
into kinetic energy, which causes each particle to overshoot its
equilibrium level and undergo a series of oscillations: the lee
waves.

     The above is suggestive, but not rigorous. (Recall that capill-
ary-gravity waves appear upstream of a fishing line). Analysis of
the wave motion and energy propagation in a moving fluid is needed
to justify the downstream appearance of the waves. Linearised wave
theory indicates that waves whose upstream phase velocity equals the
downstream flow velocity, so that they appear stationary, have a
group velocity which is less than the phase velocity and hence the
energy is swept downstream. An initial-value calculation leads to
the same result.

     While some analysis has been performed for lee waves in layered
fluids (recall the extreme case of lee waves behind topography in
one-layer flow mentioned at the end of Section 2) the theoretical
framework has been set up for fluids with both a buoyancy frequency

$$N^2(z) = - \frac{g}{\rho_o} \frac{d\rho}{dz}$$

and an upstream velocity $U(z)$ which varies with height. The use of
linearised theory and the assumption that the flow does not separate,
leads to the governing equations

$$\nabla^2 w + \ell^2(z)w = 0 \qquad\qquad (5.1)$$

$$w = Uh'(x) \qquad z = 0 \qquad\qquad (5.2)$$

$$w = 0 \qquad z = H \qquad\qquad (5.3,H)$$

$$w \to 0 \qquad (x \to -\infty) \qquad (5.4,\text{H})$$

$$w \to 0 \qquad (z \to \infty) \qquad (5.3,\infty)$$

$$w = 0(x^2 + z^2)^{\frac{1}{2}} \qquad (x \to -\infty), \qquad (5.4,\infty)$$

where

$$\ell^2(z) = (N^2/U^2) - (U''/U), \qquad (5.5)$$

w is the vertical velocity and the H version of (5.3) and (5.4) are employed if the flow is considered to take place in a channel of height H and the ∞ version if there is no upper boundary. The condition that no waves propagate upstream is expressed by (5.4). (See Miles 1969 or Turner 1973 for a derivation of these equations and a more extensive discussion on lee waves than we present here).

The solution of the infinite version of (5.1) - (5.4) is straight forward: lee-waves of non-zero amplitude and wavelength $2\pi U/N$ make up part of the response, where N is a representative buoyancy frequency and U a representative far upstream velocity. The resulting flow pattern is typified by that shown in Figure 6. The solution of the finite version brings in a new feature. Only if a representative Richardson number, Ri = N^2H^2/U^2, is sufficiently large will a lee-wave mode be possible. Explicitly, for constant N and U, if Ri < π^2 there is no wave mode which has a sufficiently large phase velocity to remain stationary against the oncoming stream. As Ri exceeds π^2 one lee-wave mode begins to exist, as Ri exceeds $2\pi^2$ a second mode is possible, and so on. We can extend to this continuously stratified situation the Froude number concept introduced in Sections 2 and 3 by writing F = U/(πNH) $\equiv$ Ri$^{\frac{1}{2}}/\pi$ (where the factor π is introduced for convenience). Then the flow is completely supercritical (no waves) if F > 1, is subcritical to the first mode if $\frac{1}{2}$ < F < 1, and so on. A typical flow pattern, though actually obtained from the non-linear theory to be described below, is shown in Figure 7. In both the finite and unbounded situations, the lee-wave amplitude is proportional to the cross-sectional area of the topography (and not just on its height). This will turn out to be an important fact to bear in mind when considering the question of separation.

Comparisons between predictions obtained from the above theoretical framework and observations in the atmosphere are scanty. The most extensive investigation is due to Smith (1976), who monitored from an aircraft the flow over a two-dimensional section of the Blue Ridge Mountain in the Appalachians, U.S.A. He concludes that the theory predicts well the wavelength of the observed lee waves but underestimates their amplitude by a factor of approximately four. Influenced by a laboratory experiment of two-layer flow over a model of the Blue Ridge, Smith conjectures that the lack of

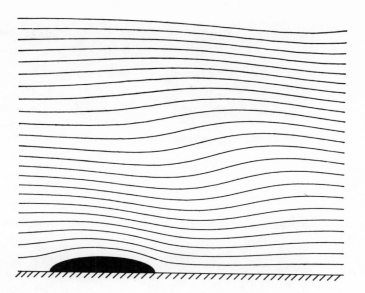

Figure 6.   Lee waves excited in a half space by a semi-elliptical
            obstacle of eccentricity 0.3, κ = N(0) h_{max}/U(0) = 0.5.
            Taken from Huppert and Miles (1969), Figure 10.

agreement *can probably be accounted for by the effects of a strong
nonlinearity lurking in the governing equations* (p.518).   In our
opinion flow separation, thereby inducing an effectively larger
cross-section in the topography, may also play a role.   We discuss
both these ideas next.

     Long (1953) pioneered nonlinear lee-wave calculations by
showing that steady, two-dimensional inviscid flow is governed
by

$$\nabla^2\delta + \tfrac{1}{2}\frac{d}{dz_0}\{\ln(\rho U^2)\}\{\delta_z - \tfrac{1}{2}|\nabla\delta|^2\} + k^2\delta = 0 \qquad (5.6)$$

where $\delta(x,z)$ is the displacement of a streamline from its far
upstream value, z_0, and $k(z)$ = $N(z_0)/U(z_0)$.   Only one assumption,
yet an important one, is made in the derivation: all streamlines
originate upstream.   Thus investigations of closed streamline
regions, or rotors as they are known by meteorologists, are
excluded, at least on a rigorous basis.   Being strongly nonlinear,
the solutions of (5.6) are difficult to obtain.   Long noticed,
however, that if far upstream the density gradient ρ' and the
dynamic pressure ρU^2 are considered to be constant, (5.6) reduces
to the linear Helmholtz equation

$$(\nabla^2 + k^2)\delta = 0, \qquad (5.7)$$

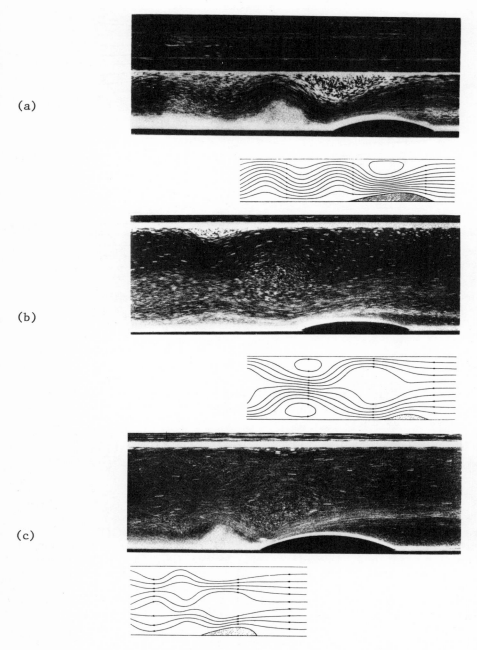

Figure 7.   Lee waves excited by an obstacle in a channel.   a)   One
            mode, b) two modes, c) three modes. Taken from Long
            (1955), Figures 7-9.

where k is a constant. Other upstream conditions which transform (5.6) into a linear equation have been suggested by Long (1958) and Yih (1965), but are not as realistic. Equation (5.7), together with the nonlinear boundary condition

$$\delta\left[x, h(x)\right] = h(x) \tag{5.8}$$

and the condition that flow can be prescribed upstream, is often called Long's model. There is considerable (unresolved) controversy as to whether the solutions to Long's model are typical of the more general equation (5.6). The discussions, conjectures and assertions enormously outweight the meagre known facts.

Long solved the system (5.7) and (5.8) together with

$$\delta(x, H) = 0, \qquad \delta(x, z) \to 0 \ (x \to -\infty) \tag{5.9,10}$$

by an inverse method. He determined a flow by imposing a non-zero value of δ over a bounded region on $z = 0$ and then chose a streamline to represent the topography. Such an inverse method is straightforward in both principle and practice, but does not allow the investigation of the change in the flow structure as the parameter k varies <u>for fixed shape</u>. Long compared his results with experiments and obtained quite good agreement, particularly when there were only a few - 0, 1 or 2 - lee-wave modes present. Figure 7 reproduces some of Long's comparisons.

In a series of papers, Miles and the author (Miles 1968a,b) Huppert & Miles (1969), Miles & Huppert (1969) instigated the direct solution of Long's model. A typical flow pattern and its variation with $\kappa = N(0)h_{max}/U(0)$ is shown in Figure 8. As κ increases, the amplitude of the lee waves increase. At a particular value of κ, say κ_c, a streamline becomes vertical somewhere. Beyond κ_c, the flow is statically unstable and closed streamlines appear, vitiating the main assumption of the model. Nevertheless, it appears that the model predicts fairly well the flow outside the closed streamline region. Huppert & Miles calculated κ_c for a variety of different topographic shapes and in particular for the whole range of semi-elliptical topography. They found that κ decreases monotonically from 1.73 for a vertical strip through 1.27 for a semi-circle to 0.67 for a semi-ellipse of infinitesimal height. Examining the response for topography of small slopes, $\varepsilon = h_{max}/L \ll 1$ where L is a horizontal length scale, which is the geophysically most relevant limit, they proved that $\kappa_c < 1$.

The form of motion and mixing that occurs beyond κ_c is of fundamental interest, though not yet well understood. Work on the generation of well-mixed regions, to be reviewed by Pollard in this volume, has not yet been linked to the present area of study.

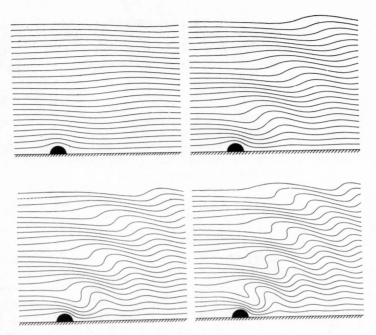

Figure 8.   Lee waves excited in a half space by a semi-circular
            obstacle for κ = 0.5,1.0, κ_c (1.27) and 1.5.   Taken from
            Huppert (1969), Figures A1–A4.

        Lee waves transport momentum downstream and the change in
horizontal momentum flux leads to a downstream wave drag, D, exerted
on the topography.   Huppert & Miles calculate the drag coefficient
$C_D = D/(\tfrac{1}{2}\rho U^2 h_{max})$ for a variety of different shapes, and some of
the results are reproduced in Figure 9.   From Figure 9a it is evident that, at least for semi-elliptical topography, the nonlinear
($\varepsilon \neq 0$) contribution to the drag can be significant.   Only one
laboratory experiment has measured drag, which is rather surprising
considering its geophysical importance.   Davis (1969) towed thin
vertical plates and also triangles through a stratified fluid.   He
concluded that the measured total drag greatly exceeded the predicted wave drag, especially so when the flow was subcritical to
two or more modes.   The disagreement was due to separation effects
and the induced turbulent wake downstream of the topography.   There
is a great need for careful quantitative drag measurements on geophysically realistic streamlined topography, which a thin vertical
barrier is not.

        We conclude this section with a discussion of some recent
work on separation behind stratified fluids, drawing heavily on the
elegant doctoral thesis of Brighton (1977).   Results on unbounded

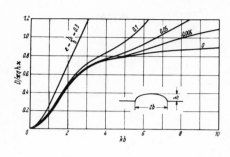
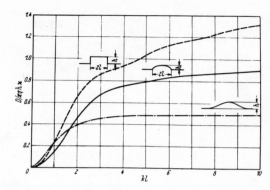

Figure 9.   a)   The non-dimensionalised wave drag on a semi-
                 elliptical obstacle in a half space. Taken from
                 Miles (1969), Figure 18.
            b)   The linearised non-dimensional wave drag on a
                 rectangular obstacle, a semi-elliptical obstacle
                 and the Witch of Agnesi. Taken from Miles (1969)
                 Figure 19.

flow have been derived from a linear boundary layer calculation
by Brighton and from a nonlinear numerical analysis using the
concept of "triple decks" by Sykes (1978). For flow over topography
of sufficiently small slopes separation is entirely absent, whether
the flow is stratified or not. For topography of moderate slopes,
if the flow is homogeneous there is the usual separation in the
lee due to boundary layer effects. As the degree of stratification
is increased, there is little change in the flow field until
$K_L = NL/U$ becomes of order unity, when separation can be suppressed.
(It is impossible to be more precise in general since the exact
value of K_L for which separation first disappears will depend heavily
on the shape of the topography). The suppression occurs because
when $K_L \simeq 1$, the wave length of the lee waves $\lambda = 2\pi\, U/N$ is
comparable to the length of the topography and there is a strong
downflow in the lee of the topography towards the lee wave trough.
For larger stratification, the lee wave length is shorter than the
topography and the suppression ceases. Brighton's calculations
diverge from Sykes', which are shown in Figure 10, at this point.
Brighton suggests that upstream separation occurs for large K_L;
Sykes suggests that inviscid lee wave rotors attached to the ground
occur.

     For flow in a bounded region the results depend much more
critically on the actual slope of the topography. First, no matter

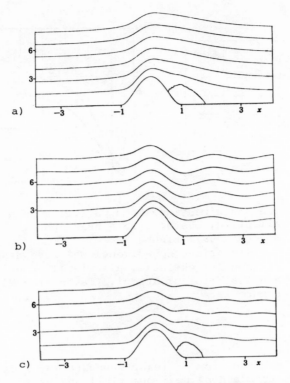

Figure 10.   Streamlines for flow in a boundary layer over a hill
              represented by $h = h_{max}\{1 - (x/L)^2\}^2 H(1 - |x|/L)$
              for $\kappa \, Re^{2/5} = 0$, 9 and 15, where Re is the Reynolds
              number.   Taken from Sykes (1977), Figures 5,6,7.

what the slope, if $K_H = NH/\pi U < 1$, the flow is supercritical and
stratification effects are minimal. For small slopes, as K_H
increases beyond 1, the downslope acceleration induced by the waves
suppresses the separation until the wavelength $\lambda = 2H/\sqrt{K_H^2 - 1} \simeq 2L$
Larger K_H (smaller wavelength) leads to separation as predicted by
the inviscid response. For moderate slopes the parameter range
(in K_H or λ) for which separation can be suppressed is considerably
reduced.

     Brighton compared his analysis which leads to the above
conclusions with an experimental study and obtained qualitative
agreement. A series of his photographs are shown in Figure 11.
He also presents the results of a series of measurements for $1 <$
$K_H < 3$ of the wavelength, which agreed well with the theoretical

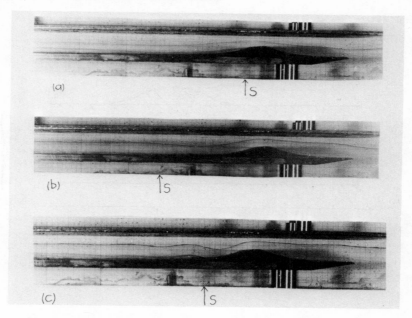

Figure 11.   Stratified flow in a channel of height 2.3 cm for
κ = 1.13, 1.4, 4.0. S marks the position of separation
Taken from Brighton (1977), Figure 4.11.

predictions, and the separation distance, whose value was virtually
indistinguishable from the measured wave length.

6.   Time-Dependent Mean Flows

Possibly the most important time-dependent mean flows occur
due to the oscillations associated with the tides. This variation
in the strength of the mean flow has an interesting effect on the
lee waves discussed in the last section. Consider the ebb tide,
flowing out of a fjord over a sill into the adjoining stratified
sea, or, on a larger scale, flowing over the continental slopes
and out to sea. Such a flow will set up a train of lee waves on
the seaward side of the topography. For sufficiently high Froude
number little mixing will take place, though the lee waves may be
of quite large amplitude. For lower Froude numbers, the lee response will be more in the form of a mixed region. Either way, the
steady disturbance set up will consist of waves whose phase velocity
shorewards is just balanced by the seawards velocity of the mean
flow. As the tide slackens, the mean flow diminishes and the phase
velocity of the waves causes them to propagate over the topography.

These concepts recently have been put forward by Maxworthy

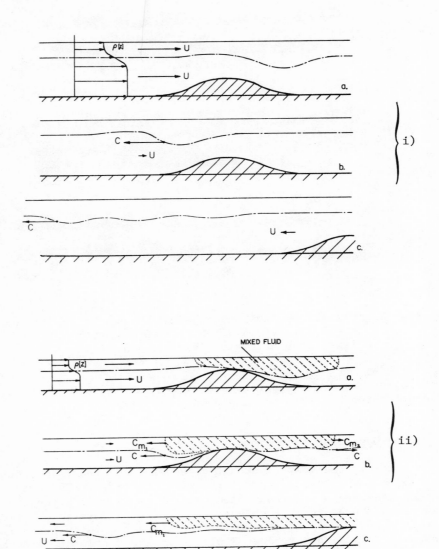

Figure 12.   The evolution of a wave train in a stratified fluid
due to the relaxation of the mean flow.
i)   The lee waves excited by the flow over topography
propagate upstream as the mean flow relaxes.
ii)   The parameters are such that the lee waves break
and form a mixed region, initially downstream of the
topography.   Taken from Maxworthy(1979), Figures
10 and 11

(1979), building on a previous theoretical discussion by Lee &
Beardsley (1974). Observations pertinent to this mechanism, which
to some extent motivated Lee, Beardsley and Maxworthy, come from
Massachusetts Bay (Halpern 1971), the southern Strait of Georgia,
British Columbia (Gargett 1976) and from Knight Inlet, British
Columbia (Smith and Farmer 1977 and Farmer & Smith 1978). We will
summarise these observations before discussing the theoretical
concepts further.

   In Massachusetts Bay, the tide runs over Stellwagen Bank,
which rises out of approximately 80 metres of water to a depth
of about 27 metres, the approximate depth of the bottom of the
seasonal thermocline. From two sets of temperature measurements
made to the west (shoreward) of the Bank, Halpern documents the
existence of a packet of internal waves during the flood tide.
The period of the individual waves was 6-8 minutes and the time of
passage of the packet was approximately 2½ hours. Halpern
calculated the first mode eigenfunction (by numerically solving
5.1), using the observed values for the velocity and density
distributions., The calculated period and structure agreed fairly
well with the observations. In a recent paper, Haury, Briscoe &
Orr (1979) used a 200 - kHz acoustic back-scattering system at
a site close to that used by Halpern to observe the wave packet
and confirm Halpern's observations. As yet no quantitative
calculation of how long it should take the waves to propagate over
the bank and appear at the observation site has yet been made to
compare with the measured values. Further, no observations have
yet been made seaward of the bank, to confirm that waves generated
during the flood tide will propagate seaward, as the proposed
explanation would require.

   In the area studied by Gargett, the Strait of Georgia is
bounded to the south by a large number of islands connected by
submarine ridges. A simplified version of Gargett's argument,
based on her observations, is as follows. During the flood phase
of the tide the southward flowing water is quite thoroughly mixed
as it flows over the ridges and through the passes. On the ebb
tide, some of this relatively homogeneous mass of water flows out
into the stratified water of the Strait and evolves in a series of
finite amplitude waves. This situation is slightly different
to that in Massachusetts Bay in that the intruding fluid is more
thoroughly mixed. The underlying principal is the same: wave
motion is the far downstream response to any source of disturbance
to a stratified fluid.

   The observations by Farmer and Smith (described in greater
detail in other papers in this volume) document the occurrence of
internal waves generated by the second sill in Knight Inlet.

If the waves are of sufficiently large amplitude, and it may be
supposed that they will be in most geophysical situations, non-
linear effects will be important in their evolution. Thus as the
tide propagates, a train of solitary waves may be expected to occur.
The same basic mechanism results at low Froude numbers, when the
fluid in the lee of the topography is quite well mixed. A series
of solitary waves will evolve, with their number depending upon
the geometry of the region, the basic stratification and the amount
of mixed fluid. Maxworthy(1979) has performed a series of experiments in a geometry like that found in Massachusetts Bay. Figure
12 taken from his paper, sketches the wave evolution described above.

REFERENCES

Brighton, P.W.M. 1977. Boundary layer and stratified flows past
      obstacles. PhD. Thesis, Cambridge University.

Davis, R.E. 1969. The two-dimensional flow of a stratified fluid
      over an obstacle. J. Fluid Mech. 36, 127-143.

Farmer, D. and J.D. Smith. 1979. Nonlinear internal waves in a
      fjord. In: Hydrodynamics of Estuaries and Fjords, ed.
      J.C.J. Nihoul. Elsevier.

Gargett, A.E. 1976. Generation of internal waves in the Strait
      of Georgia, B.C. Deep-Sea Res. 23, 17-32.

Gill, A.E. 1977. The hydraulics of rotating-channel flow.
      J. Fluid Mech. 80, 641-671.

Halpern, D. 1971. Observations on short-period internal waves in
      Massachusetts Bay. J. Mar. Res. 29, 116-132.

Haury, L.R., M.G. Briscoe and M.H. Orr. 1979. Tidally-generated
      internal wave packets in Massachusetts Bay. Nature. 278,
      312-317.

Houghton, D.D. and A. Kasahara. Nonlinear shallow fluid flow over
      an isolated ridge. Comm. Pure and Appl. Maths., 21, 1-23.

Houghton, D.D. and J. Woods. 1969. The slippery seas of Acapulco.
      New Scientist. 16, 134-136.

Huppert, H.E. and R.E. Britter. 1980. The separation of hydraulic
      flow over topography. J. Fluid Mech. (sub judice).

Huppert, H.E. 1968. Appendix to Lee waves in a stratified flow.
      Part 2. Semi-circular obstacle. J. Fluid Mech. 33, 803-814.

Huppert, H.E. and J.W. Miles. 1969. Lee waves in a stratified flow. Part 3. Semi-elliptical obstacle. J. Fluid Mech. 35, 481-496.

Lee, C.-Y. and R.C. Beardsley. 1974. The generation of long nonlinear internal waves in a weakly stratified shear flow. J. Geophys. Res. 79, 453-462.

Long, R.R. 1953. Some aspects of the flow of stratified fluids. I. A theoretical investigation. Tellus. 5, 42-58.

Long, R.R. 1954. Some aspects of the flow of stratified fluids. II. Experiments with a two-fluid system. Tellus. 6, 97-115.

Long, R.R. 1955. Some aspects of the flow of stratified fluids. III. Continuous density gradients. Tellus. 7, 341-357.

Long, R.R. 1958. Tractable models of steady-state stratified flow with shear. Quart. J. Roy. Met. Soc. 84, 159-161.

Long, R.R. 1970. Blocking effects in flow over obstacles. Tellus. 22, 471-480.

Long, R.R. 1972. Finite-amplitude disturbances in the flow of inviscid rotating and stratified fluids over obstacles. An. Rev. Fluid Mech. 4, 49-92.

Long, R.R. 1974. Some experimental observations of upstream disturbances in a two-fluid system. Tellus. 26, 313-317.

Maxworthy, T. 1979. A note on the internal solitary waves produced by tidal flow over a three-dimensional ridge. J. Geophys. Res. 84, 338-346.

Miles, J.W. 1968a. Lee waves in a stratified flow. Part 1. Thin barrier. J. Fluid Mech. 32, 549-568.

Miles, J.W. 1968b. Lee waves in a stratified flow. Part 2. Semi-circular obstacle. J. Fluid Mech. 33, 803-814.

Miles, J.W. 1969. Waves and wave drag in stratified flows. Proc. 12th Int. Cong. Appl. Mech., Stanford University, 50-76.

Miles, J.W. and H.E. Huppert. 1969. Lee waves in a stratified flow. Part 4. Perturbation approximations. J. Fluid Mech. 35, 497-525.

Sambuco, E. and J.A. Whitehead. 1976. Hydraulic control by a wide weir in a rotating fluid. J. Fluid Mech. 73, 521-528.

Smith, J.D. and D.M. Farmer. 1977. Non-linear internal waves
    and internal hydraulic jumps in a fjord. Geofluiddynamical
    Wave Mathematics: University of Washington.

Smith, R.B. 1976. The generation of lee waves by the Blue Ridge.
    J. Atmos. Sci. 33, 507-519.

Stommel, H. and H.G. Farmer. 1952. Abrupt change in width in
    two-layer open channel flow. J. Mar. Res. 11, 205-214.

Stommel, H. and H.G. Farmer. 1953. Control of salinity in an
    estuary by transition. J. Mar. Res. 12, 13-20.

Sykes, R.I. 1978. Stratification effects in boundary layer flow
    over hills. Proc. Roy. Soc. 361 A, 225-243.

Turner, J.S. 1973. Buoyancy Effects in Fluids. Cambridge
    University Press.

Whitehead, J.A., A. Leetmaa, and R.A. Knox. 1974. Rotating
    hydraulics of strait and sill flows. Geophys. Fluid Dyn. 6,
    101-125.

Yih, E.-S. 1965. Dynamics of Nonhomogeneous Fluids. Macmillan.

Yih, C.-S. and C.R. Guha. 1955. Hydraulic jump in a fluid system
    of two layers. Tellus. 7, 358-366.

BAROTROPIC AND BAROCLINIC RESPONSE OF A SEMI-ENCLOSED BASIN TO BAROTROPIC FORCING FROM THE SEA

Anders Stigebrandt

Department of Oceanography
University of Gothenburg
Box 4038, S-400, 40 Gothenburg
Sweden

INTRODUCTION

During the last few years it has become evident that the tides may provide the turbulence in at least some fjords with significant amounts of energy. Until recently it was not understood how the turbulence in the deep water below the sill level was sustained, see e.g. Gade (1970). Several possible energy sources for the turbulence in the deep water have been suggested. Among the suggested sources we may mention internal waves generated by the tides, winds and rivers flowing into the fjord. Quantitative estimates were, however, not undertaken before the present author (Stigebrandt, 1976) found that internal waves generated at fjord sills by tides possibly may sustain the turbulence. The calculation showed that just 5% of the energy released by the internal waves was needed to explain the observed mixing in the deep water in the inner Oslofjord. The present author has provided further observational evidence for the correctness of the theory (Stigebrandt, 1979a). The theory has obviously also been able to explain the mixing in the deep water in other fjords, see e.g. Perkin and Lewis (1978). The success of the theory has apparently led to an increased interest in tides in fjords in general and internal tides in particular.

The response of a fjord to barotropic forcing from the sea has both barotropic and baroclinic components. It was felt that it perhaps may be of some value to describe some of the main problems that have been described in the literature. The presentation of each case must naturally be rather brief. Anyone who intends to go deeper into the theories should consult the references cited.

THE RESPONSE OF THE WATER LEVEL IN A SEMI-ENCLOSED BASIN TO
BAROTROPIC FORCING FROM THE SEA

We are interested in the barotropic response of a semi-enclosed
basin to a water level variation outside the mouth of the basin.
Special cases of the linear problem have been much studied.   One
branch is the problem of Harbour resonance.   Here the interest
is focused on forcing periods in the range 10-200 sec.   For lower
forcing frequencies the interest has mostly concentrated on the
tidal forcing of fjords and bays and especially basins near
resonance have received attention.   The nonlinear problem (including
field accelerations and a quadratic frictional law) and tidal
choking has not been studied so much in the past although in many
respects important in a lot of fjords.

We will discuss the problem starting from the sketch in Fig.1.
A basin with surface area Y, is connected to the ocean by a channel
of length L and vertical cross sectional area A (width W and mean
depth H).   The velocity, u, in the channel is taken positive when
directed towards the basin.

The system is excited by an oscillation in the sea of the form
$h = a_0 \cos\omega_0 t$.   We will discuss the response for several special
cases of basin forms and assumptions regarding dissipation in the
system.

The Linear Regime

AI)   Y = 0, H large (Deep fjords and bays).   The free surface
has the equation $z = H + h(x,t)$ (h<<H).   Small amplitude motion and
no dissipation in the system.   The linear equations of motion then
are

$$\frac{\partial u}{\partial t} = - g \frac{\partial h}{\partial x} \tag{a1}$$

$$\frac{\partial u}{\partial x} = - \frac{1}{H} \frac{\partial h}{\partial t} \tag{a2}$$

Together these equations give

$$\frac{\partial^2 h}{\partial t^2} - gH \frac{\partial^2 h}{\partial x^2} = 0 \tag{a3}$$

We are looking for standing wave solutions.   The sought solution
thus is

$$h = a \cos(\omega_0 t+\theta)\cos(kx+\phi); \quad k = \omega/\sqrt{gh} \tag{a4}$$

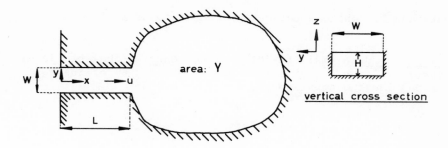

Figure 1.   Definition sketch for the barotropic response problem.

where the constants a, θ and ϕ have to be determined from the boundary conditions.   These are

$$u = 0 \quad \text{at the inner end wall } (x = L) \tag{a5}$$

$$h = a_0\cos\omega_0 t \quad \text{at the mouth } (x = 0) \tag{a6}$$

Eqs. (a4), (a5) and (a6) together gives the complete solution. This is

$$h = a_0\cos(\omega_0 t)\cos\{k(x-L) + n\pi\}/\cos(kL-n\pi) \quad n=0,1,2,3,\ldots$$

The amplitude in the semi-enclosed basin is predicted to be infinite when

$$kL - n\pi = \pi/2$$

or

$$L = \tfrac{1}{2} T(n + \tfrac{1}{2})\sqrt{gH} \quad n = 0,1,2,3,\ldots$$

Thus resonance occurs when L/λ = 1/4, 3/4, 5/4, ... where $\lambda = T\sqrt{gH}$.

In a real system dissipation of course occurs and the amplitude remains finite.   Even if dissipation does not occur inside the basin, the basin will radiate energy to the sea.   This causes two effects (i)   the response becomes finite even at the resonance frequency and (ii)   the resonance frequency changes.   The change in resonance frequency is dependent upon the aspect ratio W/L and decreases as W/L increases.   The basin appears, with respect to resonance, to be longer than L. The end correction can be found in Ippen (1966).

   AII) Y = 0, H small (friction important). (Shallow fjords
and bays). If the frictional force is modelled by a linear
frictional law the equations of motion are

$$\frac{\partial u}{\partial t} = - g \frac{\partial h}{\partial x} - \beta u$$

$$\frac{\partial u}{\partial x} = - \frac{1}{H} \frac{\partial h}{\partial t}$$

from these one may obtain the so called telegraphers equation

$$\frac{\partial^2 h}{\partial t^2} - gH \frac{\partial^2 h}{\partial x^2} + \beta \frac{\partial h}{\partial t} = 0$$

A solution to this equation is

$$h = a \, \exp(\pm \mu x) \cos(kx - \omega t)$$

where μ and k both are related to the resistance coefficient
$\beta = \omega \tan(2\alpha)$ and $\tan \alpha = \mu/k$. The details of the calculation
are given by Ippen (1966).

   The boundary conditions are the same as in case AI (Eqs.(a5) and
(a6). The boundary condition at the inner end is satisfied by
superposition of waves: an incident wave of amplitude h_1 entering
from the ocean and travelling in the positive x-direction and a
reflected wave of amplitude h_2 moving in the negative x-direction.
The amplitudes h_1 and h_2 have to be equal at the reflecting end
(x = L) in order to satisfy the boundary condition. The complete
solutions are rather messy, but they can be found, e.g. Ippen (1966).
This problem was apparently first solved by Redfield (1950) (cited
by Defant (1960).

   BI) Y ≠ 0, L<< $T_0 \sqrt{gH}$, u/ω<<L. (Frictionless). The severe
constriction, the entrance channel of length L, between the basin
and the sea requires that long barotropic waves can not be used
to describe the response of the basin to barotropic forcing from
the sea. The dynamics of the channel may be very different from
that of the basin and cannot be ignored. The role of the basin
is reduced to that of a reservoir for fluid. The oscillating mass
is essentially confined to the entrance channel.

   The condition on L is that the channel connecting the basin
to the sea is much shorter than the length of a barotropic wave
of frequency ω_0 in water depth H. The pressure gradient in the
channel may then, if u/ω<<L as explained at the end of this section

be expressed as

$$\rho g(h_0 - h_i)/L$$

where h_0 and h_i are the water levels at the outer and inner ends
of the channel and ρ is the density of the fluid. The flow in
the channel may, if non-linear and frictional effects are absent,
be described by the balance

$$\frac{\partial u}{\partial t} = g(h_0 - h_i)/L \qquad (b1)$$

The level h_i will change because of the flow through the channel.
If the water level in the basin is flat one then obtains

$$Y \frac{\partial h_i}{\partial t} = W.H.u \qquad (b2)$$

Given an oscillation $h_0 = a_0 \exp(i\omega_0 t)$ in the sea outside the
channel one obtains

$$\frac{\partial^2 h_i}{\partial t^2} + \alpha^2 h_i = \alpha^2 a_0 \exp(i\omega_0 t) \qquad (b3)$$

where

$$\alpha^2 = gWH/LY$$

The response of the basin on the outer forcing occurs at the same
frequency, thus

$$h_i = a \exp(i\omega_0 t) \qquad (b4)$$

Insertion of Eq. (b4) into Eq. (b3) gives the amplitude response

$$a/a_0 = \alpha^2/(\alpha^2 - \omega_0^2) \qquad (b5)$$

Thus there is resonance for

$$\omega_0^2 = \alpha^2 = gWH/LY$$

The acoustic counterpart to this system is the so called undamped
Helmholtz resonator.

The justification for the assumption that the water level at
the upstream end of the channel is the same as in the upstream
basin is that the Bernoulli effect on the water level upon
accelerating into the channel is small. The fall in the free
surface from the upstream basin into the hole, $\Delta\eta$, is (see also
case CI later)

$$\Delta\eta = u^2/2g$$

where u, as before, is the velocity in the channel. $\Delta\eta$ has to be small compared to $|h_0 - h_i|$ if (b1) shall be a good approximation for the momentum balance. With $\partial u/\partial t = \omega u$ (b1) gives $|h_0 - h_i| = \omega u L/g$. The condition $\Delta\eta << |h_0 - h_i|$ gives

$$u/\omega << L$$

u/ω can be identified with the excursion length, ℓ_e, for the water particles in the oscillatory motion in the channel. Thus for this theory to be valid the excursion length has to be materially smaller than the length of the entrance channel.

BII) $Y \neq 0$, $L << T_0\sqrt{gH}$, $u/\omega << L$. (Harbours). As in case BI we have a short channel compared to a long wave of frequency ω in water depth H and the excursion length is assumed to be short compared to the length of the channel. If friction is taken into account and assuming a linear friction law the following equations result

$$\frac{\partial u}{\partial t} = g(h_0-h_i)/L - \beta u$$

$$\frac{\partial h_i}{\partial t} = \alpha^2 \, Lu/g$$

Through elimination of u one gets the equation

$$\frac{\partial^2 h_i}{\partial t^2} + \beta\frac{\partial h_i}{\partial t} + \alpha^2 h_i = \alpha^2 h_0 \qquad\qquad (b6)$$

This is the equation for the Helmholtz resonator. In the present case β is believed to depend upon frictional resistance from the bottom and walls of the channel. In the acoustic analogue the coefficient is the radiation loss coefficient. Even in the present problem radiation losses from the channel to the sea are possible.

When $h_0 = a_0 \, \exp(i\omega_0 t)$

there are solutions

$$h_i = a \, \exp(i\omega_0 t) \qquad\qquad (b7)$$

Insertion of Eq. (b7) into (b6) gives the amplitude response

$$(a/a_0) = \{(1-\omega_0^2/\alpha^2)^2 + \beta^2\omega_0^2/\alpha^4\}^{-\frac{1}{2}}$$

The response is no longer infinite at $\omega_0 = \alpha$ because of friction. The resonance may even be completely hidden if β is big enough, this is the case in e.g. the Baltic-Kattegatt, a system treated in this way by Swanson (1979). He showed that a/a_0 decreases monotonically with ω_0 if $\beta > \alpha\sqrt{2}$.

Miles and Munk (cited by Ippen, 1966) treated harbours near resonance as Helmholtz resonators. They considered the dissipation mechanism in the harbour to be similar to the one in the acoustic analogue: radiation of energy from the channel mouth. They introduced a dissipation factor Q defined by

$$Q = \frac{\omega \, E}{dE/dt}$$

where E = mean energy of the oscillation

which in the earlier formulation may be identified as $Q = \alpha/\beta$. There is a vast literature on harbour resonance, one of the more recent contributions in this field is the one by Miles and Lee (1975).

The Nonlinear Regime

CI) $Y \neq 0$, $L \simeq 0$. The thin wall ($L \simeq 0$) which in this case is separating the sea from the basin has an opening through which the water can flow. Upon entering the opening the water has to accelerate and the pressure drops, see Fig. 2. (We touched upon this problem already at the end of case BI).

For low frequency oscillations the balance of forces in the accelerating phase towards the opening is established between the pressure gradient and inertial forces. The equation of motion is thus

$$\underset{\sim}{u} \cdot \nabla u = - \nabla P/\rho$$

From this equation, where $\underset{\sim}{u}$ is the three-dimensional velocity field and ρ is the pressure, the Bernoulli equation for stationary flow can be obtained. Integration of this from far upstream where $|\underset{\sim}{u}| \gg u$ and to the hole, where $|\underset{\sim}{u}| = u$, one obtains

$$\Delta\eta = u^2/2g. \tag{c1}$$

One problem arising is to relate $\Delta\eta$ to $|h_0 - h_i|$. If the flowing water leaves the hole like a jet (the flow separates from the boundaries) we have $\Delta\eta = |h_0 - h_i|$ and the whole drop in water level is found outside and upstream of the hole. Thus the water level is not symmetric around the opening. (Compare the case with a smooth bump on the bottom of a channel.) The variation in water

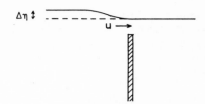

Figure 2.   The flow of water through the hole connecting the basin
            to the sea.

level associated with flow in the channel is symmetric around the
bump if separation does not occur).   The equations governing the
forced level variation in the basin are thus

$$h_0 - h_i = Q^2/(2gA^2) \text{ where } Q = uWH \qquad (c2)$$

$$Y \frac{dh_i}{dt} = Q \qquad (c3)$$

$$\frac{dh_i}{dt} = (A\sqrt{2g}/Y)(h_0 - h_i)/|h_0 - h_i|^{\frac{1}{2}} \qquad (c4)$$

The factor $(h_0 - h_i)/|h_0 - h_i|^{\frac{1}{2}}$ takes account of the sign of the volume
flux Q.   This case of response of a semi-enclosed water body to
an outer water level fluctuation was treated by McClimans (1978),
who also took the contraction effect into account.   The contraction
coefficient, ν, multiplies the vertical cross sectional area of
the hole.   In Eq. (c4) the "A" should be replaced by $\nu A (\nu \leq 1)$.

     In this case all the important dynamics in the system is situa-
ted just upstream the entrance channel (the hole) where the Bernoulli
equation determines the flow.   All the kinetic energy is dissipated
in the downstream basin through turbulence in the jet.   If the
entrance channel has a considerable length we also have to take
frictional effects (bed resistance) in the channel into account.
This is done below.

     CII $Y \neq 0$, $0 < L \ll T_0\sqrt{gH}$.   (The Baltic, the Nordåsvatnet).
When frictional effects dominate linear accelerations $(\partial/\partial t)$ the
system loses its resonator character (this also happens when $\ell_e \gtrsim L$
and the effects of damping becomes more important.   There is then
good reason to model the frictional resistance in a way more

realistic than with a linear friction law which was done in cases AII and BII. We adopt the generally accepted quadratic friction law. Because of frictional effects in the channel $\Delta\eta \neq |h_0 - h_i|$. There is a level drop, Δh, in the channel associated with friction. For channel flow we have

$$\rho g A \Delta h = \tau_b BL \qquad (c5)$$

where τ_b is the stress against the channel bed and B is the length of the wetted perimeter. With $\tau_b = \rho \kappa u^2$ we get

$$\Delta h = (\kappa BL/gA) u^2 \qquad (c6)$$

A more complete discussion of the assumption underlying this relation can be found in Stigebrandt (1979b). The relation between the total drop in water level and the velocity in the channel thus is

$$|h_0 - h_i| = (1 + 2\kappa BL/A)(u^2/2g) \qquad (c7)$$

The change in water level in a semi-enclosed basin is

$$Y \frac{dh_i}{dt} = Q + Q_f \qquad (c8)$$

where we now have added the effect of the fresh water runoff, Q_f. By combining Eqs. (c7) and (c8) one gets

$$\frac{dh_i}{dt} = \{(h_0 - h_i)/|h_0 - h_i|^{\frac{1}{2}}\}(A/Y)\sqrt{2g/(1 + 2\kappa BL/A)} + Q_f/A \qquad (c9)$$

Note that Eqs. (c4) and (c9) are similar. The Bernoulli effect and the frictional effect appears in the same way in the problem. Also note that there is no upper limit for the excursion length for Eq. (c9) to be valid.

If we want to study the response of a semi-enclosed basin upon an outer barotropic forcing of given period T it is convenient to make Eq. (c9) non-dimensional by the use of the period T and the amplitude a_0 of the water level fluctuation in the sea outside the channel. One then gets

$$\frac{dh_i}{dt} - P(h_0 - h_i)/|h_0 - h_i|^{\frac{1}{2}} - S = 0 \qquad (c10)$$

where

$$P = (A/Y)\{2gT^2/a_0(1+H)\}^{\frac{1}{2}}; \quad H = 2\kappa BL/A; \quad S = Q_f T/a_0 Y$$

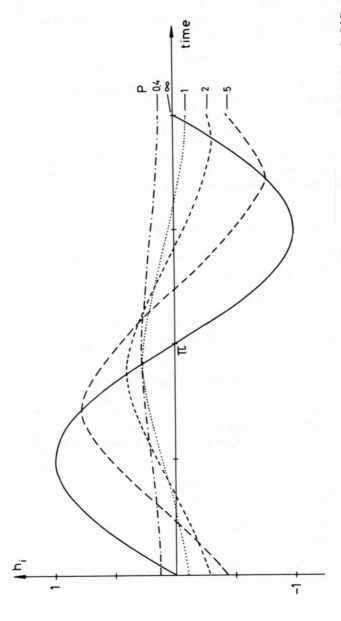

Figure 3. The solution of Eq. (c10) for varying values of the parameter P and for S = 0.067. From Stigebrandt (1979b)

Eq. (c10) was solved by the present author (Stigebrandt, 1979b) by a fourth order Runge-Kutta method for different values of P and S. Fig. 3 shows the computed water level inside the entrance channel, h_i, for different values of P and for S = 0.067. The forcing tidal elevation outside the constriction is shown by the solid line. From these curves it can be seen that the fresh water runoff to the fjord has two effects on the water level h_i. One is that the time mean value of the water level in the fjord, $\bar{h}_i$, increases with decreasing P-values. The other effect is that the periods for flood ($dh_i/dt > 0$), T_f, and ebb ($dh_i/dt < 0$), T_e, are of unequal length when S ≠ 0. These effects of runoff are more pronounced for smaller values of P.

The amplitude response or the choking coefficient, C_c, was defined by Glenne and Simensen (1963) and $C_c = (h_{i,max} - h_{i,min})/2a_0$. In Fig. 4 we have drawn C_c versus P^{-1} for S = .0.067 and in Fig. 5 we have drawn the computed phase delay, $\Delta\phi$, for the inner basin versus P^{-1} for the same S-value. $\Delta\phi$ is defined as the phase delay between high water outside and inside the constriction. Because $T_e/T_f \neq 1$ for S ≠ 0 the phase delay between the low waters is different from $\Delta\phi$. Note that P^{-1} is proportional to $\omega a_0^{\frac{1}{2}}$, the response is thus dependent upon the amplitude. In linear systems, such as the Helmholtz resonator, the response is independent of the amplitude of the input.

The asymmetry in ebb and flood periods vanishes when $\bar{h}_i$ exceeds the amplitude of the outer water level oscillation, see also Fig. 6. The basin inside the constriction is in this case rather a fresh water lake and not a fjord but still there may be tidal variations in the basin. The case when a hydraulic jump forms in the channel is also briefly discussed in Stigebrandt (1979b).

In non-linear systems the response on a single harmonic component of frequency ω_0 is spread over frequences ω_0, $2\omega_0$, $3\omega_0$, $4\omega_0$ ..... The tides in such systems show larger higher harmonics (in tidal analysis so-called shallow water constituents) than the sea outside.

Eqn. (c9) has been applied to the Baltic-Kattegatt system, see the map in Fig. 7. The water level in the Baltic is forced by changes in the water level in the sea (the Kattegatt). For low frequency oscillations the theory should be applicable ($L \ll T_0\sqrt{gH}$, $\partial u/\partial t \ll \nu_t \nabla^2 u$). In Fig. 8 observed water levels (diurnal means) in the Kattegatt (circles) and in the Baltic (horizontal mean level, continuous line) are shown. The dotted line shows the calculated water level in the Baltic for A = 0.24 $10^6 m^2$, Y = 0.37 $10^{12} m^2$, g = 9.8 m/s^2 and $Q_f = 10^9$ m^3/day and for

$$H = 2\kappa L/H = 12$$

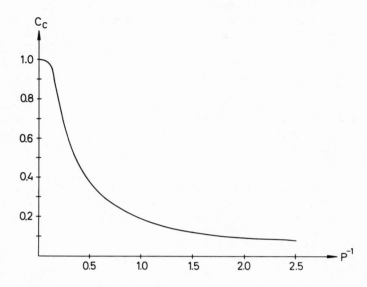

Figure 4. The amplitude response of the inner basin (the choking
           coefficient C_c) as a function of P^{-1} ($P^{-1} \sim \omega \sqrt{a_0}$).
           $S = 0.067$. From Stigebrandt (1979b).

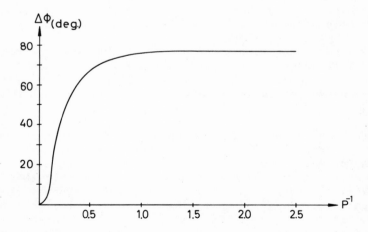

Figure 5. The phase lag for high water in the inner basin versus
           P^{-1}, $S = 0.067$. From Stigebrandt (1979b).

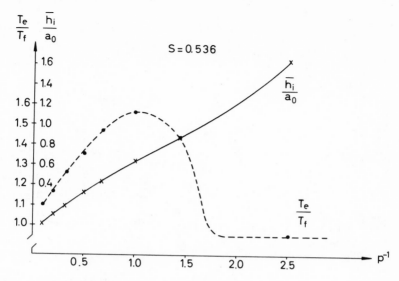

Figure 6. Normalized mean water level in the inner basin ($\bar{h}_i/a_0$) and the ratio between ebbing and flooding periods (T_e/T_t) versus P^{-1}. Here $S = 0.536$. From Stigebrandt (1979b).

This H-value gives the best fit. Thus the complicated transition between the Baltic and the Kattegatt may, with respect to barotropic forcing from the sea, be replaced by a channel with the following properties: $A = 0.24 \cdot 10^6$ m^2 and $L/H = 6/\kappa$. The deviation between calculated and measured mean water level in the Baltic is presumably caused by (i) local wind effects in the southern Baltic, affecting the water level just inside the entrance channel and (ii) the difference in air pressure between the central part of the Baltic and the Kattegatt.

The water level data and the geometrical dimensions of the Baltic-Kattegatt system were all kindly given to me by Dr. Arthur Swanson at the Fishery Board of Sweden. Dr. Swanson has treated the same system as a Helmholtz resonator (Swanson, 1979).

THE BAROCLINIC RESPONSE OF A STRATIFIED SEMI-ENCLOSED BASIN TO BAROTROPIC FORCING

In the problem of calculating the barotropic response of a semi-enclosed basin to barotropic forcing from the sea there were no conditions imposed upon the current (except at the inner end

Figure 7.   Map over the Baltic - Kattegatt system.

in case A).   In the non-linear regime (case C) however, we used the
result that the kinetic energy leaving a hole or a channel is lost
to turbulence when separation occurs (a jet).   The kinetic energy
is in this case not used for building up a barotropic pressure.

In order to study the baroclinic response of a semi-enclosed
basin to barotropic forcing from the sea we may no longer disregard
the conditions upon the velocity at topographical features.   For
simplicity we assume that the topographical feature of importance
can be taken as a step in bottom depth.   The simplest of such a step
is a vertical wall simulating a sill and we adopt this simple topographical feature for the rest of this discussion.

Internal waves constitute one part of the total response of the
semi-enclosed basin and they also influence the barotropic response.
The problem to solve the total response of the basin seems to be too
complicated in many instances and we will not here try to do that.

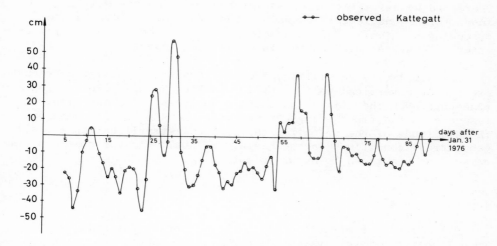

Figure 8a.   Daily mean values of the water level in the Kattegatt
             (Hornbaeck) during the period 05/02/76 – 30/04/76.

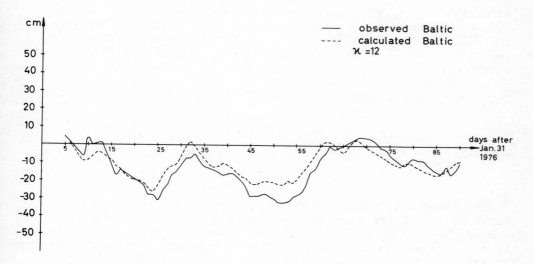

Figure 8b.   Observed (————) and calculated (–––––––) mean water
             level in the Baltic during the same period as in Fig.8a.

Therefore we assume the fluctuating transport over the sill to be known (from measurements or calculated from some theory in the preceding section). This assumption is justified if one is interested in just calculating the internal wave response.

For simplicity we will consider a fjord with gentle topography except for a vertical wall, a sill, see Fig. 9. The fluctuating transport over the sill is assumed to be known. The baroclinic response to the fluctuating current over the sill will depend on the amplitude of the velocity, the vertical stratification and also geometrical conditions in the basin. We shall discuss some typical cases that are known to occur in nature.

The Linear Regime

DI)   Two-layer stratification, the pycnocline at sill depth. In a linear two-layer system there exist two wave modes for each frequency ω: one surface mode (index b) with phase velocity $c_b = (gH)^{1/2}$ and one internal mode (index i) with phase velocity $c_i = (gd \cdot (H-d) \cdot (\rho_2 - \rho_1)/\rho_2 H)^{1/2}$. Here d $\cdot$ (H-d) and $\rho_2 (\rho_1)$ are depth and density of the lower (upper) layer, see Fig. 10.   (For a general introduction to internal wave theory see Phillips, 1977). Of course there may be an infinite number of barotropic and internal waves of the frequency ω propagating in all horizontal directions. The number of waves may be reduced to just two of each kind if the waves are forced to propagate along a narrow channel. This is the situation in many fjords and we assume that the fjord can be described by a straight channel and that the vertical wall simulating the sill is perpendicular to the axis of the channel. The orbital velocity in the internal waves, u_{1i} (u_{2i}) in the upper (lower) layer then is

$$u_{1i} = - \ \gamma\{d/(H-d)\}\cos(kx \pm \omega t) \qquad\qquad\text{(d1)}$$

$$u_{2i} = \gamma\cos(kx \pm \omega t) \qquad\qquad\text{(d2)}$$

where

$$k = \omega/c_i$$

The boundary condition at the sill on the velocity is that the horizontal velocity u vanish below sill level, z = d, and, as discussed previously, that the velocity (transport) over sill level may be regarded as known. Thus we have the following boundary conditions at the sill

$$u = \alpha \ \frac{H}{H-d} \ \cos\omega t \quad \text{(for } d < z \leq H) \qquad\qquad\text{(d3)}$$

$$u = 0 \quad \text{(for } 0 \leq z \leq d) \qquad\qquad\text{(d4)}$$

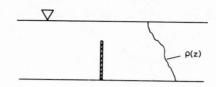

Figure 9. Definition sketch for the baroclinic response problem.

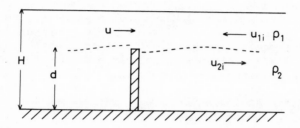

Figure 10. Definition sketch for the two layer baroclinic response
            problem.

We will now look into the possibility of satisfying the boundary
condition at the sill by a superposition of internal waves and a
barotropic wave. The internal waves do not give rise to any net
transport. The barotropic wave thus supplies all the necessary
transport. This determines the barotropic orbital velocity, u_b,
to be

$$u_b = \alpha \cos\omega t \quad (\text{at } x = 0) \tag{d5}$$

having assumed that the sill is situated at $x = 0$. The sought
combination of barotropic and internal waves thus is, from Eq.(d3)

$$\alpha - \gamma d/(H-d) = \alpha H/(H-d) \quad (d < z \leq H)$$

This gives

$$\gamma = -\alpha \tag{d6}$$

This also satisfies Eq. (d4). The internal wave satisfying the boundary condition at the sill is thus described by the orbital velocities

$$u_{1i} = \alpha(d/H-d) \cos (kx \pm \omega t) \qquad (d7)$$

$$u_{2i} = - \alpha \cos(kx \pm \omega t) \qquad (d8)$$

Internal waves propagating in both directions are thus possible. The theory presented here is more fully discussed in Stigebrandt (1976).

In a laboratory experiment it was found by the present author (Stigebrandt (1976) that only waves progressing from the sill were obtained. The reason for this was that internal waves generated at the sill were destroyed when meeting the sloping bottom in the inner reaches of the basin. With regular geometry and vertical walls in the fjord the internal waves may be reflected and a system of standing internal waves may be sustained. Whether progressive, standing or a mixture of internal waves develops thus depends upon the reflection conditions in the basin.

A densimetric Froude number, F_d, for the oscillating stream over the sill may be defined as

$$F_d = \alpha d / \{c_i (H-d)\}$$

It was found from laboratory experiments in Stigebrandt (1976) that internal waves were generated for $F_d \lesssim 1$. For $F_d > 1$ no long internal waves were generated, something like a rotor or a jet developed on the lee side.

The present author, Stigebrandt (1979a), has also reported measurements from the Oslofjord that show internal waves of tidal origin. The waves have amplitudes and phases that are very well predicted by the theory described here. The internal waves are progressive, presumably because of wave breaking against sloping bottoms.

DII) Linear stratification. The vertical stratification in many fjords is often better described as continuous rather than two-layered. An interesting special case of the continuous stratification is the linear stratification. This case is especially interesting because the internal waves which are possible in such a stratification can be described by simple analytical expressions (see Phillips (1977). An infinite number of possible linear internal wave modes exist in this stratification and the horizontal orbital velocity for the n:th mode is, if $N^2 \gg \omega^2$, given by

$$u_n = a_n \cos(n\pi z/H) \cos (kx \pm \omega t), \quad n=1,2,3,\ldots \qquad (d9)$$

where a_n is the current amplitude, k is the horizontal wave number, and $N^2 = (-g/\rho_0 \cdot \partial\rho/\partial z)$ is the buoyancy frequency squared. The dispersion relation for the waves is $\omega = kNH/n\pi$. When the water is stratified the velocity over the sill may vary with depth. We take account of this by introducing a function $g(z)$ into (d3) such that

$$\frac{1}{H-d} \int_d^H g(z)\,dz = 1$$

$$g(z) = 0 \quad (\text{for } 0 \leq z \leq d) \tag{d10}$$

The current boundary condition at the vertical section through the sill is then

$$u = \{\alpha H/(H-d)\}g(z)\cos(\omega t) \quad (d < z \leq H) \tag{d11}$$

$$u = 0 \quad\quad (0 \leq z \leq d) \tag{d12}$$

Eqs. (d11) and (d12) replace (d3) and (d4).

Again we will try to satisfy the boundary condition at the sill by a superposition of internal waves and a barotropic wave. The barotropic wave is, as in the earlier case (DI), given by Eq. (d5). The sought combination of the barotropic and baroclinic waves is thus

$$\alpha\cos(\omega t) + \sum_{n=1}^{\infty} a_n \cos(n\pi z/H)\cos(\omega t) = U(z) \quad (x=0)$$

The a_n should be determined from

$$\alpha + \sum_{n=1}^{\infty} a_n \cos(n\pi z/H) = \alpha\, Hg(z)/(H-d) \quad (d < z \leq H)$$

$$\tag{d13}$$

$$\alpha + \sum_{n=1}^{\infty} a_n \cos(n\pi z/H) = 0 \quad (0 \leq z \leq d)$$

If we choose a step velocity profile, $g(z) = \text{const.} = 1$, we obtain for a_n (making use of the orthogonality property of the functions $\cos(n\pi z/H)$

$$a_n = - (2\alpha/n\pi)\{H/(H-d)\}\sin(n\pi d/H) \tag{d14}$$

The a_n may of course be determined for other forms of the function $g(z)$. In fig. 11 we have plotted the $E(k_n)$ ($\sim a_n^2$) vs. the horizontal wave number k_n for two different values of the relative height of the sill (d/H). The general trend is that $E(k_n)$ falls off rapidly with increasing k_n for d/H in the range $0.2 < d/H < 0.8$.

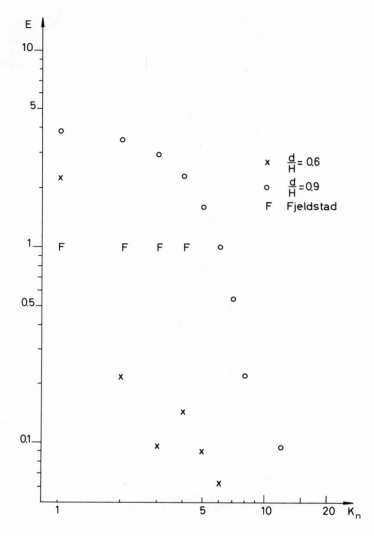

Figure 11. Line spectra for internal tides in linear stratification
for d/H = 0.6 and d/H = 0.9 respectively. The F's
show Fjeldstads results from the Herdlafjord. Arbitrary
scales. From Stigebrandt (1979b).

For values of d/H outside this range, however, the periodogram
is broad for smaller wave numbers The value of d/H gives a
specific signature in the E, k_n-space. The theory outlined here
was developed by Stigebrandt (1979b) and the theory was tested
against a classical set of measurements obtained by Fjeldstad and
cited by Defant (1961).

Fjeldstad measured the amplitude of internal waves in the Herdlafjord and he determined the amplitude of the first four vertical modes taking the real vertical stratification into account. The stratification did not differ much from linear. He found that the amplitude of the modes he determined were nearly equal. The Herdlafjord has two connections to the sea. One direct connection, which is rather narrow, in the northwestern end and with a sill depth of about 40 metres and another indirect, via the Byfjord, which is broad and deep (about 170 m). The mean depth of the Herdlafjord is about 300 m. The ratio d/H is about 0.6 and 0.9 for the deep and shallow sill respectively. The model signature of the internal waves created at the two sills should thus be rather different. The Byfjord sill should give a dominating first mode while the northwestern sill should distribute significant amounts of energy to several of the lowest modes, see Fig. 11. As already mentioned Fjeldstads analysis showed "white noise" with approximately equipartition of energy among the four modes, a signature we expect for d/H near one. Accordingly we conclude that the major components of the waves observed by Fjeldstad are generated at the Northwestern sill. Fjeldstad himself detected from simultaneous measurements along different verticals along the fjord axis that the waves propagate towards the south-east but he did not consider the generation mechanism. Unfortunately we do not know the α's for the two sills and therefore we are not able to calculate the theoretical magnitudes of the a_n:s for the two cases.

The linear theory for internal wave generation sketched here can be extended to an arbitrary continuous vertical stratification by replacing the orbital velocity functions in (d9) with those appropriate to the actual stratification.

The Nonlinear Regime

E) Two-layer stratification. Pycnocline well below the sill level. Intermittent selective withdrawal? Blackford (1978) studied the generation of internal waves by tidal flow over sills in a two-layer stratification where the interface is situated well below the top of the sill. The internal waves are mainly caused by the non-linear barotropic pressure, caused by local accelerations at the shallow sill. He assumed the barotropic pressure (the free surface) to be symmetric about the sill. The barotropic pressure gradient caused by the slope of the free surface near the sill is essentially compensated by a baroclinic pressure gradient directed in the opposite way (remembering that the system is dynamic and not static).

The most important result obtained by Blackford is that a large amplitude internal wave of twice the forcing frequency may be generated by this "Bernoulli" mechanism. As already discussed under

case CI there is Bernoulli flow just on the upstream side of the
sill. On the downstream side the flow may be characterized as
a jet. Frictional effects are thus important here and the water
level is flat from the downstream side of the sill and out into
the basin. Thus the water surface (and the barotropic pressure
field) is asymmetric over the sill and this should lead to asymmetric
waves of twice the forcing frequency. In field measurements
presented by Blackford there is in fact a pronounced asymmetry.

As the Bernoulli mechanism operates only at the upstream side
of the sill or channel when flow separation occurs on the lee
side it seems possible to describe the waves as a manifestation of
the time - dependent selective withdrawal process that must act
there. It is known (Pao and Kao, 1974) that the flow in the
process of selective withdrawal is established through the propagation
of internal waves out from the sink area. The process studies by
Blackford may probably be termed "intermittent, time-dependent
selective withdrawal".

F) High frequency internal waves generated by the jet or by an
internal hydraulic jump on the lee side of a sill or entrance channel.
Internal waves with the same frequency as the forcing tide can be
generated only if a densimetric Froude number, F_d, based upon the
stratification, the current velocity across the sill and the depths
of the two layers, is less than critical. With the definition of
F_d given in case DI and for the pycnocline at sill depth the critical
value was estimated to be about 1 in experiments reported by the
present author, see case D. If F_d is greater than critical there will
be a jet on the lee side of the sill. In this supercritical case the
jet may perhaps, for certain combinations of sill depth and stratification, undergo an internal hydraulic jump. The jump may generate
internal waves of high frequency (compared to tidal frequencies) that
travel away from the sill. The generation mechanism is in this case
thus not directly associated with the fjord sill. Farmer and Smith
(1978) have presented beautiful photos of such waves generated by
the ebb jet and propagating up the Knight Inlet. They investigated
the waves in detail and found that the waves could be described as
a train of internal solitary waves. This kind of tidally generated
internal wave has also been discussed and observed by Gargett (1976).
Hamblin (1977) observed similar waves, generated by a river flowing
into a stratified fresh water lake.

In contrast to the subcritical case it is not possible to
predict these high frequency waves (with respect to phase, amplitude
and frequency). The case of a buoyant jet flowing into a stratified
fluid was, to some extent, discussed by the present author
(Stigebrandt, 1978).

REFERENCES

Blackford, B.L.   1978.   On the generation of internal waves by
    tidal flow over a sill - a possible nonlinear mechanism.
    J. Mar. Res., 36, 529-549.

Defant, A.   1960.   Physical Oceanography, Vol. 2.   Pergamon Press.

Farmer, D.M. and J.D. Smith.   1978.   Nonlinear internal waves in
    a fjord.   In Hydrodyn. of Estuaries and Fjords, 465-494,
    Elsevier Oceanography Series 23, (J. Nihoul ed.)."

Gade, H.G.   1970.   Hydrographic investigations in the Oslofjord:
    A study of water circulation and exchange processes.   Geo-
    physical Institute, Report 24, University of Bergen, Norway.

Gargett, A.E.   1976.   Generation of internal waves in the Strait
    of Georgia, B.C.   Deep-Sea Res., 23, 17-32.

Glenne, B. and T. Simensen.   1963.   Tidal current choking in the
    landlocked fjord of Nordåsvatnet.   Sarsia, 11, 43-73.

Hamblin, P.F.   1977.   Short-period internal waves in the vicinity
    of a river-induced shear zone in a fjord lake.   J. Geophys.
    Res., 82, 3167-3174.

Ippen, A.T. (editor).   1966.   Estuary and Coastline Hydrodynamics.
    McGraw - Hill.

McClimans, T.A.   1978.   On the energetics of tidal inlets to land-
    locked fjords.   Mar. Sci. Communications 4, 121-137.

Miles, J.W. and Y.K. Lee.   1975.   Helmholtz resonance of Harbours.
    J. Fluid Mech., 67, 445-464.

Pao, H.P. and T.W. Kao.   1974.   Dynamics of establishment of
    selective withdrawal of a stratified fluid from a line sink.
    Part 1.   Theory.   J. Fluid Mech., 65.   657-688.

Perkin, R.G. and E.L. Lewis.   1978.   Mixing in an Arctic Fjord.
    J. Phys. Oceanogr., 8, 873-880.

Phillips, O.M.   1977.   Dynamics of the Upper Ocean.   Cambridge
    University Press.

Stigebrandt, A.   1976.   Vertical diffusion driven by internal waves
    in a sill fjord.   J. Phys. Oceanogr., 6, 486-495.

Stigebrandt, A.   1977.   On the effect of barotropic current
    fluctuations on the two-layer transport capacity of a constric-
    tion.   J. Phys. Oceanogr., 7, 118-122.

Stigebrandt, A.   1978.   Dynamics of an ice covered lake with through
    flow.   Nordic Hydrology, 9, 219-244.

Stigebrandt, A.   1979a.   Observational evidence for vertical diffusion
    driven by internal waves of tidal origin in the Oslofjord.
    J. Phys. Oceanogr., 9, 435-441.

Stigebrandt, A.   1979b.   Some aspects of tidal interaction with
    fjord constrictions.   To be published in J. Coastal and
    Estuarine Mar. Sci.

Swanson, A.   1979.   A simplified model of the Baltic. (in manuscript).

A LINEAR MODEL OF INTERNAL TIDES IN SILL FJORDS

Joseph R. Buckley

Seakem Oceanography Ltd.
9817 W. Saanich Rd.
Sidney, B. C.

INTRODUCTION

Internal waves with tidal periods have been observed in fjords since early in this century (Zeilon, 1914). Recent measurements in Alberni Inlet on the west coast of Canada indicated the presence of large amplitude ($\sim$10 m) internal tides in weakly stratified water of about 75 m depth (Dobrocky Seatech Ltd., unpub. man.). The model described here attempts to explain the generation and magnitude of these waves. The model will be described briefly and some preliminary results given.

Proudman (1953) showed that internal waves could be generated by the passage of gravity waves through regions of changing water depth. Rattray (1960) applied this theory to a two layered model in which internal tides in the deep ocean were generated at the continental shelf break. This model was modified by Buckley (1977) to describe the phase of the surface layer response to the baro-tropic tide in a sill fjord. A subsequent form of the deep ocean tidal model which incorporates continuous stratification {by Prinsenberg and Rattray (1975)} is also readily adaptable to a region with simplified fjord bathymetry. The model described here is similar in general form to that of Prinsenberg and Rattray. An important feature of this model is that it uses the measured vertical density structure of the water in calculating the response of the fjord.

The fjord bathymetry is parameterized by a series of long, narrow, flat-bottomed, straight channels with equal widths but usually different depths. This parameterization for Alberni Inlet is shown in Fig. 1. Region I represents the outer basin of the

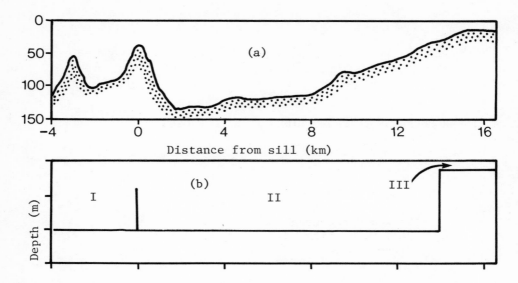

Figure 1.   (a) Bathymetry of the inner basin of Alberni Inlet.
             (b) Model bathymetry showing the three regions.

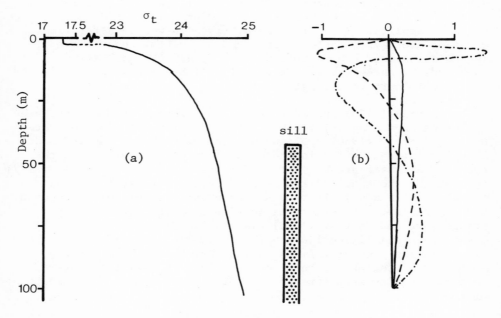

Figure 2.   (a) Sigma-t profile used for the model.
             (b) Vertical eigenfunction gradients for mode 1 (~),
                 mode 2 (-–) and mode 3 (·–).

fjord, open at the outer end to the ocean. At the junction between it and region II, the inner basin, a knife edge sill rises to within 40 m of the surface. Both regions I and II are 100 m deep. Region III represents the shallow harbour at the head of the fjord. Linearized, two-dimensional equations of motion are solved in each of these three regions. The cross-channel dimension was neglected since the channel width was less than the internal radius of deformation of the lower order modes of oscillation of the system. The plane waves resulting from this system should practically be indistinguishable from the Kelvin waves that would result from a full three dimensional solution.

Eliminating all but one of the dependent variables from the basic equations yields the following equation for the non-hydrostatic part of the pressure field:

$$\frac{\partial^2 \Pi}{\partial x^2} - \frac{\sigma^2}{N^2(z) - \sigma^2} \frac{\partial^2 \Pi}{\partial z^2} = 0 \tag{1}$$

where x is in the long-channel direction, z is positive upwards, σ is the angular frequency of the tidal wave and $N(z)$ is the measured buoyancy frequency profile. The assumption of constant depth in each region allows the pressure field to be explicitly separated into horizontal and vertical parts:

$$\Pi(x,z) = \sum_{j=0}^{\infty} P_j(x)\phi(z) \tag{2}$$

which, when applied to eq. 1 yields two series of one-dimensional equations:

$$(\frac{\partial^2}{\partial x^2} + k_j^2)P_j = 0; \quad (\frac{\partial^2}{\partial z^2} + k_j^2\{N^2 - \sigma^2\}/\sigma^2)\phi_j = 0 \tag{3}$$

The vertical equations are subject to the no flux (rigid lid) condition

$$\frac{\partial \phi_j}{\partial z} = 0 \tag{4}$$

at the surface and at the bottom. Since the buoyancy frequency profile is measured rather than analytic, the eigenvalues (wave numbers) and eigenfunctions are found numerically in a Runge-Kutta scheme.

Travelling wave solutions of the form:

$$P_j(x) = A_j \exp(ik_j x) + B_j \exp(-ik_j x)$$

are assumed for the horizontal equations in each region. In region I, the incoming amplitude A, is fixed as the barotropic wave driving the system. In region III B, is forced to be zero to simulate dis-

Table I.   Amplitudes of the horizontal wave functions calculated
            for the M_2 tide in Alberni Inlet.

Mode	Region I		Region II	
	Incoming	Outgoing	Incoming	Outgoing
0	1.00	-0.91	1.35	-1.26
1	0.00	-1.24	-0.62	-0.63
2	0.00	0.29	0.14	0.15
3	0.00	0.40	0.20	0.20

Table II.   Amplitude of isopycnal displacement 8 km from the sill
            in the inner basin of Alberni Inlet at a depth of 75 m.
            The negative signs indicate waves 180° out of phase with
            the surface tide. Amplitudes are in metres.

Mode	M_2	S_2	K_1	O_1
1	3.91	1.66	3.48	1.14
2	-3.73	-1.63	0.24	-0.30
3	-8.41	-2.55	-2.13	-1.11

Table III. Potential energy integrated from the surface to the
            bottom for the first three modes of the four largest
            tidal constituents in Alberni Inlet. Results are
            normalized with respect to the integrated energy in
            mode 1.

Mode	M_2	S_2	K_1	O_1
1	1.00	1.00	1.00	1.00
2	0.38	0.12	0	0.03
3	1.37	0.70	0.11	0.28

sipation of the incoming tidal wave on the tidal flats at the head
of the fjord. The other amplitudes are determined from the boundary and matching conditions at the junctions of the regions. At
the junction between regions I and II, the velocity fields in both
regions are set equal, above sill depth, are the pressure fields.
Below the sill depth, the velocities are set to zero. Similar
conditions apply at the junction between regions II and III. The
matching conditions are solved simultaneously in one matrix equation.

Preliminary results from this model have been calculated for
only one parameterization of Alberni Inlet, but have shown several
interesting features. The density profile used in the test was
the average of data collected by Dobrocky Seatech Ltd. (unpub.
man.) during a complete tidal cycle 21-22 Nov. 1977. This profile
is shown in fig. 2, along with the gradients of the first three
eigenfunctions calculated from it. Gradients are shown rather
than the functions since the vertical displacement of isopycnals
is proportional to the gradients.

The largest tidal constituent in Alberni Inlet is the M_2, with
a surface amplitude of 30.5 m. Amplitudes of the horizontal wave
functions for this component are shown in Table 1. These results
show that the incoming barotropic tide (mode 0) generates a baroclinic response in the three modes for which the model was solved.
The amplitudes of the incoming and outgoing waves in region II are
almost equal, indicating that the baroclinicity is generated at
the sill rather than at the harbour mouth.

The amplitude of isopycnal displacement for any mode at any
location in the inlet is proportional to the vertical gradient of
the vertical eigenfunctions, with the constant of proportionality
derived from the horizontal wavefunctions and the amplitude of the
observed tide at the sill. Table II shows the calculated amplitudes
at a location 8 km from the sill in the inner basin at a depth of
75 m for the four major tidal constituents. The amplitude pattern
of the modes is similar in all the four constituents except for the
anomalously large value for the M_2 third mode. Since the basin
length is close to 1.5 wavelengths of this mode, the large amplitude
is possibly a resonance phenomenon. Similar resonance criteria
may be invoked to account for the difference between the diurnal
(K_1 and O_1) and semi-diurnal (M_2 and S_2) components in the relative
amplitude of the second mode to the first. There is a general
tendency however for the third mode to be stronger than the second
in all constituents, possibly due to the existence of a zero crossing
of the third mode eigenfunction derivative near sill depth. Such a
zero crossing requires a large amplitude in the third mode to make
a small contribution to the solution of the no flux boundary
condition.

The potential energy of the internal waves is a function of isopycnal displacement and hence of vertical eigenfunction gradient. Therefore profiles of modal potential energy will be similar to the gradient profiles shown in fig. 2. Potential energy integrated over the water column depth gives an indication of the relative effectiveness of energy transfer from the barotropic to the various baroclinic modes. Table III shows this integrated potential energy normalized with respect to the energy of the first mode. As expected from the results in Table II, the third mode extracts more energy in all constituents than does the second.

The internal waves that this model generates are of similar amplitude to those observed. Measurements were made in only one place in the inlet so that the horizontal distribution predicted by the model may not be verified. The measurements did show a tendency to increasing amplitude with increasing depth, a tendency that the model will not support because of boundary condition eq. 4. Replacement of region II with a sloping bottomed region modelled by the methods of Wunsch (1968, 1969) should help to alleviate this discrepancy. Although the model could not reproduce all of the observed behaviour of the internal tides in Alberni Inlet, it has, in this preliminary test, given some useful insight into the gerneration of internal tides in a sill fjord and should provide a useful tool to investigate the effects of changing stratification on the distribution of baroclinic energy resulting from the passage of waves of tidal period over the sills of fjords.

REFERENCES

Buckley, J. R. 1977: The currents, winds and tides of northern Howe Sound. Ph.D. thesis. U. Brit. Col. Vancouver, Canada. 224 p.

Dobrocky Seatech Ltd. unpub. manuscript: An oxygen budget study of the deep waters in the inner basin of Alberni Inlet. A report to the department of Fisheries and Invironment, Government of Canada. Victoria, B. C. 215 p.

Prinsenberg, S. J. and M. Rattray Jr. 1975: Effects of continental slope and variable Brunt-Väisälä frequency on the coastal generation of internal tides. Deep-Sea Res. 22, 251-263.

Proudman, J. 1953: Dynamical Oceanography. Methuen & Co. Ltd., London, 349-351 pp.

Rattray, M. Jr. 1960: On the coastal generation of internal tides. Tellus 12, 54-60.

Wunsch, C. 1968: On the propagation of internal waves up a slope.
    Deep-Sea Res. 15, 251-258.

Wunsch, C. 1969: Progressive internal waves up a slope. J. Fluid
    Mech. 35, 131-144.

Zeilon, N. 1914: On the seiches of the Gullmar Fjord. Svenska
    Hyd-Biol. Komm. Skrifter 5.

TIDALLY-FORCED STRATIFIED FLOW OVER SILLS

T.S. Murty and M.C. Rasmussen

Numerical Modelling Section
Institute of Ocean Sciences, Patricia Bay
Sidney, British Columbia, Canada

INTRODUCTION

Farmer (1978) and Farmer and Smith (1978) described long, non-linear internal waves in Babine Lake and Knight Inlet on the West coast of Canada. The forcing mechanism in the case of Babine Lake is wind, whereas, for Knight Inlet, tide is the energy source. This work was started with the aim of simulating, numerically, the internal surges in these two water bodies. The model which was used in these simulations was adopted from the so-called down-welling frontal model (Simons, 1978) which will be described briefly in the next section.

Since the model used here is hydrostatic, strictly speaking lee waves cannot be reproduced. Another feature that has not been included in this model is the allowance for flow separation which, indeed, happens in nature and in the laboratory experiments. Two different models are being developed to include these features. The first one is based on the so-called Triple-deck Analysis (Sykes, 1978), which allows for non-hydrostatic effects and flow separation, but deals with a steady state problem. The second model, based on the full Navier-Stokes equations (Mason and Sykes, 1979), allows for time-dependence, continuous density variation in the vertical, non-hydrostatic effects and flow separations. The results from these two models will be reported in future publications.

THE DOWNWELLING FRONTAL MODEL

Simons (1978) started with the same set of equations as used by Cahn (1945) for geostrophic adjustment in a homogeneous fluid. For the homogeneous case, a free upper boundary was assumed, and

173

the equations were non-dimensionalized through

$$(1)\begin{cases} t = f.t' \\ x = x'/R \\ h = h'/H \end{cases} \qquad \begin{aligned} (u,v) &= (u',v')/C \\ (\tau x, \tau y) &= (\tau x', \tau y')/\rho f H C \\ C &= \sqrt{gH} \end{aligned}$$

Here, primed variables are dimensional, x is the horizontal coordinate along which the solutions are allowed to vary, f is the Coriolis parameter, h is the fluid depth, H is the reference depth, ρ is the water density, t is time and R = C/f is the Rossby radius of deformation. The wind stress components are given by $\tau x, \tau y$, and the velocity components in the x and y directions are given by u, v.

To conserve mass and momentum, the equations of motion in the x and y directions, and the continuity equation, are written in the following form after noting that

$$(2) \qquad\qquad (U,V) = (hu, hv)$$

$$(3)\begin{cases} \dfrac{\partial U}{\partial t} - V + \dfrac{\partial}{\partial x}\left(\dfrac{U^2}{h} + \dfrac{h^2}{2}\right) = \tau x \\[2ex] \dfrac{\partial V}{\partial t} + U + \dfrac{\partial}{\partial x}\left(\dfrac{UV}{h}\right) = \tau y \\[2ex] \dfrac{\partial h}{\partial t} + \dfrac{\partial U}{\partial x} = 0 \end{cases}$$

For the stratified case, a rigid lid was assumed at the surface and the Boussinesq approximation was made. The equations for the two layers are of the general form of (3). For convenience, Simons (1978) expressed the mass transport in the lower layer in terms of the mass transport in the upper layer and the total transport. Again assuming that there are no variations in the y direction, the relevant equations are

$$(4)\begin{cases} \dfrac{\partial U}{\partial t} - fV + \dfrac{(D-h)}{D}\left[\dfrac{\partial}{\partial x}\left(\dfrac{U^2}{h} + \dfrac{g\varepsilon h^2}{2}\right) - \dfrac{\tau x}{\rho}\right] - \dfrac{h}{D}\left[\dfrac{\partial}{\partial x}\left(\dfrac{U^2}{D-h}\right) - fM\right] = 0 \\[2ex] \dfrac{\partial V}{\partial t} + fU + \dfrac{(D-h)}{D}\left[\dfrac{\partial}{\partial x}\left(\dfrac{UV}{h}\right) - \dfrac{\tau y}{\rho}\right] - \dfrac{h}{D}\left[\dfrac{\partial}{\partial x}\left(\dfrac{UV-UM}{D-h}\right) + \dfrac{\partial M}{\partial t}\right] = 0 \\[2ex] \dfrac{\partial h}{\partial t} + \dfrac{\partial U}{\partial x} = 0 \end{cases}$$

Here, D(x) is the total water depth, M is the component of total mass transport along the shore, and $\varepsilon = \Delta\rho/\rho$, $\Delta\rho$ being the density difference across the interface. In (4), all the variables

are dimensional, but the primes are dropped for convenience. In
the numerical integration, a two-step Lax-Wendroff scheme was used.

DISCUSSION OF THE RESULTS

     Figure 1 shows the simulations for Babine Lake. The lower part
of the diagram shows the rigid upper surface, the interface in the
undisturbed position and the bottom topography. Note the trough
in the left side (southern side of the lake) and the main sill,
followed by a longer but shallower trough and a minor sill. In
the simulations for Babine Lake, a wind was assumed to blow uniformly
over the lake from left to right. Temporally, the wind was simulated

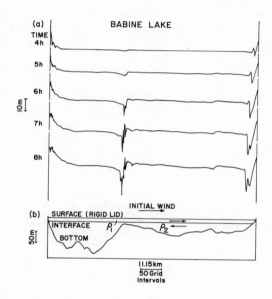

Figure 1.   Non-linear internal wave simulations in Babine Lake.

using a "ramp-function", *i.e.* initially zero, the wind increased to
a maximum in 5 hours and decreased to zero after another 5 hours.

     In the top part of Figure 1, the interface at 4,5,6,7 and 8
hours (taking zero time as the instant the wind starts to blow)
is shown. Note that the vertical scale for this part of the diagram

is five times greater than the vertical scale for the topography.
This vertical exaggeration was used to show the details of the
hydraulic jump. Three features are worth noticing: (1) a major
jump over the main sill; (2) a minor jump over the minor sill, and
(3) a major jump at the downwind end due to downwelling. At the
upwind end, some oscillation is evident. The source of this is not
confirmed, in the sense that these oscillations could be real or
could be due to numerical dispersion. Thus, it appears that jumps
are formed by flow over sills as well as by downwelling.

   The next set of experiments was run for Lake Ontario. (Note
that Simons, 1978 developed his model for this lake). It has no
sills - only a wide trough in its bottom topography (Figure 2).
On the top left-hand side of Figure 2 are shown the results without
rotation. (Unless mentioned otherwise, all the results reported
are for the non-rotating case). The plots of the interface at 10,
16, 20 and 28 hours reveal that a jump forms at the downwind end
and starts propagating into the interior. Since there are no sills,
there are no other jumps. The right side shows the interface at
10, 16, 20 and 28 hours, when the earth's rotational effects are
included. An extremely interesting result is that the jump starts
at the upwind end in this case. Thus, rotation can change the
position of initiation of the jump quite drastically. These results
are consistent with those of Simons (1978) and Houghton (1969).

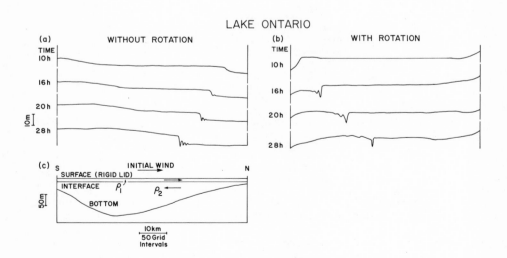

Figure 2. Simulation in Lake Ontario.

Since sills appear to generate jumps, we ran a series of experiments with two sills, Figure 3, left side bottom, shows two identical sills. The top left shows the interface at 5, 6, 7 and 8 hours. Note that, in this case, we reversed the wind direction. Identical jumps developed over the two sills and another jump was about to form due to downwelling. On the right side, the results for two sills of unequal size (the second sill is half the height of the first sill) are shown. It can be seen that the jumps are also of unequal amplitude. The jump due to downwelling, which was about to form at 8 hours, has formed by 10 hours. In Figure 3, no vertical exaggeration is used. Although the results shown in Figure 3 are somewhat obvious, these runs provide a good test of the model which seems to work quite well until the formation of the jump, and even some time after that. One interesting result which we hope to examine in future experiments is the amplification or dissipation of the jumps (or lee waves) by multiple sills, depending upon their sizes and relative distances.

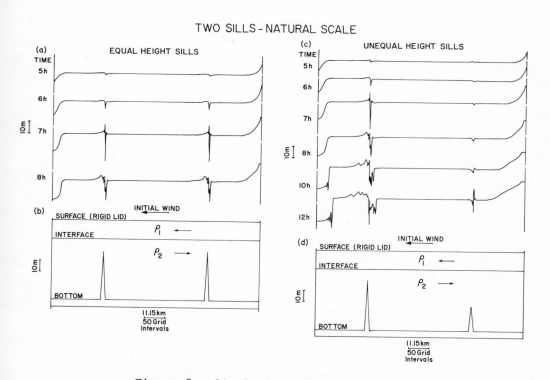

Figure 3. Simulations with two sills.

In Knight Inlet, the internal surges are forced by tidal flow
and not wind. Although we could easily have prescribed tidal flow
as a forcing mechanism, at first we attempted to simulate a laboratory experiment being done by the Coastal Oceanography group at
the Institute of Ocean Sciences, Patricia Bay (Farmer, personal
communication). In the laboratory experiment, tidal flow was
simulated by moving the sill. Our numerical experiments to simulate
the laboratory experimental results were divided into two parts.

In the first series, we used a stationary sill and supplied
energy through tidal forcing. Figure 4 shows the results for the
normal density case (left side), and a higher density gradient
case (right side). Note that no vertical exaggeration was used,
and the time-scale now is very small (a few seconds instead of
several hours). For the shape of the sill, a slightly modified
form of the "Witch of Agnesi" was used. Cyclic boundaries were used
for the high density, high tidal forcing case (closed boundaries
for all others). This avoided the excessive downwelling that would
have resulted if closed boundaries had been used here.

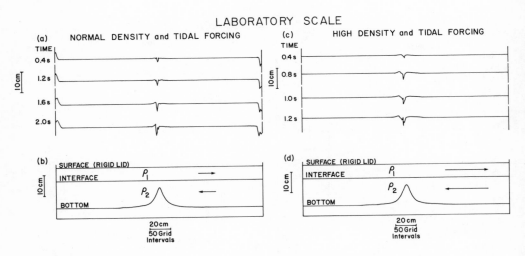

Figure 4.   Simulation of laboratory experiment.   Stationary sill.

In the second series, the sill was moved from right to left.
Figure 5 shows the initial conditions and the interface at 1, 2,
5 and 9 hours. (This simulation uses dimensions of the natural
scale and not the laboratory scale). Although the jump itself is

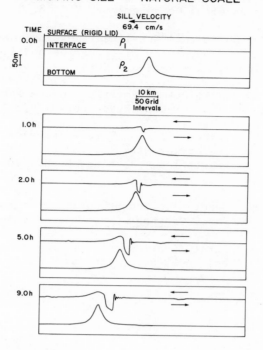

MOVING SILL - NATURAL SCALE

Figure 5. Simulation of moving sill. Natural scale.

simulated reasonably well (compared to the laboratory experiments), some features such as lee waves and flow separation are missing. As mentioned earlier, we hope to include these features through the two models under development.

ACKNOWLEDGMENTS

We thank Dr. D.M. Farmer for introducing us to this problem, Dr. T.J. Simons for providing a copy of the computer program for the downwelling front model, Mrs. Sheila Osborne for typing the manuscript and Mrs. Coralie Wallace for drafting the diagrams.

REFERENCES

Cahn, A., 1945: An investigation of the free oscillations of a simple current system. J. of Meteorology 2, 113-119.

Farmer, D.M. 1978. Observations of long non-linear waves in a
    lake. J. of Physical Oceanography, Vol. 8, No.1, 63-73.

Farmer, D.M. and J.D. Smith. Non-linear internal waves in
    a fjord. In "Hydrodynamics of estuaries and fjords". Edited
    by J.C.J. Nihoul, Elsevier, New York. 465-493.

Houghton, D.D. 1969. Effect of Rotation on the formation of
    hydraulic jumps. J. of Geo. Phys. Res., 74, 1351-1360.

Mason, P.J. and R.I. Sykes. 1979. Three-dimensional numerical
    integration of the Navier-Stokes equations for flow over
    surface-mounted obstacles. J. Fluid Mech., Vol 91, Pt. 3,
    433-450.

Simons, T.J. 1978. Generation and propagation of downwelling
    fronts. J. of Phys. Oceanography, Vol. 8, 571-581.

Sykes, R.I. 1978. Stratification effects in boundary layer flow
    over hills. Proc. Royal Soc., London, Ser. A., Vol. 361
    225-243.

GEOSTROPHIC FRONTS

Melvin E. Stern

Graduate School of Oceanography
University of Rhode Island
U.S.A.   02880

INTRODUCTION

When fresh water from a small coastal source debouches into a large mass of salt water the buoyancy force causes lateral spreading, and the Coriolis force deflects the light water to the coast. Behind the nose of the intrusion (the "bore") a geostrophic coastal current is established. The rate of propagation of the bore and the dynamics of non-linear waves on the trailing geostrophic front, are the subjects of this paper.

The effects mentioned above are illustrated in Plate I {see also Nof (1978), in which a source sink experiment in a two layer system is discussed, but with no density front}. The source region (not shown) for the light fluid is in the corner of the rotating (5 sec period) rectangular tank (13cm wide x 100cm long) and consists of a thin (2mm) walled vertical cylinder (2cm diameter) with its open mouth located 0.8cm below the parabolic free surface of the salt water whose initial mean depth was 11 cm. Dyed fresh water (0.25×10^{-3} gms/cc less than the salt water) is fed into the bottom of the cylinder by a vertical capillary tube connected to an overhead feed, the flow rate being 0.96 cm^3/sec. The cylinder acts as a buffer region, and the light fluid then rises to the free surface and then spreads laterally. In the second of the two experiments shown in Plate I, the external conditions are identical, except that the source region has been modified by placing a thin vertical plate (5cm long) parallel to the boundary, thereby providing a guiding channel for the water emerging from the top of the cylinder. The "edge effect" is somewhat different for the two geometries, but in either case the edge effect is confined to the source region, and the eddy entering on the left-hand side of frame

181

(h) is not an edge effect. The top of the tank in both experiments
is sealed by a cellophane cover; but similar waves have been seen
in identical experiments with the cover removed and having significant evaporation effects. The noteworthy features of these
qualitative* experiments are: (a) the density current emerging
from the source is deflected to the boundary, and a laminar intrusion
appears first; (b) perturbations with a horizontal scale larger than
the width of the intrusion then appear on the trailing geostrophic
front; (c) as these amplify the propagation of the bore is arrested
and the upstream flow is diverted normal to the coast and into a
geostrophic eddy; (d) the amplitudes of the latter are large compared to the width of the original current; (e) a laminar boundary
layer re-forms with a propagating bore, but new eddies are generated
downstream - and so on. Most of the time a distinctive tilt of
the amplifying wave is seen with a tilt of the crest in the opposite
direction to the boundary current; (f) the incipient frontal wave
is probably the key element in the mechanism by which the incoming
density current is mixed into the interior and by which a mean
boundary current is formed.

In order to explain the pronounced eruptions of these boundary
currents we have undertaken a theoretical study {Stern (1979)} of
a hierarchy of frontal models, the simplest of which has zero
absolute vorticity {Whitehead, Leetmaa, and Knox (1979)}, *i.e.*
the vertical component of relative vorticity in the boundary current equals the negative Coriolis parameter f. The conclusions
from this study are summarized in Fig. 1, and Fig. 1a is a schematic
perspective diagram of relatively light water flowing from a surface source region (not shown) into a deep and stationary fluid of
higher density.

A large scale "bore" (i.e. the nose of the lateral intrusion)
can propagate indefinitely along the boundary wall only if its
upstream width L is less than $g(\Delta\rho/\rho)H/(5.7f^2)^2$, where $\Delta\rho/\rho$ is the
fractional density deficit of the boundary current, H is the thick-
ness of the wall far upstream from the nose, and g is the acceleration of gravity.

In Fig. 1b there exists an initial perturbation at some distance behind the nose of the wedge, and the temporal evolution of
the former is very much dependent on the initial conditions. The
latter consists of the height of the fluid on the wall as the transverse displacement of the front from the wall.

* The geostrophic volume transport times the Coriolis parameter
divided by half the value of reduced gravity equals the square of
thickness of the fluid on the wall. This relation has been checked
(roughly) by estimating the vertical thickness of the dyed layer
10cm downstream from the source, and after ten minutes elapsed.

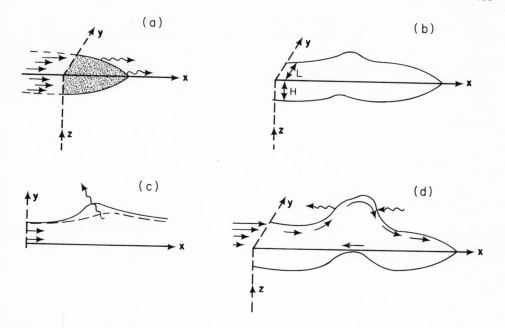

Figure 1.   (a) Surface density current emerging from a source.
Straight arrows represent currents, and wiggly arrows
are frontal propagation velocities.   (b) Propagating bore
with perturbation in the rear.   (c) Evolution of a
backward propagating wave.   (d) Blocking wave with
stationary crest and upstream propagation.

For initial conditions corresponding to the lateral bulge on
the front in Fig. 1c, the temporal displacement of the front is as
indicated by the wiggly arrow. The maximum displacement of the
front increases somewhat and reaches a limiting amplitude, but the
crest of the wave propagates upstream and the upstream half of the
wave continually steepens, thereby developing large seaward velocities. This conclusion is based on a nonlinear theory in which
the downstream wavelength is assumed to be much larger than L.
This condition is violated, however, at large time and the investigation of the breaking process requires an extension of our long
wave theory.

For the initial conditions in Fig. 1d the height of the fluid
on the wall decreases with time and approaches zero at a certain
section. When this happens the total transport thru the section

vanishes, the oncoming boundary current is "blocked", and a complete readjustment is suggested. While this phase of the dynamics is also beyond the scope of our theory, it does suggest effects similar to the ones seen in the photographs. Perhaps similar disruptions of coastal density currents can occur in the ocean.

REFERENCES

Nof, D 1978: On Geostrophic Adjustment in Sea Straits and Wide Estuaries. Jour. Phys. Oceanog. 8(5), 867-872.

Stern, M. E. 1979: Geostrophic Fronts, Bores, Breaking and Blocking Waves. Submitted to J. F. M.

Whitehead, J. A., Leetmaa, A. and Knox, R. A. 1974: Rotating Hydraulics of Strait and Sill Flows. Jour. Geophys. Fluid Dyn. 6, 101-125.

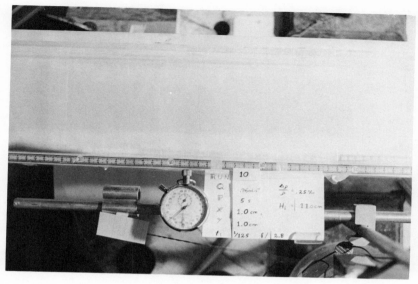

Plate I.   Dyed fresh water emerging from a small circular source
           located (but not seen) in the lower left-hand corner of
           the rotating rectangular tank, the top of which is covered
           (see text).   (a) Nose of front is 20 cm downstream from
           source and time after starting is 1 min 19 secs.

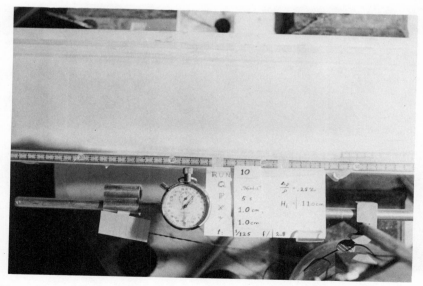

Plate I.   (b) Amplifying wave behind nose at 1 min 33 secs.

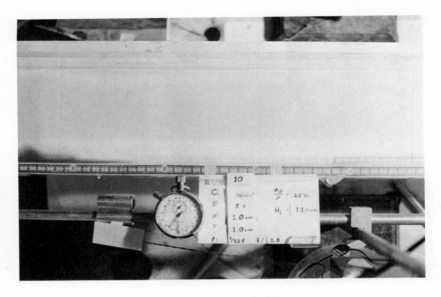

Plate I.   (c) 1 min 48 secs.

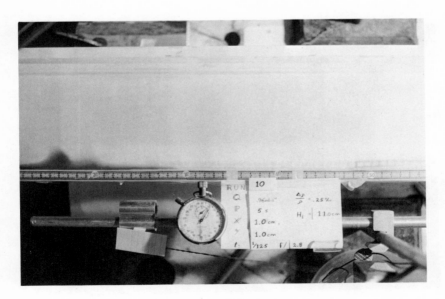

Plate I.   (d) 2 min.

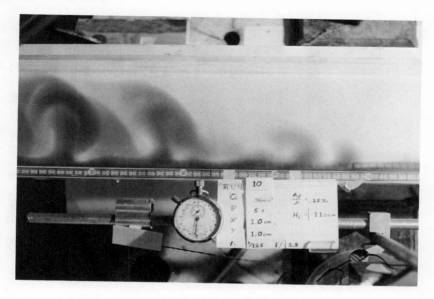

Plate I.   (e) 9 min 16 secs showing evolved eruptions of the
           coastal jet together with smaller incipient waves.

Plate I.   (f) 2 min 26 secs after start of second experiment
           (see text) with identical conditions except for source
           geometry.

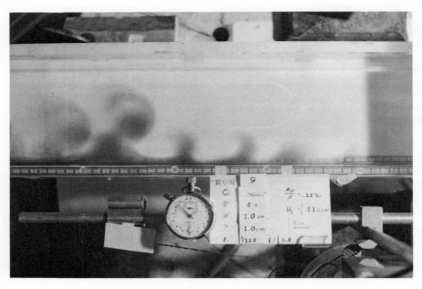

Plate I.   (g) 3 min 56 secs; note large amplitude wave is forming
           a detached vortex.

Plate I.   (h) 6 min 29 secs; a big eddy has propagated downstream
           in the lefthand side of the picture and a new cusped wave
           is seen forming on the laminar coastal current in this
           region.

Plate I. (i) 9 min 17 secs.

MODELLING OF MIXING PROCESSES IN THE SURFACE LAYERS OF THE OCEAN

R. T. Pollard

Institute of Ocean Sciences
Wormley, Godalming
Surrey, GU8 5UB, England

INTRODUCTION

Many processes that take place near the air-sea interface are common to both fjords and the open ocean. Fjord modellers need to be aware, therefore, of models that have been developed to parametrize mixing processes in the open ocean, though they may need substantial modification before being applied to relatively small enclosed bodies of water.

In my talk I discussed physical processes that can cause mixing and speculated on their relative importance. I think it inappropriate to set those speculations in print, as data from several major joint experiments (the Mixed Layer Experiment MILE 1977 in the North Pacific, the Joint Air-Sea Interaction experiment JASIN 1978 in the North Atlantic, and the GARP Atlantic Tropical Experiment GATE 1974) and individual studies (*e.g.* Thorpe, 1977, 1978) may be expected to lead to a clearer understanding of upper ocean processes in the next year or two. Also, the state of upper ocean modelling was the subject of another NATO sponsored Advanced Study Institute (Kraus, 1977) so the subject is comprehensively documented up to 1975. In this note, therefore, I shall restrict myself to a reiteration of some of the inadequacies of existing model parametrizations, and draw attention to processes of central importance to thermocline development and decay of which our understanding is rudimentary or non-existent, and model parametrizations correspondingly poor. This then is a warning to fjord modellers not to expect too much from existing models.

Two different approaches to parametrization of mixed layer processes culminated in the papers by Gill and Turner (1976) and

Thompson (1976), both of which present models of the seasonal cycle
in the upper ocean simple enough to be incorporated into large-scale
circulation models. Both papers compare several parametrizations.

Gill and Turner (1976) opt for a modification of the Kraus and
Turner (1967) hypothesis, that the rate of change of potential energy as the surface layer mixes is proportional to the energy available in the surface wind (which they assume to be constant). By
making allowance for loss of potential energy by convection in the
autumn and winter seasons, Gill and Turner simulate the hysteresis
loops shown in plots of surface layer heat content versus surface
temperature or potential energy over an annual cycle. Thompson
(1976) develops the Froude number closure hypothesis proposed by
Pollard, Rhines and Thompson (1973), based on the jumps in the
velocity and density across the base of the mixing layer.

The recipes for predicting seasonal variations in both papers
are based on rather crude assumptions. Gill and Turner point out
some weaknesses:

1. Both closure hypotheses could be improved (and many authors
    have tried, *e.g.* Price, 1979). As autumn cooling deepens the
    mixed layer, it is likely that the closure hypothesis should
    be based on conditions near the entrainment interface, rather
    than on surface input of energy or mean velocity across the
    entire surface layer. Linden (1975) showed that the rate of
    entrainment should be related to the turbulent velocity near
    the interface, not the surface input (of the stirring grid).
    Linden (1978) has also shown the wide range of profiles that
    can be obtained across the entrainment interface depending on
    the Richardson number. Measurements made during the recent
    MILE and JASIN experiments may allow improved parametrizations
    to be developed for the entrainment region, but further 'process-
    orientated' field experiments are required, particularly in the
    autumn and winter seasons. Namias (1959) found that disturb-
    ances in subsurface temperature that may persist for years may
    develop over a period of a few months, principally in the
    autumn/winter seasons. White and Haney (1978) describe the
    genesis of abnormal temperatures in the upper 400 m (cold near
    the surface, warm below 200 m) during autumn/winter 1976-1977.
    They ascribed the anomalous warming of the deep layers to con-
    vection, with advection of cold water (as well, presumably, as
    convection) causing the anomalous cooling of the surface layer.
2. A major objection to existing models is that mixing is confined
    to the "mixing layer" although there is evidence for associated
    "mixing" in the seasonal thermocline (Kitaigorodskii and
    Miropol'skii, 1970) which certainly continues to warm through
    the summer (*e.g.* Turner, 1973, p. 303). Turner ascribed the
    mixing to the intermittent breaking of internal waves propagating
    from the surface and proposed (1978, following Kitaigorodskii

et al. 1970 and Linden, 1975) that a universal form could be found for the temperature profile in the seasonal thermocline that could be used for modelling purposes, even if it gives no insight into the physics. Stabeno and Niiler (1978) found that the existence of a universal profile was not supported by data from Ocean Weather Stations Papa and Victor, and suggested that the theory for the mixing length proposed by Turner is no less difficult than the theory for the mixing processes in the thermocline itself.

Other serious contenders for "thermocline mixing" are the exchange of water with varying temperature/salinity characteristics between the mixing layer and thermocline by flow along an inclined isopycnal surface (*e.g.* Woods, 1975; Woods, Wiley and Briscoe, 1977; McPhee, 1979) or across isopycnal surfaces by double-diffusive convection (Gregg and McKenzie, 1979). These hypotheses imply that truly one-dimensional models of mixed layer/thermocline evolution are inadequate, and models must be developed which include or parametrize horizontal variability. Detailed observations of temperature and salinity variations along repeated horizontal sections will be required in the next decade before physical understanding and parametrization of horizontal variability can be achieved. The scales involved may be rather small, of order km (Pingree and Plevin, 1972; Gregg and McKenzie, 1979) to tens of km (Woods, 1975; McPhee, 1979), but scales of hundreds of km or larger (eddy scales) may also be relevant.

In fjords, horizontal variability is large, and vertical transport of properties is also possible at side boundaries and near bottom topography. Simple open-ocean one-dimensional surface mixing models may have very limited application in such situations.

REFERENCES

Gill, A. E. and J. S. Turner 1976: A comparison of seasonal thermocline models with observation. Deep-Sea Res. 23, 391-401.

Gregg, M. C. and J. H. McKenzie 1979: Thermohaline intrusions lie across isopycnals. Nature, 280, 310-311.

Kitaigorodskii, S. A. and Yu. Z. Miropol'skii 1970: The theory of the upper quasi-homogeneous layer and seasonal thermocline in the open ocean. Izv. Atm. Ocean Physics, 6, 97-102.

Kraus, E. B. (ed.) 1977: Modelling and prediction of the upper layers of the ocean, Pergamon Press, 325 pp.

Kraus, E. B. and J. S. Turner 1967: A one-dimensional model of the seasonal thermocline. II The general theory and its consequences. Tellus, 19, 98-105.

Linden, P. F. 1975: The deepening of a mixed layer in a stratified
     fluid. J. Fluid Mech, 71, 385-406.

Linden, P. F. 1978: The density step at the base of the oceanic
     mixed layer. Ocean Modelling, 18 (unpublished manuscript).

McPhee, M. G. 1979: Some thoughts and a few data on modelling and
     prediction of the upper layers of the ocean. Ocean Modelling,
     20 (unpublished manuscript).

Namias, J. 1959: Recent seasonal interactions between the North
     Pacific waters and the overlying atmospheric circulation.
     J. Geophys. Res. 64, 631-646.

Pingree, R. D. and J. Plevin 1972: A description of the upper
     100 m at Ocean Weather Station $Juliet$ (52^0 30'N, 20^0 00'W),
     Deep-Sea Res. 19, 21-34.

Pollard, R. T., P. B. Rhines and R. O. R. Y. Thompson 1973: The
     deepening of the wind-mixed layer. Geophys. Fluid Dyn. 4,
     381-404.

Price, J. F. 1979: On the scale of stress-driven entrainment
     experiments. J. Fluid Mech. 90, 509-530.

Stabeno, P. J. and P. P. Niiler 1978: Inapplicability of universal
     temperature profiles for the North Pacific's seasonal thermo-
     cline. Ocean Modelling, 14 (unpublished manuscript).

Thompson, R. O. R. Y. 1976: Climatological numerical models of
     the surface mixed layer of the ocean. J. Phys. Ocean. 6,
     496-503.

Thorpe, S. A. 1977: Turbulence and mixing in a Scottish Loch,
     Phil. Trans. R. Soc. London. A, 286, 125-181.

Thorpe, S. A. 1978: The near-surface ocean mixing layer in stable
     heating conditions. J. Geophys. Res. 83, 2875-2885.

Turner, J. S. 1973: Buoyancy effects in fluids. Cambridge Univ.
     Press. 367 pp.

Turner, J. S. 1978: The temperature profile below the surface
     mixed layer. Ocean Modelling 11 (unpublished manuscript).

White, W. B. and R. L. Haney 1978: The dynamics of ocean climate
     variability. Oceanus. 21, 33-39.

Woods, J. D. 1975: The local distribution in Fourier space-time of variability associated with turbulence in the seasonal thermocline. Mem. Soc. Roy. des Sci. de Liege. 7, 171-189.

Woods, J. D., R. L. Wiley and M. G. Briscoe 1977: Vertical circulation at fronts in the upper ocean, pp. 253-275 in A Voyage of discovery: George Deacon 70th anniversary volume edited by M. Angel. Oxford: Pergamon Press. 696 pp.

EXPERIMENTAL SIMULATION OF THE 'RETREAT' OF THE SEASONAL THERMOCLINE BY SURFACE HEATING[1]

Lakshmi H. Kantha

The Johns Hopkins University
Baltimore, Maryland 21218
U.S.A.

Towards the end of the winter-cooling period, the seasonal thermocline in bodies of water such as the upper ocean, lakes, fjords and estuaries has, in general, been eroded by the combined action of convection and mechanical mixing to such an extent that the upper mixed layer is rather deep. But with the start of the spring heating period, turbulence in the lower regions of the layer is suppressed by the stabilizing heat flux at the surface and successively shallower mixed layers are often observed to form. The process culminates in a comparatively shallow mixed layer, which as time evolves is then heated and further eroded by the wind and other mixing processes later in the spring and summer. During the winter, the seasonal thermocline advances even more rapidly under the combined action of convection due to surface-cooling and mechanical mixing. In ice-covered bodies such as arctic fjords, salt-extrusion due to ice-growth also contributes to convective erosion. The process of deepening of the thermocline has been studied extensively in recent years (for example, see Kraus, 1977) but the 'retreat' at the beginning of the spring season has not received the attention it deserves since the important contribution of Kraus and Turner (1967). The 'retreat' is germane to the problem of the determination of the mixed layer depth and surface temperatures that result later during summer heating and winter cooling because this process establishes the initial conditions for the depth at the start of the erosion cycle.

---

[1] A fifteen-minute movie on the 'retreat' process was shown during Dr. Kantha's presentation.

Kraus and Turner (1967) simulated the retreat and the advance
of the thermocline qualitatively using an oscillating grid to induce
mixing and adding fresh water at the top at discrete intervals in
increasing and decreasing amounts respectively. During the retreat
phase at least, the depth of the mixed layer D should be proportional
to the Monin–Obukhov length scale L defined as $\sigma^3/\kappa q_0$, where σ is
the rms turbulence velocity indicating the intensity of mixing, q_0
is the surface buoyancy flux and κ is von Kármáns's constant. Kraus
and Turner (1967) were not able to quantify their results, especially
since the flux was not added continuously. In this paper, the
proportionality of D to L is determined in the laboratory using
Turner-type oscillating-grid experimental set-up Fig. 1. Initially,
grid-stirring induces turbulence in a *homogeneous* but saline water
in the tank. The result is that the fluid is made turbulent over
the entire depth of the tank, although the turbulence grows weaker
as one proceeds away from the grid. When a continuous stabilizing
flux is introduced at the top, a mixed layer forms bounded below
by a buoyancy interface. The turbulence below the interface is
suppressed and decays rapidly.

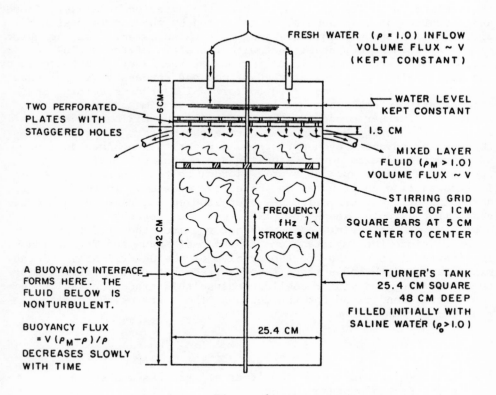

Figure 1

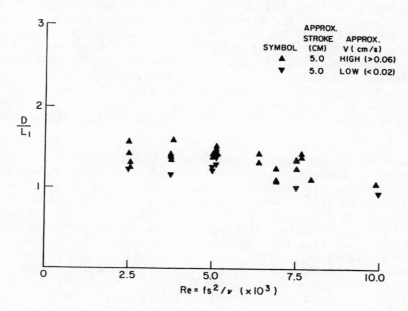

Figure 2

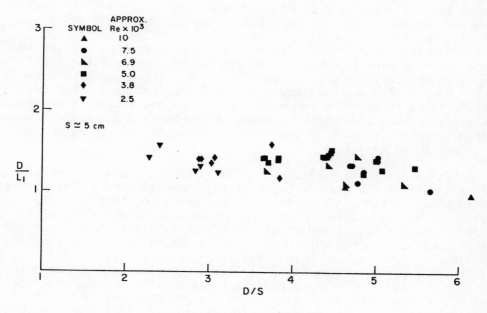

Figure 3

The buoyancy flux q_o at the surface is created by an influx of fresh water at the top and an equal outflux of brackish water from the upper portions of the mixed layer. A capping set of two per-forated plates (Fig.1) hold the fresh water until the withdrawal of the mixed-layer fluid begins and distribute the fresh water uniformly over the tank area. The fresh water volume flux V is kept constant and, since the density of the brackish water withdrawn decreases with time, the buoyancy flux decreases slowly with time. However, the decrease in q_o is seldom more than 15% and because of the weak dependence of the depth D of the mixed layer on q_o, this variation is unimportant.

In a saline but unstratified fluid, the turbulence front created by an oscillating grid has been observed (Kantha and Long, 1979 a, see also Dickinson and Long, 1978) to propagate at a speed proportional to $\sqrt{(K/t)}$, where K, the 'action' of the grid, is proportional to fs^2 (f is the frequency and s the stroke of the grid) and t is the time. Although the presence of the capping plates seems to bring the exponent down somewhat, fs^2 seems to be the fundamental quantity representing the effect of gradient and has been used to scale the results. For example, the flux q_o and fs^2 define a length scale L_1 defined by $L_1^4 = (fs^2)^3/q_o$ and the depth D of the mixed layer formed should be proportional to L_1 for a given stroke s. Note that the Monin-Obukhov length scale L can be defined as $\sigma^3/\kappa q_o$, where $\sigma = fs^2/D$ is proportional to the rms turbulence velocity at the front. $D/L = (D/L_1)^4$. Figures 2 and 3 show the variation of D/L_1 with Reynolds number, $Re = fs^2/\nu$, and D/s where ν is the kinematic viscosity coefficient.

Experiments indicate that once a mixed layer forms, if either the stirring rate is decreased by decreasing f but keeping V and q_o approximately the same or the flux q_o is increased by increase of V but keeping the stirring rate the same, another shallower mixed layer forms. The turbulence below the new interface decays rapidly. Thus, a stepwise decrease in the Monin-Obukhov scale leads to a stepwise decrease in the mixed layer depth and the halocline 'retreats' in steps!

It is difficult to extend the above results from the grid experiments to the field because mixing there is usually associated with a mean shear. A kinematic surface stress u_*^2 is invariably responsible for mixing in the upper ocean and a similar situation often exists in other bodies of fluid such as fjords, estuaries and the lower atmosphere. The Monin-Obukhov length scale L is readily defined in this case as $u_*^3/\kappa q_o$ and Kitaigorodskii (1960) in a pioneering paper was the first to suggest that the asymptotic depth D of the mixed layer formed under a surface stress in the presence of a stabilizing buoyancy flux should be proportional to L. The ratio D/L, which could be termed 'Kitaigorodskii parameter' is in general a function of L/H, where $H = u_*/f$ is proportional to the

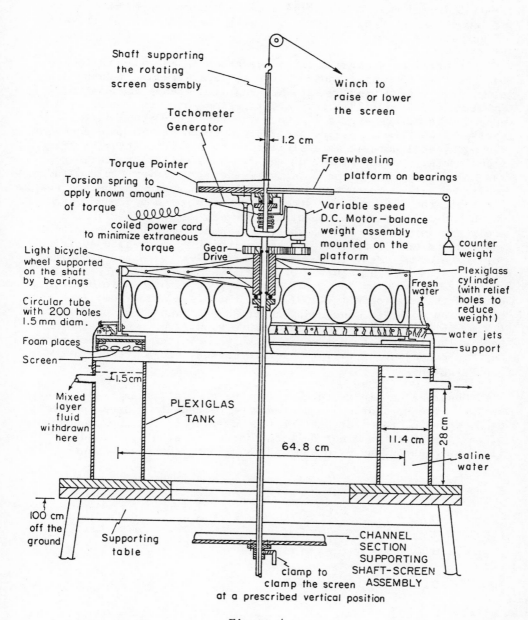

Shaft supporting
the rotating
screen assembly

Winch to
raise or lower
the screen

1.2 cm

Tachometer
Generator

Freewheeling
platform on bearings

Torque Pointer

Torsion spring to
apply known amount
of torque

coiled power cord
to minimize extraneous
torque

Gear-
Drive

Variable speed
D.C. Motor - balance
weight assembly
mounted on the
platform

counter
weight

Light bicycle
wheel supported
on the shaft
by bearings

Circular tube
with 200 holes
1.5 mm diam.

Foam places

Screen

Plexiglass
cyl inder
(with relief
holes to
reduce
weight)

Fresh
water

water jets

support

1.5 cm

Mixed
layer
fluid
withdrawn
here

PLEXIGLAS
TANK

64.8 cm

11.4 cm

28 cm

saline
water

100 cm
off the
ground

Supporting
table

clamp to
clamp the screen
at a prescribed vertical position

CHANNEL
SECTION
SUPPORTING
SHAFT—SCREEN
ASSEMBLY

Figure 4

depth of the turbulent Ekman layer. Observations (Kitaigorodskii,
1960) indicate that D/L assumes a value somewhat less than unity
at low values of L/H, that is, in the limit of negligible rotation
effects.

Experiments similar to those described above were performed
for the case of mixing induced by a surface stress. An annular
tank, of the type used by Kato and Phillips (1969) and Kantha,
Phillips and Azad (1977) for investigations of entrainment at a
buoyancy interface Fig. 4 was used. A rotating screen at the top
applied a surface stress to the fluid, which can be held constant.
The procedural details were similar to that of the stirring grid
experiments. Volume flux V of the fresh water influx at the top
was held constant and the density of the mixed-layer fluid withdrawn
monitored with a conductance probe. When this density became nearly
constant, the flux and the stress were terminated and the depth,
which had reached its asymptotic value was measured. Figure 5
shows the variation of D/L, where L is based on the buoyancy flux
towards the end of the run and the value of u_* corrected empirically
for the sidewall drag (Long and Kantha, 1979), with the Reynolds
number Re defined as $u_* D/\nu$. D/L tends asympototically to a value
of about 0.7 at high Re, a value consistent with oceanic observations
cited by Kitaigorodskii (1960). But this is considerably less than
2.8 indicated by the observations of Kondo, Kanechika and Yasuda
(1978) in a nocturnal atmospheric inversion. The nocturnal atmospheric layer may not, however, be fully turbulent over its entire
depth. Kondo, et al.(1978) observed marked intermittency in the
upper regions.

Observations indicate that for the case of mixing induced by a
surface stress, a stepwise decrease of L leads to a stepwise decrease
of D, that is, the 'retreat' of the halocline. The phenomenon is
of obvious importance not only to fjords but also other bodies of
fluid such as the upper ocean and the lower atmosphere (Kantha, 1979,

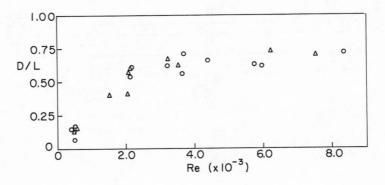

Figure 5

Kantha and Long 1979b, and Long and Kantha, 1979). In the upper
ocean, for example, shallow diurnal layers often form under conditions of strong heating at the surface and relatively weak wind
mixing. In fjords, the fresh water runoff from the river and
melting of ice during early spring in ice-covered fjords would also
tend to act like a stabilizing flux and inhibit mixing.

ACKNOWLEDGEMENTS

The author wishes to thank Prof. S.A. Kitaigorodskii for
rekindling his interest in the retreat phenomenon and Professors
O.M. Phillips and R.R. Long. This research was supported by NSF
ATM 76-04050-A1, NSF ATM 76-22284, NSF OCE76-18887, ONR N00014-75-
C-0805 and APL 600751.

REFERENCES

Dickinson, S.C. and Long, R.R. 1978. Laboratory study of the
    growth of a turbulent layer of fluid. Physics of Fluids 21:
    1698-1701.

Kantha, L.H. 1979. Turbulent mixing in the presence of stabilizing
    surface buoyancy flux - 'retreat of the thermocline'. Geo-
    physical Fluid Dynamics Lab. Report TR 79-1, The Johns Hopkins
    University.

Kantha, L.H., O.M. Phillips, and R.S. Azad. 1977. On turbulent
    entrainment at a stable density interface. J. Fluid Mech.
    79: 753-768.

Kantha, L.H. and R.R. Long. 1979b. Turbulent mixing with stabil-
    izing surface buoyancy flux. Submitted to J. Fluid Mech.

Kato, H. and O.M. Phillips. 1969. On the penetration of a turbulent
    layer into a stratified Fluid. J. Fluid Mech. 37: 643-655.

Kitaigorodskii, S.A. 1960. On the computation of the thickness
    of the wind-mixing layer in the ocean. Izv. Akad. Nauk,
    S.S.S.R., Geophysics Series 3: 425-431.

Kondo, J. O. Kanechika and Yasuda, N. 1978. Heat and momentum
    transfers under strong stability in the atmospheric surface
    layer. J. Atmos. Sci. 35: 1012-1012.

Kraus, E.B. 1977. Modelling and prediction of the upper layers
    of the ocean. Pergamon Press.

Kraus, E.B.   1977.   A one-dimensional model of the seasonal thermo-
     cline.   A laboratory experiment and its interpretation.   Tellus
     19:   88-97.

Long, R.R., and L.H. Kantha.   1979.   On the depth of a mixed layer
     under a surface stress and stabilizing surface buoyancy flux.
     Submitted to J. Fluid Mech.

TURBULENT ENTRAINMENT AT A BUOYANCY INTERFACE DUE TO CONVECTIVE TURBULENCE

Lakshmi H. Kantha

The Johns Hopkins University
Baltimore, Maryland 21218
U.S.A.

Penetrative convection is of considerable importance in geo-physical problems. For example, in the upper layers of the ocean, convective cooling of the surface aids wind-mixing in the erosion of the seasonal thermocline during the winter. On a shorter time scale, nocturnal surface cooling in bodies of water such as the oceans, lakes and fjords helps to mix the heat gained during the day over a large depth. In the lower atmosphere, the rise of nocturnal inversions during the day due to convective heat flux from the ground governs the near-surface conditions and dispersal of pollutants. The process is also important to nocturnal and winter-cooling of fjords, growth of ice in cold saline bodies of water such as arctic fjords and the arctic sea and even in spring-heating of ice-covered fresh-water lakes (Farmer 1975). Its import-ance to the growth of ice in arctic fjords, where salt extrusion by growing ice initiates and maintains convection in the water column, which in turn controls the rate of ice-growth has been emphasized by Gade et al (1974) and Perkin and Lewis (1978), while the role of similar processes in subpolar oceans on climate has been discussed by Stewart (1978). Many attempts have been made to gain a better understanding of the entrainment processes due to convection, both from laboratory simulations (Deardorff et al 1969, Heidt 1977, and Willis and Deardorff 1974 and 1979) and field observations of various pehnomena involving them (Farmer 1975, Gade et al 1974, Kloppel et al 1978, and Readings et al 1973). Of course, in the lower atmosphere, the process has been explored extensively (Kaimal et al 1976 and Readings et al 1973).

There have also been numerous attempts at theoretical modeling of penetrative convection, especially within the context of the lower atmosphere (for example see Ball 1960, Carson 1973, Deardorff

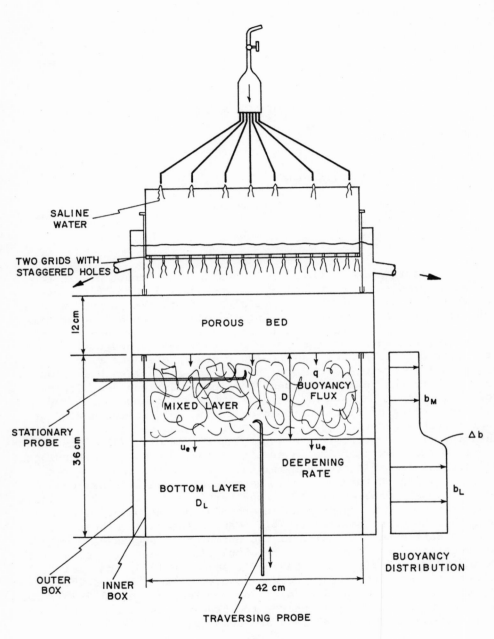

Figure 1

1979, Kitaigorodskii and Kozheloupova 1978, Long and Kantha 1978, Stull 1976, Tennekes 1973 and Zubov 1943). There has even been an attempt to model other planetary atmospheres along similar lines (Burangulov 1977). However, it is fair to say that despite the enormous attention devoted to the subject, a real understanding of penetrative convection has remained elusive. Laboratory investigations have remained the basis for proper interpretation of field observations.

Kantha (1979) has simulated penetrative convection in the laboratory using salt flux at the top of a water column (Fig. 1) instead of heat flux at the bottom as in all other experiments. This has certain inherent advantages such as the much higher Rayleigh numbers, the absence of buoyancy loss to the walls that need correcting, relatively smaller change in the initial ambient stratification over long periods and realizability of much higher Richardson number Ri_* (defined as $D \, \Delta b/w_*^2$, where D is the depth of the convective mixed layer, Δb the buoyancy change across the capping interface, and W_* the convective velocity scale defined as $(qD)^{1/3}$; q is the buoyancy flux). Experiments were done with a two-layer system, which is by far perhaps the best possible configuration for entrainment experiments. After the tank was filled with two layers up to a level above the porous bed, saline water was added at the top and its volume flux kept constant. The interface was initially sharpened by selective withdrawal. Fluorescein dye in the upper layer illuminated from below clearly marks the extent of the mixed layer. The density of the layer monitored with a conductance probe and the measured depth were used to determine q which remained nearly constant with time except at the beginning of each experiment (Fig. 2). Subscript o refers to the extrapolated initial conditions. The destabilizing salt flux gradually fills up the buoyancy difference between the mixed layer and the fluid below and also deepens the mixed layer, slowly at first but increasingly faster as time evolves until eventually it pierces the interface and plunges down at the end. A typical experiment starts at a high Ri_* of interest to winter-cooling of lakes, oceans, and fjords ($\sim$300) with Ri_* decreasing slowly with time eventually reaching low values of usual interest to the lower atmosphere (<100). Fig. 3 shows the cumulative plot of D/D_o vs $qt/D_o \Delta b_o$ in about twenty experiments. The observed entrainment rate ($u_e \equiv dD/dt$) normalized by w_* is shown plotted as a function of Ri_* in fig. 4. Despite the potential pitfalls in using the same normalizing variable (here w_*) on both axes (Hicks 1978), the observed correlation between u_e/w_* and Ri_* is believed to be accurate. The entrainment was initially high and decreased before increasing with decreasing Ri_*, the reasons for which are not clear. These initial points have been omitted in fig. 4. Figure 5 shows the variation with Ri_* of the ratio q_e/q, where subscript e denotes the entrainment flux at the interface (equal to $\Delta b u_e$).

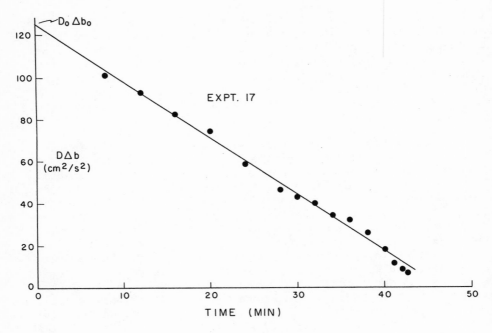

Figure 2

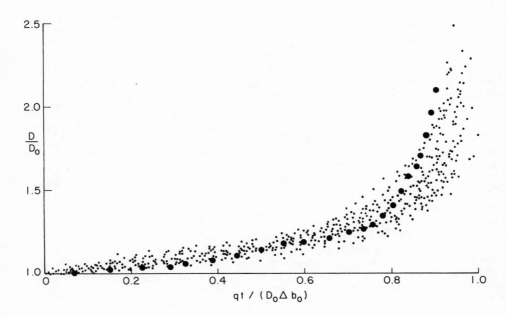

Figure 3

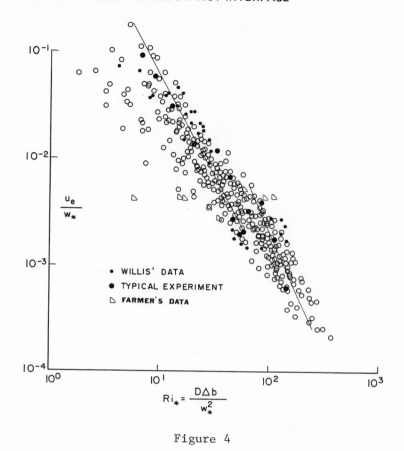

Figure 4

Theoretical modeling of penetrative convection has consisted
of specifying the value of q_e/q through suitable assumptions and
there has been a great deal of speculation concerning it. This
information is needed for the closure of the governing equations.
In the atmosphere, this ratio has been assumed to be a constant
in most models (Ball 1960 and Tennekes 1973), although recently
it has been realized that this ratio should really be a function
of Ri_* (Kitaigorodskii and Kozheloupova 1978). Stull (1976) has
summarized the values of this ratio used in models and inferred
from observations in the atmosphere. The ratio has wide-ranging
values between 0 and 1, its absolute bounds called the Zubov and
Ball limits respectively, but the present consensus is that for
most conditions in the atmosphere, the ratio is around 0.2 (Tennekes
1973). Farmer's observations also indicate a mean value that is

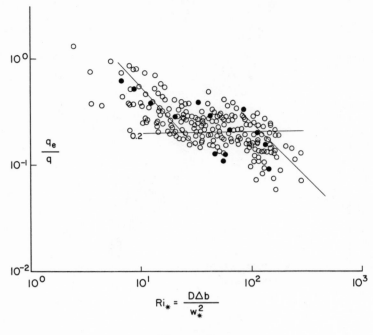

Figure 5

equivalent to this ratio being slightly less than 0.2, with significant variations around this value (Farmer 1975). The present
experiments show that the ratio is really a function of Ri_*, but at
moderate values of Ri_*, it is around 0.2 ± 0.1 (Fig. 5). At higher
Ri_*, of interest to lakes and arctic fjords, it is much smaller and
at low Ri_* it is considerably larger.

The salt water from the top enters the mixed layer in the form
of a collection of plumes situated roughly at the middle of the tank
which wanders around occasionally. Entrainment at the interface
occurs mostly in and around this plume region with the entrained
fluid swept away and mixed into the mixed layer by slow upward
motions in the regions around the plume. The process is somewhat
similar to that observed often in the atmosphere (Kaimal et al 1976
and Readings et al 1973) at the top of a convective plume. Thus the
experiment perhaps simulates a single convective cell in the field.
Careful observations of the structure of the interface at and around
the buoyant plume in the laboratory indicate that at high Ri_*,
entrainment seems to proceed by gradual sweeping away of the fluid
at the top of the interface by the plume, which keeps the interface
locally sharp although it is quite thick elsewhere, especially away
from the edges of the plume. The entrainment in this regime could

be influenced by molecular properties. In this range, qe/q is small. But as Ri_* decreases, the eddies in the plume pounding on the interface generate internal waves that propagate towards the edges and break there, the lower fluid thus entrained is mixed into the upper layer by the slow convective upward motions. At even lower values, billows due to local mean shear such as those observed by Readings *et al* (1973) in the atmosphere, appear and might contribute to the entrainment rate significantly. As Ri_* decreases further, dome-wisp pairs, postulated by Stull (1973), are brought to mind by the deep bulge in the interface and the accompanying wisps at the sides. At low Ri_* towards the end, turbulent eddies in the plume appear to impact against the interface leading to subsequent recoil of some of the denser lower fluid into the mixed layer. However, it is difficult to delineate the ranges of Ri_* where each of these processes dominate and often some appear to coexist. This suggests that the entrainment does not follow a single power law in the entire Ri_* range investigated, as this would usually mean a single dominant mechanism of entrainment in the entire range, contrary to the observations. The data, especially on the ratio q_e/q, suggests the existence of three different branches, although the scatter is too large to be certain.

These experiments shed some light on the major uncertainty in our present attempts to model penetrative convection, namely the magnitude and variation of q_e/q with Ri_*. The experiments are however exploratory and full understanding of the behaviour of q_e/q awaits more intensive studies of the structure of the entrainment interface and details of various entrainment mechanisms. The application of these results to the field is perhaps justified although the experiment appears to simulate just a single convective cell. The experiments were mostly at D_o/W value of about 0.3, where W is the width of the tank, but a few at D_o/W of about 0.45 indicated an increase in q_e/q of about 80%, suggesting significant aspect ratio effects.

This research was supported by NSF under grant no. ATM 76-2284.

REFERENCES

Ball, F.K. 1960. Control of inversion height by surface heating. Quart. J. Roy. Meteor. Soc. 86, 483-494.

Burangulov, N.I. 1977. A model of penetrative convection with applications to the atmosphere, ocean and laboratory experiments. Izv., Atmos. and Ocean. Phy. 13, 863-868.

Carson, D.J. 1973. The development of a dry inversion-capped convectively unstable boundary layer. Quart. J. Roy. Meteor. Soc. 99, 450-467.

Deardorff, J.W. 1979. Prediction of mixed layer entrainment for realistic capping inversion structure. J. Atmos. Sci. 36.

Deardorff, J.W., Willis, G.E. and Lilly, D.K. 1969. Laboratory investigation of non-steady penetrative convection. J. Fluid Mech. 35, 7–31.

Farmer, D.M. 1975. Penetrative convection in the absence of mean shear. Quart. J. Roy. Meteor. Soc. 101, 869–891.

Gade, H.G., R.A. Lake, E.L. Lewis, E.R. Walker. 1974. Oceanography of an arctic bay. Deep-sea Res.. 21, 547–571.

Heidt, F.D. 1977. Comparison of laboratory experiments on penetrative convection with measurements in nature, in Heat Transfer and Turbulent Buoyant Convection, Vol 1, ed. by Spalding and Afgan, 199–210.

Hicks, B.B. 1978. Some limitations of dimensional analysis and power laws. Boundary-Layer Meteor. 14, 567–569.

Kaimal, J.C., J.C. Wyngaard, D.A. Haugen, O.R. Cote, Y. Izumi, S.J. Caughey, C.J. Readings. 1976. Turbulence structure in the convective boundary layer. J. Atmos. Sci. 33, 2152–2169.

Kantha, L.H. 1979. A laboratory simulation of penetrative convection. Under preparation.

Kitaigorodskii, S.A. and N.G. Kozheloupova. 1978. On the entrainment rate in the regime of penetrative convection in non-stationary boundary layers of atmosphere and ocean. Izv., Atmos. and Ocean. Phy. 14, 639–648.

Kloppel, M., G. Stilke, and C. Wamser. 1978. Experimental investigation into variations of ground-based inversions and comparisons with results of simple boundary layer models. Boundary-Layer Meteor. 15, 135–145.

Long, R.R. and L.H. Kantha. 1978. The rise of a strong inversion caused by heating at the ground. Proc. of twelfth Symposium on Naval Hydrodynamics.

Readings, C.J. E. Golton, and K.A. Browning. 1973. Fine-scale structure and mixing within an inversion. Boundary-Layer Meteor. 4, 275–287.

Stewart, R.W. 1978. The role of sea ice on climate. Oceanus 21, 47–57.

Stull, R.B. 1973. Inversion rise model based on penetrative convection. J. Atmos. Sci. 30, 1092-1099.

Stull, R.B. 1976. The energetics of entrainment across a density interface. J. Atmos. Sci. 33, 1260-1267.

Tennekes, H. 1973. A model for the dynamics of inversion above a convective boundary layer. J. Atmos. Sci. 30, 558-567.

Willis, G.E. and J. W. Deardorff. 1979. Laboratory observations of turbulent penetrative-convection planforms. J. Geophy. Res. 84, 295-301.

Zubov, N.N. 1943. Arctic Ice (translated by U.S. Navy Electronics Lab., San Diego).

ENERGETICS OF MIXING PROCESSES IN FJORDS

T.A. McClimans

River and Harbour Laboratory
Norwegian Institute of Technology
Trondheim, Norway

FJORD CLASSIFICATION

Several schemes for classifying estuaries deal with the stability of the flow (Stommel, 1953, Hansen & Rattray, 1966). Pritchard (1967) gave a general geomorphologic scheme for defining fjords. Milne (1972) suggested a scheme for fjord classification (Scottish sea lochs) according to the number of sills. This gives a good picture of the qualitative nature of the type of hydrography, but fails to give a quantitative measure since no dimensions are specified.

Stigebrandt (1978) proposed a scheme for two special types of fjords (overmixed and normal) which represents extremes with respect to mixing processes, and implies topographic control. There is some question as to whether a homogeneous estuary can be considered a fjord or not. Geomorphologically many of them fall into the category termed land-locked fjords (Strøm, 1936; Glenne & Simensen, 1963).

TIDAL MIXING

In general, land-locked fjords are poorly ventilated, but when the tidal constriction has a cross sectional are A_s for which

$$A_s = 8.4A(a/gT^2)^{\frac{1}{2}}$$

where T is the tidal period, A is the surface area of the basin, a is the tidal amplitude and g the acceleration of gravity, there is a maximum of energy transfer to mixing (McClimans, 1978). Such

215

energy transport can homogenize a small land-locked fjord with a
basin depth several times sill depth (McClimans, 1973).

RIVER PLUME MIXING

Although sills often occur at narrows, there are several
examples of deep-silled narrows. In the case of a small, wide
fjord basin with mountains protecting it from wind mixing, the
narrows may act to control the mixing induced along river plume
fronts (McClimans, 1979). The narrows restrains the outflow of
brackish water, inducing a reentrainment of old river water to the
plume. For wide narrows, the river may entrain up to 13.3% seawater,
reducing to 0% when the seaward narrows is 1.84 times the river width.
These results apply as long as the fjord basin extends seaward to
ca. 10 times the river width.

WIND MIXING

A narrows is also an important topographical feature for wind
mixing. Wind mixing, however, is much more complicated than tidal
mixing at sills and river plume mixing. The narrows gives an overall control of layer thickness, but the dynamics are complicated by
wind drift and variable mixing throughout the basin. Local vertical
wind mixing (entrainment) increases the salinity of the brackish
layer. Horizontal wind mixing reduces the horizontal gradients.
Wind drift causes large scale variability which is generally
treated as stochastic. For a wide fjord, horizontal mixing has
been modeled as a homogenizing agent (Stigebrandt, 1975).

In general, wind mixing has been modeled as a statistically
stationary process without the horizontal transports which occur
in nature (Stommel, 1951; Long, 1975; McClimans, 1975; Gade &
Svendsen, 1978; McClimans & Mathisen, 1979). In the last reference
Stommel & Farmer's (1952) equations were numerically compared with
a laboratory model to look at some of the details of the recent
work on the energetics of wind mixing and the dynamics under these
special conditions. The results imply that the Kato-Phillips
formula is useful and that the effects of the control of the narrows
are as predicted by Stommel (1951), Long (1975) and Stigebrandt (1975).
Dynamically, however, the flow resistance increases greatly with
oscillating wind stress, implying Reynolds stresses and radiation
stresses through the deeper layers. These processes do not appear to
affect the energetics of wind mixing directly.

CONCLUSIONS

The energetics of mixing is usually expressed in terms of a flux Richardson number which gives the ratio of induced potential energy (buoyancy) flux to the applied energy flux. For wind induced or shear induced mixing, this is on the order of 5% to 10%. This number seems to hold also for river plume mixing; (McClimans, 1979). Stigebrandt (1976) obtained 5% for tidal induced mixing of deeper layers.

Although there are many unsolved problems, the above results indicate that topography plays a crucial role for mixing processes. No general parameter has been derived for classifying the topography, but an extension of Milne's idea may be fruitful. In each case, laboratory experiments have proven helpful in isolating the effects under study and appear to be valuable tools for the future study of combined modes.

Space does not allow a detailed description of field measurement strategies. It should, however, be clear that critical regions near the narrows, river mouths and lateral sections should replace the earlier practices of mid-fjord stations, when trying to resolve the energetics of mixing. In general it is good economy to tailor a field measurement program to the requirements of a theoretical model of the system based upon all available observations. At the River and Harbour Laboratory it is common practice to make pilot measurements in a new area before a detailed study is planned. This is in recognition of the importance of topography for the dynamics and energetics of fjords.

REFERENCES

Gade, H.G. and E. Svendsen, 1978. Properties of the Robert R. Long model of estuarine circulation in fjords. _Hydrodynamics of estuaries and fjords_. J.C.J. Nihoul, editor. Elsevier: pp. 423-437

Glenne, B. and T. Simensen, 1963. Tidal current choking in the land locked fjord of Nordåsvatnet. _Sarsia, 11:_ 43-73.

Hansen, D.V. and M. Rattray Jr., 1966. New dimensions in estuary classification. _Limnology and Oceanography 11:_ 319-326.

Long, R.R., 1975. Circulations and density distributions in a deep strongly stratified, two-layer estuary. _J. Fluid Mech. 71:_ 529-540.

McClimans, T.A., 1973. Physical oceanography of Borgenfjorden. _K. norske Vidensk. Selsk. Skr. 2:_ 1-43.

McClimans, T.A., 1975. Salinity and thickness of the brackish layer of a deep-silled fjord with narrows. Mar. Sci. Comm. 1: 305-321.

McClimans, T.A., 1978. On the energetics of tidal inlets to land locked fjords. Mar. Sci. Comm. 4: 121-137.

McClimans, T.A., 1979. On the energetics of river plume entrainment. Geophys. Astrophys. Fluid Dynamics (In press).

McClimans, T.A. and J.P. Mathisen, 1979. Stationary wind mixing in a deep-silled, narrow fjord: A comparison of laboratory observations with a numerical model. Mar. Sci. Comm. (In press).

Milne, P.H., 1972. Hydrography of Scottish west coast sea lochs. Dept. of Agriculture and Fisheries for Scotland. Marine Research/1972 No. 3. Edinburgh. Her Majesty's Stationery Office 1 - 50.

Pritchard, D.W., 1967. What is an estuary: physical viewpoint. In Estuaries G.H. Lauff, editor. Publication No. 83 AAAS, Washington. pp. 3-5.

Stigebrandt, A., 1975. Stasjonär två-lagers strømning i estuarier. River and Harbour Laboratory Report STF60 A75120. 68pp (In Swedish).

Stigebrandt, A., 1976. Vertical diffusion driven by internal waves in a sill fjord. J. Phys. Ocean. 6: 486-495.

Stigebrandt, A., 1978. Fjord problems in physical oceanography in particular internal waves, mixing and processes at geometrical constrictions. Dissertation, Dept. of Oceanography, University of Gothenburg.

Stommel, H., 1951. Recent developments in the study of tidal estuaries. WHOI, Ref. No. 51 - 33: 1-15.

Stommel, H., 1953. The role of density current in estuaries. Proc. Fifth IAHR Congress, Minneapolis pp. 305-312.

Stommel, H. and H.G. Farmer, 1952. On the nature of estuarine circulation. WHOI Tech. Report Ref. No. 52-88.

Strøm, K.M., 1936. Land-locked waters. Hydrography and bottom deposits in badly ventilated Norwegian fjords. with remarks upon sedimentation under anaerobic conditons. Skr. Vitensk. - akad. 7, Oslo.

MIXING- AND EXCHANGE PROCESSES IN A SMALL GREENLAND SILL FJORD

Tue Kell Nielsen and Niels-Erik Ottesen Hansen*

Danish Hydraulic Institute
Agern Alle 5, DK-2970 Horsholm
Denmark

INTRODUCTION

The Black Angel Mine at Marmorilik is situated in West Greenland at 71^0N. lat. Tailings from ore concentration, containing various metals, have since late 1973 been discharged into the A-fjord,† a small sill fjord (Fig.1).

The tailings outlet is placed in the basin behind the sill in order to prevent tailings solids from entering the outer fjord system. Soon after the start of the mining operations, significant metal concentration levels were detected in the fjord below the sill level (during summer) and over the entire water column (during winter).

The scope of the present investigation was to describe the hydrography of the fjord, so that possible measures to improve conditions could be evaluated.

FIELD MEASUREMENTS

Since 1972, several hydrographical and environmental investigations have been carried out in the area, resulting in a large amount

---

* Inst. of Hydrodynamics and Hydraulic Eng., Building 115,
  The Technical University of Denmark, DK-2800, Lyngby, Denmark.
† *Editor's Note: The A-fjord and the Q-fjord are Agfardlikavsâ and Qaumarujuk, discussed further in a later paper by Asmund in this volume.*

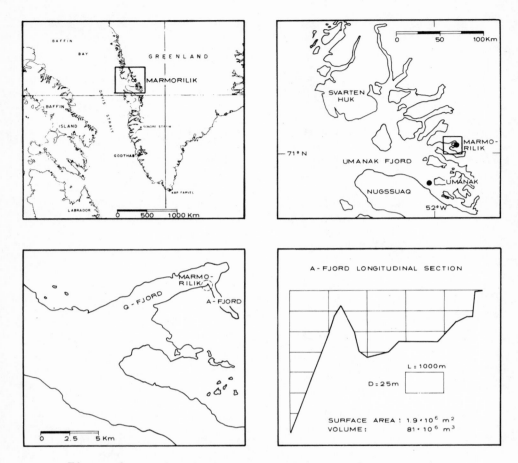

Figure 1.   A-fjord, location and longitudinal section.

of data describing distributions of salt, temperature and dissolved
metals, partly as spot observations and partly as continuous time
series.

During 1978, monthly, weekly or daily measurements of salinity
and temperature distributions were carried out, and water levels,
fresh-water runoff and wind were recorded. Vertical diffusion
coefficients were determined directly by tracer measurements, and
concentration levels of several metals and suspended solids were
measured.

The investigations involved several Danish institutions: Arctic Consultant Group, the Water Quality Institute, the Danish Isotope Centre, the Danish Hydraulic Institute, and the Ministry of Greenland.

HYDROGRAPHIC SUMMARY

During the summer and autumn, a well-defined stagnant water volume is found behind the sill and below the sill level. In the early spring, the fjord is well-mixed.

The gross salt balance of the A-fjord may be looked upon as determined by three major factors:

- A pronounced annual variation of the salinity in the outer fjord system over sill level,
- A direct fresh-water discharge of 1.1 times the fjord volume, mainly during July and August,
- Freeze-up of 0.02 times the fjord volume during January through June.

The mixing is further determined by tidal action and by local and regional wind action.

The annual net transport of salt over the sill appears to be small, so that the summer loss of salt is nearly or fully compensated by a winter inflow of salt water.

Vertical diffusion in the fjord has been determined by a weak, but measurable salt loss from the basin behind the sill, as well as by tracer measurements. Observed local and instantaneous values of the vertical diffusion coefficient were a factor 10 lower than over-all long term values, indicating that vertical transport may have taken place near the banks, and perhaps within discrete time periods.

Dissolved metals originating from the tailings outlet are accumulated in the stagnant basin volume during the summer, but are released into the outer fjord system during the winter. Larger vertical concentration gradients during the summer, and larger horizontal gradients during the winter, illustrate the annual change between vertical diffusion and horizontal mixing determining the rate of transport over the sill.

CHANGING OF MIXING PROCESSES

Several ways of reducing the current release of metals into the outer fjord system have been proposed. Among several other measures,

Table 1

Measures to keep the tailings behind the sill in A-fjord.

PERVIOUS DAM	IMPERVIOUS DAM	FLOATING SKIMMER-WALL	SALT ADDITION

MAIN EFFECTS:

PERVIOUS DAM	IMPERVIOUS DAM	FLOATING SKIMMER-WALL	SALT ADDITION
Reduces horizontal circulation	Eliminates any ingoing salt transport	Establishes a stagnant fresh-water surface layer	Establishes a stagnant salt bottom layer

OTHER EFFECTS:

PERVIOUS DAM	IMPERVIOUS DAM	FLOATING SKIMMER-WALL	SALT ADDITION
Eliminates regional wind action and reduces effect of direct wind action	Energy budget as for a lake, but with large stability	Reduces effect of direct wind action, and eliminates salt rejection from ice	No changes of energy budget, but greater stability

CALCULATED CHANGES OF METAL DISTRIBUTION:

PERVIOUS DAM	IMPERVIOUS DAM	FLOATING SKIMMER-WALL	SALT ADDITION
Larger horizontal and vertical gradients. Higher maximum levels	Very large vertical gradients. Extremely high maximum levels	Larger vertical gradients and higher maximum levels	Very large vertical gradients High maximum levels

CALCULATED CHANGES OF OUTGOING METAL TRANSPORT

PERVIOUS DAM	IMPERVIOUS DAM	FLOATING SKIMMER-WALL	SALT ADDITION
Higher during summer, lower during winter	Transport only in fresh-water runoff	Small changes	Lower during summer, higher during winter

direct changes of the natural mixing processes are being considered. This might be obtained by constructing a pervious or impervious dam upon the sill across the mouth of the A-fjord, by establishing a floating skimmer-wall at that place or by increasing the salinity of the basin water.

The nature of the initial release of dissolved metals is of great importance for the evaluation of the effect of such measures. The release will depend upon the concentration levels in the water, approaching zero for very high concentration levels. The highest obtainable concentration levels may be determined by a saturation effect, or by the future mixing in the fjord.

In the first case, a permanent reduction of the metal transport over the sill may be obtained by maintaining high concentration levels in those parts of the fjord where the initial release takes place.

If the future maximum concentration levels are determined by mixing, and not by saturation effects, the transport over the sill may only be reduced with the amount of metals accumulated in the A-fjord. In this case, it would be desirable to establish a large reservoir for accumulation behind the sill section.

ENERGY BUDGET FOR MIXING PROCESSES

To estimate the relative importance of the different mixing processes in the A-fjord the contribution to the total energy budget for each process has been calculated. The calculations are based on the methods outlined by Bo Pedersen (1979) and Ottesen Hansen (1979).

The following processes contributing to the mixing have been identified:

- tidal induced mixing between the upper and the lower layers. tidal generated internal waves.
- mixing induced by the expansion loss (Carnot loss) in the tidal current behind the steep sill at the inflow in the A-fjord (mixing of the lower layer).
- mixing induced by the action of the wind.
- mixing induced by the flow in the brackish upper layer due to fresh water discharge.
- mixing induced by the salt rejection from the ice during the freeze up.
- mixing induced by the occasional influxes of more dense water from the outer fjord - the Q-fjord.

The contributions from the different mechanisms are summarized in Table 2 for the different months of the year.

Table 2

Energy budget for each individual mechanism in kJ per month. The energy budget is defined as the gain in potential energy due to the mixing.

	Tidal exch.	Wind eff.	Fresh water run off	Salt rejec.	Inflow of more dense water	Total
JAN	?	–	–	$5 \cdot 10^5$	$3\text{--}5 \cdot 10^5$	$8 \cdot 10^5$
FEB	10^4	–	–	$5 \cdot 10^5$	$3\text{--}5 \cdot 10^5$	$8 \cdot 10^5$
MAR	10^4	–	–	$5 \cdot 10^5$	?	$5 \cdot 10^5$
APR	10^4	–	–	$5 \cdot 10^5$	?	$5 \cdot 10^5$
MAY	10^4	–	–	$5 \cdot 10^5$	?	$5 \cdot 10^5$
JUN	10^4	$2 \cdot 10^6$	5	–	?	$2 \cdot 10^6$
JUL	$5 \cdot 10^5$	$2 \cdot 10^6$	$2 \cdot 10^2$	–	?	$2 \cdot 10^6$
AUG	$5 \cdot 10^5$	$2 \cdot 10^6$	7	–	?	$2 \cdot 10^6$
SEP	$5 \cdot 10^5$	$2 \cdot 10^6$	–	–	?	$2 \cdot 10^6$
OCT	$2 \cdot 10^5$	$2 \cdot 10^6$	–	–	?	$2 \cdot 10^6$
NOV	10^5	$2 \cdot 10^6$	–	$2 \cdot 10^5$	?	$2 \cdot 10^6$
DEC	?	–	–	$3 \cdot 10^5$	$3\text{--}5 \cdot 10^5$	$6 \cdot 10^5$

REFERENCES

Ellis, D.V. and J.L. Littlepage. 1972. An oceanographic survey of Qaumarujuk and Agfardlikavsa Fiords, Marmorilik W. Greenland.

The Danish Water Quality Institute. 1973. Investigations August 1972 Qaumarujuk Fiord, Agfardlikavsa.

GF, GGU and The Ministry of Greenland. 1974. Qaumarujuk Fiord, Investigation July 1974.

GF, GGU and The Ministry for Greenland. 1975. Investigation 1974. Agfardlikavsa, Qaumarujuk.

GF, GGU and The Ministry for Greenland. 1976. Investigations 1975-76. Agfardlikavsa, Qaumarujuk.

GF, GGU and The Ministry for Greenland. 1977. Investigations 1976-77. Agfardlikavsa, Qaumarujuk.

Lewis, E.L. 1978. The movement of polluted water near Marmorilik, Greenland, its causes and possible cures.

Arctic Consulting Group.   1978.   Lake 475, Marmorilik.   Discharge
     measurements 1978.

Danish Hydraulic Institute.   1979.   Marmorilik.   Hydrographic survey
     1978.

Bo Pedersen, Fl.   1979.   A monograph on turbulent entrainment and
     friction in two-layer stratified flow.   Technical University
     of Denmark, Institute of Hydrodynamics and Hydraulic Engineer-
     ing.   In press, 392 pp.

Ottesen Hansen, N-E.   1979.   Effects of boundary layers on mixing
     in small lakes.   Paper in Hydrodynamics of Lakes, Elsevier,
     1979, Holland in press.

MIXING EXPERIMENTS IN FJORDS

Erik Buch

University of Copenhagen
Institute of Physical Oceanography
Haraldsgade 6, 2200 Copenhagen N.
Denmark

INTRODUCTION

The vertical and horizontal turbulent diffusion in the pycnocline layer has been investigated experimentally in some Nordic
fjords by means of dye diffusion experiments. The dependence of
the mixing on the oceanographical and meteorological conditions is
examined. Results from various areas are compared, namely:
  1) The Danish Belts which can be considered as a transition
      zone to the great estuarine system of the Baltic Sea.
  2) A fjord on the east coast of Sweden connected to the central
      part of the Baltic Sea.
  3) A small narrow sill fjord which is a part of the great
      fjord system around Bergen, Norway.

The experiments indicate that in the narrow areas the physical
boundaries have an influence on the vertical mixing and affect the
horizontal dispersion by generating horizontal shear. These
effects are not observed in larger areas. The vertical mixing
shows a clear dependence on stratification, vertical current shear
and wind.

Vertical Diffusion

The general equation of diffusion can be reduced to:

1) $\quad \dfrac{\partial c}{\partial t} \;=\; K \dfrac{\partial^2 c}{\partial z^2}$

assuming that

a) the advective terms can be disregarded because the observations are made in a Lagrangian reference system with
origin at the centre of the dye patch.

b) the decrease in the dye concentration is due only to the
vertical diffusion.

c) the vertical convection is negligible.

d) K is a constant.

A solution to eq. 1 was given by Kullenberg (1968):

2) $K = \dfrac{H^2}{\pi^2(t_2 - t_1)} \, ln(C_1/C_2)$

Theoretical considerations made by Kullenberg (1971) give a relation
between the vertical diffusion coefficient K and the oceanographical
environment:

3) $K = Rf \, c_d(\rho_a/\rho_o)(W_a/N)^2 dq/dz$

By plotting K against $(W_a/N)^2 dq/dz$, from experiments carried
out in open sea and coastal areas, Kullenberg found

4) $Rf \, c_d(\rho_a/\rho_o) = 9.6 \times 10^{-8}$

The present investigations in fjords give:

5) $Rf \, c_d(\rho_a/\rho_o) = 1.3 \times 10^{-9}$

Horizontal diffusion

For interpretation of the observed horizontal distribution of
dye the theory of Joseph & Sendner (1958) is used. The dye
distribution is transformed into a rotationally symmetric one, and
the horizontal diffusion velocity P is found:

6) $\sigma^2_{rc} = 6 \, P^2 t^2$

Because of the dye patch often is non-rotationally symmetric
it may be more appropriate to find the longitudinal and
transverse variances σ^2_x, σ^2_y. In the case of a

Gaussian distribution:

7) $\sigma^2_{rc} = \sigma_x \sigma_y$;

σ^2_{rc} is used for definition of the apparent horizontal

diffusivity:

Table 1

Diffusion time, diffusion velocity P (eq.5), symmetrical variance σ^2_{rc}, diffusion velocity P(eq.6) apparent diffusion coefficient K_h, diffusion scale ℓ_h, longitudinal variance σ^2_x, transverse variance σ^2_y, the product $\sigma_x \sigma_y$ elongation σ_y/σ_x.

Area and date	t sec	P(eq.5) cm sec⁻¹	σ^2_{rc} cm²	P(eq.6) cm sec⁻¹	K_h cm²sec⁻¹	ℓh cm	σ^2_{x2} cm²	σ^2_{y2} cm²	$\sigma_{x2}\sigma_y$ cm²	σ_y/σ_x
Little Belt 76 09 28	5.2×10^4	.32	1.2×10^9	.27	5.7×10^3	1.0×10^5	4.1×10^9	2.2×10^9	9.4×10^8	.23
Little Belt 76 09 29	3.5×10^4	.55	2.0×10^9	.52	1.4×10^4	1.3×10^5	3.9×10^9	9.7×10^8	1.9×10^9	.50
Arnavagen 76 10 19	1.0×10^5	.03	1.6×10^8	.05	4.0×10^2	3.8×10^4	1.6×10^8	2.2×10^7	6.0×10^7	.37
	1.3×10^5	.02	2.6×10^8	.05	5.0×10^2	4.8×10^4	5.1×10^4	5.6×10^7	1.7×10^8	.33
	1.6×10^5	.02	2.7×10^8	.04	4.2×10^2	4.9×10^4	–	–	–	–
Arnavagen 76 10 21	5.0×10^4	.14	2.0×10^8	.12	1.0×10^3	4.2×10^4	–	–	–	–
	6.5×10^4	.16	6.2×10^8	.16	2.4×10^3	$7.5\ 10^4$	–	–	–	–
Kalundborg Fjord 78 03 07	1.8×10^4	.16	1.3×10^8	.26	1.8×10^3	3.4×10^4	1.8×10^8	6.2×10^6	3.3×10^7	.18
	2.5×10^4	.42	5.0×10^8	.37	5.0×10^3	6.7×10^4	6.4×10^8	9.1×10^7	2.4×10^8	.38
Himmerfjärd 78 10 02	3.6×10^4	.23	2.1×10^8	.16	1.5×10^3	4.3×10^4	5.5×10^8	4.7×10^7	1.5×10^8	.27
Himmerfjärd 78 10 03	2.2×10^4	.55	3.1×10^8	.33	3.5×10^3	5.3×10^4	–	–	–	–
	2.7×10^4	.55	4.5×10^8	.32	4.2×10^3	6.4×10^4	1.5×10^9	1.3×10^8	4.4×10^8	.29
	4.1×10^4	.65	6.3×10^8	.25	3.8×10^3	7.5×10^4	2.3×10^9	3.4×10^8	9.0×10^8	.38
Himmerfjärd 78 10 05	2.4×10^4	.12	1.7×10^8	.23	1.8×10^3	3.9×10^4	–	–	–	–
	3.5×10^4	.11	2.1×10^8	.17	1.5×10^3	4.3×10^4	6.3×10^8	2.2×10^8	3.7×10^8	.59
	4.6×10^4	.18	5.2×10^8	.21	2.8×10^3	6.8×10^4	4.5×10^8	1.3×10^8	7.7×10^8	.17

Table 2

Parameters for prediction of variance and predicted variances

Area and date	Diffusion time (sec)	H_o cm	P(eq.19) cm sec⁻¹	K cm²sec⁻¹	a_o sec⁻¹	a sec⁻¹	b sec⁻¹	ω sec⁻¹	σ^2_x cm²	σ^2_y cm²	$\sigma_x\sigma_y$ cm²
Little Belt 76 09 27	5.2×10^4	200	.15	4.7×10^{-2}	2.3×10^{-2}	1.0×10^{-2}	1.9×10^{-2}	1.4×10^{-4}	7.1×10^9	5.1×10^7	6.0×10^8
Little Belt 76 09 28	3.5×10^4	200	.22	2.2×10^{-2}	2.5×10^{-2}	1.0×10^{-2}	5.0×10^{-3}	1.4×10^{-4}	3.0×10^9	1.1×10^6	5.5×10^7
Arnavagen 76 10 19	1.0×10^5	200	.05	6.9×10^{-3}	1.0×10^{-3}	4.0×10^{-3}	4.0×10^{-3}	1.4×10^{-4}	3.8×10^7	6.3×10^5	4.9×10^6
Arnavågen 76 10 21	3.2×10^4	200	.04	6.7×10^{-3}	5.0×10^{-3}	4.8×10^{-3}	2.8×10^{-3}	1.4×10^{-4}	8.9×10^7	9.6×10^4	2.9×10^6
Himmerfjärd 78 10 02	3.6×10^4	200	.03	2.3×10^{-2}	1.6×10^{-3}	1.3×10^{-3}	9.7×10^{-3}	4.4×10^{-4}	1.7×10^7	1.2×10^6	3.8×10^6
Himmerfjärd 78 10 03	4.1×10^4	200	.05	6.8×10^{-3}	6.2×10^{-3}	8.8×10^{-3}	3.3×10^{-3}	4.4×10^{-4}	2.5×10^8	6.3×10^4	4.0×10^6
Himmerfjärd 78 10 05	3.5×10^4	200	.04	3.1×10^{-2}	3.3×10^{-3}	3.0×10^{-3}	2.5×10^{-3}	4.4×10^{-4}	4.9×10^7	8.5×10^4	2.0×10^6

8) $K_h = \sigma^2_{rc}/4t$;

and the scale of turbulence

9) $\ell_h = 3\sigma_{rc}$.

The observations yield the following relationships:

10) $\sigma^2_{rc} \sim t^{2.2}$

11) $K_h \sim \ell_h^{4/3}$

Theoretical predictions of the horizontal variances have been carried out using the advection-diffusion model developed by Kullenberg (1971):

12) $\sigma_x^2 = (2a_o^2t^3/3 + a^2t/\omega^2)K + a_o^2H_o^2t^2/12$

13) $\sigma_y^2 = \{9b^2t/(8\omega^2) + b^2a_o t(2a_o-a)/(a^2\omega^2)\}K$

The results of the investigations of the horizontal dispersion are given in tables 1 and 2.

Conclusions

The results treated in this paper shows a considerable difference in the mixing characteristics for open areas and more narrow and enclosed areas.

The great difference between eq. 3 and eq. 4 is believed to be an effect of the physical boundaries and the stratification which is reducing the mixing due to the wind.

The relations given in eq. 10 and 11 are obtained whether the experiments are carried out in open areas or in narrow fjords.

A prediction of the dispersion by means of the above mentioned advection - diffusion model, which has been used with success in open areas by Kullenberg, gives values of the horizontal variances that are much lower than the observed ones. This suggests that some of the assumptions made in developing the model are not fulfilled in fjords. It seems very likely that horizontal diffusion and especially horizontal shear are important.

REFERENCES

Kullenberg, G. 1968. Measurements of horizontal and vertical
    diffusion in coastal waters. Rep. Inst. Phys. Oceanogr.,
    Univ. Copenhagen, 3, 50 pp.

Kullenberg, G. 1971. Results of diffusion experiments in the
    upper layer of the sea. Rep. Inst. Phys. Oceanogr., Univ. of
    Copenhagen, 12, 66pp.

Joseph, J. and H. Sendner. 1958. Uber die horizontale diffusion
    im meere. Deutsche Hydrog. Zeits. 11(2), 49-77.

TRACER MEASUREMENT OF MIXING IN THE DEEP WATER OF A SMALL, STRATIFIED SILL FJORD

Torbjörn Svensson

Department of Hydraulics
Chalmers University of Technology
Göteborg, Sweden

INTRODUCTION

The exchange of the basin water, i.e. below sill depth, in fjords is commonly envisaged as a series of inflows of heavy water over the sill with shorter or longer periods of stagnation in between. During the stagnation periods vertical mixing is the dominating process of exchange but this mixing may be very weak, especially in cases with strong vertical density gradients at or below sill depth.

No general model is available to calculate the mixing from external forcing although certain aspects may be derived from energy arguments. Thus the flux Richardson number, which is the ratio of potential energy gain to the turbulent energy production, seems to attain a constant value which is reported to be in the range 0.05-0.15. This may be expressed as :

$$Rf = K_z N^2 / \varepsilon_0 = \text{const.} \qquad (1)$$

where K_z is the vertical diffusion coefficient, $N^2 = \frac{g}{\rho} \frac{\partial \rho}{\partial z}$ is the stability of the water column and ε_0 the turbulent energy production. The distribution of ε_0 is not known and hence an empirical correlation of the type $K_z = f(N^2)$ should be sought.

DESCRIPTION OF EXPERIMENTAL BASIN

The present measurements were made in the fjord Byfjorden at the west coast of Sweden. The fjord is approximately 4 by 1.5 km and has a maximum depth of 50 m. Outside the mouth there is a

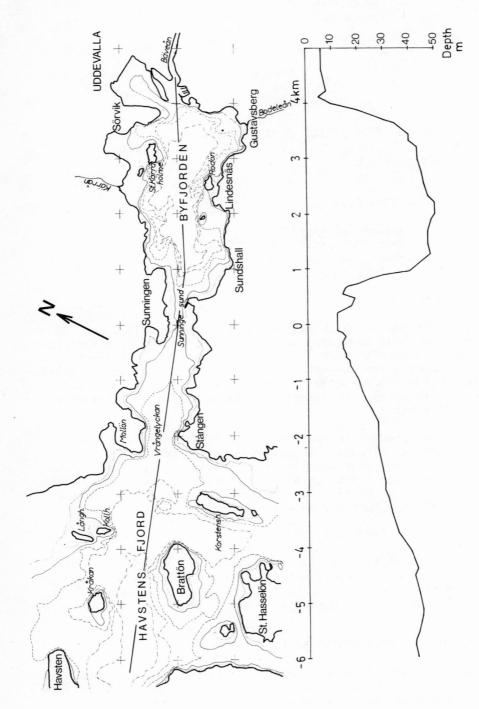

Figure 1. Plan and elevation of the Byfjorden.

fjord system surrounding the islands of Orust and Tjörn which are
adjacent to the Skagerrak. The sill at 11 m depth is situated at
the narrow entrance.

The tide is predominantly semidiurnal with a mean amplitude
of approximately 15 cm.   Meteorologically induced variations of
the water level are much larger with a difference between MHW and
MLW of 1.87 m. The fresh water supply is low in the summer,
<1 m^3/s, rising to typically 5-10 m^3/s during autumn and spring.
The stratification in the fjord is mainly maintained by the conditions outside the fjord which give rise to a strong halocline a
few meters below the sill depth. A second halocline is found at
2-4 m depth when the local fresh water supply is large.

TRACER EXPERIMENT

As part of th e Byfjorden study, (Göransson and Svensson, 1975)
a large scale tracer experiment was designed with the double aim
to simulate a proposed sewage outfall below the halocline and to
study the inflows of deep water over the sill and the vertical
mixing after the simulation phase was terminated. Rhodamine B dye
was used as the tracer, which was mixed with surface water and
continuously pumped out at 25 m depth through two horizontal
nozzles, Fig. 2. The pumping was carried out during the 101 days
period July 23 to November 1, 1970, with 22 days intermission in
all. The pumping capacity was 205 ℓ/s and the total amount of
Rhodamine injected was 127.1 ℓ trade solution.

The vertical distribution of tracer at different times is
shown in Fig. 3. The profiles are from a point in the central part
of the fjord, which is representative of the whole fjord since
differences in concentration in the horizontal were small. The
tracer profiles reveal significant inflows of water reaching the
lower part of the halocline in January and August-September 1972
and in December 1973. In the calm periods between inflows the
development of the concentration profile is characteristic for
vertical diffusion.

STABILITY OF THE TRACER

The fluorescence of Rhodamine B is known to decay in light
and to adsorb on particles in the water. In the Byfjorden,
however, the concentration of suspended solids is too low to
significantly affect the fluorescence, and the light intensity in
the deep water is practically zero, which precludes photochemical
decay. A slow decay of the tracer was found to take place and an
analysis of this was made for the calm period Jan. 26 - Aug. 21,
1972 assuming that no tracer material was transported through the

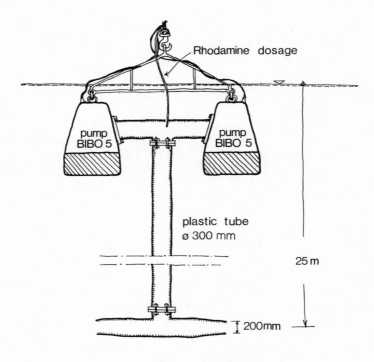

Figure 2.   Pump arrangement for tracer test in the deep water.

level of maximum concentration at 22 m depth. The change of total
mass of tracer below that level should thus represent the decay.
The measured amount of tracer below 22 m depth is shown in Fig. 4.

    The straight line in the diagram corresponds to an exponential
decay, $C=C_o \cdot \exp(-1.64 \cdot 10^{-8}\, t)$, where t is the time in sec.

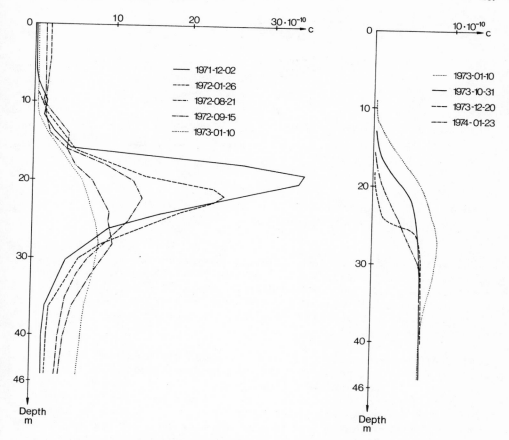

Figure 3. Vertical distribution of tracer at different times after
the pumping had stopped.

EVALUATION OF DIFFUSION COEFFICIENTS

The vertical mixing is described as a one-dimensional diffusion
process with variable diffusion coefficients, K_z, and cross-sectional area, A,

$$\frac{\partial c}{\partial t} = \frac{1}{A} \frac{\partial}{\partial z}\left(K_z A \frac{\partial c}{\partial z}\right) - kc \qquad (2)$$

where k is the decay coefficient. Eqn. (2) may be integrated from
the bottom, z_b, of the basin to evaluate the diffusion coefficients

$$K_z = \left[\frac{\partial}{\partial t}\int_{z_b}^{z} Ac\cdot dz - k\int_{z_b}^{z} Ac\cdot dz\right] / \left(A\frac{\partial c}{\partial z}\right) \qquad (3)$$

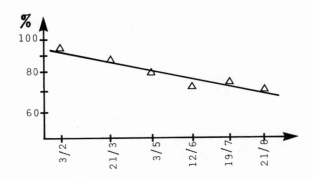

Figure 4.   Total measured amount of tracer below 22 m depth in
            January to September 1972.

This budget procedure yields, however, low accuracy where the
gradients are small, especially near the level of maximum concentration and has been supplemented by a numerical solution of
Eq. (2) starting from measured concentration profiles in the beginning of calm periods. The resulting sets of K_z for two time
periods are shown in Fig. 5.

     The diffusion coefficient has a minimum of 0.005 cm^2/s in the
halocline increasing to approximately ten times that value near
the bottom. The increasing diffusion at 13 and 15 m depth, in the
sharpest part of the halocline, is most likely the result of small
inflows over the sill into this depth interval. This is supported
by the observation that the diffusion coefficients of Fig. 5, if
applied to the salinity, would lead to a lowering of the halocline
which has not been observed.

ANALYSIS

     The resulting diffusion coefficients from the tracer experiment
are plotted against N^2 in Fig. 6 together with some determinations
based on salinity measurements during selected periods when advective inflows have been unlikely.

     The relationship may be expressed as:

$$K_z = 1.2 \cdot 10^{-4} (N^2)^{-0.6} \qquad\qquad (4)$$

A similar treatment of data for salt diffusion was made by Gade
(1970) in the Oslo fjord. In his case the exponent of (N^2) had
the value of -0.8.

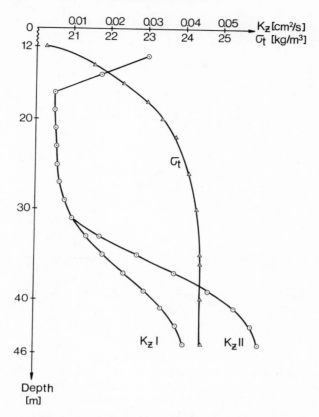

Figure 5.   Diffusion coefficients for the periods March 2 –
             August 21, 1972 (set I) and December 18, 1972 –
             June 6, 1973 (set II). Density profile during the
             first period.

It has not been possible to correlate the measured diffusion
coefficients to variations in wind velocity or the estuarine circulation and it is thus likely that the turbulence energy is supplied by the tide. A model of internal waves generated by tidal
currents at the sill and losing their energy to turbulence in the
deep water by breaking against the sloping boundaries has been
presented by Stigebrandt (1977). The model considers the total
energy conversion in the deep water by integrating Eq. (1) over the
deep water volume (below the halocline). The integral of the
energy production is set equal to the energy of the tidal induced
internal waves, ε_i, and we may make an estimate of Rf as follows:

$$Rf = \rho \int K_z N^2 dV / \varepsilon_i \qquad\qquad (5)$$

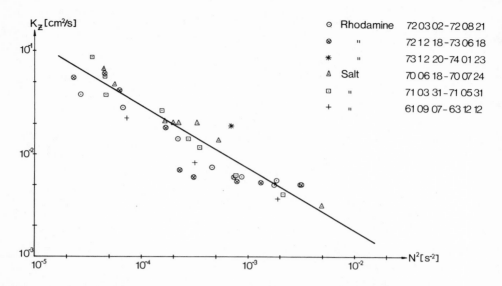

Figure 6.   Relation between vertical diffusion coefficients, K_z, and stability, N^2, in the deep water of the Byfjorden.

For the Byfjorden the energy input is estimated to 630 Watt which yields Rf in the range 0.06-0.07 for the two time periods used in Fig. 5. Similar estimates for the Oslo fjord have given Rf=0.05. Stigebrandts model thus seems to yield a fairly good estimate of the magnitude of the gross vertical mixing using Rf=0.05. The variation of diffusion coefficients in the deep water is, however, not predicted.

REFERENCES

Gade, H. G. 1970: Hydrographic investigation in the Oslo fjord, a
        study of water circulation and exchange processes.
        Geophysical Inst., Univ. of Bergen, Norway, Rept. 24, Sept. 1970.

Göransson, C-G and Svensson, T. 1975: The Byfjord: Studies of
        water exchange and mixing processes. Statens Naturvardsverk
        (Swedish Environmental Proctection Board) PM No. 594,
        Stockholm, Sweden.

Stigebrandt, A. 1976: Vertical diffusion driven by internal waves
        in a sill fjord. J. Phys. Oceanography, 6, 486-495.

ESTIMATION OF VERTICAL MIXING RATES IN FJORDS USING NATURALLY

OCCURRING RADON-222 AND SALINITY AS TRACERS

William M. Smethie, Jr.*
University of Washington
Seattle, Washington, 98195
U.S.A.

INTRODUCTION

Vertical mixing is an important process controlling the distribution of dissolved substances in the marine environment. In fjords the vertical mixing rate has a significant influence on the flushing cycle of the deep water and is very important in determining whether or not the deep water will go anoxic.

Broecker, (1965) presented evidence that radon-222 could be used as a tracer of vertical mixing in the bottom water of the ocean. Since then it has been used extensively in the ocean and in a few lakes to estimate rates of vertical mixing. Radon-222 is a radioactive inert gas with a half-life of 3.825 days. It is a member of the uranium-238 decay series and forms from the decay of radium-226.

Throughout most of the water column in the ocean, radon-222 is in equilibrium with dissolved radium-226. But in the bottom water there is a radon excess due to diffusion of radon out of the sediments and in the surface water there is a radon deficit due to diffusion of radon from the ocean to the atmosphere. Radon diffuses out of the sediments because radium-226 is enriched by up to four orders of magnitude in the sediments relative to the water. This enrichment of radium-226 is caused by the input of its parent, thorium-230, to the sediments, where it decays to radium-226. Thorium-230 forms from the decay of dissolved uranium-234 and is rapidly removed to the sediments because of its extreme insolubility.

* Present address: Lamont-Doherty Geological Observatory of Columbia University, Palisades, New York 10964, U.S.A.

In fjords, radium-226 is also enriched in the sediments which results in a radon excess in the bottom water. However the main source of radium-226 to the sediments is probably continental material weathered into the system and the enrichment may vary considerably among fjords. In the surface water there may also be a radon excess if there is fresh water inflow. The fresh water inflow will contain a component of ground water which becomes highly enriched in radon-222 as it percolates through the soil.

Vertical Mixing Rates Estimated from Radon Distributions

Vertical mixing rates, represented as a vertical eddy diffusivity, can be estimated by modelling the vertical distribution of excess radon (the radon in excess of that supported by the decay of dissolved radium). Two models were used in this study to obtain upper and lower limits of the vertical eddy diffusivity. An upper limit was obtained using what will hereafter be termed the one-dimensional model.

$$K_z \frac{d^2c}{dz^2} - \lambda c = 0 \tag{1}$$

where K_z is the vertical eddy diffusivity, c is excess radon concentration, z is height above bottom, and λ is the decay constant of radon. The underlying assumptions are steady state, no horizontal diffusive or advective transport, no mean vertical velocity, and K_z constant with respect to depth. For this model to be valid, horizontal mixing must not be rapid enough to mix radon input from the sides to the interior before the radon decays. Thus, the excess radon isopleths would parallel the bottom topography. This model gives an upper limit for K_z because it assumes the source of radon is the underlying sediments when in reality, some radon may be mixed in from the sides.

A lower limit of K_z was obtained using what will be referred to as the rapid horizontal mixing model.

$$K_z \frac{d^2c}{dz^2} + K_z \frac{dA/dz}{A} \frac{dc}{dz} + F \frac{dA/dz}{A} - \lambda c = 0 \tag{2}$$

where A is the cross sectional area of the fjord, F is the flux of radon across the sediment-water interface, and the other terms are as previously defined. The underlying assumptions are steady state, horizontal mixing infinitely rapid, no mean vertical velocity, and K_z constant with respect to z. If this model is strictly valid, the excess radon isopleths would parallel the water surface because the rapid horizontal mixing would eliminate horizontal gradients. This model gives a lower limit for K_z because the horizontal input is maximized.

The one-dimensional model was solved analytically and the rapid horizontal mixing model numerically, with appropriate boundary conditions for each. The vertical eddy diffusivity was determined by a least squares fit of the models to the data.

In Dabob Bay, a longitudinal and a transverse section of excess radon was measured. These sections revealed that the radon isopleths tended to parallel the bottom topography. Thus the rapid horizontal mixing model is unrealistic and the lower limits of K_z obtained with this model are unrealistically low.

Radon measurements were made in the bottom water of seven Fjords: Dabob Bay, Hood Canal, and Port Susan in Washington State and Lake Nitinat, Narrows Inlet, Queens Reach, and Princess Louisa Inlet in British Columbia. Generally radon increased exponentially with depth during stagnant periods. However, in Lake Nitinat, a permanently anoxic fjord, no excess radon was observed.

Values of K_z obtained from the excess radon profiles in the bottom water of fjords are given in Table 1 along with the water column stability expressed as the Brunt-Väisälä frequency squared.

Values of K_z were also estimated from excess radon profiles in the near-surface water of Narrows Inlet. Two estimates were obtained from each profile, one across the fresh water-salt water interface in the upper 5 m and one in the mixed layer between the fresh water-salt water interface and the secondary pycnocline at 35 to 40 m. The value of K_z across the fresh water-salt water interface was calculated from

$$K_z \frac{dc}{dz} \bigg|_{z_1} = (\lambda + \frac{1}{tr}) \int_{z_1}^{z_2} \tag{3}$$

$$\begin{pmatrix} \text{radon flux} \\ \text{across interface} \end{pmatrix} = \frac{\text{radon decay}}{\text{below interface}} + \begin{pmatrix} \text{radon loss due to lateral} \\ \text{exchange with water outside} \\ \text{fjord} \end{pmatrix}$$

where z is the depth of the fresh water-salt water interface, z_2 is the depth of the top of the secondary pycnocline, tr is the residence time of water in the mixed layer between the interface and the secondary pycnocline, and the other terms are as defined previously. The residence time was estimated from the tidal range and the thickness of the mixed layer to be 10.4 days. The value of K_z between the interface and the secondary pycnocline was estimated, as previously described, from the one-dimensional model with tr included to account for excess radon loss by lateral exchange,

$$K_z \frac{d^2c}{dz^2} - (\lambda + \frac{1}{tr})c = 0. \tag{4}$$

It was assumed in both calculations that sea water entering the fjord over the sill contained no excess radon. Four profiles were measured and values ranged from 0.23 to 0.47 cm²/sec across the fresh water-salt water interface and 0.9 to 3.1 cm²/sec between the interface and the secondary pycnocline (Table 2).

Vertical Mixing Rates Estimated from Salinity Changes

During stagnant periods in fjords which receive fresh water runoff, the salinity of the deep water decreases with time due to the salinity gradient maintained by the fresh water runoff. Estimates of the vertical eddy diffusivity were made from such salinity decreases in the deep water of Narrows Inlet and Princess Louisa Inlet using the method described by Gade (1970). In Narrows Inlet K_z was at a minimum of 0.045 to 0.076 cm²/sec at 44 to 50 m, which was just below the secondary pycnocline. It increased linearly with depth to a value of 0.19 to 0.25 cm²/sec at 72 m (Table 3). In Princess Louisa Inlet, K_z was at a minimum of 0.15 cm²/sec at 60 m, which is at the base of the secondary pycnocline. It increases with depth to a maximum of 2.7 cm²/sec at 150 m and then decreases with depth to a value of 0.54 cm²/sec at 170 m (Table 3).

Variation of Vertical Eddy Diffusivity, Stratification, and Energy Input

The vertical eddy diffusivity is inversely dependent on stratification and directly dependent on energy input, although the functional relationship is unknown. Welander (1968) has shown theoretically that

$$K_z \propto (N^2)^{-1}, \tag{5}$$

where N^2 is the square of the Brunt-Väisälä frequency, if the vertical turbulence is generated only by a cascade of energy from large scale horizontal turbulence. If vertical turbulence is generated only by vertical current shear, he found

$$K_z \propto (N^2)^{-\frac{1}{2}}. \tag{6}$$

From Tables 1 and 3, it can be seen that K_z for the bottom water of fjords investigated in this study varies by over a factor of 100 but stratification, expressed as N^2, varies by only a factor of 5. If K_z is plotted against N^2 on a log-log plot, the slope is far from either of Welander's predictions; this is apparently caused by different amounts of energy being available for vertical mixing in the bottom water of the different fjords. The tide appears to be the major source of this energy and the variation in the relative

Table 1

Summary of Vertical Eddy Diffusivities from Excess Radon Profiles in the Bottom of Fjords and of the Brunt-Väisälä Frequency Squared.

Fjord	Date	Station	Depth Range (m)	One-dimensional Model K_z (cm²/sec)	Rapid Horizontal Mixing Model K_z (cm²/sec)	N^2 (sec^{-2})
Dabob Bay	15 Sept 76	1	150 – 187	0.38 ± 0.03	0.22 ± 0.02	1.58 × 10^{-5}
	16 Sept 76	2	150 – 189	0.34 0.001		
	19 Apr 76	1	151 – 184	0.63 ± 0.05	0.14 ± 0.014	1.11 × 10^{-5}
Hood Canal	30 Jan 76	1	80 – 160	8.6 ± 1.6	8.6 ± 1.2	1.91 × 10^{-5}
	17 Sept 76	1	98 – 143	6.1 ± 1.3		8.08 × 10^{-6}
	20 Apr 77	1	109 – 161	36 ±26		5.76 × 10^{-6}
Port Susan	14 Apr 77	1	95 – 121	0.86 ± 0.16	0.17 ± 0.003	5.13 × 10^{-5}
	1 Jun 77	1	101 – 122	1.14 ± 0.27	0.32 ± 0.05	2.24 × 10^{-5}
Queens Reach	9 Aug 76	504	306 – 358	0.71 ± 0.14	0.12 ± 0.001	1.15 × 10^{-5}
Princess Louisa Inlet	3 Jun 76	505	135 – 176	3.9 ± 0.8	2.22 ± 0.26	1.14 × 10^{-5}
	8 Aug 76	505	149 – 177	1.73 ± 0.47	1.26 ± 0.27	1.04 × 10^{-5}

Table 2

Summary of Vertical Eddy Diffusivities from Excess Radon Profiles
in the Near Surface Water of Narrows Inlet and of the Brunt-Väisälä
Frequency Squared.

Date	Station	Depth Range	One-dimensional Model $K_z(cm^2/sec)$	N^2 (sec^{-2})
2 Jun 76	502	0 - 5	0.47	2.71×10^{-2}
		5 - 35	3.1	5.46×10^{-4}
12 Jul 76	502	0 - 5	0.31	3.07×10^{-2}
		5 - 30	1.3	1.09×10^{-3}
6 Aug 76	502	0 - 5	0.29	3.21×10^{-2}
		0 - 20	0.90	1.70×10^{-3}
8 Sep 76	502	0 - 5	0.23	3.21×10^{-2}
		5 - 35	1.8	2.31×10^{-3}

vertical eddy diffusivities can be explained assuming 1) the energy
input for vertical mixing is derived from the interaction of the tide
with sills, 2) K_z decreases as stratification increases, and 3) K_z
in the bottom water increases as the distance between the entrance
sill depth and the bottom decreases.

Princess Louisa Inlet and Narrows Inlet are both shallow
silled fjords (6 m and 11 m respectively), but K_z is 10 times
greater in the bottom water of Princess Louisa Inlet than Narrows
Inlet. This is caused by the presence of deep internal sills in
Princess Louisa Inlet that interact with the tide, and is evidenced
by the maximum in K_z at 150 m (Table 3) which is the approximate
depth of the deep internal sills.

A log-plot of K_z vs. N^2 is shown for Princess Louisa Inlet and
Narrows Inlet along with plots obtained by Gade (1970) for Oslofjord in Figure 1. The loop in the Princess Louisa Inlet plot is
caused by the rapidly varying energy input with depth in the vicinity
of the deep internal sills. At mid-depths in the fjord the slope
of the plot is approximately -1. In Narrows Inlet there are two
energy regimes, a high energy regime for the surface water and a
low energy regime for the deep water; the transition between the
two regimes occurs across the secondary pycnocline which isolates
the bottom stagnant water from the upper water. The slope of K_z
vs. N^2 is approximately $-\frac{1}{2}$. The vertical profile of K_z in Narrows
Inlet (Tables 2 and 3) reveals that K_z is a minimum at the fresh
water-salt water interface, where apparently energy input and

Table 3

Summary of Vertical Eddy Diffusivities Calculated from Salinity
Changes and of the Brunt-Väisälä Frequency Squared

Fjord	Date	Depth (m)	K (cm^2/sec)	N^2 (sec^{-2})
Narrows Inlet	7 Jun – 29 Aug 75	34	0.30	5.14×10^{-4}
		44	0.076	4.26×10^{-4}
		50	0.076	1.9×10^{-4}
		54	0.11	4.45×10^{-5}
		60	0.12	3.43×10^{-5}
		64	0.15	1.96×10^{-5}
		68	0.16	1.67×10^{-5}
		72	0.19	1.67×10^{-5}
		50–84	0.16 (avg)*	1.96×10^{-5} **
Narrows Inlet	1 Jun – 10 Jun 76	34	0.28	4.96×10^{-4}
		44	0.087	2.37×10^{-4}
		50	0.076	8.28×10^{-4}
		54	0.094	5.13×10^{-5}
		60	0.17	4.28×10^{-5}
		64	0.18	4.99×10^{-5}
		68	0.18	4.21×10^{-5}
		72	0.25	4.05×10^{-5}
		50–84	0.20 (avg)*	4.99×10^{-5} **
Narrows Inlet	10 Jul – 9 Sept 76	34	0.17	9.82×10^{-4}
		44	0.048	7.15×10^{-4}
		50	0.045	1.83×10^{-4}
		54	0.075	6.29×10^{-4}
		60	0.091	3.65×10^{-5}
		64	0.12	1.82×10^{-5}
		68	0.16	1.35×10^{-5}
		72	0.11	1.36×10^{-5}
		50–84	0.15 (avg)*†	1.82×10^{-5}
Princess Louisa	3 Jun – 10 Sept 76	60	0.15	1.01×10^{-3}
		80	0.25	1.97×10^{-4}
		100	0.30	1.85×10^{-4}
		110	0.40	1.27×10^{-4}
		120	0.98	4.53×10^{-5}
		130	1.07	3.84×10^{-5}
		140	1.70	1.38×10^{-5}
		150	2.67	9.96×10^{-6}
		160	1.38	1.56×10^{-5}
		165	0.81	1.37×10^{-5}
		170	0.54	9.06×10^{-6}
			1.5 (avg)	9.83×10^{-6} (avg)

*   Calculated assuming K_z increases linearly from 50 m to the bottom.
**   N at 64 m, the approximate midpoint of the 50–84 m depth range.
†   Bottom point ignored

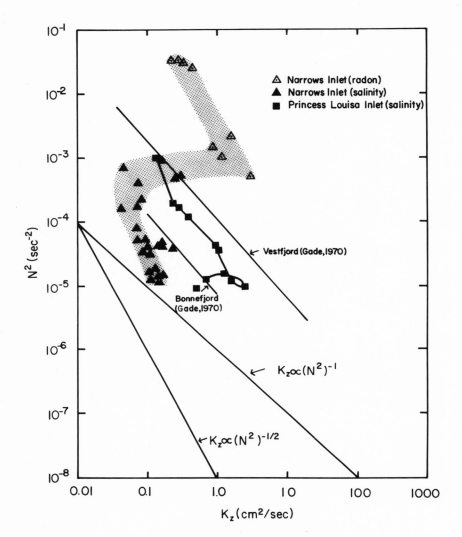

Figure 1. K_z vs N^2 for Princess Louisa Inlet, Narrows Inlet,
and Oslofjord.

stratification are both high. K_z increases to a maximum in the mixed layer between the fresh water-salt water interface and the secondary pycnocline as stratification decreases, and then decreases through the secondary pycnocline as stratification increases and energy input apparently decreases. K_z reaches a minimum at the base of the secondary pycnocline and then gradually increases with depth as stratification decreases.

There are two plots for Oslofjord, Vestfjord which is the outer fjord and Bonnefjord which is the inner fjord. The slope of both plots is -0.85. The offset of the two plots is apparently caused by different energy regimes in the two fjords, the outer fjord receiving the most energy. Gade (1970) suggested that the tide was the major energy source.

The K_z vs. N^2 log plots suggest that there may be different vertical mixing mechanisms in the bottom water of different fjords, but more data from different fjords is needed. An increase in K_z with depth in the stagnant bottom water of fjords has been observed by Gade (1970) in Oslofjord, Broenkow (1969) in Lake Nitinat and Saanich Inlet, and Devol (1975) in Saanich Inlet as well as in Narrows Inlet in this study. Since stratification generally decreases with depth in the stagnant bottom water, this may be a common feature in fjords without deep internal sills where the energy input can be expected not to vary strongly with depth.

REFERENCES

Broecker, W.S. 1965. An application of natural radon to problems in ocean circulation. In: Symposium on Diffusion in Oceans and Fresh Waters, Lamont-Doherty Geological Observatory (1964) (edited by T. Ichiye), pp. 116-145.

Broenkow, W.W. 1969. The distribution of nonconservative solutes related to the decomposition of organic material in anoxic marine basins. Ph.D. thesis, University of Washington, 207 pp.

Devol, A.H. 1975. Biological oxidations in oxic and anoxic marine environments: Rates and processes. Ph.D. thesis, University of Washington, 153 pp.

Gade, G. 1970. Hydrographic investigations in the Oslofjord, a study of water circulation and exchange processes. Geophys. Inst. U. of Bergen, Norway. Rep. No. 24. 193 + 92 pp.

Welander, P. 1968. Theoretical forms for the vertical exchange coefficients in a stratified fluid with application to lakes and seas. ACTA Regiae Societatis Scientiarum et Litterarum Gothoburgensis 1: 3-27.

MIXING INDUCED BY INTERNAL HYDRAULIC DISTURBANCES IN THE VICINITY OF SILLS

J. Dungan Smith and David M. Farmer*

Department of Oceanography
University of Washington
Seattle, Wa. 98195, U.S.A.

Many fjords along the British Columbia and Alaskan coasts are characterized by both a high freshwater discharge and a large tidal range. They also commonly have sills that penetrate well into the stratified near-surface layers, constricting the tidal flow and causing it to accelerate in the region of strong density gradient. An investigation of the nature of mixing in one such fjord, Knight Inlet, began in 1977. The observations were made with at least two vessels, using density and velocity profiling instruments together with high frequency echo sounders. A general description of the processes involved is given by Farmer & Smith (in press).

The purpose of the present paper is to describe the nature and evolution of one of the most energetic events that occurs in this environment and one of the most important in regard to the mixing of surface freshwater with more saline fluid from deeper layers. The acoustic images presented here formed the basis of extensive discussion at the workshop, since they provide striking evidence of mixing events quite different from those normally considered important in fjords. This feature, a near bottom zone of supercritical flow followed by an internal hydraulic jump, is displayed by distortion of the deep reflectors on the down-inlet (right-hand) side of the sill in Figure 1a. The dark line just beneath the surface in each image corresponds to the pycnocline. It shoals abruptly just over the sill crest and then

* Institute of Ocean Sciences, P.O. Box 6000, Sidney, B.C. V8L 4B2, Canada.

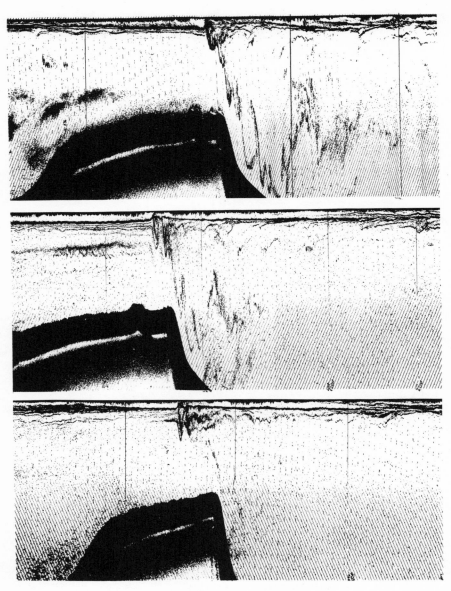

Figure 1.   Sequence showing the supercritical flow region followed
            downstream by an internal hydraulic jump during the
            maximum ebb (1a) and its subsequent decay into an up-
            stream propagating second mode internal bore (1c). For
            scale, the plateau on top of the sill is 1 km long
            and its depth is 60 m.

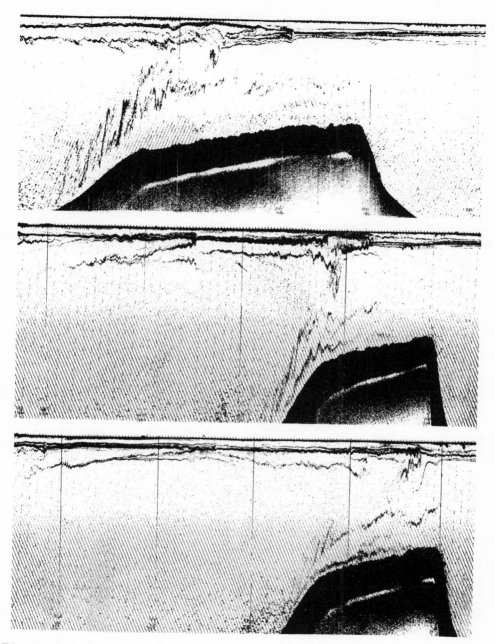

Figure 2. Sequence of images during the flood. The first image
shows the situation just before the maximum current and
the second shows the situation three hours later.

appears to split as the deeper fluid plunges down over the lee
face of the sill. The response is interpreted as being of the
second internal mode.

Figure 1a illustrates the commonly observed large amplitude
instabilities in the shear zone that separate the descending
supercritical flow from the subcritical fluid above. Little
fine structure is seen in the subcritical layer over the crest
and down-inlet side of the sill due to the considerable turbulence
produced by the disturbance, both at the base of the sharp pycnocline and at depth where the subcritical layer adjoins the supercritical one. In contrast fine layering is evident in the sharp
pycnocline beneath the surface water further down-inlet.

Figure 1b obtained sixty minutes later as the ebb slackened,
shows a decrease in amplitude of the disturbance. As in the previous
case, fine layering on the up-inlet side of the internal disturbance
and again further downstream can be seen, whereas the interior
of the feature gives the impression of a more homogeneous water
mass, in agreement with density profile data.

The situation near low slack water is shown in Figure 1c.
The supercritical flow zone has disappeared and the internal
hydraulic jump has collapsed, forming a second mode internal bore
with solitary wave-like features which travels up-inlet at the
leading edge of the mixed fluid. Less than two hours later an
analogous hydraulic feature has formed on the opposite side of
the sill due to the flood. Again a complicated, apparently
turbulent, structure develops in the pycnocline beneath the surface
mixed layer and large amplitude instabilities form on the highly
sheared interface between the deep supercritical and the intermediate subcritical flow. Figure 2b follows 2a by 3h. Fig 2c
shows the collapse just before high slack water, with formation
of a weak bore travelling down-inlet. Variations in ship speed
influence the horizontal scale in the figures.

Within $2\frac{1}{2}$ h after the change in current direction a very
turbulent, second mode disturbance developed in the pycnocline,
Fig. 3a. During the next forty minutes it thickened and became
unstable at its base, Fig. 3b. Finally by mid-ebb a feature
analogous to that in Fig. 1a had developed. A third mode
disturbance can be seen about 1 km downstream of the main one in
Figs. 2a, 2b, 3a, 3c.

Figure 4a presents an early ebb view of a larger region in
the neighborhood of the sill. At the right-hand side the finely
layered upper part of the flow is evident as is a splitting and
fattening of this segment of the pycnocline when the flood-
induced supercritical flow is approached. Downstream of this
region are numerous first and second mode internal waves. At

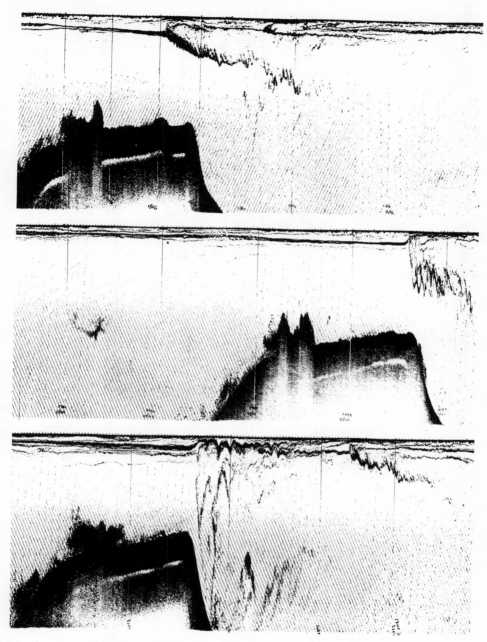

Figure 3. Sequence of figures showing development of the supercritical flow region on the down-inlet side of the sill
during the ebb

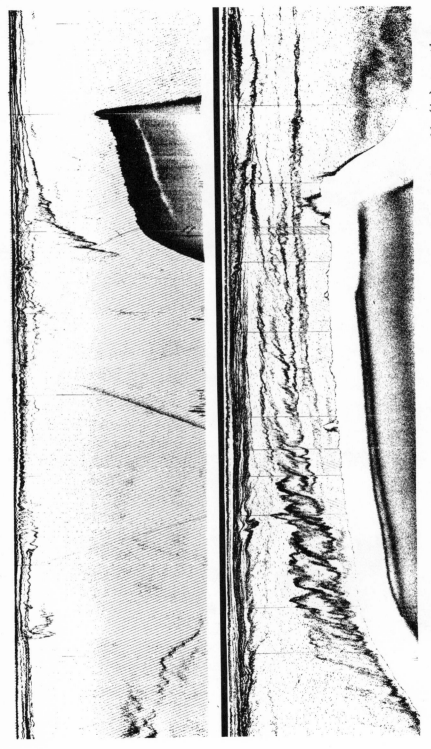

Figure 4. Flood-induced supercritical flow and an internal bore from the previous ebb (4a) together with an expanded view of the shear instabilities on top of the supercritical flow layer(4b)

the far left the up-inlet propagating internal bore can be seen followed, about a kilometer behind, by a well developed train of first mode solitary waves. The zone of mixed fluid extends between the up-inlet propagating bore and the supercritical flow above the sill.

A more detailed image of the instabilities is shown in Fig. 4b. A striking feature of all of these images are the large instabilities which must be associated with mixing. Although the amount of mixing that these cause can only be approximately estimated, it must be substantial. In fact, it is probably dominant below 10 m and certainly is dominant below 20 m as far as the near surface oceanography of Knight Inlet is concerned.

Figure 2 indicates that the flood-induced internal hydraulic disturbance is essentially steady for a three hour period and other measurements indicate that the ebb-induced disturbance is quasi-steady for at least two hours. Therefore, it is reasonable to ask if steady inviscid theories such as those of Su (1976) and Lee and Su (1977) could be used to model this disturbance. The first attempt to use such an approach revealed that an appropriately parameterized bottom boundary layer was necessary. However, even with this modification, the magnitude of mixing and momentum transfer in the interior of the water column over the sill was found to be of first order importance. To date a suitable means of parameterizing the mass and momentum exchange caused by intermediate scale internal hydraulic processes, primarily the large instabilities has not been found.

REFERENCES

Farmer D.M. and J.D. Smith. 1980. Tidal interaction of stratified flow with a sill in Knight Inlet. Deep-Sea Res. (In press).

Lee J.D. and C.F. Su. 1977. A numerical method for stratified shear flows over obstacles. J. Geoph. Res. 82, 420-426.

Su, C.F. 1976. Hydraulic jumps in an incompressible stratified fluid. J. Fluid Mech. 73, 33-47.

GENERATION OF LEE WAVES OVER THE SILL IN KNIGHT INLET

D.M. Farmer and J. Dungan Smith*

Institute of Ocean Sciences
Patricia Bay, P.O. Box 6000
9860 West Saanich Road
Sidney, B.C. V8L 4B2

It is now well known that tidally induced flow of stratified water over bottom irregularities can generate trains of internal waves whose envelope is of tidal frequency but which contain significant higher frequency components (Lee, 1972, Bell, 1975). Moreover, the existence of wave trains that might have been generated in this way has been demonstrated both by satellite imagery and direct observation in several places; recent examples include Massachussetts Bay (Haury, Briscoe and Orr, 1979) and the Andaman Sea (Osborne, Burch and Gordon, 1978). Nevertheless detailed measurements in the area of generation have been largely lacking. We describe here some observations of tidally induced flow over a sill in Knight Inlet, B.C. and identify certain similarities between these observations and recent laboratory experiments. The study of such flows is of value not only from the point of view of providing a better understanding of fjord circulation, but also because it provides a geophysical example of phenomena that are of basic fluid mechanical interest.

In common with several other locations along the British Columbia coast, trains of internal waves are frequently seen in Knight Inlet (Pickard, 1961).[1] These waves are of large amplitude, exhibit a strong downward flow at their leading edge and are highly turbulent (Farmer and Smith, 1978 and Gargett, this volume).

* Dept. of Oceanography, University of Washington, Seattle, Wa.98195
[1] A map showing the location of Knight Inlet and discussions of related aspects of the oceanography there may be found in Freeland's & Gargett's papers in this volume.

Although a wide range of responses to tidal flow over the sill
in Knight Inlet has been observed, it is now clear that the
generation mechanism, which is schematically indicated in Figure 1,
may be broadly described as follows:

1. Some time after the start of an ebb tide, flow over the
    sill generates either lee waves or an internal hydraulic
    jump. Both frequently are accompanied by intense mixing.
2. As the tide slackens, the lee waves, or the hydraulic jump,
    are no longer trapped at the sill and advance upstream
    against the decreasing flow. The former propagates as
    a train of solitary waves; whereas, the latter moves
    upstream as an internal bore.
3. The process is repeated, but in the opposite direction
    during the flood tide.

This mechanism, which has been observed in Knight Inlet (Smith and
Farmer 1977) has also been studied in the laboratory for both
uniformly stratified (Lee, 1972) and layered flows (see fig. 2 and
Maxworthy, 1979). More recent observations have revealed certain
subtleties that were not apparent in earlier data. These include
the predominance of a second mode response during the strongly
stratified summer season and also separation of the boundary layer
under appropriate conditions of stratification and tidal velocity.
This separation is clearly evident in each of the three frames of
figure 2, in which an obstacle moves at successively greater speeds
through a two layer fluid. The vortical instabilities shown in
this figure also have been observed in Knight Inlet (figure 3).

The data of concern in this report were obtained using acoustic
sounding techniques and also with a profiling current meter and CTD.
They are representative of moderate tides and high river discharge
which maintains a thin fresh layer on the surface of the inlet.
While similar results have been procured on several occasions we
will focus on two measurement sets, obtained during the strongly
stratified summer seasons of 1978 and 1979, to illustrate some of
the features associated with ebb flow over the sill. Under these
conditions the boundary layer separates near the sill crest during
the early ebb, spreading out as a sequence of billows or vortices,
(Figure 3). There is no evidence of a lee wave. However, just
before maximum ebb the separation point moves several hundred
metres downstream from the crest, the instabilities on the shear
layer decrease and become less coherent and lee waves form as
indicated in figures 1 and 2. Then as the tide slackens and the
lee waves travel back over the sill there is a retreat of the
separation point back toward the crest of the sill. The sequence
of events can be illustrated by plotting the location of the
separation point (in this case identified as depth below sill crest)
against the sectionally averaged tidal velocity over the sill. The
result (figure 4) indicates a hysteresis in the response.

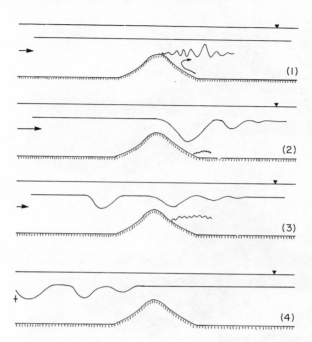

Figure 1. Schematic diagram showing certain features of the
observed response of stratified water advected from
left to right over a sill by the tide. (1) Just after
slack water: flow separates at crest of sill, shear
layer evolves as sequence of growing billows. (2) Near
maximum flow: large lee wave forms, suppressing flow
separation. (3) close to the next slack water: first
lee wave escapes upstream, subsequent lee waves continue
to control separation point. (4) Slack water: additional lee waves have escaped and form a train of waves
that travels away from the sill. The response may be
repeated for both flood and ebb tides. Other possible
responses, not illustrated here, include the formation
of a single, large breaking lee wave or hydraulic jump
early in the tide, which dominates the subsequent
response, the generation of higher mode internal lee
waves and also the generation of internal waves just
upstream of the crest when the flow is close to maximum.

Figure 2.   Laboratory model of flow over a sill in a two-layer
            fluid of total depth 10 cm and at speeds of 1.2, 1.7 and
            2.2 cm s^{-1} respectively. The interface between the upper
            layer of fresh water, dyed black, and the lower salty
            layer is deformed as lee waves develop downstream of the
            sill. Flow separation in the salty lower layer at the
            sill crest leads to vortical instability at the shear
            layer. At the highest speed the jet of salty water
            flowing over the sill is deflected downwards. However,
            the separation point remains at the crest owing to the
            small radius of curvature.

     Observations of salinity, temperature and current velocity were
obtained from a ship anchored 300 m downstream of the sill in water
of depth 100 m. The vessel was very securely moored with 2 bow and
3 stern anchors; vessel movement was monitored twice each second

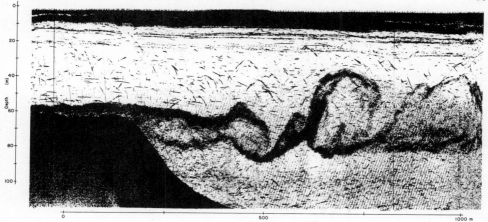

Figure 3.   Acoustic record of instabilities on the shear layer
            generated downstream of the Knight Inlet sill by the
            ebb, 0153-0213, July 30, 1978. Note vortical instab-
            ilities on shear layer emanating from separation point
            near sill crest.

using microwave positioning equipment and was rarely found to be
more than 5 cm s^{-1} during the ebb tide discussed here. From 60
profiles gathered over 6.5 hours we have generated contour plots
of salinity and the velocity component resolved along the axis
of the channel (figure 5). Cross channel flow was generally less
than 10 cm s^{-1} during the ebb and will not be discussed here.

     During the early part of the ebb most of the flow was concen-
trated in a narrow jet centered on the pycnocline and with brief
exceptions this local maximum in the current profile persisted
throughout the ebb. However, about 1540 h a second maximum at
40 m developed and thereafter dominated the current structure.
Initially this deep maximum overlay a current in the opposite
direction, indicating movement of water back towards the sill
below sill septh. But at 1615 h the lower jet started to deepen
and the current no longer reversed at depth. This structure
changed quite suddenly at 1700 h: the deep jet rose and again the
flow reversed below 60 m. From 1815 to 1900 h the jet again
descended, this time with even higher velocities (>40 cm s^{-1})
within the core. The period 1930 h to 2050 h marks the transition
from ebb to flood, with the flood beginning at depth and spreading
upwards.

     From figure 5 it is apparent that as the depth of the current
maximum increased, the isohalines also deepened. The rapid change
in current structure at 1700 was accompanied by an equally rapid
rise in isohalines below 25 m; above 25 m the isohalines were
depressed. Simultaneous traverses of the sill by a separate vessel

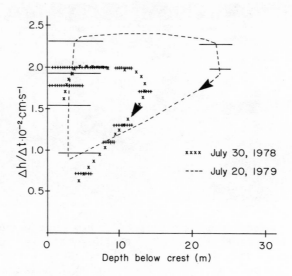

Figure 4.   Depth below the sill crest of the separation point
             plotted as a function of rate of change of sea—surface
             height (proportional to sectionally averaged tidal
             velocity). The depth of separation was determined from
             acoustic records and is indicated here as horizontal
             bars with lengths determined by the widths of the
             reflecting layer emanating from the separation point.
             Note hysterisis in response associated with formation
             of lee waves.

revealed the evolution and release of two lee waves. The times of
six consecutive traverses are indicated at the foot of figure 5.
The four most informative of these are presented as figure 6. The
first shows the situation early in the ebb when only small and short
lee waves had formed, just downstream of the sill crest, but
upstream of the anchored vessel. By traverse 3 the first lee wave
had escaped upstream and a much larger disturbance evolved downstream of the crest. The record from traverse 4 is qualitatively
the same as that from traverse 3; however by traverse 5 a large lee
wave had evolved. Here acoustic reflectors crossing the sill crest
at 25 m were depressed to 55 m in its lee. On traverse 6 this
disturbance was similar in size but had begun to move upstream
toward the sill. The large lee wave that evolved between traverses
4 and 5 again coincides with depression of the isohalines and of
the depth of the lower velocity maximum.

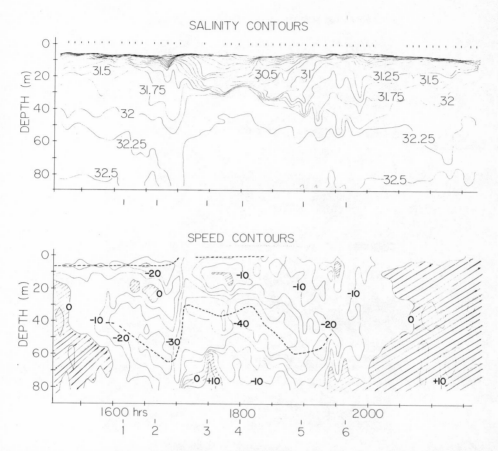

Figure 5.   Contours of constant salinity (above) and along-channel
            velocity component (below) derived from profiles taken
            300 m downstream of the sill crest on July 23, 1979.
            Flow back toward the sill (positive values) is shaded;
            numerical values denote speed contours at 10 cm s^{-1}
            intervals. Heavy dashed lines show location of speed
            maxima. The numbers at the foot of the figure refer to
            acoustic images taken from a second vessel some of which
            are displayed in figure 6.

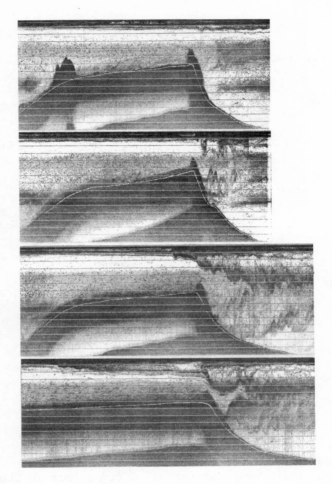

Figure 6. Acoustic images from traverses 1,2,3,5 as indicated in
           figure 5 showing the internal structure of an ebb flow
           over the Knight Inlet sill. Note the small amplitude
           and short wavelength of the internal waves on the
           pycnocline in the first traverse, relative to the larger
           ones in the second and fourth. In traverse 3 the first
           lee wave had moved upstream and a larger disturbance had
           been generated above the sill crest.

DISCUSSION

The problem of boundary layer separation of stratified flow
has recently been under active study (Brighton, 1977). It is now
known that separation of the boundary layer can be suppressed when
the wavelength of the lee wave is such that the downslope acceleration induced by the lee wave produces a favourable pressure gradient.
(Suppression of separation is not seen in figure 2 because of the
high curvature at the top of the model sill). Thus far, both
theoretical and experimental work has been confined to symmetrical
objects in uniformly stratified, unaccelerated flow which does not
readily permit application to the more complicated situation found
in Knight Inlet. Nevertheless, we expect the principles governing
the response to be similar. For example, for two dimensional hills
of moderate slope it has been suggested (c.f. Hunt, Snyder, and
Lawson, 1978) that separation is suppressed over a range of
internal lee wave wavelengths

$$2L \lesssim \lambda \lesssim 5L$$

where L is the half width of the obstacle. While the different
stratification and also the asymmetry of the Knight Inlet sill
probably changes the range over which suppression occurs, we would
still expect a transition from separated to separation- suppressed
flow as the tidal current changes.

An extensive body of theory exists for the prediction of lee
wave generation due to stratified flow over small obstacles. For
linearly stratified flow in a fluid of depth H, the wavelength is
given by

$$\lambda = 2HF(1-n^2F^2)^{-\frac{1}{2}}$$

where F is the Froude number ($F = \pi U/NH$) and n the wave mode. For
subcritical flow and given stratification N the wavelength increases
with flow speed U. We therefore expect that early in the tide the
waves will be short, but that they will increase as the flow speed
increases, eventually suppressing separation. The wave amplitude
is determined by the flow speed; but in this case the obstacle
shape also enters the problem. As an example, for an obstacle of
shape $\zeta = a^2b/(a^2+x^2)$, Lee (1972) showed the amplitude of the waves
is proportional to

$$ab \, \exp\{-(F^{-2}-n^2)^{-\frac{1}{2}}a\pi/H\}$$

Calculations carried out for this case show that the wave amplitude
initially grows quite slowly, as the flow speed increases, but then
rises rapidly as critical conditions are approached. Although this
example differs from the more complex topography and stratification
of Knight Inlet, it does suggest that the transition between small,

short waves, which cannot suppress boundary layer separation, and larger, longer waves that can, may be quite rapid. This result appears consistent with the data presented in figure 6.

The observations for the flows considered here show that only small amplitude lee waves of short length form until close to maximum ebb. A lee wave or train of waves then forms quite rapidly, suppressing the separated zone. As the waves escape upstream, the separation point advances back up the lee of the sill. The transition from separated to separation-suppressed flow is seen in figure 5 at the time each of the two lee waves form. It is clear from this figure that quite rapid fluctuations can occur in the location of the separation point during the critical stage when lee waves are being generated; similar fluctuations may have been missed in the acoustic traverses used to generate figure 4.

An interpretation of this observation based on the quasi-static lee wave theory indicated above may be partly correct, but it could hardly account for the hysterisis in the separation depth-tidal velocity diagram shown in figure 4. A possible explanation is that the flow separation itself influences the effective shape of the sill, in effect making the obstacle appear smoother to the flow. For example, it can be shown (Bell, 1975) that the smoother the obstacle, the smaller the contribution to higher frequency components of the lee wave response. Once the first lee wave has formed separation would be controlled by the waves much as described in the theoretical or laboratory results.

Of course the relative height of the Knight Inlet sill must limit the applicability of linear lee wave theory. Unfortunately there is no suitable time dependent nonlinear theory to which we can appeal; however, laboratory experiments do appear to illustrate some of the observed features. Baines (1977) carried out some tow-tank experiments primarily designed to study the validity of Long's hypothesis upstream of the obstacle, which show the formation of a strong jet on the lee, especially when there is upstream blocking and also the presence of almost stagnant regions above the jet. The Knight Inlet observations show a persistent region of small or even upstream flow throughout the ebb, centered at about 20-25 m (figure 5) which is similar to the stagnant region observed by Baines and perhaps analogous to the "rotors" observed in atmospheric flow over mountains.

The phenomena described above appear to be very important in the extraction of energy from the barotropic tidal currents in Knight Inlet and strongly influence the density and turbulence structure of the channel. It is likely that similar processes occur in other stratified fjords having significant velocities, of whatever origin, at sill depth.

REFERENCES

Baines, P.G. 1977. Upstream influence and Long's model in strat-
    ified flows. Journal of Fluid Mechanics, 82, 147–159.

Bell, T.H. 1975. Lee waves in stratified flows with simple harmonic
    time dependence. Journal of Fluid Mechanics, 67, 705–722.

Brighton, P.W.M. 1977. Boundary layer and stratified flow over
    obstacles. Ph.D. Thesis, University of Cambridge, England.

Farmer, D.M. and J.D. Smith. 1978. Nonlinear internal waves in a
    fjord. In Hydrodynamics of Estuaries and Fjords, ed. J. Nihoul,
    Elsevier, Amsterdam.

Haury, L.R., N.G. Briscoe, and M.H. Orr. 1979. Tidally generated
    internal wave packets in Massachusetts Bay, U.S.A.; preliminary
    physical and biological results. Nature, 278, 312–317.

Hunt, J.C.R., W.H. Snyder, and R.E. Lawson. 1978. Flow structure
    and turbulent diffusion around a three-dimensional hill,
    Report EPA-600/4-78-041. U.S. Environmental Protection Agency.

Lee, C.Y. 1972. Long nonlinear internal waves and quasi-steady
    lee waves. Ph.D Thesis Massachusetts Institute of Technology
    and Woods Hole Institution.

Maxworthy, T. 1979. A note on the internal solitary waves produced
    by tidal flow over a three-dimensional ridge. Journal of Geo-
    physical Research, 84, CI, 338–346.

Osborne, A.R., T.L. Burch, and R.L. Gordon. 1979. Internal waves
    in the Andaman Sea; solitons in the world ocean? Spring meeting,
    American Geophysical Union.

Pickard, G.L. 1961. Oceanographic features of inlets in the British
    Columbia coast. Journal of Fisheries Research Board of Canada,
    18(6), 907–999.

Smith, J.D. and D.M. Farmer. 1977. Nonlinear internal waves and
    internal hydraulic jumps in a fjord. Geofluiddynamical Wave
    Mathematics: Research Contributions. Appl. Math. Group.
    Univ. of Washington, Seattle.

THE HYDROGRAPHY OF KNIGHT INLET, B.C. IN THE LIGHT OF LONG'S MODEL OF FJORD CIRCULATION

Howard J. Freeland

Institute of Ocean Sciences
P.O. Box 6000
Sidney, B.C.   V8L 3H7
Canada

INTRODUCTION

This paper will discuss the circulation and associated hydrography of Knight Inlet, and will compare the behaviour with the predictions of a theory of the circulation of a two layer estuary due to R. R. Long (1975). The contents of this paper, of necessity, represent only a summary of the work completed, and a discussion in greater depth is being prepared for publication elsewhere, Freeland and Farmer (1979).

Knight Inlet was selected for intensive study for a variety of reasons, one was its geometric simplicity. The inlet has a rectangular cross-section, and is close to having a constant width throughout its length, mean width is about 3.0 km. It is generally deep, greatest depth 550 metres, but has a sill 75 km. from the head where depth falls to about 60 m. The sill is in the middle of a long, straight reach which is connected to the head by a sinuous reach. A map of Knight Inlet is shown in Figure 1 together with a longitudinal depth profile.

Most of the fresh water enters through the head, only insignificant amounts enter through the small creeks along the remainder of its length. Throughout most of the year the stratification is such that a two-layer representation seems appropriate. The general oceanography of Knight Inlet is well described by Pickard (1961) and Pickard and Rodgers (1959).

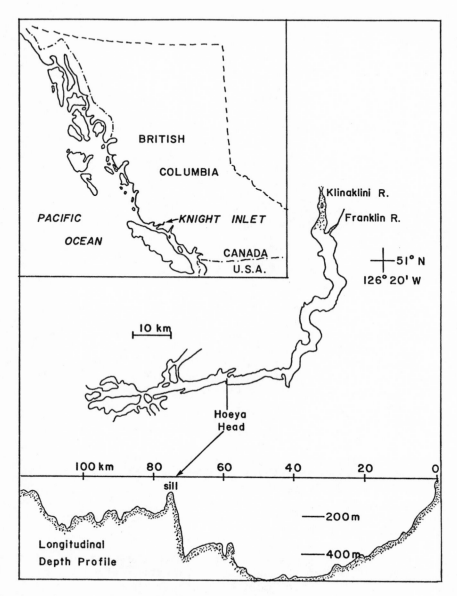

Figure 1.   Map of Knight Inlet and longitudinal depth profile of
            the inlet.   The inset locates Knight Inlet on a map
            of British Columbia, Canada.

LONG'S MODEL

Gade and Svendsen (1978) discuss a comparison of the hydrography of Sognefjord with the predictions of Long's model. Following
Gade and Svendsen we will compare observations in Knight Inlet with
a relationship between interface depth (h) and the density contrast
($\Delta\rho$) between the two layers.

$$
\frac{h}{h_o} = \left(\frac{h\Delta\rho}{h_o\Delta\rho_o}\right)^{\beta+1}\left[\frac{2F_o^2 + \beta}{2F_o^2 + \beta\left(\frac{h}{h_o}\frac{\Delta\rho}{\Delta\rho_o}\right)^3}\right]^{\frac{\beta+2}{3}}
\tag{1}
$$

In equation (1) F is the Froude number and the subscript zero
connotes values evaluated at the head of the inlet. The parameter
β measures the importance of friction; in the notation of Long
(1975) and Gade and Svendsen (1978) $\beta = s-1$. Also we can write
$\beta = 1/(1 + \mu)$, where $\mu = K/mK_n$; K is an interfacial friction
coefficient and mK_n is a measure of the importance of entrainment.
[We note in passing that the equivalent of equation (1) in Gade
and Svendsen (1978) appears to be incorrect.] For inviscid flow
$\beta = 1$, in the limit of highly viscous flow β tends to zero from
above. At some value of the Froude number, between the initial
value F and ½, the upper layer thickness is maximised, specifically
h is maximised at $F = F_m = \sqrt{\beta/(2\beta + 2)}$. As the upper layer moves
from the head towards the mouth the Froude number steadily increases.
Hence, the upper layer thickness increases until $F = F_m \leqslant$ ½ and
then decreases as F increases beyond F_m. The density contrast is
minimised, for all β, at $F = 1$ and so it is not possible to drive
the flow from sub- to super-critical conditions. Long (1975) shows,
by argument, that the flow must be controlled by the condition $F = 1$
at the mouth.

OBSERVATIONS

During the winter months the river flow is very small and a
considerable amount of mixing occurs near the head. The result is
that near-surface σ_t does not tend to zero near the head and the
estuarine circulation is not strictly fjord-like. During the mid-
summer months two problems occurred; firstly our CTD had problems
with the very sharp density interfaces and secondly, other observations suggest that the response of the inlet is in the second
internal mode (see Farmer, this volume). Hence, a second internal
degree of freedom is needed to describe summer observations. For
the purposes of a comparison with a two-layer model, we chose to
restrict attention to spring and autumn surveys.

The continuous descriptions of density versus depth acquired

from the CTD were reduced to two-layer representations by an
objective technique that conserves potential energy in the water
column and internal wave speed (1st internal mode). For four
representative cruises the parameters descriptive of the upper and
lower layer are presented in Table 1 below.

Table 1

Upper layer thickness (h) in metres; upper layer σ_t (σ_1); velocity
of upper layer (u_1) in cm/sec;
Froude number (Fr = $u_1/\sqrt{gh(\rho_2 - \rho_1)/\rho_1}$); and lower layer σ_t (σ_2).

x(km)	h	σ_1	u_1	Fr	h	σ_1	u_1	Fr
0	3.25	0.00	2.78	.0313	2.88	0.00	2.22	.0264
25	3.78	5.87	3.16	.0378	3.64	6.61	2.40	.0299
50	4.30	12.27	4.24	.0589	4.41	13.22	3.17	.0452
75	9.63	21.13	7.17	.1295	8.02	20.74	5.34	.0990
100	15.32	22.87	9.94	.2115	12.73	22.44	6.34	.1280
		$\sigma_2 = 24.313$				$\sigma_2 = 24.365$		
0	2.53	0.00	2.93	.0373	3.23	0.00	2.79	.0316
25	3.03	4.91	3.07	.0400	3.98	7.00	3.19	.0386
50	3.53	8.40	3.21	.0428	4.73	12.07	3.81	.0502
75	7.08	20.05	5.90	.1064	10.14	21.40	7.70	.1445
100	11.16	21.59	5.80	.1037	19.09	23.28	12.45	.2972
		$\sigma_2 = 24.393$				$\sigma_2 = 24.201$		

Data are presented for four surveys on dates 31/5/77, 3/11/77,
24/5/78 and 29/9/77 clockwise from the upper left and at five
locations in the inlet. The values of u_1 were calculated knowing
the river runoff (which was monitored throughout the study) and
using Knudsen's relationships.

Before making quantitative statements, based on Table 1, some
general observations applicable to all surveys of Knight Inlet may
be made.
    i)   The rate at which the upper layer deepens is much greater
        along the sinuous reach (km 50 to 100) than along the
        sinuous reach (km 0 to 50).
   ii)   The upper layer thickness at the head is about 3 m,
        considerably less than the 21 m predicted in Long (1975).
 iii)   In all of the surveys examined (21 in all) with the
        exception of one case only, the upper layer thickness
        increases steadily to the mouth; no maximum is seen.

The reason for the latter observation is evident in Table 1. The Froude numbers are remaining very low, and in fact we do not see any values of F even approaching unity as expected.

Using the data of Table 1 we plot (h/h_0) against $(\Delta\rho/\Delta\rho_0)$ as shown on Figure 2. The two sets of curves superposed are the variation predicted theoretically for each of the surveys (varying initial Froude numbers) and for two values of the friction parameter $\mu = K/mK_n$. The bold lines are for the inviscid case, $\mu = 0$, and the dashed lines are for $\mu = 5$ small viscosity. Clearly, the fit to the family of inviscid solutions is not bad; the fit to the set of viscid solutions is, however, poor. Knight Inlet appears to be in the low velocity regime of Long's model described by $h \, \Delta\rho^2$ = constant. Furthermore, the best fit is obtained in the inviscid limit.

ENERGY CONSIDERATIONS

It is interesting to examine the energetics of Knight Inlet as this sheds some light on the energy source for the mixing processes in the inlet. It is a simple matter to compute the kinetic energy and the potential energy anomaly at each of the stations listed in Table 1. One can then compute the divergence of the total energy flux over various control volumes in the inlet. We cannot measure the free surface contribution, but we can show that the ratio of potential energy changes due to the free surface divided by that due to internal modification is of order $(\rho_2 - \rho_0)/\rho_2$, in Long's model for inviscid flow. Hence, we can neglect free surface contributions if we believe interfacial friction to be small. We find then that the energy flux divergence is greater by about a factor of 20 along the straight section compared with the sinuous.

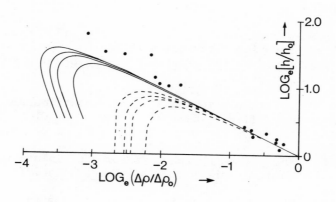

Figure 2

The contribution from kinetic effects is about 10% of that due to potential energy changes. [This is the reason why we chose to conserve potential energy in our objective scheme for reducing profiles to two-layer representations.] It is our belief that the very large inhomogeneity in energy flux divergence results from the very large amounts of turbulence generated by the tidal flow over the sill.

The wind was monitored in several locations in Knight Inlet and we conclude that:

    i)   The wind cannot do enough working on the sea surface to produce the observed energy flux divergences.

    ii) The wind is only fractionally less intense on the sinuous reach than on the straight reach, so that the observed rate of working is different by only a factor of 2.1 ± 0.5. This ratio is simply the ratio of mean cube wind speeds at the two locations.

DISCUSSION

It is clear from the energetics of Knight Inlet, and the hydrography in general, that some process that ultimately causes mixing in the fluid is considerably more active in the straight reach. We do not believe that the wind is responsible for either the inhomogeneity or the overall rate of working. It is our belief that both are a direct result of the turbulence generated by the tidal flow over the sill.

References

Freeland, H.J. and D.M. Farmer. 1979. The circulation and energetics of a deep, strongly stratified inlet. J. Mar. Res., sub judice.

Gade, H.G. and E. Svendsen. 1978. Properties of the Robert R. Long model of estuarine circulation in fjords. In J. Nihoul (editor) Hydrodynamics of estuaries and fjords, 546 pp, Elsevier, Amsterdam.

Long, R.R. 1975. Circulations and density distributions in a deep, strongly stratified, two-layer estuary. J. Fluid Mech., 71, 529-40.

Pickard, G.L. 1961. Oceanographic features of inlets in the British Columbia coast. J. Fish Res. Bd. Can. 18(6), 907-99.

Pickard, G.L. and K. Rodgers. 1959. Current measurements in Knight Inlet, British Columbia. J. Fish Res. Bd. Can. 16(5), 635-78.

TURBULENCE MEASUREMENTS THROUGH A TRAIN OF BREAKING INTERNAL WAVES IN KNIGHT INLET, B.C.

Ann Gargett

Institute of Ocean Sciences
P.O. Box 6000
Sidney, B.C. V8L 4B2

Twice each day on the turns to flood tide, a group of highly non-linear internal waves emerges from the region of a submarine sill across Knight Inlet, one of the narrow fjord-type inlets characteristic of the coast of British Columbia. Properties of the wave groups and of their generation mechanism are described by Farmer and Smith (1978) and (1979). Freeland and Farmer (1979) have suggested that mixing caused by the turbulent decay of such wave trains during propagation and final collapse is largely responsible for the observed change of mean vertical property distributions between the sill and the furthest up-inlet point reached by the waves.

Echo sounding records of the wavetrain suggest that it is indeed highly turbulent. Figure 1 is a record (courtesy of D. Farmer) from a 200 kHz hull-mounted echo-sounder, taken as the ship traversed the wave packet: notice the abrupt leading edge of the packet, the decreasing amplitude and wavelength of waves behind the leading edge, and the large incoherent structures throughout the wavetrain.

In November 1978, direct measurements of the dissipation scales of turbulent velocity and temperature fields were carried out, using the research submersible PISCES IV, stabilized for mid-water running by a set of specially designed wings, as a vehicle to carry sensors through the wavetrain. Figure 2 presents analog records of measurements during a traverse from the rear through the leading edge of one wavetrain (Fig. 1). Axial velocity (u') was measured with a heated platinum film sensor, and orthogonal velocities (v' and w') with two single-axis airfoil probes, while high frequency temperature fluctuations (T') were measured with a

277

Figure 1. Record from a 200 kHz hull-mounted echo-sounder of a train of internal waves in Knight Inlet, courtesy of D. Farmer. The leading edge of the wave group is to the right of the figure and typical wavelengths are of the order of 100-200 m near the leading edge. The superimposed arrow indicates the direction in which PISCES traversed a similar wave group which originated 12 hours previous to the one pictured here.

cold platinum film probe. Auxiliary instruments included a conduct-
ivity-temperature sensor giving mean temperature (T) and salinity
(S) at the level of the high-frequency sensors, an additional
thermistor $\sim$ 1m above, to give an estimate of mean vertical temperature gradient ($\Delta T/\Delta z$), a set of three rotor current meters for
mean forward speed and cross flows (not shown in Figure 2), a
pressure gauge (p), and three orthogonal accelerometers, mounted
just behind the high frequency sensors, for vehicle accelerations
along (PITCH) and across (ROLL and VERT) the axis of the probe
mount.

   The pressure record indicates that the submersible is able to
maintain depth fairly well until it emerges from the front of the
wavetrain; there, strong downwelling velocities act on the flat
surfaces of the stabilizing wings and force the vehicle down. As
evidenced by the narrow spike of low temperature and salinity
observed just at the front of the wave packet, this downwelling
sweeps the colder fresher near-surface layer to depth, where the
turbulent structures associated with the wavetrain mix it with
the underlying warmer and saltier water. Occurrences of the
relatively uniform temperature and salinity which are characteristic
of this underlying water are marked by arrows along the record and
allow us to identify successive wave crests behind the leading
edge, as regions where water from below is raised to the level of
the submersible. Further back from the leading edge, the absence
of intervals with T and S characteristic of the lower layer could
be due to a number of causes: a general wave-associated descent
of isopycnal surfaces, so that the submersible remains within the
upper mixing layer; a decrease in wave amplitude to the extent
that a crest no longer raises the lower layer to the level of the
submersible; a change in lower layer mean temperature and salinity
caused by the mixing occurring further forward in the wavetrain;
or simply the absence of significant wave motions in this region.

   The velocity (u', v' and w') and temperature (T') records
displayed here include frequencies from 1 to 100 Hz, corresponding
to scales of $\sim$ 1 metre down to 1 centimetre at the roughly 1 ms^{-1}
speed of the submersible. These scales cover the dissipation range
of the velocity field, allowing a direct computation of the rate
of turbulent kinetic energy dissipation by integration of the
measured dissipation spectrum. The energy-containing range of the
turbulent field (the roughly 10 m vertical and several tens of
metres horizontal scales typical of the large incoherent structures
observed on the trailing edges of the waveforms in Figure 1) cannot
be measured by the present PISCES instrumentation, because the submersible itself is moved around by the water motions at these
scales. This is most obvious in the PITCH accelerometer record of
Figure 2; where large pitching motions produced by passage through
the wavetrain vanish immediately when the submersible enters quiet

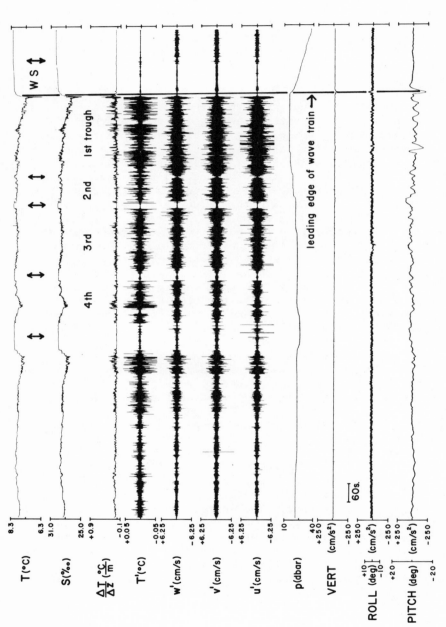

Figure 2. Chart records from turbulence and auxiliary sensors mounted on PISCES during a pass from the rear through the leading edge of a train of internal waves similar to that pictured in Fig. 1. Sensor description and interpretation of various features of this record are found in text.

water ahead of the waves. The measured velocities indicate that
the strongly dissipative turbulence associated with the wavetrain
is effectively confined to the upper part of the water column,
above and through the pycnocline: sections of record taken through
the underlying waters (arrows in Figure 2) are quiescent to the
system noise levels. The velocity records show no obvious anisotropy among the three components at these scales.

In the quantitative analysis of these and similar records
from Knight Inlet, it will be possible to examine such questions
as the degree of isotropy of the velocity field at dissipation
scales under varying conditions of mean stratification, and the
scale of separation between turbulent motions and those more
properly described as wavelike. The vertical turbulent heat flux
may be estimated from the measurements, and perhaps the buoyancy
flux, given some assumption of the relation between mean temperature
and density. Finally, the rate of turbulent kinetic energy
dissipation ε can be calculated directly from the measured velocity
fields. Since ε is often one of the major terms in a turbulent
kinetic energy balance, it will be interesting to compare its value
with those estimated for other source-sink terms such as energy
stored in the wavetrain and energy used for mixing.

REFERENCES

Farmer, D. and J.D. Smith. 1979. Generation of lee waves over
        the sill in Knight Inlet. (This Volume).

Farmer, D.M. and J.D. Smith. 1978. Nonlinear internal waves
        in a fjord. In "Hydrodynamics of fjords and estuaries". Ed.
        J. Nihoul, Elsevier, Amsterdam.

Freeland, H.J., and D.M. Farmer. 1977. The circulation and ener-
        getics of a deep, strongly stratified inlet. J. Mar. Res.
        sub judice.

FINITE ELEMENT COMPUTATION OF THE BAROTROPIC TIDES IN KNIGHT INLET, BRITISH COLUMBIA

Bruno M. Jamart and Donald F. Winter

Department of Oceanography
WB-10, University of Washington
Seattle, Washington, 98195

The accurate and efficient solution of the two-dimensional shallow water wave equations poses a challenging problem for numerical analysts. Finite difference or finite element techniques are the most commonly used numerical procedures. In the case of finite differences, the region of interest is covered by a rectangular grid, space derivatives are approximated by finite differences, and grid point values of water height and depth-averaged velocities are obtained by solving difference equations at a succession of time steps. When the finite element approach is used, the problem is first restated in terms of integrals over the domain (most often by using weighted residuals) and the integrals are replaced by sums over finite elements. The problem is thereby reduced to a system of first-order ordinary differential equations involving nodal values of the unknowns and their derivatives with respect to time: the system is then solved, as before, at a succession of time steps.

If the objective is to characterize tidal flow patterns over a time interval of a few days, the motion can be assumed to be periodic. When time-stepping procedures are used to calculate tidal motion in inlet basins of great depth D (say), excessive computer time can result from the combination of two requirements: (1) a small time step for accuracy (and, if the scheme is explicit for stability: $\Delta t \leq 2\Delta x/\sqrt{gD}$, where Δx is a representative spatial discretization interval); and (2) a large number of time steps to ensure the damping of initial transients and the attainment of a periodic state.

A procedure which provides both speed and accuracy, especially for deep inlets, has been described by Pearson and Winter (1977). The Coriolis acceleration as well as nonlinear advective transport

and boundary friction were included in their formulation. Subsequently, Jamart and Winter (1978) developed a simpler version of that
procedure for cases in which Coriolis and advective accelerations can
be neglected and the frictional force assumed linear. The resulting
problem statement is made more compact by introducing complex variable
representations of the dependent variables in the usual way. As a
consequence of the simplifications, the latter procedure is more
easily programmed and applied to common coastal features, such as
straits, bays, and inlets.

The analysis is based on the conventional set of vertically-
integrated time-dependent equations expressing conservation of
horizontal momentum and mass in two horizontal dimensions, x and y.
Boundary conditions are prescribed along the shoreline (zero normal
transport) and across the mouth of the inlet (specified elevation).
Under the assumption of a periodic solution, the dependent variables
(free-surface elevation and depth-averaged velocities) can be Fourier
decomposed, and the time-dependent equations of motion replaced by
an equivalent set of modal equations, with only x and y as independent
variables. After elimination of the Fourier coefficients of the
velocity, each modal equation for the surface elevation and the
corresponding boundary conditions can be rephrased in terms of an
equivalent variational principle. The variational problems are then
solved by means of the finite element method. The actual space-
and time-dependent currents and heights can be reconstructed by
Fourier synthesis. As mentioned above, in the procedure described
by Jamart and Winter, advective and Coriolis accelerations are
neglected and the term representing dissipation is linearized.
Even so, one of the drag-related terms in the free-surface equation
needs to be handled by an iterative procedure because it cannot be
incorporated in a variational formulation. In addition, the
calculations of the real and imaginary parts of the Fourier coefficients of the elevation are performed separately.

The approach just summarized was used to compute the barotropic
tides in Knight Inlet, British Columbia. Knight Inlet is a long,
steep-sided fjord, located on the southern mainland coast of British
Columbia, Canada; it is about 100 km long and has an average width
of about 3 km. Knight Inlet has two sills, the innermost one has a
maximum depth of about 63 m and separates a very deep basin (maximum
depth of 550 m) from the shallower ($\sim$ 200 m) outer basin. Knight
Inlet has recently been the site of extensive field programs (see
papers by Farmer, Freeland, and Smith in this volume and references
therein). For this reason, Knight Inlet seemed to be an appropriate
site for testing our model results against reliable field data.
We also wished to determine the usefulness of the linearized model
in a deep stratified fjord with shallow sills. At times of freshwater runoff, Knight Inlet becomes strongly stratified, and it
appears that a significant portion of the barotropic tidal energy
is extracted through complicated internal mechanisms associated

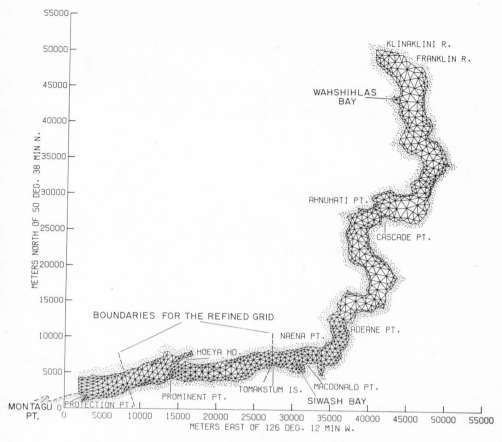

Figure 1. Finite element grid for Knight Inlet.

with flow over the sill (Freeland and Farmer, 1979). Since an
accurate parameterization of the internal energy exchange phenomena
is not yet possible, we made the working assumption that, in tidal
computations, dissipation could be modeled by assuming the "frictional" force to be proportional to the velocity $\underset{\sim}{u}$ divided by the local
time-mean water depth D:

$$\underset{\sim}{F}_r = - E \, \frac{\underset{\sim}{u}}{D}$$

The finite element grid which was used in Knight Inlet is
shown in Figure 1. Finer resolution was required in the vicinity
of the sill, and the grid used in that region is shown in Figure 2.
It appears that the model results represent fairly well the avail-

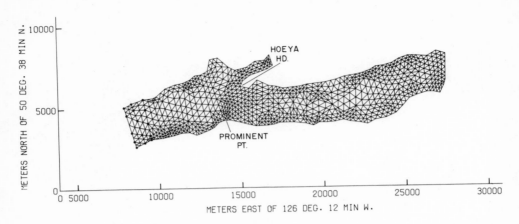

Figure 2.   Refined grid (quadrupled element density) covering part
             of the straight reach of Knight Inlet. The sill lies
             between Hoeya Head and Prominent Point.

able measured characteristics of the tidal flow in Knight Inlet.
For example, regardless of the actual dissipation process, measured
phase differences in water surface elevation can be simulated by
adjustment of E in the linear dissipation term Eu/D. Freeland and
Farmer (1979) have analyzed tide gauge records which were obtained
at three locations in the Inlet: Montagu Point (a few km west of
the model open boundary), Siwash Bay, and Wahshihlas Bay (see
Figure 1). At tidal frequency M_2, the observed phase difference
$\Delta\phi_{s-m}$ between Siwash Bay and Montagu Point fluctuates between
values of 0.75^0 (January) and 1.7^0 (June), whereas the difference
between Wahshihlas Bay and Siwash Bay remains less than 0.25^0. As
shown below, a value of $E = 10^{-2}$ m sec^{-1} seems appropriate.

Table 1

Calculated phase differences between three stations as functions of E.

E(m sec^{-1})	$\Delta\phi_{s-m}$(degrees)	$\Delta\phi_{w-s}$(degrees)
3.10^{-3}	0.41	0.07
6.10^{-3}	0.75	0.12
1.10^{-2}	1.17	0.19
$1.5\ 10^{-2}$	1.69	0.26
3.10^{-2}	3.24	0.46
5.10^{-2}	5.29	0.72

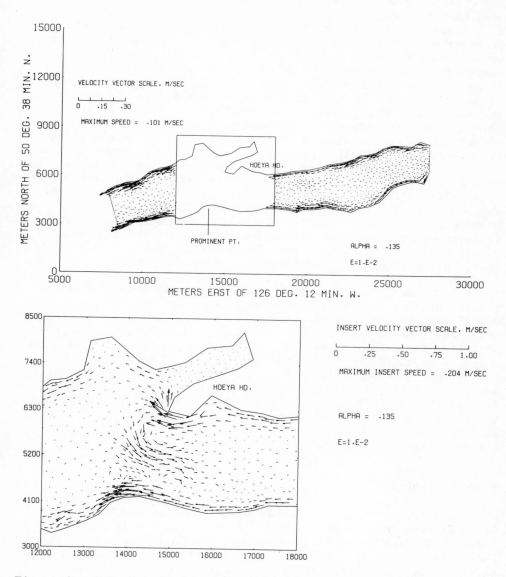

Figure 3. Velocity field in the straight reach of Knight Inlet
shortly after high tide. The lower diagram (corresponding to the inset on the upper diagram) shows detail
close to the sill. Note that the scales of the velocity
vectors are different in the two diagrams.

The depth-mean velocities, computed with the same value of E, agree
well with preliminary estimates (Freeland, personal communication)
of the barotropic tidal velocities inferred from current measurements at two stations in deep water. At the first station, close
to the western open boundary of the refined grid (Fig. 1), the
observed barotropic velocity at the M_2 frequency is approximately
14 cm sec^{-1} and precedes the elevation by 87^0; the computed magnitude and phase are about 15.5 cm sec^{-1} and 85^0 respectively. At
the second station, located in the vicinity of Tomakstum Island,
both the observed and the calculated M_2 velocity have a magnitude
of about 9 cm sec^{-1}. Due to the steepness of the fjord's sides
the dissipation term also gives rise to significant lateral shear
and phase variations that result in the formation of eddies at the
turn of the tide. This is illustrated in Figure 3 which shows the
tidal current pattern shortly after high tide, computed with the
same value of E. The computations were performed on a Cray-1
computer. For $E=10^{-2}$ m sec^{-1}, the coarse grid solution converges
(relative change $\leq 10^{-4}$ at all modes) after 10 iterations and
requires 6.7 seconds of CP time; the solution on the refined grid
takes only 2 iterations and 1.7 CP seconds.

It is our intention to describe in greater detail in a subsequent publication the computational procedure and the results
summarized above. At that time, we intend to discuss other topics
which would be of interest to applied mathematicians and numerical
analysts; these include certain complications which arise when the
Coriolis acceleration is included, questions of alternative conditions on the open boundary, and the convergence of the iterative
scheme. The tentative conclusion drawn from the work presented
here is that the simple linearized model of Jamart and Winter may
be capable of reproducing the main features of barotropic tides
in a deep fjord, such as Knight Inlet, regardless of the exact
nature of the internal dissipation processes.

ACKNOWLEDGEMENTS

This research was supported in part by the Washington Sea
Grant Program R/NE-8, and in part by the Oceanography Section,
National Science Foundation, under Grant OCE76-80720. The computing
facilities at the National Science Foundation, were made available
under Project 35371006. We are grateful to Ms. Joyce M. Lehner
for her assistance in preparing the figures.

REFERENCES

Freeland, H.J. and D.M. Farmer. 1979. The circulation and ener-
    getics of a deep, strongly stratified inlet. Manuscript
    submitted for publication.

Jamart, B.M. and D.F. Winter. 1978. A new approach to the
    computation of tidal motions in estuaries. In: Hydrodynamics
    of Estuaries and Fjords, J.C.J. Nihoul (editor), Elsevier
    Oceanography Series, 23: 261-281.

Pearson, C.E. and D.F. Winter. 1977. On the calculation of tidal
    currents in homogeneous estuaries. Journal of Physical
    Oceanography, 7(4): 520-531.

THE DISTRIBUTION OF ZOOPLANKTON COMMUNITIES IN A GLACIAL RUN-OFF

FJORD AND EXCHANGES WITH THE OPEN SEA

David P. Stone

Department of Oceanography
University of British Columbia
Vancouver, Canada

INTRODUCTION

Fjords are of particular interest to the biological oceanographer, since they can be considered discrete oceanographic units within which plankton distributions can be studied more easily than in the open sea. The relationship between each individual fjord's plankton community and that of the adjoining ocean is an additional topic of interest and is the subject of this short paper.

Knight Inlet is a glacial run-off fjord situated approximately 350 km. to the north west of Vancouver, Canada. The inlet penetrates for some 110 km. into the mainland coast mountains. Communication with adjoining marine waters is through Queen Charlotte Strait, which itself meets the Pacific Ocean 60 km. to the northwest (Fig. 1). Two shallow sills of approximately 65 m. depth partition the fjord into two basins. The outer is shallow, with a mean depth of 150 m., whilst the inner extends to over 500 m. (Fig. 2). General aspects of the fjord's physical oceanography have been discussed by Pickard (1959, 1961 and 1975), whilst recent studies have done much to elucidate features of water movement over the inner sill (e.g., Farmer this volume). Relationships between spatial and temporal patterns in zooplankton distribution, the distribution of chlorophyll, and the characteristics of hydrographic circulation in the fjord were described by Stone (1977) after ten cruises to Knight Inlet and Queen Charlotte strait between October 1974 and September 1975. During this study, an unusual planktonic community with off-shore characteristics appeared in Queen Charlotte Strait. Its composition, and possible origin and fate are discussed here.

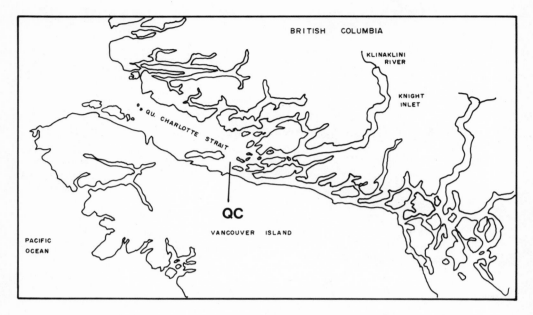

Figure 1.   Northern Vancouver Island and the study region in the
            vicinity of Knight Inlet.   The location of station QC
            is indicated.   Adapted from Pickard (1956).

METHODS

     Vertically discrete zooplankton samples were taken by horizon-
tal tows using modified Clarke–Bumpus nets of 350 μm mesh.   Each
was fitted with a calibrated flow meter.   Standard depths were at
5, 10, 30, 50, 100, 200, 300, and 500 m.   Stations were located
approximately 16 km apart along a transect from station QC in Queen
Charlotte Strait to the head of Knight Inlet.   All samples were
identified and enumerated to the species and instar level.   Salinity,
temperature, oxygen, nitrate and chlorophyll-a data were collected
concurrently.

RESULTS

     The intrusion of offshore water into the study area was first
observed in April in Queen Charlotte Strait at 100 m depth with an
increase in salinity and a fall in dissolved oxygen concentration.
These characteristics are indicative of upwelling along the outer
Pacific coast.   Such events are the results of wind-driven

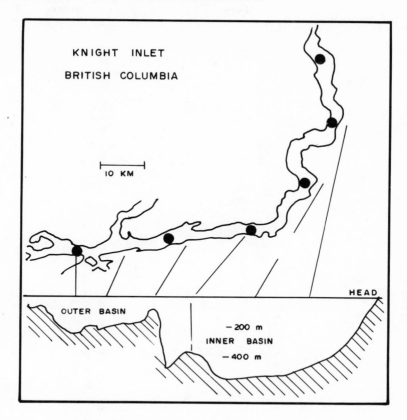

Figure 2.   The study area, Knight Inlet. The upper figure shows
            the inlet and location of sampling stations. The lower
            figure is a longitudinal profile along the inlet showing
            bottom topography. Adapted from Pickard (1956).

divergence, and occur annually at intervals throughout the summer
(Dodimead et al., 1976). An indication of intensity from month
to month at a station near the mouth of Queen Charlotte Strait can
be obtained from indices calculated from atmospheric data by Bakun
(Table I). The presence of high salinity/high density water in
Queen Charlotte Strait in April created a density gradient across
the outer sill, and an inflow occured at depth into the outer basin
of Knight Inlet. This inflow persisted throughout the summer
(although it was probably not continuous) and by July the intrusion
had begun to occupy the deeper parts of the inner basin.

    In June, three species of copepod which had been absent from
all previous samples during the study, were collected at 100 m

Table I (a)

Monthly coastal upwelling indices at a station located at 51^0N 131^0W, for the years 1972 to 1975. Units are cubic metres of water per second per 100 m length of coast. Negative values indicate onshore transport of surface waters and resultant downwelling. Abridged from Bakun's unpublished data.

YEAR	JAN	FEB	MAR	APR	MAY	JUN	JUL	AUG	SEP	OCT	NOV	DEC
							MONTH					
1972	2	-59	-25	3	2	6	17	4	38	9	-91	-7
1973	-75	-50	-3	5	-20	-15	5	40	-6	-19	2	-81
1974	5	-37	-6	-9	-3	7	8	44	3	-35	-42	-62
1975	-9	-72	2	17	-1	38	6	7	7	-24	-35	-30

Table I (b)

Mean monthly values of coastal upwelling indices at a station located at 51^0N 131^0W for the 20 year period 1948-1967. Units as in Table I (a). Abridged from Table 3 in Bakun, 1973.

JAN	FEB	MAR	APR	MAY	JUN	JUL	AUG	SEP	OCT	NOV	DEC
-64	-36	-12	-5	4	15	16	12	-3	-40	-58	-57

depth in Queen Charlotte Strait. They were *Rhincalanus nasutus*, *Calanus cristatus* and *Gaetanus intermedius*. By the following month, these immigrants were also present below 150 m in the outer inlet basin, where they remained until the study terminated in September. In July, the group of unusual immigrants had swelled to include 37 species of copepod at the Queen Charlotte Strait station. Table II shows that a large proportion of these species are not regarded as characteristic of local waters, and are more indicative of an Oregon or Californian fauna. The immigrants declined rapidly in numbers and diversity during the following months, and by September, only 18% of the original species total remained. Some species were advected into the deeper parts of the outer basin where survival was slightly better with 23% of the species remaining in September. The species which had previously been recorded in the local subarctic Pacific, like *Pleuromamma quadrungulata*, *P. ziphias*, *P. abdominalis*, *Euchirella rostra*, *Calanus cristatus* and *Rhincalanus nasutus* were those with the highest survival. In contrast, many of the more southerly species, such as *Gaetanus miles*, *Gaetanus pileatus*, *Paraeuchaeta californica*, *Metridia boecki*, *Metridia princeps* and *Heterorhabdus spinifrons* were present only in the July Queen Charlotte Strait sample. None of the immigrant species were observed in the inner basin of Knight Inlet, and none were found in the area a year later (Gardner, pers. comm.).

DISCUSSION

Speculation on the origin of the immigrant species assemblage requires a consideration of the local offshore oceanography. This is dominated by the West Wind Drift which flows across the Pacific in an easterly direction and diverges as the North American coast is approached at a latitude approximate to that of the Strait of Juan de Fuca. One branch flows northwards off the British Columbian coast as the Alaska Current, and carries a fauna which is typically sub-arctic (McGowan 1971). At the surface in winter, a current moves northwards close to the coast (Davidson Current), but this disappears for the duration of the upwelling season from April until September (Bakun 1973; Dodimead et al. 1976). A non-seasonal northward flow at approximately 200 m depth and known as the California Undercurrent has also been detected from geostrophic calculations to as far north as Vancouver Island (Reed and Halpern 1976). These observations also suggest the presence of cyclonic eddy-like features off the Strait of Juan de Fuca, where considerable modification may occur due to mixing with ambient sub-arctic water. The presence in Queen Charlotte Strait of a copepod fauna characteristic of California waters would suggest that the California undercurrent was the most likely mode of transport, and that on occasion, this current extends to at least the northern tip of Vancouver Island. The extremely high diversity of the immigrant assemblage

Table II

The Offshore Species Group: A list of all calanoid copepods thought to be characteristic of an off-shore fauna, collected in Queen Charlotte Strait (station QC) and the Outer Basin of Knight Inlet, from June until July 1975. The complete list gives occurrence in Queen Charlotte Strait only. Additional presence in the Inlet Outer Basin is indicated by an asterisk in the relevant column. A brief resume of records in the N.E. Pacific is given in remaining columns. Shih et al. (1971) reviewed all records in British Columbian coastal waters. Other authors were concerned with off-shore water adjacent to: Washington (Davis 1949), Oregon (Peterson 1972; Pearcy 1972), and California (Esterly 1905, 1906, 1911, 1913 1924: Fleminger 1964, 1967). Miscellaneous reports are given in the right-hand column. A bracketed asterisk indicates possible taxonomic difficulty.

FAMILY	SPECIES	OUTER BASIN	SHIH	DAVIS	PEARCY	PETERSON	FLEMINGER	ESTERLY	OTHERS
Calanidae	*Calanus cristatus* Kroyer	*	*	*	*	*	*		
Eucalanidae	*Rhincalanus nasutus* Giesbrecht	*	*	*	*	*	*	*	
Aetideidae	*Gaetanus intermedius* Campbell	*	(*)	(*)				*	
	Gaetanus pileatus Farran								0-35°N (Brodskii 1950)
	Gaetanus miles Giesbrecht	*	*	*	*	*	*		
	Euchirella rostrata Claus	*	*	*	*	*	*		
	Euchirella pseudopulchra Park	*	(*)	(*)	(*)	(*)	(*)	(*)	Always north of 26°N (Fleminger 1967)
	Euchirella curticauda Giesbrecht			*	*			*	
	Chirundina streetsi Giesbrecht			*	*				
	Undeuchaeta bispinosa Esterly				(*)	*		*	
Euchaetidae	*Euchaeta media* Giesbrecht					*	*	*	Central N. Pacific (Park 1968)
	Euchaeta spinosa Giesbrecht				*				
	Paraeuchaeta californica Esterly							*	
Scolecithricidae	*Scottocalanus persecans* Giesbrecht				*	*	*		
	Lophothrix frontalis Giesbrecht				*	*			
	Scaphocalanus magnus T. Scott				*	*			
	Scolecithricella ovata Farran			*	*	*			
Metridiidae	*Metridia boecki* Giesbrecht						*	*	
	Metridia princeps Giesbrecht	*						*	Oyoshio (Morioka 1972)
	Pleuromamma abdominalis Lubbock	*	*	*	*	*	*	*	
	Pleuromamma xiphias Giesbrecht	*		*	*	*	*	*	
	Pleuromamma borealis Dahl				*		*	*	
	Pleuromamma quadrungulata Dahl								
	Pleuromamma scutullata Brodskii				*				Oyoshio (Morioka 1972)
	Caussia princeps T. Scott			*	*		*	*	
Lucicutiidae	*Lucicutia bicornuta* Wolfenden				*				Central N. Pacific (Park 1968)
Heterorhabdidae	*Disseta maxima* Esterly							*	
	Heterorhabdus papilliger Claus				*	*	*	*	
	Heterorhabdus clausi Giesbrecht							*	
	Heterorhabdus spinifrons Claus							*	Kuroshio (Morioka 1972)
	Heterostylites longicornis Giesbrecht				*				
Augaptilidae	*Haloptilus oxycephalus* Giesbrecht				(*)	*			
	Centraugaptilus porcellus Johnson				*				Off California (Johnson 1936)
	Pachyptilus pacificus Johnson								Off California (Johnson 1936)
Arietelliidae	*Arietellus plumifer* Sars				*				
	Phyllopus integer Esterly							*	
Candaciidae	*Candacia bipinnata* Giesbrecht	*	*	*	*	*	*	*	

as observed in Queen Charlotte Strait, was largely due to the peculiar mixture of sub-arctic and warm temperate offshore species, and was possibly caused by such mixing processes as those described off southern Vancouver Island. The unique nature of the July Queen Charlotte Strait plankton assemblage in 1975 may partly reflect some avoidance in the past of the tedious task of taxonomically working through plankton samples. However, it seems more likely that the event was unusual. Monthly samples from Queen Charlotte Strait were available for 1974, and I found no zooplanktons of any taxon which could be considered characteristic of offshore water. There is some indication that 1975 may have been an unusual upwelling year. Bakun (1973) has described a method to estimate indices of intensity of wind-induced upwelling. Bakun's unpublished indices

for 1975 indicate that the most intensive upwelling would have been
expected in June, a month prior to the unusual plankton observation
in July. In contrast, Bakun's indices for immediately preceeding
years (1972-1974) indicate August (and September in 1972) as the
month of most intensive upwelling. The observed zooplankton
assemblage in Queen Charlotte Strait in July 1975 therefore probably
resulted from a peculiar combination of hydrographic events which
involved the California Undercurrent, the mixing of the latter with
sub-arctic water, and the occurrence of wind-induced upwelling.

REFERENCES

Bakun, A. 1973. Coastal upwelling indices, west coast of North
    America, 1946-1971. N.O.A.A. Tech. Report No. NMFS-SSRF-671.

Dodimead, A.J., F. Favorite and K. Nasu. 1976. Review of ocean-
    ography of the sub-arctic Pacific Region. 1960-1971. Bull.
    33. Int. N. Pac. Fish. Comm. 187 pp.

McGowan, J.A. 1971. Oceanic biogeography of the Pacific. Pages
    3-74 in Funnel, B.M. and W.R. Riedel (eds.), The micropaleon-
    tology of the oceans. Cambridge University Press.

Pickard, G.L. 1961. Oceanographic features of inlets in the
    British Columbia mainland coast. J. Fish. Res. Board Canada.
    18. 907-999.

Pickard, G.L. 1975. Annual and longer term variations of deep
    water properties in the coastal waters of southern British
    Columbia. J. Fish. Res. Board Canada. 32. 1561-1587.

Pickard, G.L. and G.K. Rodgers. 1959. Current measurements in
    Knight Inlet, British Columbia. J. Fish. Res. Board Canada.
    16. 635-678.

Reed, R.K. and D. Halper. 1976. Observations of the California
    undercurrent off Washington and Vancouver Island. Limnol.
    and Oceanogr. 21. 389-398.

Stone, D.P. 1977. Copepod distributional ecology in a glacial
    run-off fjord. Ph.D. Thesis. Institute of Oceanography
    and Department of Zoology, University of British Columbia,
    256 pp.

AN UNUSUAL POLYNYA IN AN ARCTIC FJORD

H.E. Sadler and H.V. Serson

D.R.E.P. Esquimalt
Fleet Mail Office
Victoria, B.C. VOF 1BO
Canada

INTRODUCTION

M. Dunbar (1957) called attention to the existence of a small circular polynya about 50 m in diameter in Cambridge Fiord in northern Baffin Island for which there was no obvious explanation. Other small polynyas are known in arctic fjords which are usually the result of turbulent mixing in areas of strong currents (Sadler 1974), but Cambridge Fiord is 100 km long, its tidal range is small, and the polynya is situated within 300 m of the delta face at the head of the fjord so that strong turbulence is very unlikely. The annual reappearance of the polynya in late winter is confirmed by a series of aerial survey photographs taken by the Royal Canadian Air Force between 1952 and 1957 and also by reports from Inuit hunters from Pond Inlet. It is first seen within about two weeks of 15 March appearing in exactly the same position each year as a circle with a diameter of about 40 m. Over a period of about a week, a lead extends from the polynya to the shore and open water is visible in the tide crack for several hundred metres either side of the shore end of the lead. The lead, unlike the polynya, changes its position from year to year, but once formed it remains fixed until general break-up (Figure 1).

1976 VISIT

On 16 May 1976, accompanied by Moira Dunbar, we spent a few hours making preliminary observations of the polynya, although with little equipment because of the limited load capacity of the DHC

Figure 1. A section of an aerial photograph of the end of Cambridge
           Fiord taken 30 April 1952. The polynya is about 60 m in
           diameter and open water can be seen all along the tide
           crack. In 1976 the outlet stream began at the upper
           edge of the polynya and ran into the upper left corner
           of the shore line. The mouth of the seasonal stream from
           the lake is shown at the bottom. The area shown is about
           1 km square.

Beaver aircraft. We found the polynya well developed, with a
diameter of 50 m and an annulus of snow and slush ice about 10 m
wide around it. Upwelling at the surface of the open water was
obvious to the eye, as was the outflow of water along the lead
towards the tide crack in the south west corner of the fjord. The
2 m wide stream ran under a raised bridge of ice, which will be
described later, and into the tide crack, in which open water
could be seen for some 500 m west of the junction. To our surprise,

we found that the water at this point was fresh to the taste.

Making a cautious approach to the edge of the firm ice, whose thickness was not measured, we lowered Knudsen sample bottles from a small hand powered winch through a hole pushed in the slush ice of the annulus. The temperature and salinity profiles were typical of arctic fjord structures, with a small halocline between 15 and 20 m (from 32.34 to 32.45 ppt) and only slight indications of fresher, warmer water above 5 m. The water depth was 48 m.

DISCUSSION

The phenomenon can be explained as the result of a fresh water spring in the bed of the fjord, the water rising to the surface in a plume under the influence of buoyancy forces amounting to 3% of the weight. The single profile obtained indicates that there can be little or no entrainment or surface spreading, while the existence of the polynya, stream and open water in the tide crack requires that the water temperature be initially about 4^0 C. The origin of such fresh water is most likely to be the deep layers of a fresh water lake and we found the probable source after taking off, when flying over a lake about 5 km from the end of the fjord. This lake has an area of about 1.5 km^2 and it is drained by a seasonal stream, shown in Figure 1, which enters the fjord about 0.6 km east of the polynya. The lake had obviously lost water since the surface run-off ended, as the ice cover sagged all around the shores and was draped over shallow spots in the bed. We found the same conditions in a stereo pair of photographs taken in 1952 and it is clear that the surface level of the lake falls by one or two metres every winter. This was confirmed by Inuit hunters from Pond Inlet (Fr. Guy Mary-Rousillière 1979). Dunbar (1957) observes that the fjord cuts into the uplifted edge of the Canadian shield to the central Baffin Island plateau and that minor faults in the gneissic rock are common.

Assuming that the vertical Richardson number is sufficiently large to avoid entrainment in the plume, it can be shown that the mechanism is sufficient to account for the existence of the polynya. In a series of photographs taken between 7 March and 21 May 1952 we found that the diameter of the polynya increased linearly from 40 m to 62 m, the mean area in March being 1800 m^2. Muller (1977), refining the methods of Walmesley and of Orvig, has estimated the net heat loss over the North Water, the large polynya in northern Baffin Bay, and assuming similar winds and air temperatures and allowing for shading by the high ground to the south west, the monthly loss of heat through the polynya in Cambridge Fiord reaches a maximum in March of 1.1 x 10^{12} J. Allowing for a drop in temperature in the polynya itself of 1^0 C (i.e. assuming a further

drop of 3^0 C in the stream and the tide crack) the maintenance of the
polynya requires a flow of about 0.1 $m^3 \cdot s^{-1}$ while the increase in
its size during the month calls for another 0.04 $m^3 \cdot s^{-1}$. A constant
flow of 0.14 $m^3 \cdot s^{-1}$ between September and the end of May would entail
a loss of 3.4×10^6 m^3 of water from the lake, equivalent to lowering
the level by about 2 m.

   Two questions remain; what controls the location of the drainage
stream and why does the polynya appear so suddenly in March? The
location of the stream is a function of the wind direction when the
first ice scum begins to form in September, the warmer fresh water
remaining unfrozen as it drifts down wind and becoming channelled
as the ice thickens. As a normal stream cuts a channel down into
an uplifting plateau, so does the fresh water cut a channel up
into the growing ice, eventually "falling" up into the tide crack.
By midwinter the stream and the polynya itself are covered by a
thin layer of ice and snow cover makes them invisible from the air,
but with the decreased radiation loss in the spring the ice cover
melts through from below, the snow cover subsides into the water
and the polynya becomes visible. The stream being narrower retains
its snow bridge longer and appears to grow out to the tide crack over
a period of a week or so.

A POSSIBLE APPLICATION

   A big problem when shipping heavy cargoes such as ore or
natural gas from ports in the arctic is to keep open an area alongside a jetty or dolphins so that ships may come alongside for
loading. It is difficult or impossible to do this using icebreakers
since the broken floes themselves become a hazard. Using conventional
heat sources such as live steam at relatively high temperatures is,
in apparent contradiction to the laws of themo-dynamics, very
inefficient. The objective is to heat a large quantity of water
evenly by one or two degrees while the effect of high temperature
sources is to heat much smaller quantities by several tens of degrees,
leading to uneven heating and large Newton cooling losses.

   We suggest that if a suitable lake can be found or created
there becomes available a large amount of "low tension" heat
energy which, because of the buoyancy forces, can be applied right
at the water surface. Assuming that it would be necessary to keep
open an area 300 x 20 m in similar climatic conditions to those in
Cambridge Fiord, the flow of fresh water at 4^0C which would be
needed over a winter season would be less than 0.7 $m^3 \cdot s^{-1}$. The
total volume used, 1.3×10^7 m^3, involves a fall in the level of
a 4 km^2 lake of some 3 m, which is not an unreasonable requirement.
The flow would begin at first freeze up so that the fresh water can
cut its own drainage channel to the tide crack. If waste heat were

Figure 2. The seaward end of the ice bridge spanning the fresh
water outlet stream. The inner surface of the arch has
bird droppings and feathers frozen on. The surface
of the stream is here covered with a layer of slush
about 1 cm thick.

to be available it could be used to heat the fresh water before
discharge, thus reducing the flow rate, while the whole process
could be made automatic by using the feed-back temperature from the
water entering the tide crack to control the flow rate and the
waste heat input.

THE BRIDGE

We mentioned earlier that just outside the tide crack the
stream ran under a peculiar ice formation. This was in the form
of a hump-backed bridge, Figure 2, about 6 m long, spanning about
3 m and with an internal height of 2 m. The ice forming the arch
was 0.5 m thick and was bent smoothly into a perfect cylindrical
section with only slight fracturing at the crown at each end. The
inside surface was smooth and had bird droppings and feathers frozen
in at some distance above the water line. A similar formation appears
on some of the photographs from other years.

A plausible explanation of this unique feature follows from
the negative coefficient of expansion of sea ice over a range of

temperatures from -2^0C to -16^0C. Using values for the volume expansion given by Doronin and Kheisin (1975) for a mean salinity of 44 ppt., a fall in the ice temperature from -5^0C to -15^0C implies linear expansion of about 0.4 m. km^{-1}. This is sufficient to buckle the thin ice over the line of weakness provided by the stream at right angles to the shore and to raise it to a height of 2 m. If the expansion rate is slow enough the ice will bend in a plastic fashion and slow rise also accounts for the frozen-in droppings and feathers.

CONCLUSION

The proposed mechanisms for the formation of the polynya, the stream and the bridge raise theoretical problems, particularly in the apparent complete absence of entrainment of salt water in the rising plume. More comprehensive measurements in the field and more investigation of the dynamics of the inflowing water are necessary before a full explanation of these phenomena can be given.

REFERENCES

Doronin, Yu. P., and D.E. Kheisin. 1975. Sea Ice. Leningrad 1975. Translation published by Office of Polar Programs and National Science Foundation, Washington, D.C. 1977.

Dunbar, Moira. 1957. Curious open water feature in the ice at the head of Cambridge Fiord. Extrait des Comptes Rendu et Rapports, Assemblée Générale de Toronto, 1957. T.IV, pp. 514 a 519. Soc. Int. Géograph. (Gentbrugge 1958).

Mary-Rousillière, G. 1979. Personal Communication.

Müller, F. 1977. Report on North Water Project activities 1 Oct. 1975 to 30 Sept. 1976. ETH Zurich and McGill U. (Unpublished Report).

Sadler, H.E. 1974. On a polynya in Makinson Inlet. Arctic, 27(2) June 1974. AINA.

SEASONAL OBSERVATIONS OF LOW-FREQUENCY ATMOSPHERIC FORCING

IN THE STRAIT OF JUAN DE FUCA

J. R. Holbrook, R. D. Muench, and G. A. Cannon

National Oceanic and Atmospheric Administration
P.M.E.L., Seattle
Wa. 98105, U.S.A.

INTRODUCTION

Although the role of local wind forcing in modifying circulation in estuaries has received much attention over the past 15
years, the importance of non-local meteorological forcing has only
recently been recognized. Studies in Chesapeake Bay (Elliott, 1978;
Wang and Elliott, 1978; and Wang, 1978), Josenfjord (Svendsen, 1977),
the Strait of Juan de Fuca (Holbrook and Halpern, 1977; Laird, Cannon,
and Holbrook, 1978; and Holbrook and Halpern, 1979) and Resurrection Bay (Muench and Higgie, 1978) have shown that coastal winds can
be highly coherent with axial estuary currents and that coastal
Ekman dynamics may play a key role in generating up-estuary near-
surface current reversals. This report discusses observations which
illustrate the effects of seasonal variability in the local and non-
local wind field on low-frequency estuarine circulation in the
Strait of Juan de Fuca.

The Strait of Juan de Fuca (Figure 1) is the principal passage connecting the Pacific Ocean to the Strait of Georgia through
the channels of the San Juan Archipelago and to Puget Sound through
the shallow sill zones of Admiralty Inlet and Deception Pass. The
Strait contains an eastern and western basin separated by a cross-
channel 60 m deep sill extending southward from Victoria. The
western basin is of U-shaped cross-section with center-channel depths
increasing to 240 m at Neah Bay. Bottom bathymetry of the eastern
basin is more complex, with deep channels (> 100m) separating numerous banks and shoals.

305

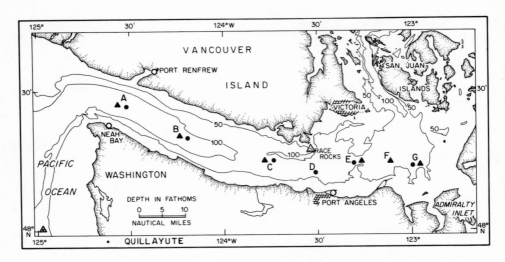

Figure 1.   Location diagram of Strait of Juan de Fuca showing bottom
            bathymetry and positions of surface (▲) and subsurface
            (●) moorings. The location at which coastal winds were
            computed by Bakun (1977) is also shown (⧌).

        Fresh water runoff enters principally via the Strait of Georgia
and Puget Sound, where circulation may be identified as a fjord
type. Prior to entering the Strait of Juan de Fuca these diluted
source waters are vertically mixed by large tidal speeds in the
channels leading into the eastern basin.

OBSERVATIONS

        3-4 month time series records of currents, winds, sea level,
atmospheric pressure and water temperature were obtained during
winter and summer experiments in the western (1976-7) and eastern
(1977-8) basins of the Strait of Juan de Fuca (Figure 1). AMF
Vector-Averaging Current Meters (VACMs) measured near-surface
(4-20m) currents and temperature from surface moorings located
along channel at sites A,B,C,E,F, and G. Modified VACMs measured
over-the-water winds at the same locations. Aanderaa current meters
measured current, temperature, and salinity at depths below 20 m
from taut-wire subsurface moorings at sites A thru G. Western and
eastern basin studies were not concurrent in time due to instrumentation limitations.

Ancillary time series of sea level records were obtained from permanent tide stations maintained by NWS at Neah Bay and Port Angeles and by Environment Canada at Port Renfrew. Winds off the Washington coast were computed at 48^0N, 125^0W from surface pressure data by Bakun (1977) and were assumed to represent the mesoscale wind field.

Annual records of Fraser River discharge and local and computed coastal winds are shown in Figure 2. During winter the Fraser River discharge, which accounted for 60-70% of the total runoff into the system, was low and wind variance was high with northward net coastal winds and weak net along-strait winds. During summer, river discharge peaked in June-July and wind variance was lower, with southward coastal winds and eastward along-strait winds modulated by a local sea-breeze effect.

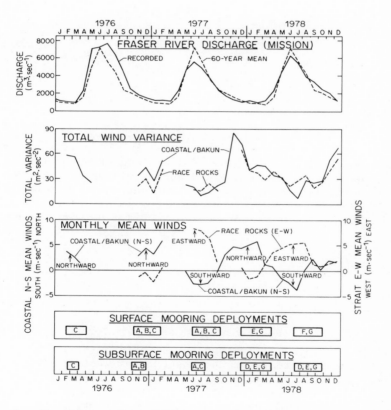

Figure 2. Monthly averaged Fraser River discharge, local (at Race Rocks) and non-local (computed by Bakun) winds during surface and subsurface mooring deployments.

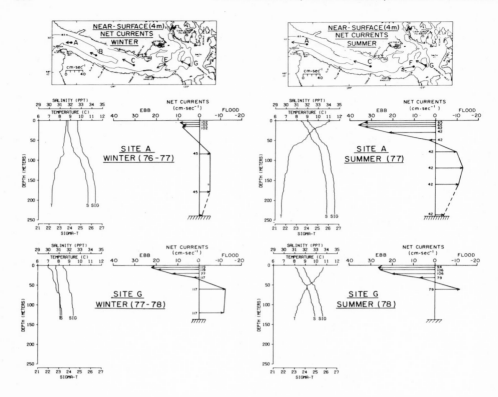

Figure 3.   Vertical distribution of net along-strait currents and
            density at sites A and G during winter and summer
            conditions. Near-surface net currents at 4m depth as
            measured from all surface moorings are shown at top.
            The level of no-net-motion occurred between 40-80m.

NET CIRCULATION

        A well developed two-layer estuarine circulation was measured
at all sites. Figure 3 illustrates the vertical distribution of
density and axial currents at the eastern and western ends of the
Strait during winter and summer. The level of no-net-motion occurred between 40-80m. During winter the net 4m flow decreased with
distance seaward from 31 cm/sec at site C to 10 cm/sec at Site A.
This decrease was due to the eastward reduction in amplitude and
frequency of near-surface current reversals generated by coastal
winds (Holbrook and Halpern, 1979). Net near-surface flow was less
variable (20-22 cm sec^{-1}) in the eastern Strait. During summer

vertical stratification and longitudinal density gradients were
stronger and, led to a more vigorous induced gravitational circulation, with near-surface currents ranging from 27-40 cm sec^{-1}.

LOW FREQUENCY VARIABILITY

Currents in the Strait of Juan de Fuca were extremely variable
in time. Current variance fell naturally into three energetic bands
which accounted for 86-90% of the total variance: subtidal
(>0.025 cph), diurnal (0.033-0.043 cph), and semidiurnal (0.076-
0.085 cph). Although diurnal and semidiurnal tidal fluctuations
accounted for most (65-88%) of the along-strait variance, significant variance (up to 27%) was found at subtidal frequencies during
winter. Some 80-85% of the along-strait subtidal variance occurred
at frequencies >0.0067 cph.

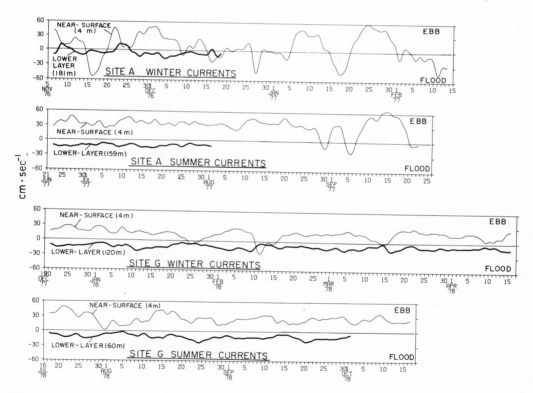

Figure 4.   Subtidal variability sites A and G are shown. Tidal and
higher frequency fluctuations have been removed by applying a A$^2_{25}$A$_{24}$ Godin low-pass filter.

Examples of subtidal variability are shown in Figure 4. Tidal
and higher frequency fluctuations have been removed by applying a
A_{25}^2 A_{24} Godin low-pass filter (Godin, 1973) which has a half-ampli-
tude frequency of 0.0143 cph. Between June and August, the near-
surface and lower layer currents exhibited less variability than
during September-April. The longer period fluctuations were generally 180^0 out of phase between the surface and deep layers. The
largest subtidal current fluctuations occurred at 4m at site A

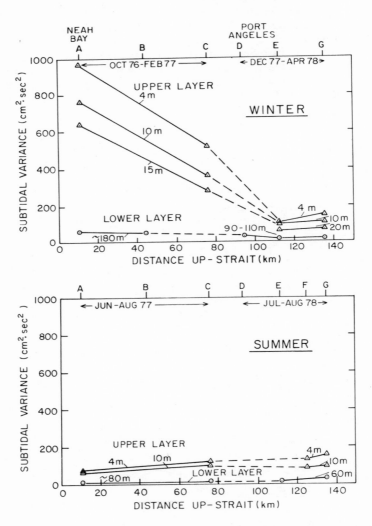

Figure 5.   Along-strait distribution of subtidal variance during
            winter and summer. Subtidal variance was primarily con-
            fined to the upper 35-40m of the water column in the west-
            ern strait during non-summer months.

where periods of subtidal current reversal (up-strait flow) accounted for 35% of the filtered record, lasted from 3-10 days, and had maximum up-strait speeds of 25-55 cm·sec^{-1}.

In the western strait, near-surface subtidal variance decreased with distance up-strait and with depth during winter (Figure 5). Subtidal along-strait variance at 4 m was 778 cm^2 sec^{-2} near the entrance at site A and 470 cm^2·sec^{-2} (40% less) 63 km up-strait at site C. It was greatest near-surface and decreased with depth at a rate of 25-30 cm^2·sec^{-2}·m^{-1} between 4-10 m and by 15-20 cm^2·sec^{-2} ·m^{-1} between 10-15 m. The subtidal variance was primarily confined to the upper 35-40 m of the water column. During summer in the western Strait and throughout the year in the eastern Strait, subtidal variance below 50 m was less than 200 cm^2·sec^{-2} and decreased gradually with depth.

Holbrook and Halpern (1974) have shown that subtidal fluctuations were poorly correlated with local wind forcing, but were well correlated with coastal wind forcing. Coastal winds along a 020-200^0 T axis explained (were significantly coherent with) 71 and 44% of the along-strait, near-surface current variance at sites A and C, respectively. Maximum near-surface current reversals at site C lagged site A by 62 hours, indicating an up-strait phase velocity of 28 cm·sec^{-1}. Periods of deceleration in the along-strait flow, prior to maximum reversal, were coincident with SSW coastal winds and rising sea level at Neah Bay and Port Renfrew (Figure 6). Figure 6 shows low-passed filtered time series of local winds at site B, near-surface and deep at sites A, B and C, vertical current shear between 4 and 10 m at sites A and C, computed geostrophic coastal winds (Bakun, 1977) off the Washington coast, corrected (with Quillayute atmospheric pressure) sea level at Port Renfrew and Neah Bay, and atmospheric pressure at Quillayute. Vertical lines are drawn during periods of maximum up-strait current reversal at Site A. Periods of deceleration in the along-strait flow, prior to maximum reversal, are coincident with SSW coastal winds and rising corrected sea level at Port Renfrew and Neah Bay. This suggests an Ekman flux mechanism may be responsible for the generation of these reversals which slowly propagated up-strait. Maximum up-strait reversals were accompanied by a reversal in the vertical current shear. Temperature measurements at site A, B, and C (not shown) demonstrated that warmer, fresher coastal water was advected up-strait during the reversals with a time lag giving an up-strait water particle velocity of 2-26 cm·sec^{-1}.

Although coastal low pressure systems with southwesterly winds occur more frequently and persist longer during winter than during summer, occasionally a low pressure system will dominate the coast long enough during summer to generate a response in the Strait. Such a system was observed in August 1978 (Figure 7). SSW winds associated with this system generated a near-surface up-strait current

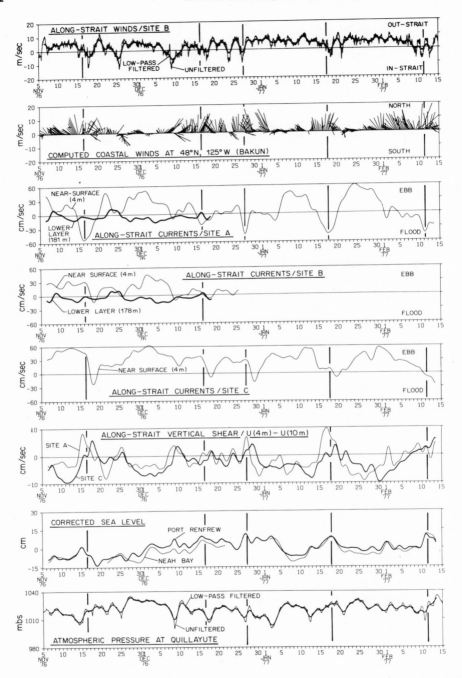

Figure 6. Low-passed filtered time series of winds, currents and
            vertical shear.

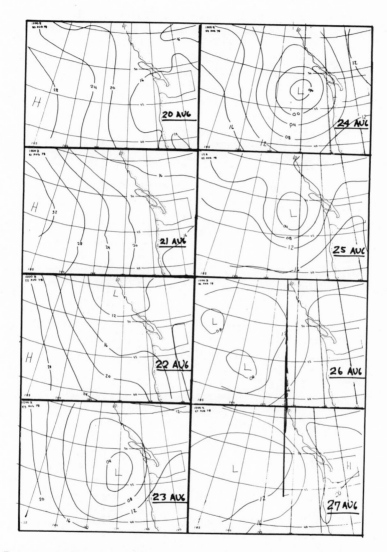

Figure 7. Surface pressure maps showing stationary low-pressure system centered 300–400 km offshore between 23–25 August 1978.

Table 1.   Summary of subtidal variance coherent with local and non-
           local winds. SSW–NNE coastal wind accounted for much of
           the large current variance observed near-surface during
           the winter in the western Strait. Local winds explained
           a smaller amount. Throughout the year, a background
           level of variance of 100–200 cm$^2 \cdot$sec^{-2} was directly
           related to neither local nor coastal winds.

$$\left. \frac{\text{COHERENT VARIANCE}}{\text{TOTAL VARIANCE}} \right\} \%$$

WINTER

	A	C	E	G
COMPUTED COASTAL WINDS Along 020–200°T At 48°N, 125°W	553 ⎱ 71% 779 ⎰	207 ⎱ 44% 469 ⎰	13 ⎱ 25% 50 ⎰	0 ⎱ 0% 94 ⎰
ALONG–STRAIT LOCAL WINDS	121 ⎱ 16% 779 ⎰	52 ⎱ 11% 469 ⎰	0 ⎱ 0% 50 ⎰	3 ⎱ 3% 94 ⎰

SUMMER

	A	C	F	G
COMPUTED COASTAL WINDS Along 020–200°T At 48°N, 125°W	3 ⎱ 6% 47 ⎰	0 ⎱ 0% 58 ⎰	0 ⎱ 0% 96 ⎰	0 ⎱ 0% 151 ⎰
ALONG–STRAIT LOCAL WINDS	0 ⎱ 0% 47 ⎰	7 ⎱ 12% 58 ⎰	– ⎱ – 96 ⎰	– ⎱ – 151 ⎰

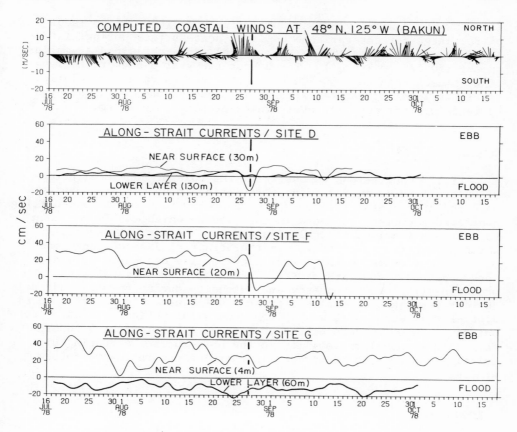

Figure 8. Southerly coastal winds associated with the offshore low-pressure system (Figure 7) generated an along-strait current reversal 125 km up-strait at site F. Maximum reversal at sites D and F occurred approximately 4 and 4½ days respectively, after the beginning of the strong southerly coastal winds.

reversal which was observed 4-6 days later in the eastern strait (Figure 8). The vertical line drawn at the time of arrival of the maximum reversal at site D illustrates the phase lag of this response relative to the coastal winds. The reversal was measured at site F 39 hours later than at site D, giving an up-strait phase velocity of 21 cm·sec^{-1} in the eastern basin.

SSW–NNE coastal winds accounted for most (200–550 $cm^2 \cdot sec^{-2}$) of the large current variance observed near-surface during winter in the western Strait (Table 1). Local winds explained a smaller amount (50–125 $cm^2 \cdot sec^{-2}$). Throughout the year a background level of subtidal variance (100–200 $cm^2 \cdot sec^{-2}$) had no apparent relation to either local or coastal winds. This background low-frequency variance was not coherent over scales greater than 13 km.

SUMMARY

Field observations demonstrated the role of non-local meteorological forcing in modifying circulation in an estuary opening onto the continental shelf. Winds conducive to coastal down-welling generated a current response in the estuary which was larger than the net gravitational circulation and lead to current reversals which propagated up-estuary. In the Strait of Juan de Fuca, the variance associated with coastal wind forcing decreased with distance up-strait and with depth. Since southerly coastal winds occurred with lower frequency and intensity during summer, subtidal current reversals were rare between June–August. In contrast local along-strait winds were weakly coherent with subtidal current fluctuations. Thus the seasonal nature of non-local forcing accounted for major seasonal variability in the near-surface estuarine circulation and must be considered in planning comprehensive future studies.

ACKNOWLEDGEMENT

This work was funded by the U. S. Environmental Protection Agency and managed by NOAA's Puget Sound Project office at the Marine Ecosystems Analysis (MESA) Program.

REFERENCES

Bakun, A. 1977: Personal Communication

Elliott, A. J. 1978: Observations of the meteorologically induced circulation in the Potomac Estuary. Est. Coastal Mar. Sci. 6, 285–299.

Godin, G. (1973): The analysis of tides. University of Toronto Press, Toronto, Canada. 264 pp.

Holbrook, J. R. and D. Halpern 1974: STD measurements on the Oregon coast, July/August 1973. CUEA Data Rept. 12, University of Washington, Seattle, 397 pp.

Holbrook, J. R. and D. Halpern 1977: Observations of near-surface currents, winds and temperature in the Strait of Juan de Fuca during November 1976-February 1977. Trans. of the American Geophysical Union, 53 (12), 1158 (Abstract).

Holbrook, J. R. and D. Halpern 1979: Observations of over-the-water winds and their correspondence to shore winds in the Strait of Juan de Fuca. In preparation.

Laird, N. P., G. A. Cannon and J. R. Holbrook 1978: Winter intrusions of coastal waters into the Strait of Juan de Fuca estuary. Trans. of the Amer. Geophys. Union, 59 (4), 305 (Abstract).

Svendsen, H. 1977: A study of the circulation in a sill fjord on the west coast of Norway. Marine Science Communications, 3 (2), 151-209.

Wang, D. P. 1978: Sub-tidal sea level variations in the Chesapeake Bay and relations to atmospheric forcing, J. Phys. Oceanogr. 9 (2), 413-421.

Wang, D. P. and A. J. Elliott 1978: Non-tidal variability in the Chesapeake Bay and Potomac River: Evidence for non-local forcing. J. Phys. Oceanogr. 8, 225-232.

Muench, R. D. and D. T. Higgie 1978: Deep water exchange of Alaskan subartic fjords. Estuarine Transport Processes, (B. Knirfoe, ed.), Univ. of S. Caroline Press, Columbia, S. C., 239-267.

VERTICAL STRUCTURE OF FLUCTUATING CURRENTS IN THE STRAIT OF JUAN DE FUCA

R. D. Muench and J. R. Holbrook

Pacific Marine Environmental Laboratory
NOAA, ERL, Seattle
Wa. 97105 U.S.A.

INTRODUCTION

The Strait of Juan de Fuca has been the focus for an intensive circulation study from 1976-79. During 1977-78 this study focused upon the eastern half of the Strait, which is the principal connection between the Strait of Georgia via channels through the San Juan Islands, Puget Sound via Admiralty Inlet and the Pacific Ocean via the western Strait of Juan de Fuca (Figure 1). Simultaneous observations of winds, currents, sea level and water parameters such as temperature and salinity were obtained to allow examination of the kinematics and dynamics of large-scale tidal and non-tidal flow. This paper utilizes time-series current observations to summarily discuss the vertical distribution of time-varying flow in the eastern Strait.

FIELD PROGRAM

Time-series current data were acquired at six locations (Figure 2) using Aanderaa current meters in a taut-wire mooring configuration for observations below 20 m. Vector-averaging current meters were suspended from surface arrays were used to obtain curent data above 20 m. The moorings were deployed during winter (December 1977-April 1978) and summer (May-September 1978); this paper utilizes results only from the winter moorings. Hydrographic surveys were conducted during the deployment periods using a profiling CTD to map the internal distribution of temperature, salinity and density. Wind data were acquired from recording anemometers mounted on the surface floats and from shore stations located about the periphery of the eastern Strait.

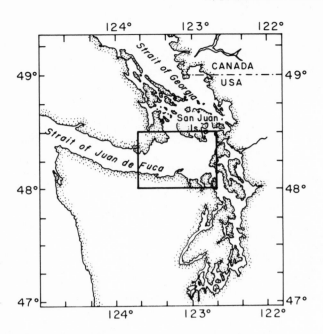

Figure 1.   Geographical location of the Strait of Juan de Fuca
region.   Study area is defined by the rectangle.

OBSERVATIONS

Representative vertical distributions of temperature, salinity
and density (as sigma-t) at Station 48 are shown for the start and
finish of the winter mooring period (Figure 3).   Vertical stratification in density did not fluctuate appreciably through the
winter at this location.   Salinity clearly exerted a dominant control over density; this was most pronounced in December 1977 when
the vertical distribution of temperature was nearly uniform despite
a pronounced density gradient.   There was no persistent vertical
density discontinuity, rather, there was a uniform top-to-bottom
gradient with smaller transient step-like features superposed.

Low-pass filtered currents observed in the upper and lower
layers during winter are summarized in Figure 2.   Stations 40, 41
and 43, were located off the mouths of Haro, Rosario Straits and
Admiralty Inlet, respectively, and reflect the influence of water
exchange through those channels.   We confine our attention to
stations 48, 49 and 42 since we are concerned primarily with flow
processes in the central eastern Strait and these stations were the

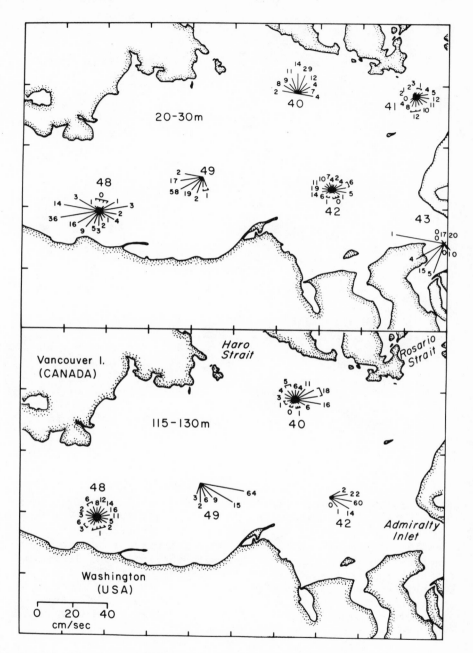

Figure 2. Low-pass (35 hour) filtered currents in surface and deep
          layers (upper and lower figures, respectively) during
          December 1977–April 1978.

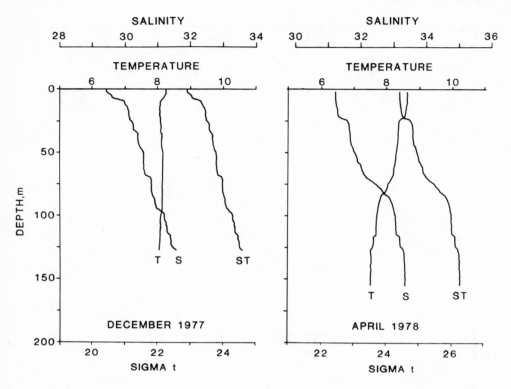

Figure 3.   Vertical profiles of temperature, salinity and density
            (as sigma-t) during December 1977 and April 1978 at
            Station 48.   Station location is shown on Figure 2.

most centrally located relative to the flow axis.   In the figure
lengths of lines indicate mean current speeds, and orientations of
lines indicate direction sectors in which currents were flowing.
The small numbers at the ends of lines are percentages of observations during which currents flowed in the indicated directions.
Large numbers are station identification numbers.   The upper layer
observations show a net outward (westerly) flow with superposed
scatter including reversals to inward (easterly) flow.   Axial flow
speeds were of order 20 cm sec $^{-1}$.   Each record also exhibited a
marked cross-channel flow component.

The near-bottom observations define a deep net inflow with
superposed directional fluctuations (Figure 2).   This supports the
concept of a mean "estuarine" flow in the Strait of Juan de Fuca,
with a net surface outflow and deep inflow (cf Cannon, 1978).

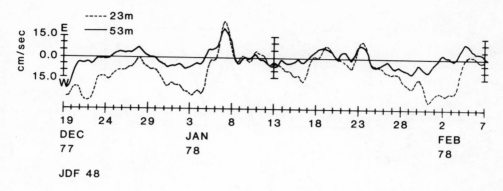

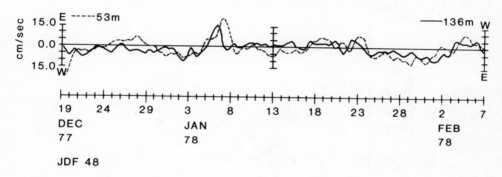

Figure 4.   Representative segment of east-west low pass filtered
            current time series at Station 48 comparing flow within
            surface layer and surface to deep layer flow (upper and
            lower figures, respectively). Lower graph has deep
            currents reversed in sign to more clearly show the
            correlation.

Intermediate depth current observations obtained over the same
period (not shown) suggest that the depth of no net motion fluctuates
and can fall anywhere between 40 and 80 m (cf. Holbrook, *et al.*
elsewhere in this volume).

    Axial flow, approximated by east-west flow, in the eastern
Strait of Juan de Fuca was vertically correlated within the upper
layer (Figure 4). Low frequency (below about .013 cph) fluctuations
were well correlated between the upper and lower layers but were
approximately 180^0 out of phase between these layers; e.g. an east-

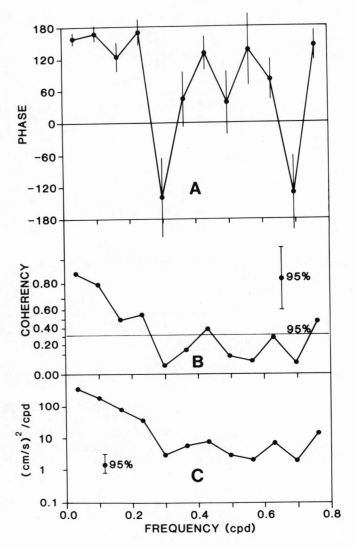

Figure 5. Phase (A), coherency (B) and energy cross-spectra (C) comparing surface and deep layer east-west flow at Station 48 during December 1977-April 1978.

ward pulse in the lower layer coincided with a westward pulse in the upper layer, and vice versa. Higher frequency events were not correlated between the upper and lower layers. The visual correlation observed on the time series plots was substantiated by coherence analyses carried out on upper and lower layer east-west

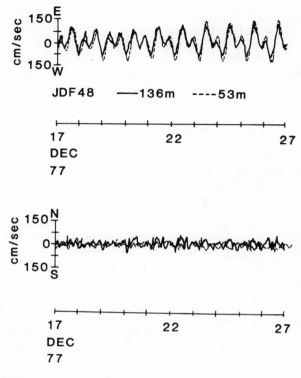

Figure 6. Representative segments of unfiltered east-west (along channel) and north-south (cross-channel) current components (upper and lower figures, respectively) at Station 48.

flow components (Figure 5). Vertical coherencies at frequencies below about 0.013 cph were above the 95% significance level and had phase differences of 160-170^0. At frequencies above 0.013 cph coherencies were not generally significant at the 95% level. This was probably due primarily to lack of variance energy in this band and a consequent low signal to noise ratio.

About 90% of the current variance energy fell in tidal frequency bands. Axial components of the tidal currents were vertically highly coherent and in phase through the depth range of our observations (Figure 6). Cross-strait tidal components contained far less energy than the axial components and were at times 180^0 out of phase between the upper and lower layers. While the axial currents contained a dominant semidiurnal tidal signal, the cross-strait currents contained primarily higher tidal harmonics and

little or no semidiurnal signal. Visual inspection of the unfiltered
current time series (Figure 6) suggests that the M_4 component was
dominant in cross-channel flow and that it was this component which
exhibited vertical phase reversal. Because of the low cross-channel
signal to noise ratio however, coherency analyses yielded ambiguous
results.

DISCUSSION

Two observed points deserve comment here: 1) the phase difference between upper and lower layer low-frequency axial current
fluctuations; and 2) the vertical phase reversal in higher order
cross-channel tidal harmonics.

The phase reversal between upper and lower layer axial currents
can be examined in light of probable forcing. Vertical reversal
in mean flow is a consequence of the estuarine circulation mode,
with an outflowing upper, dilute layer underlain by an inflowing
layer of marine source water. A pulse of seaward flow in the upper
layer might lead to an inward pulse in the lower layer by the same
mechanism - vertical entrainment flow coupled with volume continuity.
The cause of the low frequency fluctuations is however not clear.
It is shown by Holbrook *et al* (elsewhere in this volume) that local
wind forcing accounts for little of the observed flow variation,
but that coastal winds to the west may exert a significant influence.
Information on conditions to the west along the coast are presently
insufficient to rigorously examine these factors.

The cause of vertical phasing of the higher order cross-channel
tidal components is likewise not readily apparent. Forrester (1970)
observed a similar phenomenon in the St. Lawrence estuary and developed a theoretical model which explained the phasing by application
of wave theory to a layered fluid. His results depended upon a
vertical phase lag in the axial tidal component, which was not the
case in the Strait of Juan de Fuca. Moreover, the Strait of Juan
de Fuca is not vertically layered, but rather stratified more or
less uniformly with depth. Mean surface to bottom Vaisala frequency
was of order 0.013 cph, or two low for resonant interaction with
tidal components.

Analysis of these features is complicated both by an extremely
complex topography and by spatial variations in vertical stratification. Application of either analytical or numerical models to
obtain, for example, normal modes for internal motions at other than
a fixed vertical location becomes a non-trivial problem. These and
the problem of remote forcing for low frequency motions are critical
to an understanding of the system, and will be addressed in the
future.

ACKNOWLEDGEMENT

This work was funded by the U. S. Environmental Protection Agency and managed by NOAA's Puget Sound Project Office of the Marine Ecosystems Analysis (MESA) Program.

REFERENCES

Cannon, G. A. (Ed.) 1978: Circulation in the Strait of Juan de Fuca: some recent oceanographic observations. NOAA Tech. Rept. ERL 399-PMEL 29, 49 pp.

Forrester, W. D. 1970: Geostrophic approximation in the St. Lawrence estuary. Tellus XXII(1), 53-65.

Holbrook, J. R., R. D. Muench and G. A. Cannon 1979: Seasonal observations of low-frequency atmospheric forcing in the Strait of Juan de Fuca (elsewhere in this volume).

PHYSICAL OCEANOGRAPHY OF THE NEW ZEALAND FJORDS

B.R. Stanton and G.L. Pickard*

D.S.I.R., New Zealand Oceanographic Institute
Post Office Box 12-346, Wellington North
New Zealand

INTRODUCTION

The south-west coast of the South Island of New Zealand is
indented by a group of fjords extending from Milford Sound in the
north to Preservation Inlet in the south. The fjords are the
drowned lower reaches of valleys formerly occupied by glaciers.
They have typical fjord features, being narrow, steep sided and
having one or more submarine sills separating their relatively
deep basins from the Tasman Sea. The mountainous nature of the
land and its exposure to prevailing westerly weather systems results
in high rainfall and large freshwater runoff into the fjords.

The features and all 14 main fjords and most of the side arms
were studied in December 1977. Earlier data are available for some
of the fjords. Some fjords are simple single passages while others
are complex and inter-connected.

Water samples were obtained with N.I.O. Bottles; salinities
were determined by inductive salinometer and dissolved oxygen by
Winkler titration. Temperatures were determined with reversing
thermometers and bathythermograph.

FJORD STATISTICS

Fjord lengths range from 10 to 44 km, mean widths from 0.7 to

*Department of Oceanography, University of British Columbia,
2075 Wesbrook Mall, Vancouver, British Columbia V6T 1W5.

2.3 km, outer sill depths from 30 to 145 m, mean depths from 100
to 280 m and maximum depths to 420 m. Some fjords have inner sills.

Average annual precipitation at two stations near fjord heads
is 6200 mm and 5300 mm respectively and rainfall generally has
maxima in November and March. Rain plus snow melt give maximum
freshwater input during November to February. Only one river, the
Cleddau into Milford, is gauged. A considerable freshwater input
(gauged) to the head of Doubtful comes from Lake Manapouri through
a hydroelectric power station. In the Doubtful/Thompson Sounds
system, this outflow forms a major proportion of the whole freshwater input with little seasonal variation.

Tides are semi-diurnal with marked inequality and ranges are
from 1.6 to 2.3 m at springs.

SALINITY AND TEMPERATURE

There was generally a thin, surface layer of low salinity
water to about 10 m depth, a strong halocline and then relatively
uniform deep water. Most temperature variations were in the upper
10 m.

In six fjords the surface salinity varied from 5 to 10 $^0/_{00}$ at
the head rising to 25 to 33 $^0/_{00}$ at the mouth, while in the others
the salinity ranged from 26 to 32 $^0/_{00}$ at the head to 30 to 34 $^0/_{00}$
at the mouth. None of the fjords had the well-mixed surface layer
over a sharp halocline found in many American fjords but in many
there was an upper moderately low salinity layer in which the salinity increased slowly with depth to 2 to 5 m with a main halocline
started at the surface as nearly as could be determined. There was
a good negative correlation between the mean salinity in the upper
10 m layer and the fresh-water runoff into the fjord estimated from
watershed areas and Cleddau River runoff which was assumed typical
for the coast.

In the deep zone the six northern fjords showed a salinity
maximum of about 35.06 $^0/_{00}$ at around 100 m the salinity then
decreasing to about 35.02 $^0/_{00}$ at the bottom. In the southern ones
the salinity increased steadily to the bottom where the salinity
was 35.0 $^0/_{00}$ at mid-coast decreasing to 34.3 $^0/_{00}$ at the most
southerly fjord, Preservation. Small variations of salinity (less
than 0.1 $^0/_{00}$) between basins and side arms were observed in some
fjords.

Surface temperature showed a range of 1^0C or less along each
fjord and most values were between 12 and 14^0C. The vertical
profiles were usually a mirror image of the salinity profiles with a

thermocline coinciding with the halocline. In some cases there was a thin surface layer (2 to 4 m) of cool water. Below the thermocline, the temperature decreased steadily; the range from upper to deep water was 2^0 to 3^0C for the northern fjords and 1 to 2 C for the southern ones.

Transverse sections for temperature and salinity in the upper 20 m were worked across Thompson, Bradshaw and Doubtful Sounds but no systematic variations were observed.

DENSITY

The density range from surface to bottom was much the same for the six northern fjords with maximum values (for σ_t) of about 26.7 kg m^{-3}. The deep water values decreased southward in the southern group to 26.2 kg m^{-3} in Preservation, except for Chalky, immediately north of Preservation, which had a maximum value of 26.6 kg m^{-3}.

T,S CHARACTERISTICS

The T,S characteristic curves were similar for the six northern fjords showing the major effect of salinity changes on density changes in the upper 50 to 100 m while below that depth temperature changes had the major effect. In the southern group, salinity changes predominated in determining density changes at all depths. The set of curves for the fjords show the trend toward lower temperatures and salinities toward the south, particularly from mid-coast.

T,S characteristics offshore of the middle and north fjord coast showed a marked salinity maximum of 35.1 ^{0}/$_{00}$ at 100 m depth, decreasing through 34.7 ^{0}/$_{00}$ at 400 m. Other offshore data indicate that this maximum is less marked or disappears further south.

DISSOLVED OXYGEN

Dissolved oxygen values at the surface were at or above saturation at values of 6 to 7 mL/L, decreasing in the deep water to 5 to 6 mL/L. Three side arms showed lower values of 4.5, 4.2 and 3.0 mL/L and in Deep Water Basin, a lagoon-like basin at the head of Milford, a value of 0.8 mL/L was measured at the bottom. Offshore values (not measured in 1977) are generally between 5 and 6 mL/L from the surface to 400 m.

CONCLUSION

The distribution of water properties suggests that sub-surface water exchange between the fjords and the Tasman Sea is free. The

inner sills give rise to small variations of deep water properties
between basins. The water properties in Chalky, the next most
southernmost fjord, suggest that vertical mixing is more important
than in other fjords, possibly due to its having the greatest
exposure to the strong southwest winds.

REFERENCES

Batham, E.J. 1965. Rocky shore ecology of a southern New Zealand
    fiord. Trans. Roy. Soc. N.Z. Zool. Vol. 6, 215-227.

Garner, D.M. 1964. The hydrology of Milford Sound. pp. 25-33
    in Skerman, T.M. (Ed.), "Studies of a southern fiord",
    Bull. N.Z. Dept. Scient. Ind. Res. 157, 25-33.

Jillett, J.B. and S.F. Mitchell. 1973. Hydrological and biological
    observations in Dusky Sound, south-western New Zealand, pp. 419-
    427 in Fraser, R. (Comp.), "Oceanography of the South Pacific
    1972", N.Z. National Commission for UNESCO, Wellington.

Stanton, B.R. 1978. Hydrology of Caswell and Nancy Sounds, pp.
    73-84 in "Fiord Studies: Caswell and Nancy Sounds, New Zealand"
    G.P. Glasby, (Ed.), New Zealand Oceanographic Institute Memoir
    79.

A WAY TO MEASURE THE RESIDUAL FLOW IN THE LIM FJORD - DENMARK

Torben Larsen

Institute of Civil Engineering
University of Aalborg
DK-9000 Aalborg, Denmark

INTRODUCTION

The Lim Fjord is in the northern Jutland in Denmark. As well as the rest of the Danish landscape the fjord was formed in the glacial time. The length of the fjord is 170 km, the surface area is 1500 square km and the average depth is 6 m. Maximum depth is 28 m. (See Figures 1 and 2).

During historic times the fjord has always been connected with the Kattegat to the east. But at least from the Viking age the fjord has been separated from the North Sea to the west. In 1825 a flood in the North Sea opened a channel between the fjord and the sea. After some decades with unstable conditions the channel stabilized its course at Thyborøn. This dramatic event forced major changes in salinity and the biological environment of most of the fjord.

Due to the small water depth in the fjord only insignificant vertical salinity gradients occur in the main course. Therefore, the transport of soluble matter depends only on the residual flow and the dispersion. For most of the fjord the residual flow dominates the transport.

The environmental problems in the fjord are most significant in the secondary fjords where the residual current does not reach. All the secondary fjords have the typical, undesirable eutrophic state, which also can be observed in a number of Danish fjords. The influence of the residual flow on nutrient level in the secondary fjords is indirect but important. The residual current in the fjord is responsible for the concentration level in the main

333

Figure 1.   The Lim Fjord in Denmark

course and defines the boundary concentration level for the secondary fjords. On average over the year the residual flow runs eastward with a value of 50–100 m^3/sec.

THE BACKGROUND FOR MEASUREMENT OF THE RESIDUAL FLOW

The water exchange in the Lim fjord depends primarily on the
water levels in the North Sea and the Kattegat, and on the wind.
But also the water levels in the North Sea and the Kattegat depend
on the wind, and former investigations, The Lim Fjord Committee

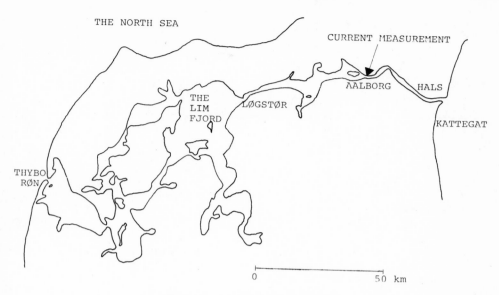

THE NORTH SEA

CURRENT MEASUREMENT

AALBORG   HALS

THE
LIM   LØGSTØR
FJORD

KATTEGAT

THYBO
RØN

0   50 km

Figure 2.   The Lim Fjord

(1976), show that the residual current is well correlated to the
wind vector along the fjord. The correlation is surprisingly good
for moderate to strong winds, but for light winds this empirical
correlation seems not to hold. The physical reason for this is the
well known fact that the friction factor between wind and water surface during light winds is not well defined. In northwest Europe
light winds are often connected with high pressure, easterly wind
direction and high temperature, and occur typically in the spring
and summer. Those conditions are also the most critical from the
point of view of water quality.

Current measurement in the Lim Fjord using conventional automatic current meters is difficult during the summer due the large
amount of floating seaweed. A daily check of the current meter is
necessary but even so the record is often interrupted. In the
following is presented a simple method to find the residual flow
from observation of the current direction.

THEORETICAL RELATION BETWEEN TIDAL EXCURSION AND DURATION

As a rough approximation current measurements at Aalborg
(Fig. 3) shows that the one dimensional current in the fjord can
be simplifed as (see Fig. 4):

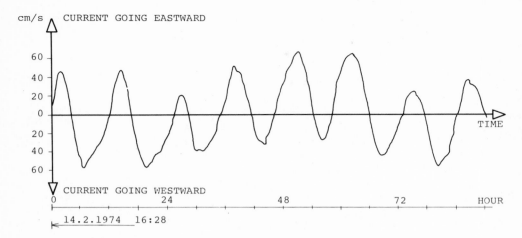

Figure 3.   Selected part of current measurements at Aalborg.

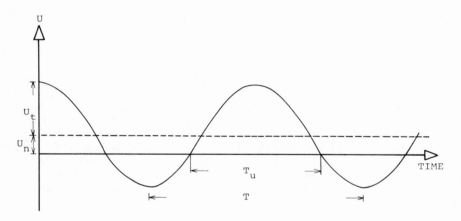

Figure 4.   Theoretical tidal current and residual current.

$$U = U_t \cos(2\pi t/T) + U_n \qquad\qquad (1)$$

where:-

U is actual current volocity (average over cross-section)

U_t is amplitude of tidal current (average over 28 days is 0.44m/s near Aalborg)

T is tidal period (M_2-component 12.42 hour)

t is time

U_n is residual current

We now define the current excursion V as (Fig. 4)

$$V = \int_{t_o}^{t_o + T_u} U \, dt \qquad (2)$$

where T_u is the period in which the current is going the same direction.

Combining (1) and (2) gives

$$V/U_t T = \sin(\pi T_u/T) - (\pi T_u/T)\cos(\pi T_u/T)/\pi \qquad (3)$$

As a practical approximation (3) can be developed as:

$$V/U_t T = 1.60(T_u/T)^{2.63} \qquad (4)$$

It is now postulated that V can be determined from (4) using observed values of T_u.

This version of the method neglects the influence of the variation of the water level during the tidal period. For the Lim Fjord at Aalborg this is a reasonable approximation. The tidal range is 0.3 m and the average depth is 11 m where the current measurements were carried out.

CALIBRATING THE METHOD BY MEASUREMENTS

During 1971 the Royal Danish Hydrographic Office had an automatic current meter placed under one of the bridges crossing the Lim Fjord at Aalborg, and during 1974 the author continued the measurements on the same position. It should be interposed that most of the velocity measurements were inapplicable because of the flowing seaweed. From the intact measurements 72 periods have been chosen to calibrate the method. The periods have been chosen from days with moderate to light winds. In Fig. 5 the results are shown. A curve fitting gives

$$V = 4.84 \times 10^{-4} T_u^{2.08} \text{ (V in cm and } T_u \text{ in sec)} \qquad (5)$$

of if we insert T = 12.42 hour and U_t = 0.44 m/s

$$V/U_t T = 1.16(T_u/T)^{2.08} \qquad (6)$$

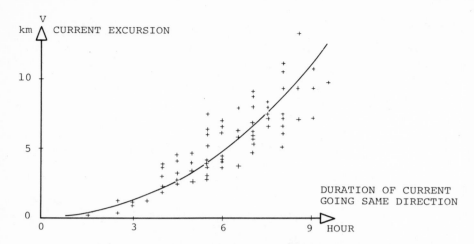

Figure 5.   Connection between current excursion and current duration.

        Comparing the theoretical equation (4) with the empirical
equation (6) shows:

T_u(hours)	$V/U_t T$	
	(4)	(6)
2	0.01	0.02
4	0.11	0.08
6	0.26	0.24
8	0.47	0.50
10	0.74	0.90

        The agreement seems satisfactory. It should be mentioned that
statistical tests to distinguish two different equations for the
relation (6), one for eastgoing currents and one for westgoing
currents, were carried out, but the tests did not show significant
differences.

COMPARING MEASURED RESIDUAL CURRENT AT LØGSTØR BY ESTIMATED
RESIDUAL CURRENT AT AALBORG

        Fig. 6 illustrates, by an example, the results of estimating
the residual current by the method described in this paper compared
to some direct measurements carried out in another connection, the
Lim Fjord Committee (1976), near Løgstør, which is 50 km from

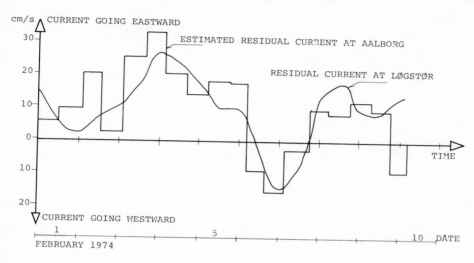

Figure 6.   Comparison between measured and estimated residual current.

Aalborg.   Due to fresh water inflow to the Lim Fjord between Løgstør and Aalborg and due to the effect of the storing capacity, the residual flow cannot be expected to be exactly the same in Aalborg and Løgstør if a small timescale, like a tidal period, is used. Whatever that can be said in the accuracy of the method, it has been shown to be the only practical and economical way, for the moment, to observe the residual flow in the Lim Fjord during longer periods.

ACKNOWLEDGEMENTS

The Research Committee of University of Aalborg has supported this work.   The writer thanks his students Mr. Niels Einar Jensen and Mr. Ervin Pedersen for handling the large amount of data.

REFERENCE

The Lim Fjord Committee 1976:   "The Lim Fjord Investigation 1973-75" in Danish.

CONTROLLING IMPACT OF CHANGES IN FJORD HYDROLOGY

Torkild Carstens and Henrik Rye

The River and Harbour Laboratory
The Norwegian Institute of Technology
Trondheim, Norway

One of the early large post-war hydro-power developments in Norway, Rossaaga, brought to our attention the effect of changes in the runoff regime on the dynamics of the fjord receiving this freshwater runoff. The immediate and obvious consequence of increasing the winter discharge was that the formerly ice-free fjord developed an ice cover. To avoid this in other fjords several methods have been proposed and studied at the River and Harbour Laboratory.

We want to know how much the temperature and salinity must be raised artificially at the head of the fjord in order to avoid freezing inside the mouth of the fjord.

The heat source is the sea-water in the fjord. Its heat is made available by two entrainment processes, one which is natural in the sense that is is not controlled, and one which is artificial or man made.

The heat sinks are the atmosphere and the precipitation on the fjord surface. When only the large terms are retained the heat balance becomes

$$\rho_2 c_p Q_e (T_2 - T_f) = A(q_n + L\dot{M} - q_e) \qquad (1)$$

Q_f - fresh-water discharge m^3/s

Q_e - volume flux of artificial entrainment

q_e - $\rho_2 C_p w_e (T_2 - T_f)$ Naturally entrained heat flux per unit area
cal/s m^2

q_n - heat loss per unit area to the atmosphere cal/s m^2

ρ_i - water density of upper (index 1) and lower layer (index 2)

C_p - specific heat cal/g^{0}C

L - latent heat cal/g

w_e - vertical entrainment velocity m/s

$\dot{M}$ - rate of snow precipitation g/s m^2

T_f - freezing point of upper layer

T_i - sea-water temperature

A - desired ice-free area m^2

For (1) to hold there must be unhindered exchange within the upper layer of thickness h_1 so that the entire heat reservoir $\rho_1 C_p (T_1 - T_f) h_1 A$ can be depleted. Vertical salinity gradients therefore must be weak.

From (1) Q_e can be determined if q_n, $\dot{M}$ and q_e are known. For the heat loss function q_n we have tested several options in the literature and settled for the one proposed by Ryan et al. 1975. For the entrainment velocity w_e we used Kato and Phillips' formula,

$$w_e = 4.9 \times 10^{-9} \frac{W^3}{\frac{\Delta\rho}{\rho} g h_1} \tag{2}$$

which requires prior knowledge of h_1. In (2) W is the wind speed, $\Delta\rho$ is the density jump and g is the acceleration of gravity. For the time being we can predict h_1 only for fjords with restricted mouth, for which $h_1 = \frac{3}{2} h_{mouth}$

$$w_e = 4.9 \times 10^{-9} W^3 \left[\frac{b}{\beta S_2 g Q_f}\right]^{2/3} \tag{3}$$

where $\beta = \frac{1}{\rho_o} \frac{d\rho}{dS} \overset{\sim}{=} 8.10^{-4} \, {}^o/_{oo}{}^{-1}$ and b is the width at the mouth.   An
example is shown in Fig. 1 (Stigebrandt et al. 1976)

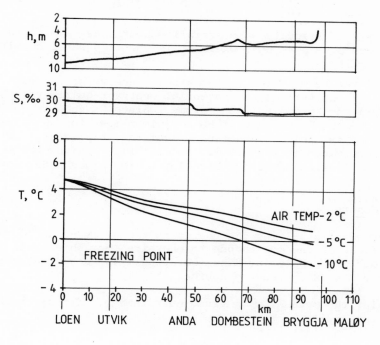

Figure 1

OPTIONS OF FORCED MIXING

The forced mixing required to bring about initial dilutions
of only 5 or 10, not to mention 20 or 30 times the freshwater
discharge, calls for substantial volume fluxes. For instance, to
dilute 100 m^3/s 10 times, it is necessary to supply 1000 m^3/s of
sea-water. Such quantities cannot economically be pumped through
pipes. They must be free flows, driven by *jet or plume entrainment*.

Effluent Control

This method of providing initial dilution is similar to sewage
outfalls. The mixing energy is supplied in the tailrace from the
powerhouse, either as kinetic energy in a high-velocity jet, or
as potential energy in a submerged outfall.

Reduced scale hydraulic models are readily adapted to such

studies, and the VHL experience with outfall experiments for hydro-
power plants was recently reported by MATHISEN and SÆGROV 1978.
The energy loss is simply

$$\Delta E = \frac{U_o^2}{2g} + \frac{\Delta \rho}{\rho}(H + \frac{D}{2}) \tag{4}$$

where U_o = exit velocity, H = outfall submergence, D = outfall
diameter and $\Delta \rho = \rho_2 - \rho_f$, where ρ_f is the freshwater density.

The first term in (4) is the kinetic energy of the exit jet,
while the second term is the potential energy required to submerge
the outfall and create a rising plume. Since ΔE goes to infinity
with both U_o and H, there will be a minimum loss for any desired
dilution. This minimum cannot be calculated at present, but is
readily determined in a hydraulic model. Fig. 2 gives the results
of an investigation of ΔE for Nordfjord. (Mathisen & Saegrov 1978).

Ambient Control

Instead of supplying the energy to the effluent as in the case
described above, one may choose to energize the ambient water. The
10 year old air bubbler at Rana (Carstens 1970) does that with
compressed air as the energy source.

Again, the mechanism is plume entrainment, with air bubbles
forcing the mixing.

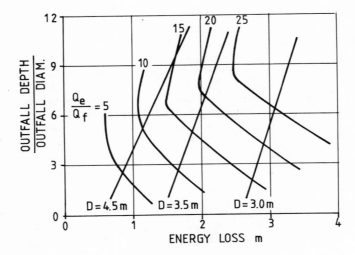

Figure 2

The loss function (4) gives directly the energy loss per m^3 of water passing through the outfall.

The lack of a model law for upscaling of experimental results with air bubbles precludes the use of convenient small-scale tests to determine the loss function

$$\Delta E = \frac{Q_a}{\rho_w g Q_f} \ (P_o \ell n(1 + \rho_w gH/P_o) + P_f)\eta \qquad (5)$$

where Q_a = air flux at atmospheric pressure P_o, H = submergence of openings, P_f = pressure drop in air pipe and η = efficiency of compressor.

For complete mixing by an air bubbler the vertical jet must penetrate the upper layer. That is, the horizontal surface velocity U_s of the deflected half-jet directed upstream must equal or exceed the surface velocity U_1 of the upper layer (Berge, 1964). This bubble-induced surface velocity is a function of Q_a and H, given by Taylor, 1955.

$$U_s = C(g\frac{Q_a}{b})^{1/3} \left(1 + \frac{\rho_w gH}{P_o}\right)^{-1/3} \geq U_1 \qquad (6)$$

According to Bulson 1961, C= 1.46 for a single air pipe. With two pipes we obtained a coefficient C = 1.9 in some coarse field tests. Better two-dimensional tests in a dock in Trondheim by Traetteberg, 1967, yielded C = 1.71 for the optimum spacing of the pipes. Since Q_a is proportional to C^3, the savings from the use of two pipes and C = 1.71 in (5) amount to 38% compared with one pipe.

If Q_a determined by (6) delivers an initial entrainment Q_e which exceeds that determined by (1), the control is shifted from the ambient to the effluent.

To reduce Q_a we may redesign the outlet so as to reduce U_s, keeping in mind that $Q_a \sim bU_s^3$.

It may be argued that the energy cost should be less when the power is used to accelerate ambient water to a high velocity rather than accelerating the effluent to a high velocity. Actually, our tests in still water (Saegrov, 1977) indicate that ambient control using air bubbling consumes less energy than effluent control. In the case of RANA the loss (6) is 0.6 - 1.1m and roughly the same as for a well-designed outfall.

Jets

Another idea is to use big propellers to drive a row of vertical jets. Energy-wise this would probably be the cheapest method, however, capital costs would run high.

Combined Control

If part of the fresh-water is released in the normal way at the surface while the remaining is released as a submerged jet, we have a combined system. This option may become attractive for topographic and other reasons, but it has not yet been tested.

REFERENCES

Berge, H., 1964. Teknisk Ukeblad No 24, pp 1-8 (In Norwegian).

Bulson, P.S., 1961. The Dock & Harbour Authority - May 1961, pp 15-22.

Carstens, T., 1970. ASCE Journ. Waterways and Harbours Div. WW pp 97-104.

Kato, H. & O.M. Phillips, 1962. J. Fluid Mech. 37, pp 643-655.

Mathisen, J.P. S. Saegrov, 1978. Sixteenth Conf. on Coastal Eng., Hamburg.

Stigebrandt, A., I. Schei, J.P. Mathisen, T. Audunson, 1976. Forced mixing of seawater into the fresh water discharge from the planned hydroelectric power plant in Loen, Nordfjord. The River and Harbour Lab., Trondheim. Rep. No. STF60 F76096 156 pp. (In Norwegian).

Saegrov, S., 1977. The River and Harbour Lab., Trondheim. Rep. No. STF60 F78015.

Taylor, G.I., 1955. Proc.Roy.Soc A vol 231, pp 466.

Traetteberg, A., 1967. River and Harbour Lab., Trondheim. Unpubl. report. Project 600185.

WATER MOVEMENTS TRACED BY METALS DISSOLVED FROM MINE TAILINGS
DEPOSITED IN A FJORD IN NORTH WEST GREENLAND

G. Asmund

Grønlands Geologiske Undersøgelse
Øster Voldgade 10, DK-1350
KØBENHAVN-K, Denmark

MINING AND TAILINGS

The location of the studied area on the west coast of Greenland is seen in Fig. 1. A mining company has, since the fall of 1973, produced a zinc and lead concentrate, and 1400 tons per day of tailings with a chemical composition of approximately:

50%	carbonate, marble and dolomite
50%	FeS_2, pyrite
1%	Zn, mainly as ZnS, sphalerite
0.5%	Pb, mainly as PbS, galena
0.01%	Cd

This is pumped out in the fjord with sea water which is also used as flotation media. The tailings line ends at a depth of 42 m at location T in Fig. 1, see Asmund 1976.

THE DISSOLUTION PROCESS

The first pre-mining environmental investigation in February 1974, only 4 months after mining had started, disclosed unexpectedly high values of dissolved zinc, lead and cadmium. At the 2 stations in the fjord where the dumping occurs, the concentrations found were as listed in Table 1.

These findings gave rise to concern by the Ministry of Greenland, which is the environmental controller, and resulted in half-yearly environmental investigations on the site. In the latest years

347

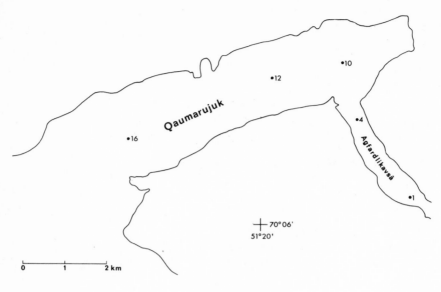

Figure 1.

Table 1.   µg/ℓ dissolved metals, Feb. 1974, GGU *et al.* 1974.

	Station 1			Station 4		
Depth	Zn	Cd	Pb	Zn	Cd	Pb
2 m	353	2	176	304	2	186
20 m	625	2	165	341	2	217
bottom	292	2.5	207	615	4	248

a major investigation by the mining company (Greenex) has thrown
light on the dissolution process. The sulphide minerals are themselves insoluble in sea water, but after oxidation they are somewhat soluble. It was therefore believed that the source of the
dissolved metals were sulphates formed by *in situ* oxidation at the
fjord bottom. However, the investigations have shown that the major
source of the dissolved metals is some compounds in the tailings,
formed during the mining and flotation, that are soluble but not
dissolved in the tailings water.

When these compounds, presumably basic sulphates, enter the
marine environment, and the pH drops from the 9-10 in the tailings
water to 8 in sea water, some of them dissolve. The dissolved
metals zinc, lead and cadmium act as perfect tracers for water
originating from the bottom of Agfardlikavsâ and show where and when
currents occur and how much mixing there is in the fjord system.

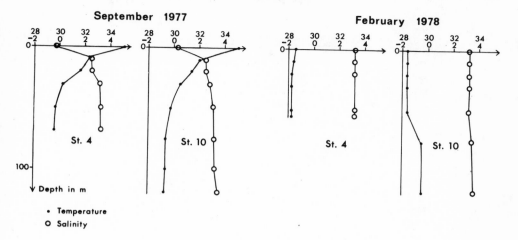

Figure 2.

HYDROGRAPHY

This fjord system starts with the 0.5×4 km^2 Agfardlikavsâ which reaches a depth of 80 m and has an average depth of 35 m. It is separated from the fjord Qaumarujuk by a sill with maximum depth 27 m. Qaumarujuk is 2×10 km^2 with a fairly uniform depth ranging from 100 to 200 m. There is no sill between this and the next fjord, in which the metals are diluted to undetectable concentrations. Temperature and salinity curves for the two innermost fjords are shown in Fig. 2 for a typical winter (ice covered) and a typical summer situation. Dissolved oxygen has been determined several times always with the result that the water was close to saturation at all depths.

SAMPLING METHODS AND ANALYSIS

Water samples were obtained by a Hydro-Bios reversible plastic water sampler. Temperature was read on a reversible thermometer, oxygen determined by the Winkler-method and salinity by chloride titration. Polyethylene bottles (1ℓ capacity) were filled for heavy metal analysis. Experience has shown that it makes no difference whether the samples are shipped as taken or they are filtrated, acidified and frozen before shipment to the laboratory. Metals were determined by anodic stripping voltammetry with graphite or glassy carbon electrodes coated with a mercury film. The pH was adjusted to 4.5 by titration with dilute suprapur nitric acid.

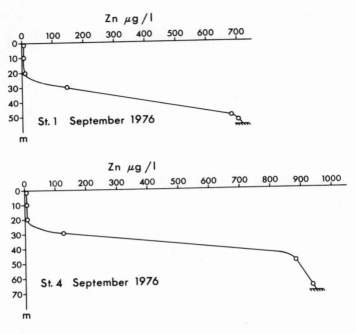

Figure 3.

The metals Zn, Cd and Pb are found in almost constant ratio
in water affected by the mine tailings:

3<µgZn/µg Pb<1.5   average 2

50<µgZn/µg Cd<200   average 100

For that reason only the zinc profiles are discussed in the following.

Agfardlikavsâ ice free situation

The zinc profiles for September 1976 are shown in Fig. 3 as
typical for the situation where the ice has been absent for three
months. It is seen that the concentration of zinc can reach up to
1000 µg/ℓ where the water is in direct contact with the tailings
depot. The surface layers, from 0 to 20 m, are much lower in dissolved metals due to the unhindered action of wind, tidal forces and
fresh water run-off that continuously renew the water above sill
depth. The steep gradient between high and low metal concentrations
at 30 m corresponds to the depth of the sill between Agfardlikavsâ
and Qaumarujuk (27 m).

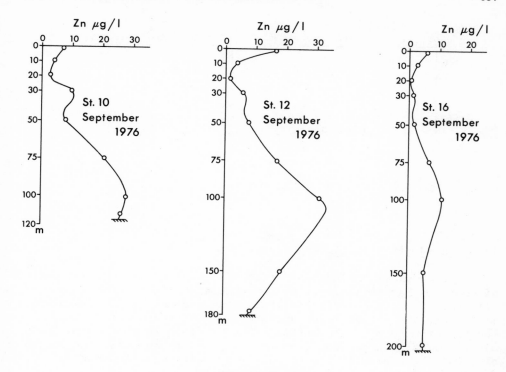

Figure 4.

Qaumarujuk ice free situation

The metal profiles in Qaumarujuk in the ice free situation, as illustrated in Fig. 4, are not fully understood. A tentative explanation is, that the fresh water run-off from the bottom of Agfardlikavsâ ($95 \cdot 10^6 m^3$/year flowing from June to September) mixes with saline metal-rich water in the surface of Agfardlikavsâ and continues as a light surface current over the sill and out through Qaumarujuk. This current is compensated for by an opposite directed current at the depth of 20 m. There is a distinct maximum of dissolved metal at just below 30 m in Qaumarujuk. This is also explained as an effect of the sill. Water with high metal content may occasionally escape over the sill and afterwards stay at that depth. The very marked maximum at 100 m in Qaumarujuk is not explained by any other observation. Obviously water from the bottom of Agfardlikavsâ passes over the sill and down to 100 m in Qaumarujuk. This maximum is never present during the investigations in March when the ice cover is around 80 m.

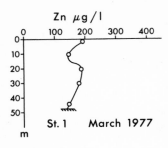 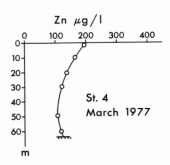

Figure 5.

Ice covered situation Agfardlikavsâ

Fig. 5 shows the concentration of dissolved zinc as a function
of depth in March 1977. It is remarkable that the highest concentrations are in the surface although the sources of the dissolved
metals are the bottom layer and the plume of suspended tailings
running from 40 m to the bottom. The water in this situation is
highly mixed as also seen from the temperature and salinity curves,
Fig. 2. The mixing is caused partly by salt rejection during
freezing and partly by density currents from Qaumarujuk.

Qaumarujuk ice covered situation

The metal rich water thus brought up to the surface in
Agfardlikavsâ are now much more accessible to the upper layers of
Qaumarujuk. As seen from Fig. 6 the water in Qaumarujuk above sill
depth (27 m) is highly affected by metals. Below sill depth the
water is unaffected.

CONCLUSION

Zinc, cadmium and lead released from mine tailings at the bottom of the fjord Agfardlikavsâ can be used as tracers for water
movements. In the summer fresh water run-off produces a strong
stratification in both Agfardlikavsâ and Qaumarujuk. The bottom
layers in Agfardlikavsâ are stagnant and enriched in dissolved
metals. During ice cover this stratification is broken down in
Agfardlikavsâ where the mixing then is complete. In Qaumarujuk a
new stratification is formed with metal-rich water above the depth
of the sill between the two fjords, and unaffected water below sill
depth.

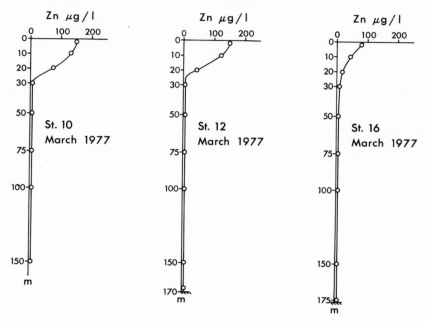

Figure 6.

REFERENCES

Asmund, G., Bollingberg, Haldis J. and Bondam, J. 1976: Continued
    environmental studies in the Qaumarujuk and Agfardlikavsâ fjords,
    Mârmorilik, Umanak district, central West Greenland. Rapp.
    Grønlands Geol. Unders. 80, pp. 53–61.

GGU *et al*. 1975: Recipientundersøgelse 1974, Agfardlikavsâ,
    Qaumarujuk. 108 pp. Copenhagen.

Several Reports, Greenex Ltd., Landemærket 10, DK-1119, Copenhagen,
    Denmark.

EXCHANGE PROCESSES ABOVE SILL LEVEL BETWEEN FJORDS AND COASTAL WATER

Harald Svendsen

Geophysical Institute, University of Bergen
Bergen, Norway, N-5014

INTRODUCTION

During the period from September 1972 to November 1975 the Geophysical Institute of the University of Bergen carried out an investigation in six fjords, Hylsfjord, Sandsfjord, Saudafjord, Erfjord, Jøsenfjord and Jelsafjord, in Ryfylke in the southern part of Norway, Fig. 1. The investigation was based on measurements of temperature, salinity, oxygen, current, freshwater runoff, water level and wind.

This paper presents some results from the study of the circulation in the intermediate layer, i.e. between the brackish layer and the sill depth. The results from a comprehensive investigation of the circulation in the Jøsenfjord (Fig. 1b) in June 1974, indicated that the mean circulation in the intermediate layer was strongly influenced by the wind conditions offshore (Svendsen 1975, 1977 and Svendsen and Thompson 1978). In a 6 day period dominated by winds towards north at the coast, outflow of water occurred in the lower part of the intermediate layer in the fjord compensated by inflow of water from the upper layer of the fjord to the upper part of the intermediate layer. In this period the salinity decreased and the temperature and oxygen content increased in the whole layer. A sudden change of the wind conditions, to winds towards south, reversed the circulation and the salinity increased and the temperature and oxygen content decreased.

From these observations some interesting questions arise: Is the relation between the wind at the coast and the circulation of the intermediate layer which existed in the Jøsenfjord in June 1974 typical, i.e. is the circulation of the intermediate layer

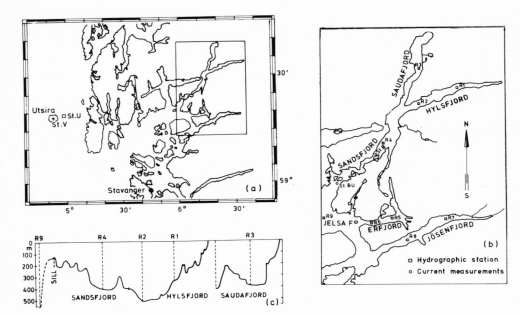

Figure 1.   (a) Map of a part of the southwest coast of Norway.
             (b) Map of the Ryfylkefjords. (c) Longitudinal section
             of the Sandsfjord, Hylsfjord and Saudafjord.

mainly governed by wind conditions at the coast? Is the two-layer
circulation, which was observed in the Jøsenfjord, representative
of the circulation of the intermediate layer?

        Although no current measurements were made in the intermediate
layer of the fjords during the three year period (except June 1974
in the Jøsenfjord) we may infer the answer to the first question
using the relationship between current and hydrographic parameters
found in the Jøsenfjord in June 1974.

        The salinity, temperature and oxygen content of the water were
observed once a month at selected depths at fixed positions in the
mentioned fjords. In the intermediate layer the variation of the
hydrographic parameters followed the same pattern in all the fjords
in the observation period. This result implies that the mean circulation in the intermediate layer was the same in the six fjords.
Salinity and temperature were also sampled at St. U at the coast,
and wind measurements were made on an island (Utsira), St. V, 7 km
off the coast, Fig. 1a.

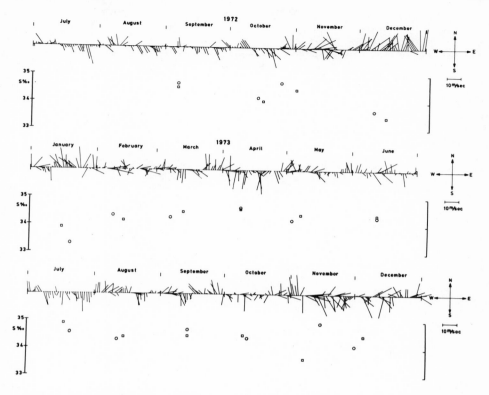

Figure 2.   Daily mean values of the wind at St. V (N: Longshore
            component towards north) and the integrated mean of the
            monthly observations of the salinity from four depths
            (30, 50, 75 and 100 m) at St. U (0), and from six depths
            (30, 40, 50, 60, 75 and 100 m) at St. R4 ( ), September
            1972-December 1973.

     Figures 2, 3 and 4 show the wind (daily mean) at St. V
(Utsira), Figure 1a, and the integrated mean of the monthly observations of the salinity from four depths (30, 50, 75 and 100 m) at
St. U at the coast and from six depths (30, 40, 50, 60. 75 and 100 m)
at St. R4 in the Sandsfjord.

     These figures indicate a correlation between periods of wind
to the north along the coast and decreasing salinity in the intermediate layer of the fjords. Therefore, winds to the north should
coincide with inflow in the upper part of the intermediate layer
and outflow in the lower part, and winds to the south with currents
in the opposite directions.

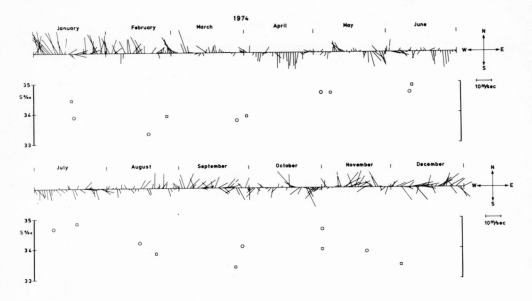

Figure 3.   Caption as in Figure 2, 1974.

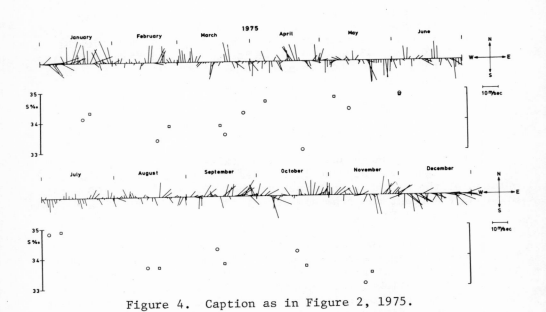

Figure 4.   Caption as in Figure 2, 1975.

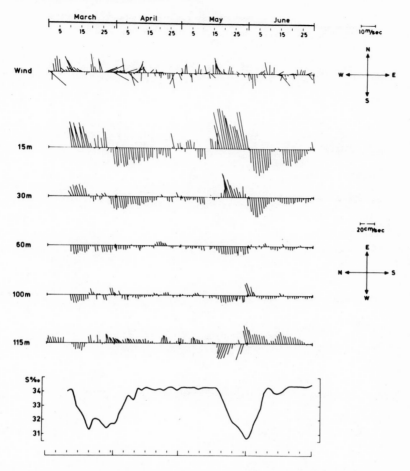

Figure 5. Daily mean values of the wind at St. V (N: Longshore
           component towards north), current at St. B (E-component
           up-fjord) and salinity at 30 m at St. B, March-June 1976.

        To gain more confidence in the relationship between hydro-
graphic parameters and currents in the intermediate layer, current
measurements were made at five depths (15, 30, 60, 100 and 150 m)
at St. B and at the sill (115 m) at St. BU in the Sandsfjord from
March to October 1976. One of the current meters (30 m) carried
a conductivity cell. Except for a few days the mean circulation
took place as a two-layer system, Figures 5 and 6. The figures
show daily mean values of the wind at St. V, current at St. B
and salinity at 30 m.

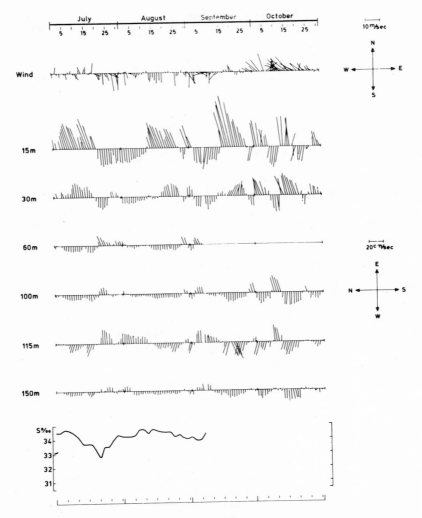

Figure 6. Daily mean values of the wind at St. V (N: Longshore
component towards north), current at St. B (E-component
up-fjord) and salinity at 30 m at St. B, July–October
1976.

The shear zone was situated most of the time within the depth
interval 30–60 metres except in April and June when the zone was
within the interval 100–115 m.

The assumed relation between the direction of the circulation in the intermediate layer and the wind direction at the coast seems well supported by the observed mean current at St. B. In about 70% of the periods when the mean wind at the coast has a component towards the north the flow is up-fjord in the upper part and down-fjord in the lower part of the intermediate layer at St. B. The salinity at 30 m decreases in these periods. The reversed circulation, with increasing salinity at 30 m, appeared in about 80% of the periods in which the wind has a component towards the south.

The results above indicate that it is possible, with tolerable accuracy, to ascertain the direction of the mean circulation and the salinity development in the intermediate layer by studying only the direction of the mean wind at the coast.

REFERENCES

Svendsen, H. 1975: A study of the circulation in a sill fjord on the west coast of Norway. Lecture given at the XVI General Assembly of IUGG, Grenoble 1975.

Svendsen, H. 1977: A study of the circulation in a sill fjord on the west coast of Norway. Mar. Sci. Commun., 3(2), 151-209.

Svendsen, H. and Thompson, R. O. R. Y. 1978: Wind-Driven circulation in a fjord. J. Phys. Oceanogr., 8(4), 703-712.

PARTICULATE MATTER EXCHANGE PROCESSES BETWEEN THE ST. LAWRENCE ESTUARY AND THE SAGUENAY FJORD

Jean-Claude Therriault, Rejean de Ladurantaye
and R. Grant Ingram†
Pêches et Océans Canada
Division d'océanographie
C. P. 15,500
Québec, P. Q. Canada   G1K 7X7

INTRODUCTION

Since the first suggestion by Drainville (1968) of occasional penetration of waters from the St. Lawrence estuary into the deep layer of the Saguenay fjord, a number of studies have both confirmed the existence of this exchange process (Therriault and Lacroix 1975; Yeats and Bewers 1976; Sundby and Loring 1978; Chanut and Poulet 1979) and indicated its tidal periodicity (Seibert *et al*. 1979). This circulation pattern, implying a regular intrusion of near surface estuarine waters, rich in particulate organic carbon, into the lower layer of the Saguenay fjord is the probable cause of the very high standing stock of zooplankton observed in the deep waters of the fjord as compared to those of the estuary. Considering the potential ecological importance of such an enrichment mechanism for the Saguenay Fjord, it was most important to estimate the rate at which this intrusion occurred and the physical factors that control it. This paper summarizes an attempt that has been made to characterize the water masses involved in this exchange and to describe the tidal circulation at the entrance sill and within the outer basin, in order to calculate a budget for the particulate organic matter in the lower fjord region.

†Marine Sciences Centre
 McGill University
 3600 University Street
 Montréal, Québec
 H3A 2T8

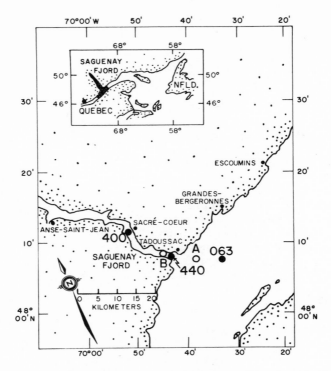

Figure 1.   Station location in the Saguenay fjord and the St.
             Lawrence estuary: open circles are used to indicate
             the location of the current meter moorings.

MATERIALS AND METHODS

        Field studies were carried out in August and October of 1978.
On both occasions vertical profiles of temperature, salinity,
chlorophyll, nutrients and particulate organic matter were taken
at two anchor stations (stns. 440 and 400) in the lower part of
the Saguenay fjord and at a nearby site. (stn. 063) in the St.
Lawrence Estuary (Fig. 1). The fjord stations were separated from
station 063 by a shallow sill of ∿ 20 m deep. An Aanderaa current
meter (sampling interval: 1 min) was moored on the shallow entrance
sill (stn. A) during both cruises (Fig. 1). An additional current
meter (stn. B) was moored in the outer basin at a depth of 68 m
during the August program. Hourly vertical current profiles (30 s
sampling rate) were also taken at each station during the October
cruise. A detailed account of the methods used to collect and
analyse the data will appear in another publication (manuscript in
preparation).

RESULTS AND DISCUSSION

　　During both cruises, the distribution of chlorophyll a(Chl a), particulate organic carbon (POC), nutrients, phytoplankton counts and sigma-t isopleths over the tide cycle showed a consistent variation pattern indicating the penetration of near surface waters (10-30 m) from stn. 063 into the deep layer at stn. 440 during the flood. Over the same time interval, the current meter data showed the presence of tidal flow in the lower layer above the sill directed into the fjord. The inward flowing currents were found from the start of the flood, at LW+1, to high water (HW). A net upstream current ∿ 5 cm s^{-1} was observed over the 5 day mooring period at 68 m in the outer basin. By examining density values for the August sampling period, a sigma-t range of 22.5-23.0 was found to be characteristic of the water masses showing elevated chlorophyll a and particulate organic carbon concentrations. Such characteristics were observed simultaneously throughout the deep layer of stn. 440 and within the 10-30 m deep layer at stn. 063 during the flood. T-S diagrams support this conclusion and confirm both the similarity of these water masses and their estuarine origin. The mechanism responsible for the periodic introduction of dense estuarine water at sill depth during the flood is related to the generation of a high amplitude progressive internal tide in the area of stn. 063 (Forrester 1974; Therriault and Lacroix 1976).

　　This penetration of estuarine waters, which can yield, at times, large fluxes of particulate organic material into the deep layer of the fjord is thought to be of significant ecological importance for enhancing the productivity of the deep waters of the Saguenay. As a result of this, it was decided to calculate a budget for the particulate organic carbon in the outer basin of the Saguenay. In order to estimate the fluxes across the sill, a simple model of conditions was constructed using observed current meter data at the sill and POC concentration at station 063. The model results were the following:

Sampling period		Inflowing water mass	Outflowing water mass	Net Flux
August (R ∿ 2.8 m)	Water volume	7.36×10^8 m^3	8.01×10^8 m^3	-1536 m^3 s^{-1}
	POC content	332 tonnes	261 t	72 (per tide cycle)
October (R ∿ 4.2 m)	Water volume	8.90×10^8 m^3	9.82×10^8 m^3	-2130 m^3 s^{-1}
	POC content	141 t	91 t	50 t (per tide cycle)

R = tidal range

The calculated net outward volume flux of 1536 and 2130 m^3 s^{-1} for August and October respectively, were comparable with the known river discharge $(1 - 2 \times 10^3$ m^3 $s^{-1})$ which supports the reliability of our model results. The important point to stress here is the observation of a net input of POC into the deep water of the Saguenay fjord over the tide cycle. Using the model results and assuming a basin volume of 2.30×10^9 m^3, below sill depth, we have estimated that a complete replacement of the waters of the first basin should occur every 1.5 to 7 days, depending on the value of the residual inward volume used to make the calculations and the tidal range. These estimates are considerably less that the 65 day replacement time calculated by Seibert et $al.$ (1979). However these authors also mentioned the possibility of replacement occuring in as little as 3 days, which is consistent with our findings.

REFERENCES

Drainville, G. 1968: Le fjord du Saguenay: Contribution à l'océano-
    graphie. Nat. Can. 95, 809-855.

Chanut, J. P. and S. A. Poulet 1979: Distribution des spectres de
    taille des particules en suspension dans le fjord du Saguenay.
    Can. J. Earth Sci. 16, 240-249.

Forrester, W. D. 1974: Internal tides in the St. Lawrence estuary.
    J. Mar. Res. 32, 55-56.

Seibert, G. H, R. W. Trites and S. J. Reid 1979: Deepwater exchange
    processes in the Saguenay fjord. J. Fish. Res. Bd. Can. 36,
    42-53.

Sundby, B. and D. H. Loring 1978: Geochemistry of suspended part-
    iculate matter in the Saguenay fjord. Can. J. Earth Sci. 15,
    1002-1011.

Therriault, J. C. and G. Lacroix 1975: Penetration of the deep
    layer of the Saguenay fjord by surface waters of the St.
    Lawrence Fjord by surface waters of the St. Lawrence Estuary.
    J. Fish Res. Bd. Can. 32, 2373-2377.

Therriault, J. C. and G. Lacroix 1976: Nutrients, chlorophyll,
    and internal tides in the St. Lawrence Estuary. J. Fish. Res.
    Bd. Can. 33, 2727-2757.

Yeats, P. A. and J. M. Bewers 1976: Trace metals in the waters
    of the Saguenay Fjord. Can. J. Earth Sci. 13, 1319-1327.

ZOOPLANKTON VARIABILITY IN SKJOMEN, NORTHERN NORWAY, AND EXCHANGE WITH THE OUTER FJORD

Norma Jean Sands and Harald Svendsen†

Marine Biological Station
University of Tromsø
Tromsø, Norway, N-9001

Skjomen is a narrow fjord, about 20 km long and 1 to 2 km wide, in northern Norway (68° 18'N, 17° 20'E). It has a maximum depth of 150 m and a sill depth of 60 m separating it from the larger Ofot-fjord. Skjomen is surrounded by steep mountains. One major, and several minor, rivers empty into the fjord. In 1970 a hydro-electric power plant was built on the major river. An investigation lasting three years was conducted in the fjord just prior to the building of the plant (Loeng 1978, Strømgren 1974). After the plant started operations, in 1977, another three year investigation began. The present study, to examine short term zooplankton variability, is part of this later investigation.

Zooplankton samples were taken from two stations, one inside Skjomen at 150 m depth, another just outside the sill in Ofotfjorden at 350 m. Current measurements from 10 current meters, hung from a fixed buoy, were made at a point between these two stations inside the sill and at 150 m. Zooplankton sampling consisted of five sets of hauls per station taken within a 36-40 hour period in October 1977, January, March and July 1978. In October 1977 each set of hauls consisted of three replicate hauls taken with a vertical Juday net from bottom-100 and 100-0 m. In 1978 the sets consisted of four replicate hauls, three preserved for counting and one dried for biomass estimation. The depth intervals used in 1978 were bottom-100, 100-50 and 50-0 m.

The zooplankton community of the area is dominated by copepods most of the year. *Calanus finmarchicus* and *Oithona similis* are by

†Geophysical Institute, University of Bergen, Bergen, Norway.

far the most dominant. In Ofotfjorden *C. hyperboreous* and *Microcal-anus pusillus* are the next most common while in the inner-most part of Skjomen (35 m) *Pseudocalanus elongatus* is the third most abundant species. Larvae are common in all three areas at various times of the year. The zooplankton communities of inner Skjomen, mid Skjomen, and the Ototfjord station are quite similar and no species common to one area is not found at least in low numbers in the other two areas suggesting free exchange.

Biomass and abundance estimates over the 36-40 hour period within the various months show no significant correlation with periods of light and dark so that no pattern of vertical migration could be established. However, between months there is a significant increase in biomass and abundance day round in spring and summer over fall and winter. In the winter *C. finmarchicus,* a relatively large copepod, is found almost exclusively below 100 m while the small copepod, *O. similis,* is the major constituent in the upper waters. In spring mature *C. finmarchicus* come to the surface (upper 50 m) leaving smaller copepods and ostracods to dominate in the lower waters. This causes a reduction in biomass in the deeper water. In summer the new, young *C. finmarchicus* migrate to the lower waters but the abundance of other species in the upper layers keeps the biomass high there.

Coefficients of variation (standard deviation/mean x 100) were calculated for species components (various copepod stages, species, and groups of species). Non-parametric tests were used to test for differences between day and night hauls and between hauls 24 hours apart. No significant difference could be found. There was also no significant difference between replicate hauls (taken minutes apart) and sets of hauls (taken several hours apart). The coeff-icient of variation found between both replicate hauls and sets of hauls falls within ranges found by many authors for replicate hauls (20-45%). However, when abundance estimates were calculated from only one haul instead of from three replicate hauls, the coefficients of variation between sets were significantly higher in most cases (mean 54%). This higher variation results in higher confidence limits for abundance estimates making comparisons more difficult. Thus in studies involving comparisons of zooplankton abundances over time or space, replicate sampling is worthwhile.

In an attempt to explain the variation in abundance estimates found in Skjomen, tests of correlation between abundance and current measurements were made. The currents in Skjomen are strongly affected by tides at most depths. The water movement is dominated by oscillating currents of semi-tidal period. These are generated at the sill and generally penetrate down to the bottom but not up through the pycnocline. At most times of the year the pycnocline lies between 2 and 10 m. The surface water movement is controlled mostly by the winds. In July the surface currents were up-fjord,

in the other three months they were down-fjord. The current pattern
from surface to bottom could change from a one layer system to a
multi-layer system with the change in tidal direction. Often
several shear currents would occur in the upper 50 m above sill
depth. The strongest currents were found in the upper 50 m.

Water displacement between sample taking was calculated from the
current measurements and correlated with corresponding changes in
abundance of zooplankton. In many cases the greatest differences
in zooplankton catch occurred with the greatest water displacement
but over all rankings of the two did not correlate. This could
indicate that for periods with small displacement, other factors
play a more important role.

A closer look at individual species changes during the October
and January periods show several different patterns. In October
the two *Oithona* species both increase in the deep water while
decreasing in the upper layers (<100 m). This could either be due
to vertical migration of a relatively stable population or due to
opposing currents carrying individuals into one layer and washing
them out of the other. However, the increases occurred with an
inflow of water into Skjomen while *Oithona* is more numerous in
inner part of the fjord than outside in Ofotfjorden. Also *Microcalanus*, which is more numerous in Ofotfjorden, and *Pseudocalanus*,
which is more numerous in inner Skjomen, increased and decreased
in a similar manner at both depth intervals. Obviously species
behavior is also important in the distribution of these species.

In January most of the copepods increased and decreased in a
similar manner at all three depth intervals. In the deeper waters
(>50 m) numbers increased while the flow was out of the fjord,
but numbers also increased in the upper 50 m at the same time where
net flow was into the fjord.

The existing currents near the mouth of Skjomen plus the similarity in zooplankton composition together suggest continuous
exchange of species between Skjomen and Ofotfjorden. The complex
pattern of water flow between the two fjords makes correlation
between species abundance and currents difficult. The multi-layer
flow allows for several areas (depths) of slack water where species
would not be displaced. The tidal influence on the currents also
causes less net water displacement so that plankton displacement
is not as great as would be if one only considered single direction
flow over 24 hours. To test the effect of tidal pulses more samples
within a 24 hour period should be taken. Smaller depth intervals
in the upper 50 m where there was often more than one direction of
flow would also be useful, one dividing point being at the pycnocline since water flow above this is generally wind driven and below,
tidal driven.

REFERENCES

Loeng, H.   1978.   Hydrografi og strømforhold i Skjomen 1969-1973.
    Radgivende Utvalg for Fjordundersøkelser Skjomenprosjektet
    Rapport nr. 1 - Oslo.   88 pp.   111 figs.

Strømgren, T.   1974.   Zooplankton investigations in Skjomen 1969-
    1973.   Astarte 7:   1-15.

MEAN FIELD DISTRIBUTIONS OF A DISSOLVED SUBSTANCE IN THE VICINITY

OF BRANCHES IN A FJORD SYSTEM

Paul F. Hamblin and Eddy C. Carmack*

National Water Research Institute
P.O. Box 5050
Burlington, Ontario L7R 4A6
Canada

INTRODUCTION

A characteristic feature of many fjord and fjord—like lakes is their complicated plan, consisting of a system of branching arms generally with a river entering at the head of the arm; a typical example is the Trondheim Fjord (McClimans, 1978). Modelling of the distributions of nutrients, pollutants, sediments, larvae and other water quality parameters requires a knowledge of turbulent mixing between water masses in the vicinity of channel junctions. A method of determining the appropriate mixing coefficients, part-icularly for the large spatial and temporal scales, is to deduce them from the distribution of a naturally occurring material substance. In the present study simple analytical methods were used to construct solutions of the advective—diffusion equation in the vicinity of junctions of arms which are subsequently applied to the conductivity distribution within a riverine layer in a fjord lake.

PHYSICAL CONSIDERATIONS

In this study we restrict our attention to time scales sufficiently long to permit steady conditions to obtain, and focus only on the far-field portion of river plume. We assume that the plume is spread across the fjord or that the internal Rossby radius

* National Water Research Institute Branch, Pacific and Yukon Region, 4160 Marine Drive, West Vancouver, B.C. V7V 1N6, Canada.

of deformation is greater than the fjord width and that we may vertically average the concentration of the river water.

Vertical averaging introduces the dispersion effect or enhanced horizontal diffusion due to vertical diffusion acting upon vertical shear, and due to entrainment of saline water into the brackish layer. An expression for the shear dispersion effect has been developed for a riverine layer in a fjord lake by Hamblin and Carmack (in preparation) analogous to that for dispersion in a circular pipe of Gill and Sankara (1971). The shear dispersion for time scale sufficiently long to allow the establishment of an equilibrium between shear and vertical diffusion is $D^2\bar{U}^2/30 \, k_v$, where D is layer depth, k_v is vertical diffusivity and $\bar{U}$ is the vertically averaged flow. The presence of vertical shear then gives rise to the anisotropy of the diffusion processes. Other contributions to the dispersion effect in fjords have been studied by Smith (1978).

THE DIFFUSION EQUATION MODEL

Based on the above considerations we adopt the following advection-diffusion model for the concentration field of a conservative substance, $c(x,y)$.

$$U(x,y) \frac{\partial c}{\partial x} + V(x,y)\frac{\partial c}{\partial y} = \frac{\partial}{\partial x}(K_{11}\frac{\partial c}{\partial x} + K_{12}\frac{\partial c}{\partial y}) + \frac{\partial}{\partial y}(K_{12}\frac{\partial c}{\partial x} + K_{22}\frac{\partial c}{\partial y})$$

where U is the component of flow parallel to the x-axis and similarly V along the y-axis.

$$K_{11} = K_s\cos^2\theta + K_N\sin^2\theta,$$

$$K_{12} = (K_s - K_N)\sin\theta\cos\theta,$$

$$K_{22} = K_s\sin^2\theta + K_N\cos^2\theta,$$

and K_s is the dispersion coefficient (sum of horizontal diffusivity and dispersion) in the streamwise direction, K_N is the diffusivity normal to the flow (both assumed constant) and θ is the angle between the streamlines and the x-axis.

Exact solutions to this equation have been obtained by Hamblin (1979) by the Wiener-Hopf, Boussinesq and Schwartz-Christoffel methods for branching channels meeting at junction angles of 0^0 and 180^0 and for various cases of channel breadths and flow velocities.

APPLICATION TO A FJORD LAKE

Kootenay Lake is a long (100 km), narrow (4 km), deep (150 m

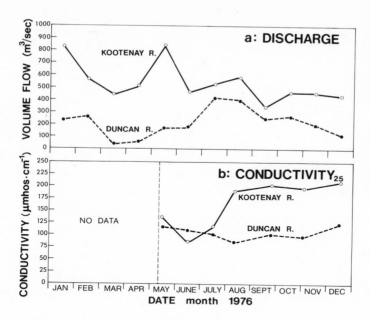

Figure 1. Time history of (a) discharge (m^3/s) and (b) conductivity (μmho/cm) of the Duncan and Kootenay Rivers.

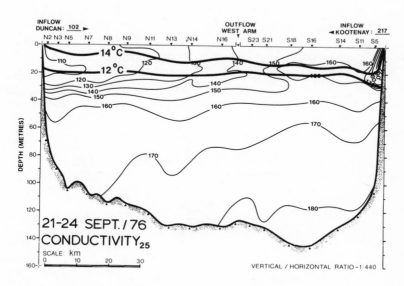

Figure 2. Vertical distribution of conductivity (referenced to 25^0C) and temperature (^0C) along the north-south axis of Kootenay Lake, September 21-24, 1976. (Salinity X $10^6 \simeq 0.8$ conductivity).

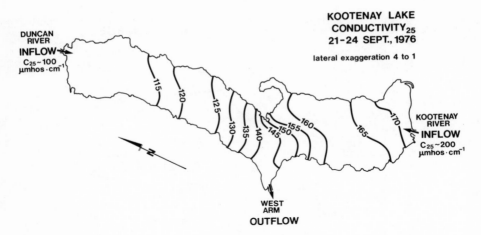

Figure 3.   Vertically averaged conductivity between the 12 to 14⁰C
             isotherms, 21-24 September 1976. Note that the cross-
             lake scale is exaggerated by a factor of 4.

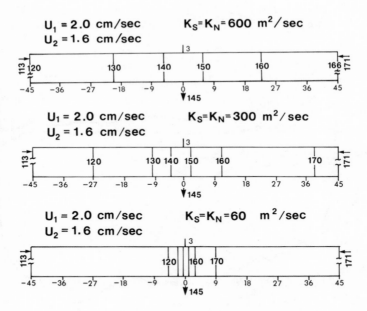

Figure 4.   Theoretically determined distributions of conductivity for
             horizontal diffusivities of (a) 600 m^2/s (b) 300 m^2/s,
             (c) 60 m^2/s. Note that the y-scale is exaggerated by a
             factor of 3.

maximum depth) fjord lake in British Columbia. During the late summer and fall of 1976 measurements of conductivity profiles at approximately one month intervals in Kootenay Lake in conjunction with daily inflows of the two principal rivers entering at opposite ends of the lake established that both the conductivities and river inflows were relatively steady in late summer, see Fig. 1. The cross-sectional distributions of conductivity and temperature along the longitudinal axis of the lake (Figure 2) show that the river interflows occur between the 12 and 14^0C isotherms. A horizontal map of conductivity averaged over the above temperature interval is presented in Figure 3. The slight cross-lake asymmetry of the conductivity isopleths near the sources of river water reveals the effect of a Coriolis deflection of the incoming river water (cf. Hamblin and Carmack, 1978). From the river discharge values and estimates of the cross-sectional area of the riverine layer the average flows were taken as 1.6 cm/s in the north arm and 2.0 cm/s in the south arm for the purposes of applying the model. Conductivities of 171 µmho cm^{-1} and 113 µmho cm^{-1} were assumed at the south and north ends of the lake, respectively. Distributions of conductivity were computed and are displayed in Figure 4 for three strengths of the dispersion coefficient and for the assumption of horizontal isotropy of dispersion. The distribution for a horizontal dispersion coefficient of 300 m^2/s best matches the observed conductivity field. When the strength of the mixing is reduced by a factor of five a narrow frontal zone is predicted in the vicinity of the outflow; for a mixing coefficient of twice the above value the isopleths are too widely separated along the axis of the lake. It may be shown that it takes approximately two weeks to achieve to steady conditions using a dispersion coefficient of 300 m^2/s.

We may appeal to the shear dispersion effect dicussed above to estimate vertical diffusivity. If we assume a linear vertical shear in the riverine layer of $2\bar{U}/D$, and that the shear dispersion term is much greater than the horizontal diffusivity, then according to the theory outlined above the average vertical eddy diffusivity over the 10 m depth riverine layer is 5 x $10^{-6}m^2$/s.

CONCLUSION

A relatively simple procedure has been briefly outlined for deducing the mixing coefficients between different river water masses from naturally occurring distributions of a conservative substance which is applicable under certain restrictions to fjords and fjord-like lakes of complex plan. In consideration of a more complex system of branching fjords than discussed here Hamblin (1979) has established criteria for treating the system as a one-dimensional branching network.

REFERENCES

Gill, W.N. and R. Sankara. 1971. Dispersion on a non-uniform slug
    in time-dependent flow. Proc. Roy. Soc. Series A, No. 1548,
    Vol. 322, pp 101-117.

Hamblin, P.F. 1979. An analysis of advective diffusion in branching
    channels. Manuscript submitted for publication.

Hamblin, P.F., and E.C. Carmack. 1978. River-induced currents in
    a fjord lake. J. Geophys. Res., Vol. 83: No. C2 pp 885-899.

McClimans, T.A. 1978. Fronts in fjords. Geophys. Astrophys.
    Fluid Dynamics, Vol. 11, pp 23-24.

Smith, R. 1978. Coriolis, curvature and buoyancy effects upon
    dispersion in a narrow channel. Hydrodynamics of Estuaries
    and Fjords. Ed. J.C.J. Nihoul, Elsevier, Amsterdam.

PELAGIC PRODUCTIVITY AND FOOD CHAINS IN FJORD SYSTEMS

J.B.L. Matthews and B.R. Heimdal

Institute of Marine Biology
University of Bergen
N-5065 Blomsterdalen
Norway

FJORD AS BIOTOPES

The fjord systems of the world are all quite young in evolutionary time, yet they offer opportunities that are unique among coastal systems. Their common origin as glaciated coastal valleys and the common hydrographic conditions documented elsewhere in this volume give them a number of characteristic features. From a marine biological point of view the most important of these are:
a) Topography
    i) A deep basin connecting directly or indirectly with the open sea at one end over a relatively shallow sill, typically one half to one tenth basin depth.
    ii) A length many times greater than the width.
b) Climate and hydrography
    iii) Pronounced seasonal change, particularly in temperature and fresh-water run-off, as found at middle and high latitudes.
    iv) Development of a halocline, typically in summer and autumn, between brackish surface water and fully saline subsurface water.
c) Hydrodynamics
    v) Vertical mixing (see Pollard, 1980, this volume) at certain times of year, particularly the autumn.
    vi) Estuarine circulation driven by the freshet.
    vii) Tidal flow acting as a hydraulic pump.
    viii) Density or wind-driven currents bringing water into the fjord over the sill or taking it out at the surface, in both cases with a corresponding counter-current. When the horizontal density difference is great enough, the inflowing water displaces the water in the deep basin.

Table I

Maximum values of phytoplankton number (cells · 10^6 · l^{-1})

Place	Reference	Year	M	A	M	J	J	A	S	O	N
Norway											
Ullsfjord	Heimdal 1966	1963	0.03	0.5	0.04		0.3	0.2	0.04	0.02	0.01
Balsfjord	Eilertsen 1979	1977	0.9	2.1	2.8	0.7	0.4	3.1	0.9		
Skjomen	Schei 1973 a	1970	0.04	0.8	0.2	0.7	0.2		0.8		0.01
"	" 1973 a, b	1971	0.01	1.1	0.3	1.7	4.0	0.1	0.1		0.01
"	" 1979	1978	0.4	1.9	0.5	0.3					
Trondheimsfjord	Hegseth 1977	1975	3.5	17.0							
Nordåsvann	Hope 1943 in Tangen 1974	1941					1.5	1.0	2.2	0.4	
"	Tangen 1974	1969					132.5	308.1	42.3	11.5	4.9
Hardangerfjord	Hjelmfoss 1957 in Tangen 1974	1955		0.7			0.3		0.1	0.1	1.4
"	Øverland 1959 in Tangen 1974	1956					2.6	1.5	2.7		
Søndeledfjord	Østergren 1977	1975		15.0	5.2		0.4	1.0	1.7	6.9	1.4
Drøbak	Hasle & Smayda 1960	1957	1.1		0.3	0.7	0.5	1.4	4.0	6.4	0.3
I. Oslofjord Langvikbukta	Johannessen 1978	1975	16.0	56.7	98.7	97.8	61.1	137.6	7.0	9.3	12.2
I. Oslofjord, Nakkholmen	"	"		39.4	15.4	39.9	17.5	55.9	8.3	9.9	5.7
Sweden											
Byfjord	Söderström et al. 1976	1973	19.7	0.9	16.5		4.5		4.7		
Canada											
Howe Sound	Stockner & Cliff 1976	1974			26.0						
Strait of Georgia entrance to Howe Sound*	Stockner et al. 1979	1976	<0.4	13	1	2	4				0.7

* Maximum number recorded in the surface layer of the Strait of Georgia in 1976 and 1977 was 17.9·10^6·l^{-1}

In the Norwegian language there is a useful distinction between fjords *sensu stricto* and smaller, more enclosed systems known as "polls" (cf. Gaarder & Bjerkan, 1934). The former have sills at 100 m or more, a deep basin or basins at several hundreds of metres, and a total length of tens or even a hundred or more kilometres. Polls have shallow sills at only a few metres depth, basins perhaps 100 m deep, usually rather less, and a length of only a few kilometres. Though there are of course intermediates in size, the important difference is the depth of the sill relative to the pycnocline: Fjord: Sill depth greater than depth of pycnocline.
                  Poll: Sill depth less than depth of pycnocline.
As the word fjord has achieved an accepted meaning in English, so it is suggested that "poll" may also be a useful introduction for a geographical feature for which English does not possess a suitable noun.

FJORDS AS STOCK HOLDERS

Fjords have traditionally provided fish in good quantity to the local populations in Norway, Scotland, British Columbia and Greenland. A briefer tradition is that of fjords as happy hunting grounds for marine biologists. But is this just an impression that fjords are especially rich in marine life in comparison with the open continental shelf? There are the sprat and mackerel fisheries of western and southern Norway which in summer move into the fjords; there are locally renowned fisheries for local herring stocks, Loch Fyne for example, and for early year classes of Atlanto-Scandian herring; and there are the salmon fisheries of British Columbia, Scotland and Norway, highly seasonal affairs catching the fish as they return from the open sea to spawn in fresh water. Compared with the fisheries off the Washington coast, in Queen Charlotte Sound, in the Minch, on Haltenbanken and in Lofoten for example, the catch from the fjords does not seem proportionately great. As far as marine biology is concerned, may it not be that variety is an important attraction of the fjords, as well as the comparative calm of fjord conditions (cf. Brattegard, this volume)? What really are the pelagic stock levels in fjords?

There are as yet no entirely satisfactory methods for measuring biomass in samples taken from the sea and any attempt to review information on standing stock must be qualified by the additional serious reservation that the data have been compiled from disparate sources, the result of different methods of sampling and analysis as well as a reflection of conditions at different seasons or depth zones. This summary of information is therefore accompanied by the cautionary rider that the original studies must not be forgotten when assessing comparability.

Table II

Maximum values of chlorophyll a (mg · m⁻³)

Place	Reference	Year	F	M	A	M	J	J	A	S	O	N	D
Norway													
Trondheimsfjord	Hegseth 1977	1975	0.4	5.9	13.0	6.7	5.9		1.8	4.3	7.5	1.3	
Lindåspolls	Lannergren 1978	1974		10			1.2						
Fauskangerpoll	Sjøvoll pers. comm.	1976	0.6	8.4	6.0	5.1	2.6	2.6	1.4	2.1	1.1		
Korsfjord	Heimdal unpulb.	1978	1.0	0.6	11.2	1.7	1.7		1.2	2.8	1.9	0.5	
I. Oslofjord, Langvikbukta	Johannessen 1978	1975		82.9	41.8	26.9	27.9	52.0	38.7	71.3	8.1	5.3	0.3
I. Oslofjord, Nakkholmen	"	1975	0.3	11.2	44.3	4.9	13.6	7.1	19.8	8.0	6.3	3.6	0.5
Iceland													
Reydarfjørdur	Thórdardóttir pers.comm.	1978					6.9						
Scotland													
Loch Creran	Tett & Wallis 1978	1970–1976	----- 37 -----			------- 7 -------					<1		
Canada													
Strait of Georgia	Parsons et al. 1969b	1967	2–6										
"	Heath 1977	1975				--- 8 ---	------- 8 -------			8	---------		
Saanich Inlet	Takahashi et al. 1977	1974–1975			22	--- 3 - 15 ---			9				
Knight Inlet	Stone 1977	1975	0.5	0.6	6.0	2.0	4.8	12.6	4.6	12.7	<1		
U.S.A.													
Puget Sound	Winter et al. 1975	1966–1967	------ 15 ------			------							

The earliest estimates of standing stock were counts of cells per unit volume of water, a method which can be refined at the time of analysis by multiplying species counts by the estimated mean volume or weight per cell; it is obvious that there is no simple or constant relationship between number of cells and biomass. There seems to be no geographical pattern in the general level of maximum recorded values (Table 1). The highest values in Norway occur in the vicinity of urban areas, namely in Trondheimsfjord, Nordasvann and Inner Oslofjord, and Söderström *et al.* (1976) found a three-point trend towards higher numbers in eutrophicated (polluted) environments on the Swedish west coast. Stockner *et al* (1979) consider that increased primary production in the Strait of Georgia is due in part to eutrophication. It may be concluded that, where such high numbers have been recorded, conditions favour small forms, increase productivity, or decrease grazing pressure – or perhaps that sampling in accessible areas is more frequent than in remote areas, so increasing the chance of recording true maximum values.

An alternative measure of phytoplankton stock, chlorophyll a concentration, has been rather more widely applied in various fjord systems. The method, correctly applied, can produce consistent results for comparison, though chlorophyll is by no means a perfect measure of biomass (see Sakshaug, 1978). Chlorophyll a can reach consistently high levels in the studied fjords (Table II); by comparison 8 mg m^{-3} was a usual level in blooms in the northern North Sea in 1961 and 1962 (Steele and Baird, 1965). The high levels in inner Oslofjord, coupled with the high cell count, indicates that there really can be a large biomass in the area.

Corresponding samples of zooplankton stock are perhaps more easily evaluated since biomass can be measured directly, though the sampling characteristics of the various types of nets that have been used can be very different. There has also been a number of different methods used to estimate zooplankton biomass. The summary of published results (Table 3) is presented in terms of mg dry weight $\cdot$ m^{-3}, on the basis where necessary of simple conversions, as follows: 1 ml displacement volume = 1 g wet weight = 0.2 g dry weight = 0.1 g C. In most cases the biomass is that of the epi-plankton, samples having been taken from 100 m to the surface. It is not possible to relate these values directly to the biomass in the whole water column.

FJORD COMMUNITIES

The pelagic fauna and flora of many fjord systems are well known. A review of such investigations lies outside the scope of this article, so only a few representative studies are referred to here. Most of them stem from that initial period of inquisitive-ness, recording the species found in an area no-one had looked

Table III

Examples of zooplankton standing stock (recalculated where necessary for comparison)

Place	Reference	Year	Month	D[†]	Quantity g dry wt/m²
Norway					
Skjomen, Inner	Strömgren 1974 a	1970	June	35	8
" , Outer	"	1970	June	150	30
Lindåspolls	Haug 1972	1971	July	80	3*
Korsfjord	Matthews & Bakke 1977	1971	August	650	40
Hardangerfjord, Inner	Gundersen 1953	1951	July	300	2
" , Outer	"	1950	July	300	7
Åkrafjord	"	1951	July	500	11
Canada					
Strait of Georgia	Parsons et al. 1969 a	1966	June	100	26
Chile					
Baker Inlets	Brown et al. 1975	1970	March	–	3.2
Middle Inlets	"	1970	March	–	32

* Fixed material
† Depth of sampled water (m)

closely at before. Gran (1902), Nordgaard (1905), Jørgensen (1905), and Runnström (1932) did this on the coast of Norway, Marshall (1949) on the west coast of Scotland, Digby (1954) in East Greenland, Jespersen (1934) in West Greenland, Stockner and his associates (see Stockner *et al.*, 1979) and Stone (1977) in British Columbia.

Without exception these investigations refer to "net" communities, organisms often found dead together in a sample (*pace* Fager, 1963), and with particular emphasis on, if not bias towards, diatoms and adult calanoid copepods. Organisms which are of the same order of size may be competitors for a common resource, most obviously food, but they are less likely to interact directly as, for example, predator and prey. Information relating organisms to each other quantitatively across the whole size spectrum is still lacking, though the stimulus of work by Sheldon *et al.* (1972) and others has begun to focus attention on particle size distribution, as yet at the lower end of the spectrum, in some fjord systems.

In Oslofjord and Korsfjord net phytoplankton was responsible for most of the primary production during periods of bloom (Johannessen, 1978; Heimdal, unpubl.). Although ultraplankton could predominate at other times, its contribution to total annual production was small in both fjords. In Lindaspolls, Skjoldal & Lannergren (1978) found that nanoplankton predominated and that larger forms seemed not to be extensively grazed when they did occur. In Fauskangerpoll, ultraplankton was responsible for 50% of primary production in surface layers, increasing with depth in the euphotic zone (Sjövoll, pers. comm.).

Size fractioning of zooplankton has been little practised hitherto in open fjord systems. McLean (1979) working in the enclosed Lindaspoll system has found that 33% of the total dry weight sampled over a period of a year by a 180 μm mesh Juday net passed through a 1 mm mesh, 70% through 0.5 mm; the small copepod *Oithona* was prominent in the samples. In other years *Pseudocalanus* was the dominant form (Ellingsen 1973). Small cyclopoids are the dominant copepods in Nordasvann (Wiborg, 1944; Johannessen, 1972). Eriksson (1978) found *Oithona* and *Pseudocalanus* to be prominent in Fauskangerpoll. By contrast, the zooplankton in larger fjords is characteristically dominated by *Calanus* (e.g. Lie, 1967; Strömgren, 1974 a). Fosshagen (this volume) has clearly shown the distinction in community between a "*Calanus* fjord", Jøsenfjord, and a "non-*Calanus* fjord", Hylsfjord.

The size characteristics of the plankton community seem not to change drastically along the length of a fjord, though concentrations may become both heavier and more patchy as a consequence of water movements in constricted areas (Wiborg and Bjørke, 1968), and smaller forms may become relatively more numerous in the innermost reaches (Lie, 1967). The poll community is in general an

Table IV

Values of daily primary production (mg C · m⁻² · day⁻¹)

Place	Reference	Year	F	M	A	M	J	J	A	S	O	N
USSR												
Barents Sea Fjord	Propp 1976							343				
Norway												
Balsfjord	Eilertsen 1979	1977		260	1250	1230	695	360	920	235		
Straumsbukta	Throndsen & Heimdal 1976	1974			684			2300		380		
Lindaspoll	Lännergren 1976	1973			———		150–900	———			<300	
Fauskangerpoll	Sjøvoll pers. comm.	1976	139	1947	1001	930	737	1103	547	324	226	
Korsfjord	Heimdal unpubl.	1978	162	68	1993	1635	374	313		684	429	15
I. Oslofjord, Langvikbukta	Johannessen 1978	1975	1	695	3759	5612	2638	1025	1642	1667	35	42
I. Oslofjord, Nakkholmen	"	"	8	791	4530	2767	1483	957	1769	504	157	140
Sweden												
Byfjord	Söderström et al. 1976	1972							892	528	308	
Kungsbackafjord	Olsson & Olundh 1974	1970			180	190	793	675	500	725	675	
Greenland												
Godthaabfjord, entrance	Steemann Nielsen 1958	1953–1956		600	455	425	1200	700	750	300	100	
Godthaabfjord, inner part	"	"			1240		390		750	100	60	
Canada												
Strait of Georgia*	Parsons et al. 1970	1965–1968	250	470	550	1200	320	270	310	200	230	120
Strait of Georgia, Fraser River plume	Parsons " 1969 b	1967		140	1135	2132	353					
Indian Arm	Stockner & Cliff 1979	1975–1976			—— 6640 ——							
Howe Sound	Stockner et al. 1977	1973–1974			—— 4893 ——			5712				
USA												
Valdez Arm	Goering et al. 1973	1971–1972					———— 4000 ————					

* Maximum value recorded was 2562 mg C · m⁻² · day⁻¹

ultraplankton - microzooplankton community, the fjord community
chiefly a net phytoplankton - macro/megazooplankton community.

There has been much discussion of the existence of alternative
food chains, where the smaller size fractions lead to medusae and
ctenophores instead of to fish as in the case of the larger phytoplankton and zooplankton fractions (see Landry, 1977). Large
stocks of hydromedusae and rapid population growth have been
recorded in inner fjord systems (Huntley & Hobson, 1978) and
ctenophores are a common component of the plankton in such areas
(Parsons *et al*. 1970). In Norway the bathypelagic scyphomedusan,
Periphylla periphylla, is characteristic of the deep fjord basins
at a considerable distance from the open coast (Fosshagen, pers.
comm.) where to a large extent it must be isolated from oceanic
stocks. The indigenous pelagic fish in inner fjord systems are
small; the myctophid, *Benthosema glaciale*, is more abundant there
than further out towards the coast (Gjösæter, pers. comm.) and the
landlocked herring stock of Lindaspolls are smaller and grow more
slowly than the conspecific coastal stocks (Lie *et al*., 1978).
These differences in community structure have not been quantified
in terms of community energetics.

FJORDS AS PRODUCTIVE SYSTEMS

Estimates of primary production exist for a number of fjords
in British Columbia, Scandinavia and Greenland (Table IV). In most
cases incubations have been taken preferentially around the time of
the spring bloom, so the chances are that most measurements are high
but not necessarily maximal. Since production is related to the
level of the producing stock (as opposed to productivity which can
be defined as the production achieved by unit stock of phytoplankton
per unit volume or surface area), areas of high stocks in Scandinavia
tend to show high production as well. Other areas, particularly
in inner reaches of the British Columbian fjord system (Howe Sound
and Indian Arm), for which there are not equivalent biomass data,
show that other fjord systems are also potentially very productive.

By their nature and definition, fjords are subject to marked
seasonal change in climate and hydrography. The daily values
obtained under bloom conditions cannot therefore be expanded to
give annual primary production estimates. There are, however, a
number of areas where such estimates have been made on the basis
of year-round measurements (Table V). The various authors have
placed reservations on the precision of these results, since light,
vertical movement and other factors can change so much more
frequently than it is possible to run production measurements.
Nevertheless, annual production seems to be commensurate with the
general level of production in coastal waters, estimated by Ryther
(1969) at 100 g C . m^{-2} . yr^{-1}. The possibility seems to exist,
however, for quite remarkably high levels to occur, where boundary

Table V

Estimates of annual phytoplankton production

Place	Reference	Year	Primary production, g C · m⁻² · yr⁻¹
General			
Coastal Zone	Ryther 1969	1969	100
Upwelling areas	"	"	300
Norway			
Balsfjord	Eilertsen 1979	1977	110
Lindaspolls	Lännergren 1976	1973	90-100
Sweden			
Kungsbackafjord	Olsson & Olundh 1974	1970	100
Greenland			
Godthaabfjord, entrance	Steemann Nielsen 1958	1953-1956	98
" , inner part	"	"	95
Canada			
Howe Sound, entrance	Stockner et al. 1977	1973	300
" " , entrance	"	1974	516
" " , inner stations	"	1973	118
" " , " "	"	1974	163
Port Moody Arm	Stockner & Cliff 1979	1975-1976	532
Indian Arm and the Narrows region	"	"	260
Strait of Georgia	Parsons et al. 1970	1965-1968	120
Strait of Georgia, Fraser River Plume	Stockner et al. 1979	1975-1977	149
Strait of Georgia, sheltered boundary waters of inlets	"	"	511
USA			
Puget Sound	Winter et al. 1975	1966-1967	465
Port Valdez	Goering et al. 1973	1971-1972	150
Valdez Arm	"	"	200

conditions, land run-off, or "eutrophication" give particularly nutrient-rich conditions. Peculiar conditions of vertical stability can give rise to unseasonal blooms, even in November and December in south-western Norway (Braarud, 1974; Nygaard, pers. comm.), which contribute to the annual production, though it is doubtful whether such blooms effectively stimulate secondary production.

No method yet exists for the rapid and exclusive measurement of secondary production (see Edmondson and Winberg, 1971), but estimated values of the production: biomass ratio (turn-over rate) have been used to convert a few values of standing stock to production, in all cases with reservations arising from the use of published values for turn-over. The production of some other populations which are prominent in their particular communities has been computed from data on their population dynamics using techniques of systems identification (see Parslow *et al*., in press). Such data as are available for fjord systems are summarized in Table VI. The periods over which production has been estimated are very different and few estimates cover the whole production period.

The information on carnivorous production is yet more scanty. Large stocks of medusae and ctenophores are a commonplace occurrence and population growth can be high under suitable conditions (Huntley and Hobson, 1978) but this has not yet been quantified for more than single species over a short space of time; *in situ* fish production has not yet been satisfactorily estimated in fjord systems, partly for reasons to be considered in the next section. There are also considerable stocks of smaller planktonic carnivores which can be studied alongside *Calanus* etc. and which represent an extra, carnivorous link in the food chain still available to plankton-eating fish. Assuming P/B ratios are no higher for carnivores than for herbivores (carnivores tend to have the longer life span and to be less subject to predation) information on carnivore stocks can indicate production levels which will err if anything on the high side. *Eukrohnia hamata* (Chaetognatha) and *Euchaeta norvegica* (Copepoda) have the highest stock levels of all macroplanktonic carnivores in Korsfjord. (Matthews & Bakke, 1977), but to them must be added a large number of less abundant species and the carnivorous component of many apparent omnivores. It has been estimated that carnivores comprise about 5% of the macroplankton of Korsfjord (Matthews & Bakke, *op. cit*.), probably equivalent to rather less than 5% in terms of production.

On the basis of the above values some cautious conclusions may be drawn, propositions which can be more rigorously tested in future studies:
1. The enormous variation in primary production in inner systems (polls) seems to be associated with low ecological efficiency, at least in the case of the larger size fraction. High

Table VI

Estimates of zooplankton production

Place	Reference	Year	Period	D	Method	Quantity g dry wt. m^2	Comments
Norway							
Lindaspolls, Inner	Ellingsen 1973	1971	Apr-Sept	80	P/B	0.4	Small copepods
" , "	"	1971	"	80	"	0.16	*Sagitta elegans*
" , Outer	"	1971	"	80	"	0.4	Small copepods
" , "	"	1971	"	80	"	0.12	*Sagitta elegans*
Korsfjord	Matthews 1973	1971–72	one year	650	R.&gr.r.	0.8 –1.25	*Meganyctiphanes norvegica*
"	Calc.fr. Matthews et al. 1978	1972–73	"	650	"	800	*Calanus finmarchicus*
Sweden							
Kungsbackafjord	Olsson & Ölundh 1974	1969–70	May–Oct	16	P/B & B.I.	1.6–2.8	Copepods
Canada							
Strait of Georgia	Parsons et al. 1969a	1967	Feb–May	100	B.I. & gr.r.	15	Chiefly *Calanus plumchrus*
" "	" 1970	1965–67	one year	20	" "	17.3	Herbivores
Saanich Inlet	Heath 1977	1975	July–Nov	150	B.I.	5.6	*Euphausia pacifica*
"	Huntley & Hobson 1978	1975	June	50	"	0.55	*Phialidium gregarium*

* P/B = Ratio of production to biomass (see Edmondson & Winberg 1971), using published turnover rates

R = Observed increase in number of recruits per unit time

gr.r = Growth rate derived from changes in size distribution per unit time

B.I. = Observed increase in biomass per unit time

D = Depth of sampled water column (m)

efficiency may be achieved in outer areas, particularly where advection can increase herbivore stocks.
2. Herbivore biomass and hence secondary production may be markedly affected by predation.
3. Carnivore production can be relatively high when secondary production is low, either seasonally or geographically.

FJORDS AS PART OF LARGER SYSTEMS

Some of the Korsfjord figures, notably the *Calanus* production in relation to primary production, are impossible values for a static water column. Much of the *Calanus* production can only be explained as advective production. Another example of the phenomenon is demostrated by the stock of a single generation of *Meganyctiphanes norvegica* as represented in the samples from a mid-fjord water column in Korsfjord (Fig. 1). The recorded biomass follows a normal recruitment pattern in the first summer but declines drastically the following winter and early spring. The same generation returns in its second summer and crashes again early the following year. Young and parents, even indications of grand-parents, fluctuate in unison. Production estimates for Korsfjord (see Table VI) are based on actual appearances, i.e. both periods of increase in Fig. 1 count as production, while the reality for the component generations in the population would show a virtual cessation of production by the early summer of the second year, at an age of about 15 months.

Stömgren (1974 b) has discussed the differential effect of fjord circulation on the various developmental stages of *Calanus finmarchicus* ; Sands (in press) has concluded that only one of the five species of chaetognath recorded in Korsfjord actually breeds there successfully; Matthews & Pinnoi (1973) found that *Pasiphaea sivado* seemed not to be able to maintain a population in Korsfjord without recruitment from outside; and Stone (1977) has demonstrated the incursion of an offshore copepod community into Knight Inlet. On a more local scale, Wiborg & Bjørke (1968) have found aggrega-tions of *Calanus*, particularly at the head of fjords or in narrow sounds, which appear to be the result of water movements acting on stocks with preferred vertical distributions.

These examples illustrate the open-ended nature of fjord communities, which nevertheless seem to maintain individual or group characteristics as discussed above and summarized in Fig. 2.

It is tempting to explain these differences in a) the phytoplankton and herbivores in terms of stratification and water exchange in surface layers, and b) the relative abundance of planktonic nectonic, demersal, and benthic carnivores in terms of the length of the water column.

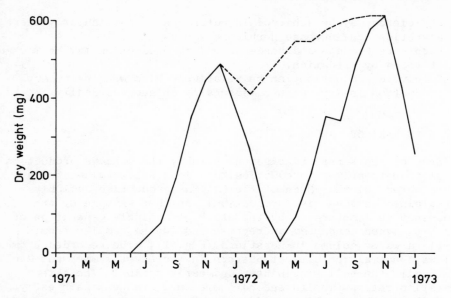

Figure 1. The biomass of the 1971 generation of *Meganyctiphanes*
           norvegica under 1 m² in Korsfjord. The broken line
           indicates the supposed biomass of the total year
           class including that part which disappeared from the
           sampling area in spring.

    Upwelling areas are generally considered to provide good
conditions, of nutrient supply and vertical water movement, for
diatom growth. Such conditions are found in many coastal areas
and often near the head of fjords. Production seems to be higher,
however, in the outer fjord areas where upwelling appears to be
limited to entrained water from the counter current. For high
levels of primary production in spring the really significant
event may be the seasonal breakdown of vertical stability and mixing
of water throughout the water column in the previous autumn and
winter. The newly advected deep-basin water is only available as
a growth medium when it is lifted to the surface a year later.
On the other hand, the restricted conditions in more enclosed
systems, particularly in "polls", and especially when overlaid
with brackish, almost fresh, water, gives "ultraplankton conditions".
The different phytoplankton communities support herbivores of
different sizes.

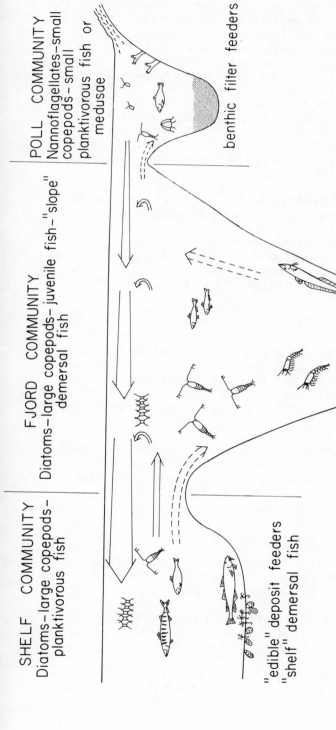

Figure 2. A schematic representation of a fjord system from continental shelf (left) to fjord basin (centre) to "poll" (right). The near-surface arrows represent the freshet, augmented by entrained water (curved arrows) and compensated for by the counter current at middle depth. The broken arrows indicate the periodic inflow of water replacing old resident water in the fjord and poll basins. Stagnation is indicated in the basin of the poll whose sill breaks the pycnocline. Note the increased aggregation and decreasing diversity towards the head of the fjord.

Further development of the food chain may be dependent on the depth of the water column. In shallow inlets benthic herbivores like filter-feeding bivalves and tunicates can abound. In a shallow sea on the continental shelf fall-out and settling can support a deposit-feeding benthos and epibenthos on which demersal stocks of fish can prey. In a deep fjord basin mesopelagic organisms - myctophids, *Periphylla*, *Beroe*, *Pasiphaea* etc. - provide an extra link in the food chain, thus reducing the food available to the benthos and demersal fish. Such fish as do occur are "slope fish" like ling and cusk. (There are of course shallower areas in fjords and appropriate fish, cod and plaice for example, are found there.) In the polls anoxic deep water may effectively eliminate benthos and demersal fish.

However, this may be, fjords play an important part in the spatial cycle of life in many high-latitude coastal areas, productive regions all of them. Their most obvious function of use to man is that of a nursery ground for pelagic fish stocks. Fjords as physical systems cannot function without specific external hydrographic and meteorological conditions. Interaction is equally essential in maintaining the biological functioning of fjords.

ACKNOWLEDGEMENTS

We are most grateful to those people mentioned in the text whose personal communications have greatly added to the body of information in this paper, and particularly to the many former students whose theses contain important results which are not yet widely accessible in the published literature. It is hoped that they will become so.

We believe the information has not been used out of context, but we take responsibility for the assumptions and calculations made in order to summarize the data, compare results, and draw conclusions.

REFERENCES

Braarud, T. 1974. The natural history of the Hardangerfjord.II. The fjord effect upon the phytoplankton in late autumn to early spring 1955-56. Sarsia, 55, 99-114.

Brattegard, T. 1980. Why biologists are interested in fjords. Fjord Oceanographic Workshop, Institute of Ocean Sciences, Patricia Bay, Sidney, B.C.

Brown, R.G.B., F. Cooke, P.K. Kinnear and E.L. Mills. 1975. Summer seabird distribution in Drake Passage, the Chilean fjords and off Southern South America. Ibis, 117, 339-356.

Digby, P.S.B. 1954. The biology of marine plankton copepods of Scoresby Sound, E. Greenland. J. Anim. Ecol., 23, 298-338.

Edmondson, W.T. and G.G. Winberg. 1971. A manual on methods for assessment of secondary productivity in fresh waters. IBP Handbook No. 17, Blackwell, Oxford, pp. 358.

Eilertsen, H.C. 1979. Planteplankton, minimumfaktorer og primær-produksjon i Balsfjorden, 1977. Cand. real. thesis, Univ. Tromsø, pp. 130.

Ellingsen, E. 1973. Kvalitative og kvalitative zooplankton-undersøkelser i Lindaspollene. Mars 1971-april 1972. Cand. real. thesis. Univ. Bergen, pp. 100.

Eriksson, L. 1978. Holoplankton i Fauskangerpollen. En undersø-kelse av vertikalvandring under døgnet. Cand. real. thesis Univ. Bergen, pp. 153.

Fager, E.W. 1963. Communities of organisms. In The Sea.: Ideas and observations on progress in the study of the sea. Vol. 2 M. N. Hill(ed.), Interscience, N.Y., pp. 415-437.

Fosshagen, A. 1980. How the zooplankton community may vary within a single fjord system. Fjord Oceanographic Workshop, Institute of Ocean Sciences, Patricia Bay, Sidney, B.C. (This volume)

Gaarder, T. and P. Bjerkan. 1934. Østers og østerskultur i Norge, Grieg, Bergen, pp. 96.

Goering, J.J., W.E. Shiels and C.J. Patton. 1973. Primary Product-ions, pages 251-278 in D.W. Hood, W.E. Shiels and E.J. Kelley, editors, Environmental Studies of Port Valdez, University of Alaska, Institute of Marine Sciences occasional publication #3, 496 pp.

Gran, H.H. 1902. Das Plankton des Norwegischen Nordmeeres. Rep. Norw. Fish Mar. Invest., 2 (5), 1-222.

Gundersen, K.R. 1953. Zooplankton investigations in some fjords in western Norway during 1950-1951. Fisk. Dir. Skr. Ser. Hav-unders., 10, 1-54.

Hasle, G.R. and T.J. Smayda. 1960. The annual phytoplankton cycle of Drøbak, Oslofjord. Nytt Mag. Bot., 8, 53-75.

Haug, A. 1972. Biomasse-undersøkelser av zooplantkon i Linda-spollene i tidsrommet mars 1971 - april 1972. Can. real. thesis University of Bergen, pp. 98.

Heath, W.A.   1977.   The ecology and harvesting of euphausiids in the
     Strait of Georgia.   Ph.D. thesis, Univ. British Columbia, pp.187.

Hegseth, E.N.   1977.   Artssammensetning, kjemisk sammensetning og
     minimumsfaktorer for vekst hos planteplankton under første
     varoppblomstring i Trondheimsfjorden, 1975.   Cand. real.
     thesis, Univ. Trondheim, pp. 195.

Heimdal, B.R.   1966.   Planteplanktonet i en nord-norsk fjord,
     Ullsfjord i Troms, fra november 1962 til desember 1963.
     Cand. Real. thesis, Univ. Oslo, pp. 215.

Huntley, M.E. and L.A. Hobson.   1978.   Medusa predation and plankton
     dynamics in a temperate fjord, British Columbia.   J. Fish.
     Res. Bd. Can., 35, 257-261.

Jespersen, P.   1934.   The Godthaab Expedition 1928.   Copepoda.
     Medd. Grønland, 79 (10), 1-166.

Johannessen, P.   1972.   Undersøkelser i Nordasvannet 1969-1970.
     Hydrografi, planktoniske copepoder, og en kort oversikt over
     meduser og ctenophorer.   Cand. real. thesis, Univ. Bergen,
     pp. 159.

Johannessen, T.   1978.   Primærproduksjonen pa to stasjoner i indre
     Oslofjord (Langvikbukta og Nakkeholmen).   Cand. real. thesis,
     Univ. Oslo, pp. 245.

Jørgensen, E.   1905.   Hydrographical and biological investigations
     in Norwegian fiords.   B. Protistplankton.   Bergen Mus. Skr.,
     7, 49-151.

Landry, M.R.   1977.   A review of important concepts in the trophic
     organization of pelagic ecosystems.   Helgol. wiss. Meeresun-
     ters., 30, 8-17.

Lännergren, C.   1976.   Primary production in Lindaspollene, a
     Norwegian land-locked fjord.   Botanica Marina, 19, 259-272.

Lännergren, C.   1978.   Phytoplankton production at two stations
     in Lindaspollene, a Norwegian land-locked fjord, and limiting
     nutrients studied by two kinds of bio-assays.   Int. Rev. ges.
     Hydrobiol., 63, 57-76.

Lie, U.   1967.   The natural history of the Hardangerfjord. 8.
     Quantity and composition of the zooplankton.   Sept. 1955-
     Sept. 1956.   Sarsia, 30, 49-74.

Lie, U., O. Dahl and O.J. Østvedt.   1978.   Aspects of the life
     history of the local herring stock in Lindaspollene, western
     Norway.   Fisk. Dir. Skr. Ser. Havunders., 16, 369-404.

Marshall, S.M. 1949. On the biology of the small copepods in Loch Striven. J. Mar. Biol. Ass. U.K., 28, 45-122.

Matthews, J.B.L. 1973. The succession of generations in a population of *Meganyctiphanes norvegica*, with an estimate of energy flux. Int. Counc. Explor. Sea. Ann. Mtg., 1973, Plankton Ctee L:19, 1-6.

Matthews, J.B.L. and J.L. W. Bakke. 1977. Ecological studies on the deep-water pelagic community of Korsfjorden, western Norway. The search for a trophic pattern. Helgol. wiss. Meeresunters., 30, 47-61.

Matthews, J.B.L. and S. Pinnoi. 1973. Ecological studies on the deep-water pelagic community of Korsfjorden, western Norway. The species of *Pasiphaea* and *Sergestes* (Crustacea, Decapoda) recorded in 1968 and 1969. Sarsia 52, 123-143.

McLean, E.S. 1979. Zooplankton production and respiration in Lindaspollene, western Norway: Biochemical determinations on mixed populations. Cand. real. thesis, Univ. Bergen, pp. 94.

Nordgaard, O. 1905. Hydrographical and biological investigations in Norwegian fiords. A. The greater forms of animal plankton. Bergens Mus. Skr., 7, 23-48.

Olsson, I. and E. Ölundh. 1974. On plankton production in Kungsbacka Fjord, an estuary on the Swedish west coast. Marine Biology, 24, 17-28.

Østergren, I. 1977. Fytoplanktonundersøkelser i Søndeledfjorden og i kystvannet ved Risør, April 1976. Cand. real. thesis, Univ. Oslo, pp. 174.

Parslow, J., N.C. Sonntag and J.B.L. Matthews. In Press: Technique of systems identification applied to estimating copepod population parameters. J. Plankton Res.

Parsons, T.R., R.J. Le Brasseur, J.D. Fulton and O.D. Kennedy. 1969a: Production studies in the Strait of Georgia. Part II. Secondary production under the Fraser River plume, February to May 1967. J. Exp. Mar. Biol. Ecol., 3, 39-50.

Parsons, T.R., K. Stephens and R.J. Le Brasseur. 1969b. Production studies in the Strait of Georgia. Part I. Primary production under the Fraser River plume, February to May 1967. J. Exp. Mar. Biol., 3, 27-38.

Parsons, T.R., R.J. Le Brasseur and W.E. Barraclough. 1970. Levels of production in the pelagic environment of the Strait of

Georgia, British Columbia: a review. J. Fish. Res. Bd. Can.,
    27, 1251-1264.

Propp, M.V. 1976. Ecological system of a fjord-type gulf of the
    Barents Sea. Sov. J. mar. Biol., 2, 95-102.

Runnström, S. 1932. Eine Uebersicht über das Zooplankton das
    Herdlaund Hjeltefjordes. Bergens Mus. Arb., (7), 1-67.

Ryther, J.H. 1969. Photosynthesis and fish production in the sea.
    Science, 166, 72-76.

Sakshaug, E. 1978. The influence of environmental factors on the
    chemical composition of cultivated and natural populations
    of marine phytoplankton. Dr. philos. thesis, Univ. Trondheim,
    pp. 82.

Sands, N.J. In press. Ecological studies on the deep-water pelagic
    community of Korsfjorden, western Norway. Population dynamics
    of the chaetognaths from 1971 to 1974. Sarsia, 65.

Schei, B. 1973a. Planktonundersøkelser i Skjomen, en arm av
    Ofotfjorden, mars 1970-April 1971. Cand. real. thesis, Univ.
    Tromsø, pp. 182.

Schei, B. 1973b. Planteplanktonundersøkelser i Skjomen mai-
    november 1971. Tromsø Mus. Prelim. Rep., No. 4, pp. 17.

Schei, B. 1979. Planteplanktonundersøkelser i Skjomen mars-mai
    1978. Prelim. Rapp. radg. utv. fjordunders., 2/1979, pp. 1-16.

Sheldon, R.W., A. Prakash and W.H. Sutcliffe Jr. 1972. The size
    distribution of particles in the ocean. Limnol. Oceanogr.,
    17, 327-340.

Skjoldal, H.R. and C. Lännergren. 1978. The spring phytoplankton
    bloom in Lindaspollene, a land-locked Norwegian fjord. II.
    Biomass and activity of net- and nanoplankton. Marine Biology,
    47, 313-323.

Søderstrøm, J., B.Rex, M. Rex and E. Hildenwall. 1976. Byfjorden:
    Marinbotaniska undersøkningar. Statens Naturvardsverk, PM
    684, pp. 218.

Steele, J.H. and I.E. Baird. 1965. The chlorophyll a content of
    particulate organic matter in the northern North Sea. Limnol.
    Oceanogra., 10, 261-267.

Steemann Nielsen, E. 1958. A survey of recent Danish measurements
    of the organic productivity in the sea. Rapp. B. -v. Cons.
    perm. Int. Explor. Mer., 144, 92-95.

Stockner, J.G. and D.D. Cliff. 1976. Effects of pulpmill effluent on phytoplankton production in coastal marine waters of British Columbia. J. Fish. Res. Bd. Can., 33, 2433-2442.

Stockner, J.G. and D.D. Cliff. 1979. Phytoplankton ecology of Vancouver Harbour. J. Fish. Res. Bd. Can., 36, 1-10.

Stockner, J.G., D.D. Cliff and D.B. Buchanan. 1977. Phytoplankton production and distribution in Howe Sound, British Columbia, a coastal marine embayment-fjord under stress. J. Fish. Res. Bd. Can., 34, 907-917.

Stockner, J.G., D.D. Cliff and K.R.S. Shortreed. 1979. Phytoplankton ecology of the Strait of Georgia, British Columbia. J. Fish. Res. Bd. Can., 36, 657-666.

Stone, D.P. 1977. Copepod distributional ecology in a glacial run-off fjord. Ph.D. thesis, Univ. British Columbia, pp. 256.

Strömgren, T. 1974a. Zooplankton investigations in Skjomen 1969-1973. Astarte, 7, 1-15.

Stromgren, T. 1974b. Zooplankton and hydrography in Trondheims-fjorden on the west coast of Norway. K. norske Vidensk. Selsk. Mus. Miscellanea, 17, 1-35.

Takahashi, M., D.L. Seibert and W.H. Thomas. 1977. Occasional blooms of phytoplankton during the summer in Saanich Inlet, B.C., Canada. Deep-Sea Res., 24, 775-780.

Tangen, K. 1974. Fytoplankton og planktoniske ciliater i en forurenset terskelfjord, Nordasvatnet i Hordaland. Cand. real. thesis, Univ. Oslo, pp. 449.

Tett, P. and A. Wallis. 1978. The general annual cycle of chlorophyll standing crop in Loch Creran. J. Ecol., 66, 227-239.

Throndsen, J. and B.R. Heimdal. 1976. Primary production, phytoplankton and light in Straumsbukta near Tromsø. Astarte, 9, 51-60.

Wiborg, K.F. 1944. The production of zooplankton in a landlocked fjord. The Nordasvatn near Bergen in 1941-42. Fisk. Dir. Skr. Ser. Havunders., 7, 1-83.

Wiborg, K.F. and H. Bjørke. 1968. Utbredelse av raudate i kyst- og fjordstrøk sør for Bergen i mai-juni 1968 og muligheten kommersiell utnyttelse av dyreplankton. Fiskets Gang, 42, 727-730.

Winter, D.F., K. Banse and G.C. Anderson. 1975. The dynamics
    of phytoplankton blooms in Puget Sound, a fjord in the north-
    western United States. Marine Biology, 29, 139-176.

HOW THE ZOOPLANKTON COMMUNITY MAY VARY WITHIN A SINGLE FJORD SYSTEM

Audun Fosshagen

Institute of Marine Biology
N-5065 Blomsterdalen
Norway

INTRODUCTION

This zooplankton investigation in southwestern Norway was started as a part of a monitoring program before a hydroelectrical power plant was built. Field investigations were started in 1972 and finished in 1975. The investigated area consists of three fjord branches, connected with a main fjord which leads to the open sea (Figure 1).

The branches have deep basins (320 to 670 m) and all have a sill depth of about 100 m where they open into the main fjord. Jøsenfjorden has two large rivers near the head. Erfjorden is little influenced by freshwater. There is little freshwater run-off directly into Hylsfjorden, mainly from small rivers, but there is considerable freshwater inflow from outside the fjord. Three different water masses occur:

1. An upper layer, 0 to 20 m, of brackish water - with marked vertical gradients in temperature and salinity in summer and autumn. Major seasonal and short term changes take place.
2. An intermediate layer, 20 to 100 m, where conditions are less variable than in the upper layer. Water movements are mainly caused by differences in density between corresponding depths outside and inside the sill.
3. Basin water, 100 m to the bottom, which is characterized by rather stable conditions with small ranges of temperature and salinity (6.5^0 to 8^0C, c.35 $^0/_{00}$).

The deep water in these fjords was always well aerated. (3.6 - $6.0m\ell/\ell$). Complete physical data are presented elsewhere (Svendsen and Utne, in press).

399

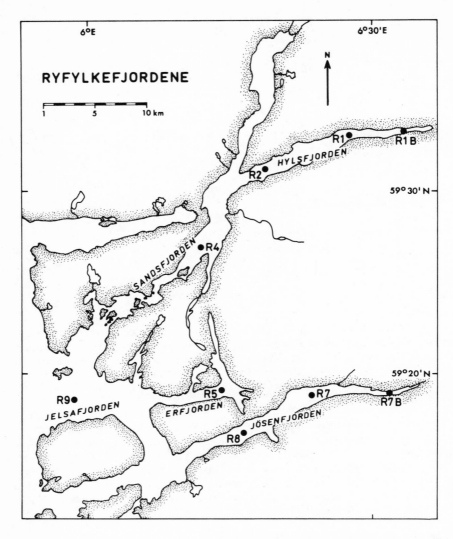

Figure 1. Map of the investigated area with zooplankton stations.

        Zooplankton were collected at 9 stations on 30 cruises (10 per
year), mainly by vertical hauls with a Juday closing net, mesh size
180 μm, from the bottom to 100 m and from 100 m to the surface.

        A little over 500 samples have been analysed into species or
groups of species. Half the sample of each net haul was used for
determining the biomass as dry weight (mg m^{-3}) in the whole water
column.

The seasonal variation in biomass (Table 1) was more pronounced in the outer area (R5 to R9) than in the inner area (R1 to R4). The maximum usually occurred in the autumn. The total annual biomass was about four times larger in Jøsenfjorden than in Hylsfjorden. Many differences have been noted in species composition and abundance between the different areas of the fjord system.

A brackish water community was conspicuous in the Sandsfjord system (R1 to R4), restricted to the uppermost 3-5 m, with the copepod *Eurytemora affinis*, the cladocerans *Evadne nordmanni* and *Podon* spp. and rotifers as the most important members. *E. affinis* was found nowhere else.

The following species and groups were found in greater abundance in the Sandsfjord system than elsewhere: Bivalve larvae, polychaete larvae, the copepods *Aetidius armatus* and *Oncaea* spp. and the ostracods *Conchoecia* spp.

In the outer area Cyphonautes larvae, *Calanus finmarchicus*, *C. hyperboreus* and *Euchaeta norvegica* were more abundant than in the inner area. The much larger biomass in the outer area than in the inner was mainly due to *C. finmarchicus*.

The following taxa were relatively abundant in the spring of 1974: bivalve larvae, polychaete larvae, cirripede larvae, echinoderm larvae, Hydrozoa, *Evadne nordmanni* and *Fritillaria borealis*. This assemblage seems to have been associated with a massive inflow of deep water which lifted up the old bottom water, relatively warm and rich in nutrients, to the upper water masses, at least as far as 5-10 m depth.

In August of the same year, an occurrence of the pteropod *Spiratella retroversa* in particular seems to have been connected with water transport into the fjords in the intermediate layer. In July the species was found only in the main fjord but in August it was abundant in the inner part of all branches. An inflow of water, with high temperature, low salinity and a low content of nutrients, seems to have taken place in the upper part of the intermediate layer, while water flowed out of the fjord near the sill depth. The properties of the inflowing water indicate that it must have had its origin near the surface.

An inflow of deep water restricted to the main fjord took place in the spring of 1975. The result was a strong increase in the number of *Conchoecia* spp., *Microcalanus* spp., *Spinocalanus brevicaudatus* and *Calanus hyperboreus*, all deep water species, which must have been introduced from outside the fjord from the North Sea.

Table 1. Biomass (dry wt. mg m^{-3}) from 30 divided vertical hauls
          at 3 stations in Ryfylkefjordene with a Juday closing net
          during 1973-1975 [from Fosshagen (in press)].

Date		R1		R5		R7	
		D_1	D_2	D_1	D_3	D_1	D_4
1973							
Jan	18-19	0.25	1.45	0.35	11.30	0.68	27.11
March	15-16	2.89	1.35	11.16	2.31	6.32	0.80
April	12-13	1.75	0.95	10.79	4.51	6.00	0.94
May	10-11	5.16	2.20	6.54	8.55	11.50	3.63
June	18-20	5.20	2.71	18.07	21.90	15.83	4.77
July	19-20	3.92	0.81	25.33	20.70	16.17	5.49
Aug	16-17	4.62	3.02	26.08	31.55	21.72	17.05
Sept	11-12	5.54	2.73	10.01	42.10	3.71	14.21
Oct	11-12	6.57	2.84	11.42	34.55	7.68	24.25
Nov	8-9	1.11	2.23	0.53	36.39	1.93	16.85
1974							
Jan	17	0.06	0.99	3.62	13.17	5.47	14.85
Feb 28-Mar 1		0.81	0.74	1.20	1.79	3.13	0.84
April	4-5	13.39	0.55	18.21	3.71	16.44	1.61
May	9-10	15.10	3.71	23.58	7.50	52.17	3.67
June	11-12	5.06	3.97	9.16	9.10	7.04	5.02
July	16-17	10.07	1.98	11.97	8.63	22.70	4.27
Aug	22-23	38.65	4.14	33.23	22.56	96.43	15.80
Sept	26-27	12.64	1.59	12.78	22.87	3.37	32.26
Oct	29-30	2.31	4.17	5.57	51.45	1.28	38.89
Dec	5-6	5.37	3.61	2.37	31.52	1.15	21.82
1975							
Jan	23-24	0.40	1.75	1.86	24.16	1.29	16.35
Feb	25-26	1.95	1.65	9.49	4.59	46.85	0.75
April	4-5	11.04	1.54	37.13	5.84	14.81	1.46
May	8-9	10.70	4.85	38.06	7.35	20.02	2.84
June	5-6	3.36	6.17	20.52	14.28	15.17	2.61
July	10-11	3.67	5.32	47.29	36.82	25.62	7.93
Aug	21-22	12.43	4.40	88.20	47.60	105.60	27.90
Sept	18-19	1.67	3.12	21.04	33.30	30.03	29.14
Oct	23-24	0.56	2.48	2.53	23.65	0.68	27.18
Nov	19-20	0.73	1.79	6.01	32.47	1.30	25.16

Depth ranges D_1, D_2, D_3, and D_4 for the hauls are 100-0m, 300-100m,
245-100m, and 275-100m, respectively.

At times when strong and steady winds were blowing out of the fjord plankton might be concentrated at the head of Jøsenfjorden. This was particularly striking in August 1975, when large concentrations of *Oikopleura* spp. and *Calanus finmarchicus* were found.

At the head of Hylsfjorden no such concentration of plankton could be observed at the same time, although the two fjords have the same directions and probably are exposed to much the same forces of wind. The reason for the differences in the abundance of the plankton seem to be found in the thickness of the brackish water layer of the fjords and in the topography of the fjords.

In Jøsenfjorden this upper layer usually has a thickness of about 1-2 m while the same layer in Hylsfjorden has a thickness of 3-5 m. Because of the large runoff of freshwater and the open mouth of Jøsenfjorden, the transport out of the fjord will be large, and a resulting countercurrent will bring in saline water and water rich in plankton to the head of the fjord. The wind blowing out of the fjord may increase this transport.

The runoff of freshwater into Hylsfjorden proper is about 20% of that of Jøsenfjorden and from several small rivers. However, most of the brackish water in the fjord seems to have been transported into it from the outside. As the mouth of the Sandsfjord system is narrow this will probably hamper the outflow of freshwater.

The wind alone does not seem to be able to start a surface current and a resulting countercurrent in the saline water below the halocline in Hylsfjorden to bring in plankton to the head of the fjord.

The productivity in the Sandsfjord system seems to be lower than that of the outer parts, (R5,R9), judging from the low biomass of zooplankton. One of the reasons may be found in the prominent brackish water layer. A brackish water community becomes established, indicating a long residence time for the water in the system. Because of great stability, the thick layer of brackish water may act as a lid covering the saline water and water rich in nutrients. Wind mixing will be hampered and nutrients prevented from reaching the upper zone.

The turbidity of the upper layer in Hylsfjorden was greater than that of Jøsenfjorden. Measurements of the Secchi disc depth in Hylsfjorden showed this depth to be about half that of Jøsenfjorden. Observations by scuba divers confirmed that the visibility was much poorer in the brackish water than in the saline water underneath, probably due to particles and "yellow substance".

Possibly because of unfavorable light conditions the number
of phytoplankton cells at 10 m and 20 m depth in Hylsfjorden were
lower than that at the same depth in Jøsenfjorden.

With a very sharp halocline in Hylsfjorden, April to November,
often with steep gradients in salinity from 4 $^0/_{00}$ to 30 $^0/_{00}$ in
one metre, it may be expected that very few zooplankton organisms
are able to pass this barrier. A result may be a self sustaining
community in the brackish water layer. It may be hypothesized
that shortage of food is one of the reasons for the relatively low
biomass, particularly in the water column below the halocline in
Hylsfjorden. This could happen both because sufficient nutrients
are not available in the most productive zone, and because few of
the plankton organisms in the brackishwater layer are available as
food for the organisms in the saline water below.

Gelatinous zooplankton, for example jellyfish and ctenophores
was more important in the plankton of the Sandsfjord system than
elsewhere. Dominant species in the Sandsfjord system were small
copepods like *Oithona* spp., *Microcalanus* spp., *Acartia* spp.,
Temora longicornis, Paracalanus parvus, Pseudocalanus sp. and *Oncaea*
spp. Of these only *Microcalanus* spp. and *Oncaea* spp. seemed to be
more abundant here than in the outer area.

It has previously be found by underwater observations by
HAMNER *et al*. (1975) that *Oncaea* spp. predominantly lives on the
surface of gelatinous zooplankton.

Chaetognaths, particularly *Eukrohnia* spp., were found in deep
water in much the same abundance in all fjords. This could mean
that chaetognaths, as predators, are playing a more important role
in the Sandsfjord system than elsewhere. A marked abundance, however, of large copepods such as *Calanus finmarchicus, C. hyperboreus*
and *Euchaeta norvegica* were present in the outer area.

The abundance of plankton-eating fish, mainly sprat and herring,
is not well documented for these fjords. However fishermen state
that Jøsenfjorden is a much better fjord for sprat than Hylsfjorden.
In the inner part of Jøsenfjorden there is also every winter a good
fishery on cod.

As a whole the two main pathways of trophic organization summarized by LANDRY (1977) seem to be represented in these two fjords:
in Hylsfjorden small species of phytoplankton → small copepods →
jellyfish and chaetognaths seem to be the dominating pathway while
in Jøsenfjorden large diatoms → large copepods → is more prevailing.

The first type of foodchain exists in a stable physical environment with low nutrients and the other in a turbulent environment
with an upward transport of nutrients.

As both the pathways, though not exactly as described by Landry (*op. cit.*), exist in the investigated area, they may indicate that the conditions, particularly in Hylsfjorden, are characterised by stability with low turbulence and, consequently, low nutrients.

REFERENCES

Fosshagen, A. (in press). Dyreplankton i Ryfylkefjordene 1972-1975. Rapp. radg. utv. fjordunders. Oslo.

Hamner, W. M., L. P. Madin, A. L. Alldredge, R. W. Gilmer and P. P. Hamner 1975. Underwater observations of gelatinous zooplankton: Sampling problems, feeding biology and behaviour. Limnol. Oceanogr. 20(6), 907-917.

Landry, M. R. 1977. A review of important concepts in the trophic organization of pelagic ecosystems. Helgolander Wiss. Meeresunters, 30, 8-17.

Svendsen, H. and N. Utne (in press). Fysisk-oseanografisk under-søkelse i Ryfylkefjordene 1972-1975. Rapp. radg. utv. fjordunders. Oslo.

A PRELIMINARY EXAMINATION OF ZOOPLANKTON SPECIES GROUPINGS AND ASSOCIATED OCEANOGRAPHICALLY DEFINED REGIONS ALONG THE BRITISH COLUMBIA MAINLAND COAST

Grant A. Gardner

Ocean Ecology Laboratory*
Institute of Ocean Sciences
P.O. Box 6000
Sidney, British Columbia
V8L 4B2, Canada

INTRODUCTION

Although the physical and hydrographic characteristics of most British Columbia fjords are well known (e.g. Pickard and Stanton, this volume; Pickard 1961), the biological characteristics have never been comprehensively surveyed. Several papers have considered biological problems in geographically restricted areas, and others have concentrated on the Straits of Georgia and Juan de Fuca due to their convenient locations and commercial importance. Prior to the present study, however, only one geographically extensive work existed (LeBrasseur 1954). Despite the importance of the coast-line, many areas had never been sampled, there was little information on species distribution patterns and still less on differences in plankton community composition along the coast. Even LeBrasseur had taken few deep water tows and dealt only with broad taxonomic groups.

The study summarized here represents a first attempt to study coastal scale patterns in plankton community composition. The basic premise of the study is that broad grouping patterns can be identi-fied in both the plankton and hydrography of British Columbia main-land fjords and near shore waters. Although Pickard (1961) coarsely segregated fjords on the basis of runoff regime, no biological

* Current address: Department of Biology, Memorial University of Newfoundland, St. John's, Newfoundland, Canada, A1B 3X9.

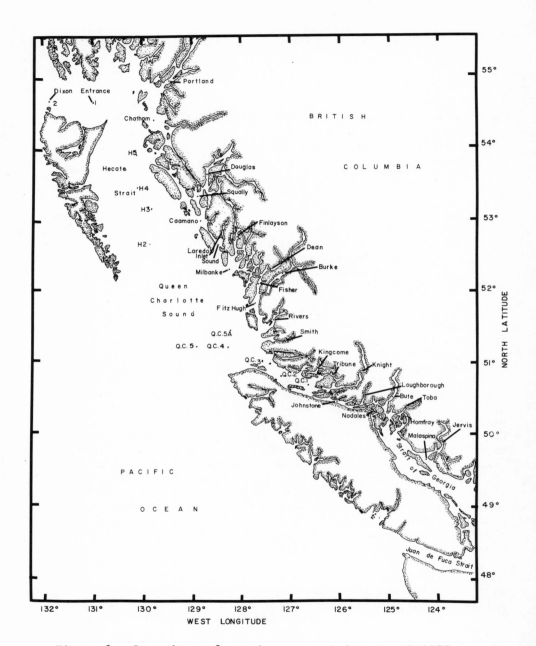

Figure 1.   Locations of stations occupied in April 1977.

grouping had ever been attempted. In this paper, zooplankton
community composition is used as an index on which such groupings
might be based. The results provide a framework which allows geographically isolated studies to be put into context with respect
to the coast as a whole.

Methodology

    The data discussed were collected in April 1977. The thirty-
eight stations (Fig. 1) divide equally among fjords proper, con
necting channels and near shore waters. At each station, temperature
and salinity data were collected from standard depths, and a vertical
net haul (50 cm mouth diameter; 350 μm mesh) was taken from near
bottom to the surface. A closing net was used to collect a discrete
vertical sample from below sill depth at stations separated from
connecting waters by a relatively shallow sill. Replicate hauls were
taken in Knight Inlet and Dixon Entrance. Plankton samples were
completely sorted and plankters normally identified to species or
genus; although subsampling was usually necessary, subsamples smaller
than 1/32 were rarely required.

    To examine grouping patterns in the hydrographic data, temper-
ature/salinity values were plotted for each of five depth horizons
(10-30m, 50-100m, 125-200m, 225-300m, 325-bottom). Plotting the
data in this manner facilitated interpretation of relationships
within the set of stations. Envelopes enclosing discrete groups of
data points were then related to geographic groupings. Biological
grouping were extracted using a clustering method previously proven
useful with time series plankton data (Gardner 1977).

RESULTS AND DISCUSSION

    Near surface temperature/salinity plots show three regions,
defined primarily by temperature differences (Fig. 2; Table 1).
The intermediate temperature group further divides into six on the
basis of salinity differences: however, some of the resulting groups
are only marginally distinct, and one, group 3, splits further into
10, 20 and 30m envelopes. The net result is a clearly defined
southern group (1a, b), an equally clear northern group (6a, b)
and an intermediate group. Within the intermediate group there is
a separate open water group of stations from Queen Charlotte Sound
and Hecate Strait (4), three groups containing either one or two
stations each (5,7,8), and a poorly separated pair (2,3) containing
the remainder of the stations. Separation into northern, southern
and intermediate groups holds for all depth horizons; however, the
infrastructure of the intermediate group varies and the only consistent grouping is of Burke, Dean and Fisher Channels and Fitz Hugh
Sound within group 3.

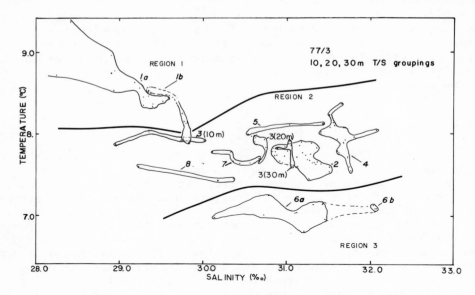

Figure 2.   Temperature/Salinity envelopes in the 10-30m horizon.

Table 1

Temperature/Salinity groupings in the 10-30 m depth horizon.

Group	Stations
1a	Malaspina, Jervis, Homfray, Toba, Bute
1b	Nodales, Loughborough
2	Johnstone, Q.C. 1, Q.C. 2, Smith, Douglas, Milbanke, Finlayson, Laredo Sound, Laredo Inlet, Caamano, Squally
3	Fitz Hugh, Burke, Dean, Fisher
4	Q.C. 3, H.5, H.4, H.3, H.2, Q.C. 4, Q.C. 5, Q.C. 5A
5	Rivers
6a	Chatham, Portland, Dixon 1
6b	Dixon 2
7	Tribune, Kingcome
8	Knight

    Altough the results of clustering split and vertical haul
data together are not presented here, two points arising in the
analysis support the clustering method. In all but one case, split
and complete vertical hauls from the same station clustered together
in 'natural' groups. (In this context, a 'natural' cluster is

comprised of stations which are geographically close and likely
exposed to similar hydrographic regimes). In addition, replicate
hauls clustered at relatively high levels of similarity.

It should be emphasized that references to 'high' or 'low'
levels of similarity should not be interpreted as statistical
judgements. The use of cluster analysis in the current context,
as in Gardner (1977), is inferential rather than for rigorous
hypothesis testing. When this and similar techniques are used
carefully and in full awareness of their pitfalls they are extremely
valuable in biological detective work. Thorough discussion of
multivariate techniques may be found in Cooley and Lohnes (1971),
Williams (1971) and others.

Clustering only the complete vertical hauls yields six 'natural'
clusters at an arbitrary similarity level of 0.32 (Fig. 3). The
groups to note are a southern fjords group (1), an open water group
(3), and two intermediate groups (2,4). Within the intermediate
groups, cluster 2 contains stations not directly connected with
open water. A subgroup of cluster 2 contains Burke, Dean and
Fisher Channels – the majority of one of the physically definable
subgroups. The group containing Caamano Sound through Portland
Inlet is primarily composed of stations exposed directly to open
water. The exception are Chatham Sound, a shallow northern station,
Laredo Inlet, a mid-coast inlet connected to Laredo Sound by a
constricted passage, and Squally Channel, protected from open water
by a short channel. There is no obvious reason for these stations
to be included in cluster 4. The other subgroup of cluster 4,
containing the Dixon Entrance stations, corresponds closely to the
hydrographically defined northern group.

Similar geographic grouping of stations have consequently been
derived from both biological and physical oceanographic data.
The grouping patterns are characterized by southern, northern,
open water and intermediate fjord groups. Further division of the
intermediate fjords group is possible and certain subgroups recur
in both in both the physical and biological groupings. The degree
of similarity implies strong links between the physical regime and
the composition of the zooplankton community. This result is not
unexpected, but there was previously no indication of either the
geographic scale of the groups or the strength of the linkage.
Note that the observed correspondence does not imply that
temperature/salinity characteristics are responsible for community
composition. As has been pointed out by several authors (e.g. Bary
1963, Gardner 1977), T/S relationships help define water bodies
which may differ in any number of biologically important parameters.
It is this array of parameters, in conjunction with geographic and
temporal factors, which determine composition.

The results presented here were repeated in a second cruise
(October 1977), and so although data are still sparse it appears

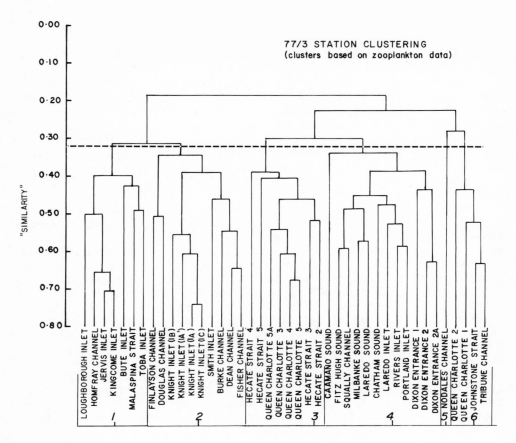

Figure 3.   Clustering of stations based on vertical haul data.
All clusters are based on a similarity level of 0.32.

that the geographical scale of the groupings is persistant. The
mechanisms creating patterns on this scale are complex and will be
discussed elsewhere (Gardner, in prep.); however, one of the key
factors is likely the interplay between fresh water runoff and
intruding oceanic water. Runoff is variable and both contributes
to the composition of intruding water and drives estuarine circulation. The composition of intruding water will vary both spatially
and temporally. Temporal variation can have subtle long term
effects on zooplankton communities (Gardner 1977), and it now appears
that spatial variation has similar, but more pronounced, effects.

The first approximation to a coastal scale study has shown
broad areas of similar physical and biological characteristics.

Knowledge of the grouping patterns will be sueful in the efficient design of future studies. The data collected suggest many areas for further research, and represent the only baseline zooplankton data covering the length of the mainland coast of British Columbia. This and companion studies will greatly enhance our understanding of plankton ecology in and near British Columbia fjords, and will provide a foundation on which we may build more general studies of plankton community ecology.

REFERENCES

Bary, B. McK. 1963. Distributions of Atlantic pelagic organisms in relation to surface water bodies. pp. 51-67 in Dunbar, M.J., ed., Marine Distributions. Roy. Soc. Can. Spec. Pub. No. 5.

Cooley, W.W. and P.R. Lohnes. 1971. Multivariate Data Analysis. John Wiley and Sons, Inc., N.Y., Lond., Sydney, Toronto. 364 pp.

Gardner, G.A. 1977. Analysis of zooplankton fluctuations in the Strait of Georgia, British Columbia. J. Fish. Res. Board Can. 34: 1196-1206.

LeBrasseur, R.J. 1954. The physical oceanographic features governing the plankton distribution in the British Columbia inlets. M.A. Thesis, University of British Columbia. 52 p.

Pickard, G.L. Oceanographic features of inlets in the British Columbia mainland coast. J. Fish. Res. Board Can. 18: 907-999.

Williams, W.T. 1971. Principles of clustering. Ann. Rev. Ecol. Syst. 2: 303-326.

GADOID POPULATIONS OF WESTERN SCOTTISH SEA LOCHS AND THEIR EXCHANGES WITH WEST COAST STOCKS

Adrian Cooper*

Dunstaffnage Marine Research Laboratory
P.O. Box NO. 3
Oban, Argyll, PA34 4AD, Scotland

INTRODUCTION

The opening of the Scottish Marine Biological Assocation Laboratory at Dunstaffnage in 1969 stimulated extensive surveys of the ecology of fish populations in the sea lochs and associated inshore areas around Oban. Gordon (1977) showed that the inshore areas were nursery grounds for young Norway pout. (*Trisopterus esmarkii*, Nilsson 1855), Whiting (*Merlangius merlangus*, Linnaeus 1758) and poor-cod (*Trisopterus minutus*, Linnaeus 1758) respectively, the populations consisting of mainly 0-group and 1-group fish. Since there was no evidence of spawning it was assumed that the young juveniles were recruited from adult populations located in the open sea to the west of Mull. In this paper the movement of juvenile whiting, Norway pout and poor-cod from the offshore adult stocks to the inshore nursery grounds is described. The likelihood of a return of maturing fish to the adult stock is assessed and the ecological significance of these migrations are discussed.

MATERIALS AND METHODS

Fig. 1 shows the area studied. Five sampling sites were chosen to compare inshore and offshore environments, bearing in mind the suitability of the sea-bottom for trawling and the availability of ship-time. Fish populations at these sites were sampled at approximately 3-weekly intervals from November 1974 to April 1977 using:

* Present Address: Fisheries Ecology Research Project, Dept. of Zoology, U.W.I., Mona, Kingston 7, Jamaica.

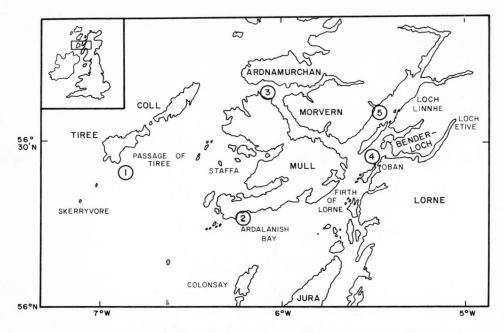

Figure 1.   Chart showing the area studied.   Sampling sites and their
             depths: (1) Tiree Passage, 80 m; (2) Ardalanish Bay,
             77 m; (3) Bloody Bay, 50 m; (4) Firth of Lorne, 47 m;
             (5) Loch Linnhe, 90 m.

1.   a Plymouth 2 m ring trawl to capture planktonic stages;
2.   a Gourock No. 1 mid-water trawl to sample pelagic juveniles
     which escape the plankton net;
3.   a wing (bottom) trawl with a headline of 58 feet and a footrope
     of 76 feet to capture demersal fish.

The sampling gears, their performance and operation have been described by Cooper (1979).   Planktonic larvae and pelagic juveniles
caught by the mid-water trawl were preserved in 4% formalin and
returned to the laboratory for identification.   When very large
numbers of demersal juveniles were caught the total number of each
species was estimated and a random sample of 200 fish deep-frozen
and returned to the laboratory.   The total length of each fish was
measured to the nearest mm and otoliths were collected from each
cm length group for age determination according to the method of
Williams and Bedford (1973).

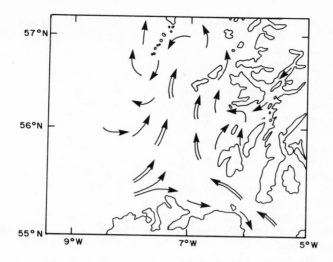

Figure 2. Surface circulation to the west of Scotland.

RESULTS

    Planktonic larvae of whiting and poor-cod were found only at
the three most offshore sites (Tiree Passage, Ardalanish Bay and
Bloody Bay) between April and July with maximum abundance in May.
No planktonic larvae of Norway pout were caught during the present
study and in surveys from 1970 to 1977 only 14 planktonic larvae
have been positively identified, 11 of which were taken in 1974
(Gordon, pers. comm.).

    Juvenile whiting were first found inshore in June/July at
lengths of 2-7 cm and were simultaneously caught in bottom and mid-
water trawls. The abundance of demersal fish increased rapidly at
inshore sites to a maximum of over 8000/h trawling in September and
October. At Ardalanish Bay and Tiree Passage abundance increased
more slowly, reaching a maximum only in May of the 1-group at Tiree,
where catch rates never exceeded 1000/h. Whiting therefore aggreg-
ated inshore for a short period as 0-group and early 1-group fish.

    In poor-cod and Norway pout the pelagic nektonic stage did
not last as long as in whiting. Both species were caught at all
sites from July, with pelagic and demersal juveniles occurring simultaneously.   In Norway pout and poor-cod the abundance of demersal
juveniles at all sites increased rapidly, probably due to an increasing proportion of the population being retained by the net

i.e. incomplete recruitment. In contrast to whiting, Norway pout
and poor-cod abundance was at all times greater at Tiree Passage
than at any other site.

From July to November inshore juvenile fish of all species
were longer and presumably older than fish caught offshore. This
difference was most obvious in pelagic fish caught by the mid-water
trawl, which has a slightly finer mesh size.

Following the peak of abundance, when the fish were first
completely sampled by the bottom trawl in October, there was a
gradual decline in abundance of all species at all sites. Regress-
ions were calculated for this reduction in abundance with time at
each site and for each species. In all cases a large and highly
significant part of the variance of abundance was explained by
regression on age. In all species the reduction in abundance was
most rapid inshore, and also more rapid at the shallow inshore site
of the Firth of Lorne (47 m) than at the deep inshore site of
Loch Linnhe (90 m).

DISCUSSION

The results indicate that planktonic whiting and poor-cod were
restricted to offshore sites and Norway pout larvae were not caught.
Recruitment to inshore areas occurred after the planktonic phase
and was presumably an active migration. These observations can be
explained to some degree by the hydrography of the area. Tulloch
and Tait (1959), Craig (1959) and most recently Ellett (in press)
have deduced circulation patterns for the west coast of Scotland
based on the limited information available. Although all authors
stress the sparseness of the data and lack of long term measurements
it seems fairly well established that three main water types are
important in the study area (Fig. 2):

1.   northward flowing Atlantic water from the north-east coast of
     Ireland;
2.   northward flowing Irish Sea water from the North Channel;
3.   Coastal water formed locally as a result of rainfall run-off.

Evidence from the distribution of Caesium 137 (Jefferies, *et al.*,
1973) which is derived from Windscale nuclear power station discharges into the Irish Sea, suggests that the dominant influences
in the Tiree Passage area are Irish Sea water and possibly Coastal
water, although Craig (1959) found water of near-oceanic salinity
forming a tongue directed inwards towards the south Minch and Firth
of Lorne. The absence of Norway pout larvae in the study area could
be explained if juvenile populations were derived from spawning
grounds known to exist to the north of Scotland (Raitt, 1965) or

from other areas to the north of Tiree Passage. Poor-cod spawn in Tiree Passage, but although mature adult and young larval whiting were found there, it is thought that the area is an adult feeding ground with spawning occurring to the south (Cooper, 1979). Little information is available concerning circulation in the outer Firth of Lorne area, although Craig (1959) found net water movements to be inshore near the bottom and offshore at the surface, which is to be expected considering the large freshwater surface run-off. Although wind-stress and tides undoubtedly confuse the general pattern of net flow, this circulation pattern would effectively restrict planktonic larvae in the surface waters to offshore sampling sites while reinforcing the active immigration of juveniles on the bottom.

In Norway pout and poor-cod it appears that the inshore migration was a simple dispersal with fish in greatest abundance near the breeding grounds, whereas in whiting there was a greater density of fish inshore for a limited period in the 0-group. Russell (1976) pointed out that different horizontal distributions can result from differences in the vertical distribution if water currents vary with depth. This is normally considered to be important only in the planktonic stage, but Harden Jones (1968) notes that whereas larvae must drift, adults can. The association of whiting over 2 cm long with the Scyphomedusa *Cyanea cappillata* between May and October (pers. obs., Bailey 1975) may delay the adoption of the demersal habit, thus affecting the horizontal distribution.

Following the peak of abundance of demersal fish inshore in October of the 0-group, abundance declined more rapidly than offshore. Predation and other natural mortalities undoubtedly accounted for part of this reduction in abundance. The most important predators of gadoids on inshore nursery grounds are probably older whiting, spur-dogs (*Squalus acanthias*) and possibly mackerel (*Scomber scombrus*) in summer. Reef-dwelling fishes such as cod (*Gadus morhua*), pollack (*Pollachius pollachius*) and saithe (*Pollachius virens*) may feed on the species studied but are unlikely to be important because of their low catch rates. Since trawl records suggest that predators are more common offshore and there is no evidence of mass mortalities it is concluded that a proportion of the population return to the offshore spawning grounds. Inshore catch-curves therefore represent survival and emigration while offshore curves represent survival and immigration from inshore areas.

Inshore nursery grounds are not unknown for gadoids, and their ecological significance is thought to be threefold. Nikolsky (1963) argued that the separation of nursery and adult feeding grounds was an adaptation towards abundance, since there may not be enough food to support both adults and immatures on the same ground. Secondly, estuaries in general are thought to be areas of abundant food supply, much of it unconsumed by resident species (Barnes, 1974). Finally, several authors have considered that estuaries allow the escape of juveniles from predators *e.g.* McErlean, (1973) and Gunter (1938). This requires further consideration since it introduces the question of why the maturing fish leave the area at all, when there is such an abundance of younger fish as food. Part of the explanation for this is thought to lie in the preference by older fish for deeper water. For example, in this study mean lengths of fish at the deeper sites were frequently greater than at shallow sites and in winter fish length appeared to decrease (Cooper, 1979). This was thought to be due to the earlier migration of larger fish. In fact in many marine species the physiological tolerances of adults and for spawning are more restrictive, or different for larvae and juveniles, so emigration may be related directly to maturation. In addition the availability of a limited food supply may be important in determining the timing of emigration.

REFERENCES

Bailey, R. S. 1975: Observations on diel behaviour patterns of North Sea gadoids in the pelagic phase. J. mar. biol. Ass. U. K. 55, 133–142.

Barnes, R. S. K. 1974: Estuarine Biology. 1st ed., 76 pp. London: Edward Arnold.

Cooper, A. 1979: Aspects of the ecology of gadoid fish of the west coast of Scotland. Univ. Stirling, Ph.D. Thesis.

Craig, R. E. 1959: Hydrography of Scottish Coastal Waters. Mar. Res. No. 2, 30 pp.

Ellett, D. J. (in press): Some oceanographic features of Herbridean waters. Proc. R. Soc. Edinb. Ser. B. (in press).

Gordon, J. D. M. 1977a: The fish populations in inshore waters of the west coast of Scotland. The biology of the Norway pout (*Trisopterus esmarkii*). J. Fish Biol. 10, 417–430.

Gordon, J. D. M. 1977b: The fish populations in inshore waters of the west coast of Scotland. The distribution, abundance and growth of the whiting *Merlangius merlangus* (L.). J. Fish Biol. 10, 587–596.

Gunter, G. 1938: Seasonal variations in abundance of certain estuarine and marine fishes in Louisiana with particular reference to life histories. Ecol. Monogr. 8, 313–346.

Harden Jones, F. R. 1968: Fish Migration. 1st ed. 325 pp. London:
    Edward Arnold.

Jefferies, D. F., Preston, A. and Steele, A. K. 1973: Distribution
    of Caesium 137 in British coastal waters. Mar. Pollut. Bull.
    4, 118-122.

McErlean, A. J., O'Connor, S. G., Mihursky, J. A. and Gibson, C. I.
    1973: Abundance, diversity and seasonal patterns of estuarine
    fish populations. Estuar. Coast. mar. Sci. 1, 19-36.

Nikolsky, G. V. 1963: The ecology of fishes. Translated by L.
    Birkett. 1st English Ed., 352 pp. London: Academic Press.

Raitt, D. F. S. 1965: The stocks of Trisopterus esmarkii (Nilsson)
    off north-western Scotland and in the North Sea. Mar. Res.
    No. 1, 24 pp.

Russell, F. S. 1976: The eggs and planktonic stages of British
    marine fishes. 1st. ed., 524 pp. London: Academic Press.

Tulloch, D. S. and Tait, J. B. 1959: Hydrography of the north-
    western approaches to the British Isles. Mar. Res. No. 1,
    32 pp.

Williams, T. and Bedford B. C. 1973: The use of otoliths for age
    determination. In: T. B. Bagenal (Ed.): Ageing of fish.
    pp 114-123. Old Working: Unwin Bros.

PHYTOPLANKTON BIOLOGY OF SAANICH INLET, B.C.

Louis A. Hobson

Department of Biology
University of Victoria
Victoria, British Columbia
Canada

Determination of budgets for organic carbon in fjords is difficult because of, in part, variance in the spatial and temporal distributions of phytoplankton. For example, Hobson (Unpub. Results) found large variations in distributions of *in situ* chlorophyll a fluorescence "believed to be a crude estimate of phytoplankton biomass" in Saanich Inlet, a fjord in southern Vancouver Island, British Columbia, Canada, during the spring and summer months of 1974. These variations in biomass suggested large variation in primary production, the most important source of organic carbon to the fjord, which because of physical constraints could not accurately be assessed by the carbon-14 technique (Steemann-Nielsen 1952). Therefore, it was assumed that a mathematical model relating primary production to environmental variables would be the technique of choice to determine production values in the fjord. Many of these models are available (e.g. Fee 1969), and all contain expressions for the responses of photosynthesis to varying irradiance (e.g. Jassby and Platt 1976). Most of these expressions contain two parameters, a, the response of light-limited photosynthesis to changes in irradiance, and P_{max}, the maximum rate of photosynthesis independent of irradiance. This latter parameter probably controls the bulk of primary production in temperate waters from spring to fall seasons, and little is known about its seasonal behaviour. Therefore, I carried out a study of the variability of P_{max} and its relationships to environmental variables before using a model to predict primary production in the fjord.

The results of the first part of this study showed that P_{max} increased from a minimum in early spring to maximum values in summer

and then decreased to a minimum in fall of 1975 and 1976 (Fig. 1).
These results are similar to those of Platt and Jassby (1976)
and Taguchi (1970), which the former authors attributed to variation
in seawater temperature. However, the potential for other variables
to affect P_{max} is great because P_{max} is a ratio of photosynthetic
rate to cell biomass, usually measured as chlorophyll a, which can
be influenced by nutrients (Hobson and Pariser, 1971), irradiation
(Steeman-Nielsen 1975), and daylength (Hobson et al. 1979).

The second part of this study was an attempt to discover
environmental variables covarying with P_{max} to suggest cause and
effect relationships between selected variables and P_{max}. This
attempt was made with a multiple regression analysis (Nie et al.
1975) attributing variance in P_{max} to concentrations of chlorophyll
a, NH_4-N, NO_3-N, PO_4-P, $Si(OH)_4-Si$, and protein, seawater temperature,
daylength, and pH. Coefficients of determination (r^2) were
considered to be significant when they exceeded 0.05. Two and four
different regression equations were generated for data from 1975
and 1976, respectively, to predict P_{max}. These equations are:

1975, 1: $P_{max} = \left[4.17-2.59(\{NO_3\}) + 6.25(\{NH_4\}) + 0.47\left[\{Si(OH)_4\})\right]\right]$

$$\pm 4.53 \qquad r^2 = 0.79$$

2: $= \left[-43.73 + 3.02(D.L.) + 0.61(\{Si(OH)_4\}) - 0.44(Chl)\right.$

$$\left. + 5.89(\{NH_4\}) - 1.96(\{NO_3\})\right] \pm 4.22 \qquad r^2 = 0.84$$

1976, 1: $P_{max} = \left[-20.04 + 1.21(D.L.) + 0.0032(Pr) + 0.049(\{Si(OH)_4\})\right.$

$$\left. + 0.26(T)\right] \pm 1.76 \qquad r^2 = 0.90,$$

2: $= \left[-29.18 + 1.77(D.L.) + 0.33(T) + 0.071(\{Si(OH)_4\})\right.$

$$\left. + 0.18(Chl)\right] \pm 1.71 \qquad r^2 = 0.91$$

3: $P_{max} = \left[-16.69 + 1.39(D.L.) + 0.073(\{Si(OH)_4\})\right] \pm 1.56$

$$r^2 = 0.86,$$

4: $= -15.59 + 1.31(D.L.) \pm 1.98 \qquad r^2 = 0.81$

Terms in these equations that are not self-evident are daylength in
hr from sunrise to sunset (D.L.), chlorophyll a concentrations in
mg m^{-3} (Pr), and temperature in ^{0}C (T). All nutrient concentrations
are expressed as μM. Irradiation at the sea surface was not
included in regression analyses because of its close relationship
to surface seawater temperature. Examination of these equations
shows that in 1975, either gave equally good predictions of P_{max}

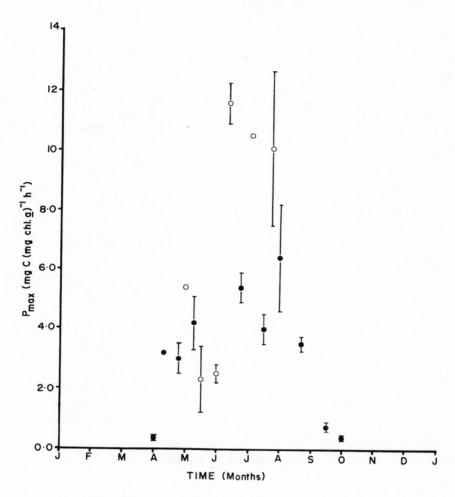

Figure 1. Variations in hourly rates (24-hr x̄) of light-saturated photosynthesis during 1975 (open circles) and 1976 (closed circles). Photosynthetic rates were measured with carbon-14 as a tracer in 24-hr in situ experiments. Vertical bars give confidence limits at the 95% probability level. Symbols without bars have unknown confidence levels.

for the period mid-April to mid-June. For 1976, none of the
equations predicted the low P_{max} measured in early April, while
from mid-April to mid-June, number 3, including daylength and
silicate concentrations gave predictions close to measured values.
During summer months, either number 1 or 2, including daylength,
silicate concentrations, temperature, and a measure of phytoplankton
biomass, gave closest predictions of measured values. However,
quantitatively choosing among these equations would be impossible
because of their large standard error at the 95% confidence level.
It is not surprising that these errors are large because regression
analysis assumes linearity between independent and dependent variables, and it is likely that variables affect P_{max} via non-linear
mechanisms.

Results of the multiple regression analyses suggest that
environmental control of P_{max} is complex and varies within and
between seasons. This variance makes the validity of production
models in neritic ecosystems dubious without concurrent measurements
of P_{max}. The results also consistently indicated that daylength and
silicate concentrations were related to P_{max}, but whether or not
these correlations were due to cause and effect relationships can
only be determined by further study.

REFERENCES

Fee, E.J. 1969. A numerical model for the estimation of photo-
    synthetic production, integrated over time and depth, in
    natural waters. Limnol. Oceanogr., 14: 906-911.

Hobson, L.A. and R.J. Pariser. 1971. The effect of inorganic
    nitrogen on macromolecular synthesis by Thalassiosira fluv-
    iatilis Hustedt grown in batch culture. J. exp. mar. biol.
    6: 71-78.

Hobson, L.A., F.A. Hartley and D.E. Ketcham. 1979. Effects of
    variations in daylength and temperature on net rates of
    photosynthesis, dark respiration, and excretion by Isochrysis
    galbana Parke. Plant Physiol., 63: 947-951.

Jassby, A.D. and T. Platt. 1976. Mathematical formulation of the
    relationships between photosynthesis and light for phytoplankton.
    Limnol. Oceanogr., 21: 540-547.

Nie, N.H., C.H. Hull, J.G. Jenkins, K. Steinbrenner and D.H. Bent.
    1975. Statistical package for the social sciences, 2nd. Ed.
    McGraw-Hill Co., N.Y., N.Y. xxiv + 675 pp.

Platt, T. and A.D. Jassby. 1976. The relationship between photo-
    synthesis and light for natural assemblages of coastal
    marine phytoplankton. J. Phycol., 12: 421-430.

Steemann-Nielsen, E.   1952.   The use of radio-active (C^{14}) for
    measuring organic production in the sea.   J. Cons. Int.
    Explor. Mer, 18:   117-140.

Steemann-Nielsen, E.   1975.   Marine photosynthesis.   Elsevier Sci.
    Publ. Co., Amsterdam.   vi+141 p.

Taguchi, S.   1970.   Seasonal variations of photosynthetic behaviour
    of phytoplankton in Akkeshi Bay, Hokkaido, with special
    reference to low photosynthetic rate in summer associated
    with large percentages of dwarf cells.   Bull. Plankton Soc.
    Japan, 17:   65-77.

ECHO - SOUNDING AND ECOLOGY

B. McK. Bary

Department of Oceanography
University College
Galway

INTRODUCTION

It is probably agreed that underwater scattering of biological origin at low frequencies is attributable to fish and at high frequencies, not only to fish, but to zooplankton as well. There is a range of frequencies over which both may occur, but within which zooplankton scattering is spasmodic and not clearly defined. This has been the cause of some confusion. Bary and Pieper (1970) presented results gained from several years of echo sounding in Saanich Inlet that had been accompanied by towing of an instrumented High Speed Sampler (the Catcher; Bary and Frazer, 1970) and of mid-water trawls, (Barraclough, pers. comm.).† Of two quantitative studies at U.B.C., one, carried out by Beamish (1971) using a special hydrophone array, showed that individual euphasiids, 2 to 3 cm long, when subjected to 104 kHz pulses from above and the side, caused echoes that were recorded. The back scattering cross section was calculated to be 1.4×10^{-4} cm^2, four-fifths of which was attributed to compressibility contrast between the animals and seawater and one-fifth to density contrast. Pieper (1971) studied quantitatively the volume back scattering at high frequencies attributed to zooplankton in Saanich Inlet.

Between them, these several approaches indicated that earlier hypotheses (e.g., Hersey and Backus, 1962) over respective roles of fish and zooplankton in underwater scattering were essentially

† Dr. E. Barraclough, Dept. of Fisheries and Oceans, Pacific Marine Biological Station, Nanaimo, B.C., Canada.

correct – namely, that zooplankton scattering becomes progressively
more pronounced as frequency increases from, say, 50 towards 200 kHz.
Fish may be recorded at all frequencies, but apparently differentially, over the range of 12 to 200 kHz. At low frequencies – i.e.,
below about 40 kHz – only fish appear to have been recorded in
Saanich Inlet. Certainly there is no evidence from the Saanich
observations that zooplankton scattering occurs at 12 kHz – and
indeed, it is unlikely that zooplankton scattering has been recorded
anywhere at that frequency, despite the diffuseness in records that
has suggested zooplankton was the cause (Hersey and Backus, 1962).

     With these points in mind, some of the interesting ecological
information obtained from prologed echo-sounding studies, more
especially in Saanich Inlet, can be outlined. Features such as
zooplankton – fish relation; behaviour with respect to artificial
lights; and diel vertical migration of fish and zooplankton, will
be briefly discussed. Some topographic and physical oceanographic
features and their effects on the distribution of scatterers are
also illustrated.

     The water in Saanich, as is known (e.g., Herlinveaux, 1962),
is atypical in structure – e.g., low oxygen below about 100 m
depth. A result is that organisms may not attain a low, "preferred",
light intensity at their day depth. Because of this, results from
Saanich Inlet are examined in relation to records from other seas
to determine their universality.

UNDERWATER SCATTERING AND FREQUENCY

     Until the question of the sort of organisms causing scattering
at frequencies between, say, 5 and 200 kHz was reasonably answered,
then ecological interpretations based on scattering studies were
arguable. Simultaneous recordings at 12, 44, 107 and 197 kHz in
Saanich Inlet clearly show two broadly distinct forms of scattering.
One, which is diffuse, amorphous, and usually in one or more "layers"
occurs at 107 and 197 kHz, but only occasionally and then weakly at
44 kHz; it is not recorded at 12 kHz. The other type is recorded by
the narrow-beam, high frequency sounders as more or less angular,
line traces: if specimens are numerous a diffuse record may be
produced. A diffuse record is common when obtained using the
broad-beamed, low-frequency sounder.

     Concurrent sampling has been carried out, using instrumented
plankton samplers and trawls, towed above, below and within these
two general forms of scattering. Zooplankton biomass, from individual tows or from stratified series of tows made with the Catcher,
is not consistently related to scattering being caused by organisms
other than zooplankton, viz., fish. In contrast, the vertical

distribution of zooplankton, from the stratified tows made relative to scattering at 197 kHz, confirms that the diffuse "zooplankton scattering" is associated consistently with very high no/m^3 of zooplankton, chiefly euphausiids. "Fish scattering" at this frequency may or may not be associated with high numbers of the zooplankton.

Both biomass and no/m^3 collected in Saanich Inlet attest to the extraordinary quantity of organisms associated with the scattering at 197 kHz there; values up to 7 or 8 g /m^3 or 400 to 600 specimens/m^3 (of only the larger organisms) can be obtained. Additionally, vertical series of tows show that the broad categories of organisms dealt with tend to stratify differentially, by day and night; the categories were mainly "euphausiids", "amphipods" and "decapods".

VERTICAL MIGRATION

Diel vertical migrations of zooplankton and fish are recorded by echo-sounders.

A day-night pattern for migrating fish has been recorded at 12 kHz in Saanich; it is applicable elsewhere, in general terms. Some fish migrate, some do not; reasons for this are speculative, even when only a single species is involved. There is also a pattern to the migration. Frequently, 4 "stages" are apparent (Bary, 1967). By correlating the onset of the stages with the principal surface light changes, viz., sunset or sunrise, dawn or dusk, it appears they occur in a regular daily and seasonal relationship to these light changes. The stages arise from different, successive alterations in speeds of migration and they can be related to arbitrarily drawn under-water "isolumes". The relationships of ascending and descending scattering to the light intensity, represented by the "isolume", differ, with the descent approximately 1.4 times faster than the ascent. The stages and the different rates of migration appear to be world-wide phenomena; they are found in many 12-kHz records of migrating fish, whether from turbid coastal water or clear oceanic water (Bary, 1967). They were usually present in Saanich Inlet, and occurred on the one occasion the light intensity was measured at the day depth of the scattering. On this occasion, the intensity was three orders of magnitude higher than found at the day-light scattering depth in Santa Barbara Channel (Boden and Kampa, 1965). (Kampa's, 1970, comments and discussion of the photo-environment and scattering are relevant and important.)

According to Ringelberg (1964) changes of suitable magnitude from the light intensity to which organisms have become adapted will stimulate directed movements. Briefly, when the rate of change is rapid, the amount and direction of movement increases, and vice versa. It is from these reactions that the stages originate (Bary, 1967). It is also because of these reactions that organisms in Saanich Inlet (and at least at times these must adapt to high light intensities at their day depth) respond to changing light with the same stages as those organisms found at much lower light intensities in more "normal" environments.

   Zooplankton scattering appears to undergo similar stages in migration to the fish scattering, but even in Saanich the picture may be obscured by, for example, multiple layers. To date, however, insufficient work has been done on the plankton scattering to determine the speeds in the stages, either for Saanich or else- where, and on the relationships between fish and zooplankton migra- tions. It is important that the zooplankton migratory stages be related to the major surface light phenomena - and subsequently to precise underwater measurements. It is necessary, of course, to collect and identify the scatterers.

FISH AND ZOOPLANKTON BEHAVIOUR

   Zooplankton scattering usually migrates separately from fish scattering and more usually at a shallower depth. Occasionally, fish may not migrate or may migrate at shallower depths than the zooplankton. Reasons for these different relationships are not apparent.

   Behavioural features can be examined with the aid of underwater lights. For example, fish at the surface at night can be confused, and then may no longer respond, after repeated, alternating, brief periods of high-intensity light and darkness. Organisms in scatt- ering at its day depth avoid underwater T.V. lights, with the zoo- plankton re-grouping at shallower and the fish migrating to deeper levels than the light. At night, the underwater light, when lowered in steps, "drove" organisms from the surface to deep levels - even into the deep water in Saanich in which they rarely occurred.

   Schooling of fish, the times of day and under what conditions it occurs, whether it is persistent in some species and not in others, and so on, are facets of practical and economic importance and of interest to behaviourists. The Saanich records suggest that schooling may periodically be a group-feeding reaction. It is not known whether the information from Saanich Inlet parallels that from elsewhere, even from other British Columbia waters. Echo- sounders, especially at high frequencies and with narrow beams, and used from moored vessels, can provide continuous records on schooling that are unobtainable otherwise.

SCATTERING AND OCEANOGRAPHIC PROCESSES

Underwater scattering, especially from zooplankton, may provide interesting information to other scientific disciplines. For example, an abrupt change in the depth of scattering may indicate the crossing of a front between water masses (Davies, 1977). Or, wavelike fluctuations in depth can indicate internal wave systems. Or, the distribution of scattering where it abuts onto a shallow obstruction may indicate water movements over or round it. A turbulent wake, resulting from a current passing a large obstruction, can be detectable for some distance.

Scattering not attributable to organisms can arise as a result of density layering, caused by temperature and/or salinity discontinuities. These do not appear to be detectable at 12 kHz, but are frequent at 44 kHz, and above. Frequent attempts to collect organisms from this type of scattering have failed.

Finally, in certain sedimentary conditions gas may be produced which bubbles forth; in Saanich there is periodic release of bubbles from the bottom that appears to be triggered by the approach to low tide. The bubbles are obvious on echo traces, when the ship (or sounder) is stopped, but when underweigh a bubble often produces a "comet" trace almost indistinguishable from a fish echo.

CONCLUDING COMMENT

Despite the many reports and publications on underwater scattering, the technique has barely begun to be exploited as an ecological tool, even in Saanich Inlet where prolonged, intensive studies have been carried out. It is the writer's opinion that Saanich Inlet offers probably a better locale than anywhere else to further studies on underwater scattering, and on the behaviour, composition, and distributions of the organisms composing it. It is considered, also, that the unusual water structure in the inlet may be an advantage rather than a hindrance in the study of some – if not many – of the features discussed. Further, in Saanich and other coastal waters of British Columbia, there appears to be a future for combining observations of scattering of biological origin, particularly at high frequencies, and the study of certain physical phenomena such as internal waves, turbulent wakes and movements of water about obstructions.

REFERENCES

Bary, B. McK., 1967. Diel vertical migrations of underwater scattering, mostly in Saanich Inlet, B.C. Deep-Sea Res., 14 35-50.

Bary, B. McK. and E.J. Frazer. 1970. A high-speed opening-closing
    plankton sampler (Catcher II) and its electrical accessories.
    Deep-Sea Res. 17, 825-835.

Bary, B. McK. and R.E. Pieper. 1970. Sonic-scattering studies in
    Saanich Inlet, British Columbia: a preliminary report. Proc.
    Int. Symp. on: Biological sound scattering in the ocean,
    Maury Center for Ocean Science, Rep. NIC 065: 601-611.

Beamish, Peter. 1971. Quantitative measurements of acoustic
    scattering from zooplanktonic organisms. Deep-Sea Res. 18(8),
    811-822.

Boden, B.P. and Elizabeth M. Kampa. 1965. An aspect of euphausiid
    ecology revealed by echo-sounding in a fjord. Crustaceana,
    9(2), 155-173.

Davies, I.E. 1977. Acoustic volume reverberation in the eastern
    tropical Pacific Ocean and its relationship to oceanographic
    features. Deep-Sea Res., Note, 24, 1049-1054.

Herlinveaux, R.H. 1962. Oceanography of Saanich Inlet in Vancouver
    Island, British Columbia. J. Fish. Res. Bd., Can. 19(1),
    1-37.

Hersey, J.B. and R.H. Backus. 1962. Sound scattering by marine
    organisms. No. 13 in: The Sea: Ideas and Observations on
    progress in the study of the sea, M.N. Hill, Ed., Interscience
    I, 498-539.

Kampa, Elizabeth, M. 1970. Photoenvironment and sonic scattering.
    Proc. Int. Symp. on: Biological sound scattering in the ocean,
    Maury Center for Ocean Science, M.C. Rep. 005: 51-59.

Pieper, R. 1971. Quantitative back-scattering from zooplankton in
    Saanich Inlet. Ph.D. thesis, Univ. of British Columbia,
    Vancouver, Canada.

Ringelberg, J. 1964. The positively phototactic reaction of
    Daphnia magna Straus: a contribution to the understanding of
    diurnal vertical migration. Netherlands J. Sea Res., 2(3),
    319-406.

VARIABILITY IN SEA-LOCH PHYTOPLANKTON

Paul Tett and Brian Grantham

Scottish Marine Biological Association
Dunstaffnage Marine Research Laboratory
P.O. Box 3, Oban, Argyll, Scotland

INTRODUCTION

Waters in temperate latitudes generally show marked seasonality in phytoplankton crop, production and composition. Only in coastal waters, however, is it possible to sample intensively or long enough to assess the extent to which the seasonal cycle varies from year to year. But because of the efects of runoff, tidal mixing, and geography on the vertical and horizontal stability of the water column it seems likely that phytoplankton crop and production are more variable, and show more patchiness, in coastal waters, and especially in fjords, than in the open sea. The seasonal cycle might thus be relatively indistinct in fjords. The aim of this paper is to resolve spatial and temporal variability in sea-loch phytoplankton biomass into several components so as to begin to understand which processes in phytoplankton ecology can be modelled deterministically and which must be treated stochastically.

METHODS

Phytoplankton standing crop in Loch Creran, Scotland, was estimated during 1972 through 1979 by fluorescence measurement of extracted chlorophyll. Creran is a small fjord opening into a larger fjord, the Firth of Lorne on the west coast of the Scottish Highlands. Both the loch (mean depth 17 m) and its entrance sill (5 m) are shallow. Tett & Wallis (1978) describe the loch and the method of chlorophyll measurement. As reported in that paper, only values for depths between 1 and 8 m are analyzed: this zone is the seawards moving brackish layer, which in Creran is only weakly separated from a deeper, saltier layer.

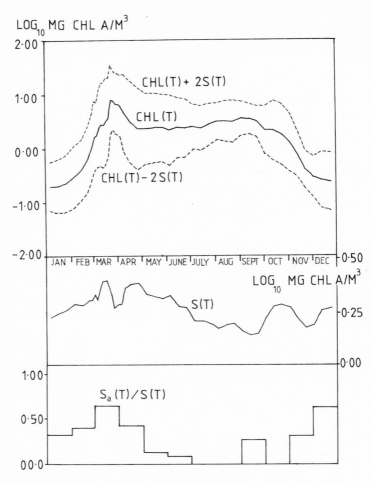

Figure 1.   The general annual cycle of chlorophyll standing crop in
            Loch Creran, and some components of variation about the
            general cycle. The envelope defined by mean chlorophyll
            concentration ($CHL(T)$, $\log_{10}$ mg chl a m^{-3}), plus or minus
            two standard deviations ($S(T)$, $\log_{10}$ mg chl a m^{-3}) of
            chlorophyll concentration at time T, includes 95% of all
            measurements made between 1 and 8 m depth in the main
            basin of Loch Creran. One of the components of variation
            with the envelope is 'between-year' variation and the
            variance due to this ($S_a^2(T)$) is shown as monthly-averaged
            proportion of total 'within-envelope' variance $S^2(T)$.

Analysis of variance was used to partition components of variation. The model used resolves each log-transformed chlorophyll measurement into components due to the general annual cycle ($CHL(T)$), and four sources of variation about this general cycle. The sources, and their corresponding variances, are

(a)   variation between years ($S_a^2(T)$)

(b)   variation on the scale of the loch ($S_b^2(T)$)

(c)   small scale patchiness ($S_c^2(T)$)

(d)   errors in measurement ($S_d^2(T)$)

The variances $S_a^2(T)$ through $S_d^2(T)$ sum to $S^2(T)$, the overall variance of all measurements made in time period T between 1972 and 1979. The periods used for the analysis were generally of 10 days, but 3 day periods were used for the season of the spring phytoplankton increase.

RESULTS

A total of 1,832 measurements were analyzed, and the results in figure 1 confirm the findings of Tett & Wallis (1978) concerning the features of the general annual cycle. Table 1 summarizes the variance components, averaged over all periods and years, and relates each to $S^2(T)$. Several components show strong seasonality. Of these the most interesting is $S_a^2(T)$, 'between-year' variation, which is greatest is spring and least in late summer (figure 1). $S^2(T)$ is also least in late summer.

DISCUSSION AND CONCLUSIONS

The analysis reported here is incomplete. Not all our data has been analysed in respect of each component of variation. In addition assumptions involved in the model need to be checked. Several results are however so clear that it is unlikely that a more complete analysis will change the following conclusions.

(1)   Table 1 shows that variance due to $CHL(T)$ is nearly three times $S^2(T)$: thus an annual cycle is discernible above 'noise' due to patchiness and year-to-year variation. The most prominent features of the cycle are a spring increase a stable period during midsummer, and an autumn dieback. See Tett & Wallis (1978) for more details.

(2)   Most of the 'within-envelope' variance $S^2(T)$ comes in roughly equal parts from 'between-year' ($S_a^2(T)$) and 'loch-scale' ($S_b^2(T)$) variability.

(3) $S_a^2(T)$, the variances due to differences between years, is greatest during the spring, when { d $CHL(T)/dT$ } is most, and when small changes in the timing of the spring increase can have large effects on phytoplankton standing crop. $S_a^2(T)$ is least in summer. Is this because summer weather differs least from year to year than that of other seasons, or are the processes regulating phytoplankton growth in July and August such as to be independent of weather?

Table 1

Components of variation amongst chlorophyll measurements in Loch Creran. The units of variance are ($\log_{10}$ mg chl a m^{-3})^2, and each variance is also expressed as a proportion of total 'within-envelope' variance $S^2(T)$. When variance is given as a function of time (T), time advances in periods of 3 or 10 days. Note that $S^2(T) = S_a^2(T) + S_b^2(T) + S_c^2(T) + S_d^2(T)$ is an assumption of the model used.

Source of variance	Symbol	Variance	Proportion
1. Total variation (all samples)	S^2	0.33	3.8
2. General annual cycle	$S^2 - S^2(T)$	0.24	2.8
3. Total variability within envelope in each period	$S^2(T)$	0.087	1.00
4. Variability between years	$S_a^2(T)$	0.037	0.43
5. Loch-scale variation (within periods and years	$S_b^2(T)$	0.038	0.44
6. Small scale patchiness	$S_c^2(T)$	0.009	0.11
7. Measurement error	$S_d^2(T)$	0.002	0.02

REFERENCE

Tett, P. and A. C. Wallis. 1978. The general annual cycle of chlorophyll standing crop in Loch Creran. J. of Ecol., 66, 227-239.

AMINO ACID FLUXES IN MARINE PLANKTON COMMUNITIES CONTAINED IN

CEPEX BAGS

J. T. Hollibaugh

Institute of Marine Resources, A-018
University of California, San Diego
La Jolla, California, 92093, U. S. A.

INTRODUTION

Our group is interested in the general problem of the tropho-dynamics of bacterioplankton in marine ecosystems. An important aspect of the activities of bacteria is their role as mediators of nutrient transformations (Figure 1). We have attempted to quantify their contribution to the nitrogen flux via animo degradation using the plankton communities in CEPEX CEEs 78-2 and 78-3 (Menzel & Case 1977) as experimental systems. The data and discussion which follow give the results of that effort.

I would like to acknowledge the assistance of several individuals who contributed to this work. A. L. Carruthers, J. A. Fuhrman and F. Azam provided assistance with the collection and processing of the microbiological samples. I am grateful for the efforts of the CEPEX staff and summer students, particularly R. Brown and F. Whitney, for collection the inorganic nutrient and productivity data. This research was funded by a grant from the International Decade of Ocean Exploration, NSF OCE77-26400 to Dr. F. Azam.

MATERIALS AND METHODS

The data reported here were taken to study the role of bacteria in marine plankton cummunities as part of the 1978 CEPEX Foodweb I experiment. Integrated samples were taken from 5 depth intervals (0-4m, 4-8m, 8-12m, 12-16m, 16-20m) from CEE 78-2 or 78-3. Chlorophyll a NO_3-, NH_4+, alkalinity, and dissolved organic carbon (DOC) were determined by the methods described in Strickland and Parsons (1972). Primary production was estimated by the standard radiocarbon method in 4 h *in situ* incubations (1000 to 1400 local time).

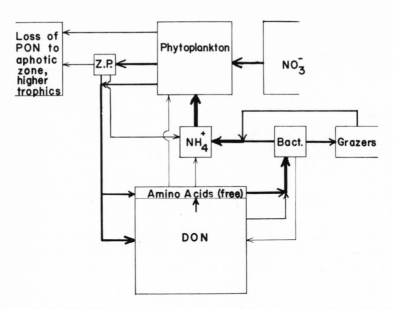

Figure 1.   Generalized nitrogen cycle expected in the euphotic
            zone of coastal temperate water columns. Arrows
            denote directions of fluxes and approximate magnitudes,
            the area of the enclosed boxes gives approximate pool
            sizes. Open boxes indicate major surces and sinks
            of nitrogen.

     Microbiological samples were analyzed as follows. Bacteria in
the samples were enumerated by the acridine orange direct count
method of Hobbie *et al*. (1977). A visual estimate of cell size
distribution (percent of population in each of three classes) was
used to estimate the biovolume which was converted to biomass
carbon by the factor 100 fg $C/\mu m^3$ (Ferguson & Rublee 1977). The
turnover rates of D-glucose and L-leucine were determined in 0.5
to 1 h incubations by the method of Azam and Holm-Hansen (1973)
using $D-6-^3H$-glucose and $L-4,5-^3H$-leucine (New England Nuclear).
The filters (0.45 μm pore size Millipore or 0.6μm pore size Nucleo-
pore) were radioassayed after clearing in either Scintiverse
(Fisher Scientific) or Aquasol II (New England Nuclear) liquid
scintillation cocktails in a Beckman LS 100 C liquid scintillation
spectrometer. A correction for quenching and counting efficiency
was made using the external standard ratio and appropriately
quenched standards. A self absorption correction of 0.5 determined
using $D-6-^3H$-glucose, was applied when ^3H was being radioassayed.
DOC and primary amines were determined on samples vacuum filtered
through baked (450^0C for at least 4h) Reeve Angel 984 glass fiber
filters in baked all-glass filtration apparatus at 120 torr pressure

differential. The samples were immediately frozen in baked glass screw-cap tubes. The caps of the tubes were lined with concentrated nitric acid cleaned teflon sheets. Primary amines were determined as glycine equivalents by the method of North (1975).

RESULTS

Amino acid fluxes were estimated from the data as follows. If both the size of a substrate pool and the rate of turnover of that pool are known, the flux of substrate can be calculated simply by the equation:

$$\text{Flux} = \text{pool size} \times \text{turnover rate}$$

Because of the possible interference of non-amino acid primary amines (amino sugars, peptides, aliphatic and aromatic amines), the primary amine values were multiplied by 0.5 to estimate the size of th e dissolved free amino acid (DFAA) pool (North, 1975). The turnover rate of ^3H labelled L-leucine was used to estimate the turnover rate of the DFAA pool. Competition experiments indicated that all of the amino acids tested inhibited the uptake of L-leucine to some extent, suggesting that amino acid uptake may be relatively non-specific at the community level, unlike glucose uptake. Furthermore, others (Crawford et al. 1974; Wright, 1974; Hobbie et al, 1968) have found that the turnover times of a variety of amino acids and glucose were within an order of magnitude of each other when measured in the same water sample, with the turnover time of L-leucine lying mid-range. We have compared the turnover rates of L-leucine, L-arginine and L-glutamate in samples taken from the Scripps Institute of Oceanography pier and found them to be within a factor of three of each other. The turnover rates of L-leucine and D-glucose in our CEPEX samples were significantly correlated and of similar magnitude (n = 51, r^2 = 0.22, p < .001, leucine turnover rate = 10.1 + 0.53 glucose turnover rate). The total amino acid flux (F_{AA}) was thus calculated from the primary amine concentration (1^0 amine) and the turnover rate of L-leucine (leu f/t) as:

$$F_{AA} = (1^0 \text{ amines}) \times 0.5 \times \text{leu f/t}$$

Amino acid fluxes ranged from 0.09 to 2.42 µM (glycine equivalents) day^{-1}. The mean amino acid flux in euphotic zone samples (0-8m) was greater than the mean flux in aphotic zone samples (8-20m) but the difference was not significant. When the parameters were compared by linear regression analysis, significant correlations (p < 0.01) were found between: glucose and leucine turnover rates; glucose turnover rate and amino acid flux; leucine turnover rate and primary production, numbers of bacteria, bacterial biomass; and amino acid flux and primary production, bacterial biomass. The correlation between amino acid flux and primary production had a higher significance (0.001 < p < 0.01) than the correlation between leucine turnover rate and primary production (0.01 < p < 0.005). DOC concentration was not correlated with any other parameter.

Amino acid-N flux was compared to ammonia-N flux as the degradation of amino acids by bacteria leads to the production of ammonia (Hollibaugh, 1978; Wright, 1974). At steady state, the ammonia flux should reflect the amino acid flux if the input due to amino acid degradation is a significant portion of the total ammonia input. Under non-steady state conditions, changes in the ammonia pool size should reflect the amino acid flux. It is difficult to measure the steady-state turnover rate of the ammonia pool directly, however, estimates of the rate of loss from the pool can be made on the basis of primary production estimates in euphotic zone samples. Pool size can be measured directly and the flux through the pool can thus be estimated. This was done based on the following assumptions:

1. 50% of amino acid-N flux is excreted as ammonia, 1 g-at N/mole of amino acid.
2. Ammonia input from other sources is negligible.
3. 12-hr day for primary production.
4. Phytoplankton C/N = 6/1.
5. 50% of phytoplankton N comes from ammonia (Harrison and Davies 1977; Harrison, Eppley and Renger, 1977; Eppley, pers. commun.).

The results of these calculations are presented in Fig. 2. Linear regression analysis of this data gave the equation:

Regeneration from amino acids = 0.45 + 0.46 × ammonia flux
into phytoplankton

with a significant correlation ($n = 21$, $r^2 = 0.82$, $p < .001$). Primary production was correlated with ammonia concentration, but at a much lower significance level ($n = 24$, $r^2 = .28$, $p < .05$).

For the non-steady state condition to apply, the fluxes into and out of the ammonia pool must not be equal. This condition is most simply realized in the aphotic zone where there is no uptake by phytoplankton. Input functions include ammonia regeneration zooplankton excretion, and diffusion. In the deep water of the CEEs (12-16 m and 16-20 m layers) ammonia gradients were not large so that the perturbations from molecular diffusion were probably small.

Ammonia production rates in the 12-16 m and 16-20 m samples were calculated from the ammonia concentrations observed one

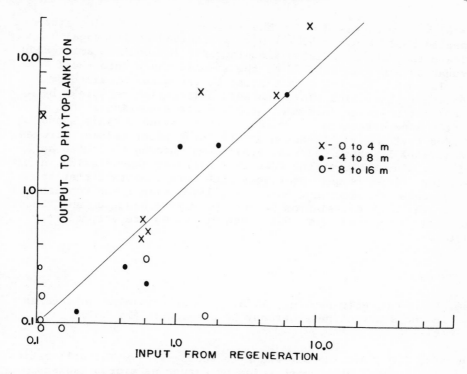

Figure 2.   A comparison of the expected rates of turnover of the
ammonia pool in euphotic zone samples due to phytoplankton uptake (ordinate) and ammonia release
accompanying amino acid degradation (abscissa).   The
diagonal line has a slope of 1.0.   See text for details.

sampling day before, on the same day as, and one sampling day after
an amino acid flux observation.   The rate calculated from the
interval prior to the flux observation was compared to the rate
for the subsequent interval to guard against disruptions due to
uncontrolled mixing factors.   Interval pairs for which the sign
of the change in pool size differed were discarded.   The average
rate of change of the ammonia pool was calculated for the remaining
pairs from the change in ammonia concentration observed over the
whole interval.   This value was then compared to the expected
ammonia input calculated from the amino acid flux as described
above.   A significant relationship between these two numbers was
observed when they were submitted to linear regression analysis

(n = 8, r^2 = 0.5, p < 0.05). The slope of the regression line (expected pool change = 0.07 + 0.37 x observed pool change) indicate that the calculated ammonia regeneration from amino acid degradation accounts for 37% of the ammonia input into these water layers. This value is similar to the value obtained by comparing regeneration via amino acid degradation to uptake by phytoplankton in the (assumed) steady state situation in the euphotic zone (regeneration = .45 + .46 x uptake). This supports the idea that a large fraction of the total water column nitrogen flux (37-46% by these calculations) moves through the amino acid pool. It also supports the idea that nitrogen regeneration in the water column is quantitatively much more important than has been supposed (Sheldon and Sutcliffe, 1978; Ryther and Dunstan, 1971). The high correlation between primary production and the amino acid flux indicates that these processes are very tightly coupled.

REFERENCES

Azam, F. and O. Holm-Hansen. 1973. Use of tritiated substrates in the study of heterotrophy in seawater. Mar. Biol. 23: 191-196.

Crawford, C.C., J.E. Hobbie and K.L. Webb. 1974. The utilization of dissolved free amino acids by estuarine microorganisms. Ecology 55: 551-563.

Harrison, W.G. and J.M. Davies. 1977. Nitrogen cycling in a marine planktonic food chain: nitrogen fluxes through the principal components and the effects of adding copper. Mar. Biol. 43: 299-306.

Harrison, W.G., R.W. Eppley and E.H. Renger. 1977. Phytoplankton nitrogen metabolism, nitrogen budgets, and observations on copper toxicity: controlled ecosystem pollution experiment. Bull. Mar. Sci. 27: 44-57.

Hobbie, J.E., C.C. Crawford and K.L. Webb. 1968. Amino acid flux in an estuary. Science 159: 1463-1464.

Hobbie, J.E., R.J. Daley and S. Jasper. 1977. Use of Nuclepore filters for counting bacteria by fluorescence microscopy. Appl. Environ. Microbiol. 33: 1225-1228.

Hollibaugh, J.T. 1978. Nitrogen regeneration during the degradation of several amino acids by plankton communities collected near Halifax, Nova Scotia, Canada. Mar. Biol. 45: 191-201.

Menzel, D.M. and J. Case. 1977. Concept and design: Controlled experimental ecosystem pollution experiment. Bull. Mar. Sci. 27: 1-7.

North, B.B. 1975. Primary amines in California coastal waters: utilization by phytoplankton. Limnol. Oceanogr. 20: 20-27.

Ryther, J.H. and W.M. Dunstan. 1971. Nitrogen, phosphorous, and eutrophication in the coastal marine environment. Science 171: 1008-1013.

Sheldon, R.W. and W.H. Sutcliffe, Jr. 1978. Generation times of 3 H for Sargasso Sea microplankton determined by ATP analysis Limnol. Oceanogr. 23: 1051-1055.

Strickland, J.D.H. and T.R. Parsons. 1972. A practical handbook of seawater analysis. Bull. 167 (2nd ed.). Fish. Res. Bd. Can., Queen's Printer, Ottawa. 310 p.

Watson, S.W., T.J. Novitsky, H.L. Quinby, and F.W. Valois. 1977. Determination of bacterial number and biomass in the marine environment. Appl. Environ. Microbiol. 33: 940-946.

Wright, R.T. 1974. Mineralization of organic solutes by hetero-trophic bacteria, p. 546-565. In R.R. Colwell and R.Y. Morita (eds.), Effect of the Ocean Environment on Microbial Activities. University Park Press, Baltimore.

BIOCHEMICAL COMPONENTS AS INDICATORS OF SEASONAL CONDITION OF DEEP-WATER ZOOPLANKTON

Ulf Båmstedt
University of Göteborg
Tjärnö Marine Biological Laboratory
Post Office Box 2781, S-452 00 Strömstad
Sweden

The seasonal variation in the main biochemical composition of 5 megaplankton species from Korsfjorden, western Norway, reveals a general pattern of accumulation of lipids during autumn and utilization of the energy reserves during the winter and spring. The proportion of both protein and ash show seasonal variations more or less inversely with the lipid proportion. The average composition and seasonal range of the 5 species are summarized in Table 1, together with the corresponding weight-specific energy content, derived by using standard energy-equivalents for each component.

The onset of breeding is usually well correlated with the lipid and energy cycles; the copepods *Chiridius armatus* and *Euchaeta norvegica* both spawn most intensly in mid-winter and mid-summer when the energy content is relatively high, and the euphausiid *Meganyctiphanes norvegica*, with a life span of two years, breeds in the beginning of the year when the lipid content is at its maximum. The chaetognath *Eukrohnia hamata* has a more protracted breeding period which is probably most intense in autumn, and the mysid *Boreomysis arctica* breeds the year around, but probably most intense in autumn when the energy content is high. For *Euchaeta norvegica* and *Meganyctiphanes norvegica* there is a significant positive relationship between both body dry-weight and lipid proportion, and body dry-weight and weight-specific energy content during autumn - early winter, and this positive relationship disappears in early spring, thus indicating that the gradual build-up and later elimination of reproductive products are partly responsible for this. However, the overall decrease in lipid content during late winter - early spring, which is common for all

species and stages, is indubitably a consequence of the feeding
condition; the standing stock of presumptive prey organisms decreases
usually from December on to a minimum in March - April.

The seasonal cycles of the main biochemical components have been
compared between the different species and stages by means of
correlation analysis and this shows that stages within a species
have more similar seasonal trends than different species. It also
shows that the copepod *Chiridius armatus* has the most different
seasonal variations among the 5 species investigated and this might
be an indication of a somewhat different trophic position compared
with the 4 other species.

Since ATP (adenosine triphosphate) is a key factor in the
metabolism of organisms seasonal changes in the metabolic rate
may be inevitably correlated with changes in the ATP content.
Table 2 shows the seasonal variation in ATP concentration of 4
megaplankton species from Korsfjorden, western Norway. They all
show the highest ATP concentration in spring and an indication of
a second increase in autumn. Incomplete data for other species
indicate that the spring maximum is a general characteristic.
Values for herbivorous species indicate that they may reach the
peak at the time of spring phytoplankton-outburst, thus earlier
than omnivorous and carnivorous species.

These results have much in common with previously published
results on respiration and excretion of boreal zooplankton, where
a pronounced seasonal peak in spring is of general validity, and it
is therefore suggested that the seasonal variation in ATP concentration indicates inherent seasonal variation in metabolic rate.

Growth involves synthesis of proteins whereby RNA (ribonucleic
acid) is of great importance. As a rule, the growth rate decreases
with increasing size, both intra- and interspecifically, and this
is also true for the RNA concentration. A multi-specific relationship between individual body dry-weight and RNA concentration of
zooplankton from Korsfjorden, western Norway, sampled over 1 year,
reveals a relationship where an increase from 1 to 100 mg body
weight involves a 2/3 reduction of the RNA concentration. Much of
the variation around the regression line is due to the intermingling
of several species and single species therefore show considerably
closer correlation between RNA concentration and body weight.

Comparison between the relationships body weight / growth
rate and body weight / RNA concentration of megaplankton organisms
reveal almost identical regression coefficients, thereby indicating
a close linear relationship between growth rate and RNA concentration.

Table 1

Main biochemical composition and energy content (mean value with the seasonal range in parenthesis) of 5 zooplankton species from Korsfjorden, western Norway, sampled over 13 months.

Species	Stage	Biochemical composition (percentage of body dry weight)					Energy content (J/mg dry weight)
		Ash	Chitin	Carbohydrate	Lipid	Protein	
Chiridius armatus	female	7.4 (6.4–9.5)	7.0 (5.7–7.4)	2.0 (1.6–2.2)	24.2 (12.7–30.7)	46.1 (35.6–55.6)	25.7 (23.4–26.7)
Euchaeta norvegica	female	7.0 (4.4–11.2)	4.9 (3.7–6.4)	1.7 (1.0–2.2)	38.3 (26.4–47.3)	40.6 (35.6–46.5)	28.0 (25.2–29.9)
E. norvegica	male	6.2 (4.4–7.8)	5.2 (4.4–6.4)	1.4 (0.8–1.9)	37.5 (32.5–42.9)	47.7 (43.4–55.4)	28.0 (26.7–29.0)
E. norvegica	c-v	6.6 (4.4–11.8)	4.7 (3.7–9.3)	1.7 (0.9–2.4)	37.2 (25.9–43.4)	42.5 (38.5–48.3)	27.9 (24.8–29.3)
E. norvegica	egg	3.3 (3.2–3.3)	1.0 (0.9–1.0)	2.5 (1.7–3.8)	54.9 (49.9–57.5)	41.2 (36.4–48.8)	31.5 (30.6–31.9)
Boreomysis arctica	female	14.5 (11.8–16.9)	2.9 (2.8–3.2)	1.9 (1.2–2.7)	34.2 (26.8–43.1)	37.7 (31.9–44.3)	25.5 (24.3–27.6)
B. arctica	male	15.5 (13.2–16.9)	3.8 (2.3–5.4)	1.7 (1.1–2.4)	31.5 (24.4–40.3)	39.7 (32.0–46.7)	24.9 (23.4–26.8)
B. arctica	juv.	14.2 (10.9–18.6)	2.7 (2.3–3.2)	2.0 (1.3–4.0)	34.1 (21.8–44.6)	34.7 (25.2–41.4)	25.8 (23.2–28.2)
Meganyctiphanes norvegica	female	11.4 (10.1–12.6)	3.4 (2.8–4.4)	1.7 (1.4–2.1)	18.2 (9.8–26.9)	61.6 (53.1–76.5)	23.7 (22.2–25.4)
M. norvegica	male	11.6 (10.3–13.5)	3.5 (2.9–4.4)	1.6 (1.0–2.1)	17.7 (9.2–27.8)	61.8 (51.5–72.3)	23.6 (21.8–25.4)
M. norvegica	juv.	12.6 (11.4–14.4)	3.1 (2.5–3.8)	2.0 (1.7–2.5)	14.5 (6.8–19.6)	62.6 (53.4–75.0)	23.0 (21.6–23.8)
Eukrohnia hamata		18.3 (14.6–21.9)	0.1 (–)	1.5 (1.2–1.7)	32.1 (24.6–39.8)	39.1 (36.2–43.3)	24.4 (22.9–26.3)

Table 2

ATP concentration (μg/mg dry weight) of 4 zooplankton species from
Korsfjorden, western Norway, sampled over 15 months.

| | | Sampling month | | | | | | | | |
Species		Oct.	Nov.	Jan.	Feb.	Apr.	May	July	Sept	Dec.
Euchaeta	n	6	13	15	25	20	20	17	16	18
	$\overline{X}$	1.30	1.54	4.15	6.86	7.49	5.96	3.48	6.12	4.24
norvegica	s	.84	.75	.70	1.63	1.20	.76	.70	.99	1.02
Boreomysis	n	9	13	14	2	27	0	0	6	0
	$\overline{X}$	2.08	.49	1.56	3.46	6.48	–	–	2.70	–
arctica	s	.98	.53	.76	.19	1.58	–	–	.50	–
Meganycti-	n	11	11	15	14	5	5	19	15	12
phanes	$\overline{X}$	.73	.97	1.49	2.84	8.47	4.12	2.35	3.75	2.77
norvegica	s	.28	1.27	.30	.13	.93	.97	.71	1.80	1.95
Eukrohnia	n	4	5	15	20	9	4	0	5	0
	$\overline{X}$	3.11	.44	.23	1.42	2.48	1.31	–	3.72	–
hamata	s	.73	.25	.05	.54	.67	.19	–	1.02	–

   Since growth is mainly restricted to the period when food is
abundant this should occur from spring to early winter for omnivores
and predators. Table 3 summarizes results from RNA measurements on
11 megaplankton species from Kosterfjorden at the Swedish west
coast. With a few exceptions the highest RNA concentrations are
found in April - June, thus in the beginning of the food-surplus
period.

   The referred results show that the seasonal condition of
zooplankton populations, which is determined by factors like feeding
intensity, reproductive state and growth rate, these in turn governed
by environmental factors, is inevitably reflected by biochemical
and energetical parameters. Measurements of these parameters may
be further refined to yield quantitative estimates of condition
parameters.

Table 3

Average RNA concentration of 11 zooplankton species from Kosterfjorden, western Sweden, sampled over 8 months. Included sampling months are indicated by asterisks and the positive and negative above indicate the sampling month with maximum (+) and minimum (-) RNA concentration.

Species	Stage	Sampling month F M A M J J A S	Dry Wt (mg)	RNA $\bar{X}$ (μg/mg DryWt) (range)
COPEPODA				
Calanus hyperboreus	c-V, c-VI	∓ * * * * Ī*	1.3	5.5 (0.9-7.5)
Chiridius armatus	female	∓ Ī * * * * *	0.5	19.1 (12.3-29.0)
Euchaeta norvegica	c-V, c-VI	* Ī * ∓ ∓ * *	2.4	6.5 (2.8-12.7)
Metridia longa	female	* * * ∓ Ī* *	0.2	19.6 (4.0-41.9)
MYSIDACEA				
Boreomysis arctica	mixed	* Ī* ∓ ∓	17.7	6.5 (3.0-10.9)
EUPHAUSIACEA				
Meganyctiphanes norvegica	mixed	∓ * Ī *	37.6	13.2 (6.7-18.5)
Thysanoessa inermis + raschii	mixed	Ī * Ī* *	4.8	15.0 (7.5-53.3)
DECAPODA NATANTIA				
Pasiphaea multidentata	mixed	* Ī ∓ * * *	89.3	16.3 (11.1-20.4)
AMPHIPODA				
Parathemisto abyssorum	mixed	* * * ∓ * * * * Ī* *	2.4	28.7 (5.9-51.5)
CHAETOGNATHA				
Eukrohnia hamata	mixed	Ī * * * ∓	3.8	9.9 (5.8-13.5)
POLYCHAETA				
Tomopteris helgolandica	mixed	Ī* * * ∓ * * *	2.9	38.9 (18.3-70.3)

DEEP WATER RENEWAL IN FJORDS

H. G. Gade and A. Edwards†

Geophysical Institute
University of Bergen
Bergen, Norway, N-5014

1. INTRODUCTION

The notion of an isolated but easily studied body of seawater, removed for long periods from the complexities of advection or too much diffusion, is an attractive one. Early discussion of deep water renewals in fjord basins appears in the literature in connection with such stagnant water. Particular attention was given to stagnant basins with intermittent renewals - wherever discovered. Advective type water exchanges were linked to the prevailing wind conditons. They were most often considered to be a result of wind blowing upper, lighter water masses out of the fjord and thus forcing heavier water to rise. In this way, water appearing at sill level might be heavy enough to replace the resident water of the fjord basin (Gran & Gaarder, 1918; Strom, 1936; Braarud & Ruud, 1937).

Although this is still an accepted mechanism for renewal in some shallow silled fjords, later contributions draw attention to changes of the coastal water structure, often in response to longshore winds causing coastal upwelling. Important advances in the description and understanding of relevant processes were made in papers by Eggvin (1943); Gade, (1970, 1973); Linde, (1970); and Helle (1978). Recent interest has also focussed on the importance of the tides, particularly their currents, for the exchange processes in fjords. As well as causing transports in and out of the fjord, the tides are often the main source of energy for mixing and cir-

† Scottish Marine Biological Assocation
  P.O.B. 3, Oban, Argyll, Scotland.

culation of basin water. It has been found that the mixing process is enhanced where baroclinic modes of the tidal wave are prevalent (Stigebrandt, 1976, 1979). The role of penetrative convection has been pointed out (Saelen, 1950; Nutt & Coachman, 1956). Furthermore, the characteristics of the deep water in a number of fjord basins have been extensively dealt with in many papers, among which are Saelen (1950, 1967), Beyer (1954), Pickard (1961, 1963, 1967, 1971), Gade (1963); Edwards & Edelsten (1977) and Reid *et al.*, (unpubl. man.).

2.   THE NATURE OF DEEP WATER RENEWAL OF FJORD BASINS

2.1   Steady state and intermittent circulation in fjord basins, renewals

Because of the highly stratified nature of most fjords, the estuarine circulation is often thought to be determined by the entrainment of sea water into the brackish layer. This concept may be adequate for a rough description of circulatory features of the brackish layer, but fails for a proper understanding of the circulation below the main pycnocline. A closer inspection reveals that there are indeed diffusive processes which can be dealt with by eddy coefficients at all levels in the fjord, in most cases quite satisfactorily by overall coefficients, functions of depth and time only (Gade, 1970; Stigebrandt, 1976).

If such eddy diffusion were the only effective process in a fjord basin - in otherwise stagnant conditions - one would observe a continuous lowering of density at all levels until the water column was completely uniform, determined by the upper boundary condition. Apart from any possible microstructure, such a process is reflected in the sign of $d^2\rho/dz^2$ which under such conditions is invariably negative. This decrease of density is the basis of renewal of the deep water by advection of water from outside into the basin. The advection takes the form of a density current developing from the sill and is conditioned by the adjacent water above sill depth being denser than the resident water in the basin.

Provided the density stratification of the coastal water adjacent to such a fjord remains perfectly steady, sea water will enter and descend to the greatest depths at all times (Fig. 1). The flow of sea water will then be dominated by the vertical eddy exchange processes in the basin water;

$$\nabla \cdot (\rho \underline{v}) = \frac{\partial}{\partial z}\left(K_z \frac{\partial \rho}{\partial z}\right)$$

where $\underline{v}$ is the velocity vector.

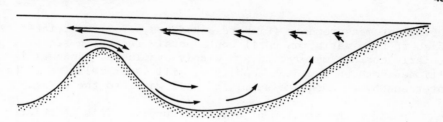

Figure 1. Schematic diagram of deep water renewal in a fjord basin.

Deep water renewal by steady state advection, as implied above, would in many cases be sufficient to keep the basin water well aerated.

Such steady conditions are, however, rarely met in nature. Normally, seasonal and other fluctuations of the density field in the adjacent water interfere with the inflow of sea water. If the density of the water available above sill depth is reduced, the current weakens and may for periods vanish completely. In this case the basin water becomes stagnant. Otherwise, increased density of the adjacent water may accelerate the intrusion of sea water to the fjord even to the extent that the inflow reaches the maximum corresponding to overmixed conditions (see 4.4).

If the fluctuations of the external conditions are sufficiently strong relative to changes in the basin, the inflow of sea water to the fjord basin will be repeatedly interrupted, causing the deep water renewal to be intermittent.

The alternating behaviour of sill fjords between active and stagnant phases, being caused by intermittence of the deep water renewal, occurs on a variety of time scales. The shortest of these are to be found in the tidal motion which often overrides the steady state currents of the sill region. Other fluctuations are tied in with the weather and may contain elements coupled to the change between sea and land breezes or to events such as the passing of low pressure areas. Intermittence is often seen on a time scale of several weeks and appears to be caused by correspondingly ex-tended periods of predominant winds. Important also are seasonal changes which may lead to periodicity of a year or to intermittence of stagnant intervals of multiples of a year.

2.2 High frequency forcing - tidal and other short term processes affecting the deep water renewal

The intermittence of renewal on long time scales has been long recognized (see 2.3), but the earlier difficulty of making

sufficient measurements (Pickard & Rodgers, 1959) restricted study
of short term variation until continuously recording sensors, particularly current meters, became widely available. Renewal is conditioned both by the high density of sill water relative to basin
water and by an adequate supply of sill water to the basin.

Consider the problem of adequate supply. This is related to
the flows over the sill, which have barotropic and baroclinic components, whose interaction determines the manner in which the renewal is made intermittent. In a deep silled fjord, the barotropic
component associated with tides, wind and meteorological disturbance
may be insufficient to reverse the baroclinic flow over the sill
and inflow will continue as long as the density condition is satisfied. If the barotropic flow varies, the baroclinic flow will
be modulated and the intensity of renewal will vary accordingly.
Over shallower sills there is a relevant hydraulic control on the
size of purely baroclinic flow, which is limited such that a suitably defined Froude number does not exceed unity (Stommel & Farmer,
1953; Kullenberg, 1955). The barotropic component is not so limited
and indeed tends to increase as shallower sills are considered.
Consequently, the shallower the sill, the more likely is it that
bartropic flow will switch off inflow if directed out of the fjord
and augment it if directed inwards. As a result, inflows are pulsed
rather than modulated as over deeper sills. However, because of
the barotropic augmentation, such systems, if compared to purely
baroclinic ones, can carry more renewing water into the basin and
effectively increase the carrying capacity of the sill (Stigebrandt,
1977).

From this viewpoint there is no difference in principle between
the influence of tides, wind and other meteorological disturbances.
All cause barotropic flow which modifies the inflow. Their relative
importance may be judged from the long term means of the modulus
of the flows. Usually the tides predominate, but for short periods
the other factors may be more important.

It is sometimes difficult to distinguish tidal modulation of
inflow from internal basin oscillations (Skreslet & Schei, 1976)
which interfere with measurement in the flow path at the basin bottom near the sill. If a water property is measured simultaneously,
the periods of renewal rather than oscillation may be easily identified (Fig. 2). In the simplest cases, where the transit time over
the sill is short, modulation is diurnal or semidiurnal, as the
local tide (Skreslet & Loeng, 1977; Edwards & Edelsten, 1977).
Fig. 2 is an example from Loch Eil (D. J. Edelsten, pers. comm.)
which also indicates a spring - neap variation. Springs enhance
renewal by increasing supply and by their effect on sill mixing
(see below). Over longer sills, the springs inflow peak may be
delayed in its passage to the basin (Cannon & Laird, 1978) and even
arrive during neaps.

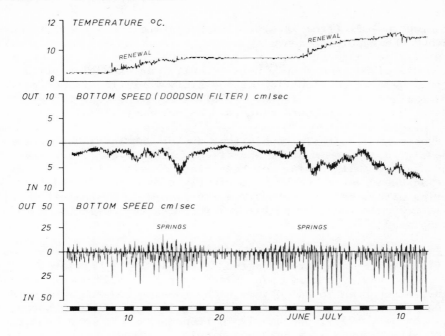

Figure 2.   Tidally pulsed inflows measured at the bottom of a basin
            (Loch Eil 1976, station X, depth 50 m, Edwards *et al*.
            1980).   The lowest record shows semidiurnal pulsing.
            The central filtered record shows an increase in the
            intensity of renewal during spring tides when (top record)
            bottom temperature changes most rapidly.

        The influence of the wind is twofold. It causes baroclinic
flow which in the sill region can import renewing water, and it
changes the fjord water level – causing barotropic currents which
diminish or augment (Ozretich, 1975) inflow over the sill according
to their direction. Occasional examples suggest that a down-fjord
wind has increased inflow (Johannessen, 1968) or renewal (Gade, 1968;
Bell, 1973) by accelerating the estuarine circulation, but as flow
can be multilayered (Pickard & Rodgers, 1959) it is not possible to
generalize the relation between wind direction and renewal. The
wind is probable more significant in its effect on the adjacent
density field, for satisfaction of the density condition (Gade, 1976;
Skreslet & Loeng, 1977; Helle, 1978).

        A varying freshwater run-off affects renewal in three ways. It
drives baroclinic and barotropic currents at the sill and it alters
the density of inflow if there is mixing at the sill associated with
tidal or other currents. Without this mixing, the effect of an

increasing run-off (from zero) is to accelerate the baroclinic currents and the supply of renewing water (Gilmartin, 1962; Bell, 1973).
The supply increases to a maximum, beyond which increasing barotropic
outflow diminishes the supply until finally, at exceptionally high
run-off, there is only outflow at the sill, no inflow (blocking),
and renewal is prevented (Beyer, 1976). The steady and transient
states of such systems were discussed by Welander (1974).

Even with a suitable supply from any of the above mechanisms,
the density condition must be satisfied. It may fail because adjacent density falls during coastal stratification fluctuation or
if the adjacent water, being sufficiently dense, is yet modified
by mixing with lighter water during its passage over the sill. In
this latter case the renewal rate varies with the mixing or with
the supply of light water.

Examples of tidal variation of inflow because of adjacent density fluctuations are rare (Anderson & Devol, 1973; Collias &
Barnes, 1966; Reid et al., 1976; Seibert et al., 1979), but inflow
rate has often been found to vary with changing coastal longshore
wind on time scales of a few days (Helle, 1978; Skreslet & Loeng,
1977; Svendsen, 1977), presumably in response to offshore convergence or divergence. When adjacent density is varying, it is not
necessary that it should always be greater than the basin density,
only that it should sometimes be greater, and that fluctuations in
inflow and density should be in such phase that there is a net inward flux of denser water.

Sills are regions of increased current, larger shears and
tidal jets, and mixing is greater there than in the main fjord.
Important contributions to this theme have been made by Drinkwater
& Osborn (1975) and McClimans (1978). Mixing of upper water into
the inflow reduces inflow density and inhibits renewal. Where this
effect is important, the density will vary on many time scales,
perhaps seasonally, but in any case the basin will tend to renew
at times of low run-off (Bell, 1973; Matthews & Quinlan, 1975;
Ozretich, 1975). Cannon (1975) thought that water entering Puget
Sound during neaps spends more time in the sill region than during
springs and so emerges from the sill too light to cause renewal at
neaps. An alternative mechanism, not reported in the literature,
is that inflow modification by mixing could be minimal in neaps -
which would thus promote renewal. It sometimes happens that significant quantities of run-off are mixed down so that an inverse
relation between sill salinity and preceding run-off can be demonstrated (Edwards & Edelsten, 1977). Temperature effects on density
are small and run-off consequently controls the renewal through its
influence on the density condition. In Loch Etive the run-off over
the preceding few weeks is relevant, and the stagnation-renewal
mechanism is thus capable of short term response, faster than seasonal (Fig. 3).

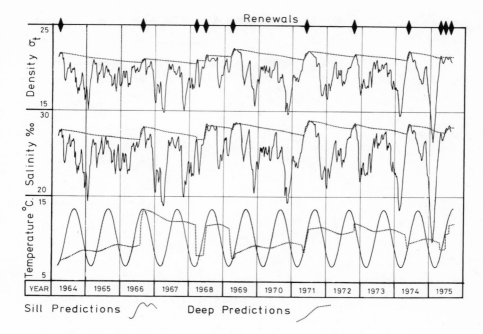

Figure 3.   Application of Welander's model to Loch Etive. Hindcast
records of the properties of sill and inner basin bottom
water show that renewals occurred at irregular intervals
whenever the sill density exceeded the bottom density.
Sill water responds on short time scales to changes in
the local run-off.

2.3 <u>Low frequency forcing – seasonal and other long term processes</u>
<u>affecting the deep water renewal</u>

Striking cases of intermittence of deep water renewal to a
fjord basin are seen in connection with seasonal changes of the
density structure offshore. In Norwegian fjords such changes are
recognized as coupled to the monsoonal nature of the major wind
field, being predominantly northerly in the summer and southerly
in the winter. The resulting coastal divergence or convergence
shows up clearly in the levels of the isopycnals of the coastal water
such as are evident from the 10-year averages presented in Fig. 4.
The diagram also suggests that the fluctuations are coupled to the
amount of fresh water in the coastal current such as derives from
the Baltic outflow and runoff from southern Norway (Helland-Hansen
& Nansen, 1909). Thus the first salinity maximum may be attributed
to winter retention and is felt at all levels down to at least 100 m.
The secondary maximum coincides with the peak of northerly winds in

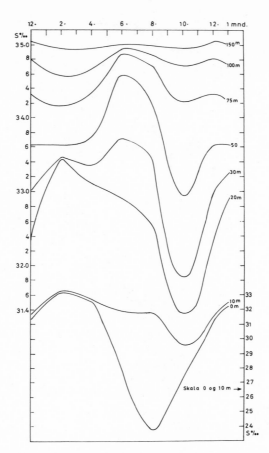

Figure 4.   Average salinity isopleths at the coastal station
            "Sognesjoen", for the period 1947-1956.

May-June, a time the local runoff is considerable although not maximal. The abrupt change in August followed by a minimum salinity
in October reflects the generally increased low pressure activity
with southerly winds and heavy rains in the autumn. The coincidence
of phases at all depths is clear indication of surface convergence
as the predominant process.

     A brief glance at Fig. 4 suggests a distinction between shallow
silled fjords which should experience influxes in the late winter
and deeper silled fjords where the major influxes are to be expected
in the middle of the summer. The first is undoubtedly the case for
most fjords in southern Norway with sills less than 40 m. In fjords
with sill levels deeper than 100 m inflows are often seen as an almost continuous process beginning in January or February extending

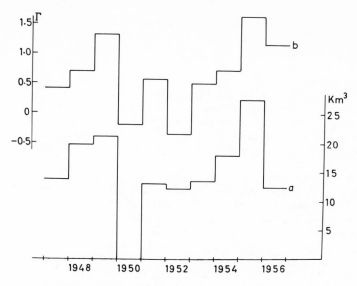

Figure 5.   Yearly volumes of inflow to the Bergen Byfjord (lower
diagram) and associated strength of northerly winds measured by the wind parameter T.   Reproduced with permission of H. Helle, 1978.

throughout the summer, only interrupted by isolated intervals of
passing cyclones.

A number of British Columbia fjords behave in the same way.
Typically for the shallow-silled fjords, Indian Arm (sill depth 26 m)
exhibits annual intrusions during January-March.   The bottom water
appears, however, to remain stagnant for several years at a time
(Gilmartin, 1962).

A similar feature is found in Alaskan subarctic fjords.   Winter
retention in the form of snow is likely to be responsible for the
salinity maximum above 60 m in the late winter whereas below 200 m
the maximum values follow 6 months later (Royer, 1975, 1979; Hoskin
et al., 1978).   The conditions of the source water for deep water
renewal in Alaskan subarctic fjords are further discussed by Muench
& Heggie (1978).   The seasonal variation of the density field in the
Gulf of Alaska is here assumed to be primarily determined by the
changing of the predominant circulation of the lower atmosphere.   It
is pointed out that intrusions to shallow silled fjords take place
in March-April, whereas the deeper-silled fjords experience renewals
later in the year.   In fjords with sill depths below the depth of
minimum annual density variation in the Gulf of Alaska, about 150 m,
deep water renewal may occur as late as September.

    The pattern discussed above is also seen in the Oslofjord
(sill depth 19.5 m) although not entirely for the reasons given.
The monsoonal winds of the area have directions predominantly out
of the fjord in the winter and into the fjord in the summer and the
upper water masses of the fjord react quite clearly to the longitudinal surface stress. Local upwelling in the fjord, especially
outside the sill area, contributes to increasing the density of the
water available for inflow across the sill. The effect is obvious
in sections obtained during strong winds (Fig. 5) and as a seasonal
trend it is reflected in the ice-free condition of the fjord.

    A number of fjords, of not the majority, respond to seasonal
forcing of the coastal oceanographic climate. Wherever the annual
variation is large in comparison with variations on shorter time
scales, the annual cycle is likely to show up in the deep water,
indicating partial or complete renewal of the basin water.

    Such is the case in Puget Sound where the bottom water is re-
placed in the late summer due to coastal upwelling response to
northerly winds. At other times of the year renewals appear to be
triggered by other mechanisms (Cannon & Ebbesmeyer, 1978).

    The influence of longshore winds on renewal of the deep water
of the Bergen Byfjord has been examined in detail by Helle (1978).
Helle compared the volumes of the individual annual inflows, as
determined by Linde (1970) for a 10-year period, with a longshore
wind anomaly and was able to show a clear correlation, as evident
in Fig. 6. Furthermore, direct current measurements over the sill
show inflows of dense water directly coupled to the northerly wind
component at a nearby coastal station.

    The effects of wind on exchange between coastal and fjord water
have also been dealt with by Svendsen (1977), Svendsen & Thompson
(1978), and in an article by Svendsen appearing in the present
volume. Attention has been given to northerly winds and associated
upwelling along the Norwegian west coast. Simultaneous current
meter records document intrusions of relatively dense water into
the fjord.

    It is not obvious that fjords will always respond with basin
water renewal to seasonal forcing of the density structure of the
adjacent coastal water. The necessary condition that the water
available for inflow has a density equal to or greater than that of
the bottom water of the basin may still not be satisfied. With the
steadily decreasing density of the resident water, however, renewal
must eventually take place, although possibly with intervals of many
years. In Fig. 7 the Bonnefjord basin of the Oslofjord exhibits an
example of stagnation from 1962 to 1966, when another complete renewal was observed. The oxygen values in the 400 m level in the
innermost basin of the Nordfjord (Fig. 8) are not equally conclusive,

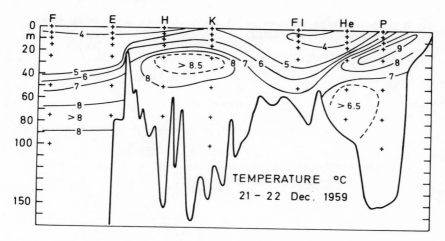

Figure 6.   Example of convergence of surface water in the northern
            end of the Oslofjord (St. Fl) under southerly gales.

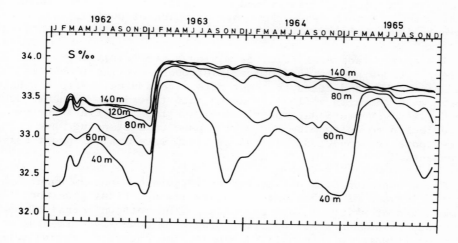

Figure 7.   Isopleths of salinity for the Bonnefjord basin of the
            Oslofjord 1962-1966.

but suggest intervals between the recorded major renewals of up to
10 years.

3.   VERTICAL DIFFUSION IN FJORD BASINS

On the basis of observed fluxes of water properties it is justified to assume that turbulent diffusion happens at all levels in

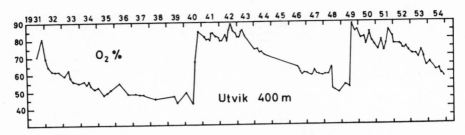

Figure 8.   Isopleths of oxygen percentage during 30 years at the
            "Utvik" station in the innermost basin of Nordfjord.
            Reproduced with permission of O. H. Sælen , 1967.

a fjord basin. The state of turbulence will vary from place to
place and also with depth. There is also a time variation which
may be a function of the external and internal physical conditions
such as the stability of the stratification or the existence of ice
cover, or dynamically conditioned by variation in the turbulence
generation. In both cases there are examples of periodic and non-
periodic fluctuations.

    The energy made available for mixing comes mainly from two
sources - tides and meteorological distrubances. The tides are the
most dependable and often the most effective source. With the barotropic wave there is a flux of energy into the fjord. Except for
fjords with very serious constrictions, most of the tidal energy
is reflected so as to make the wave behave as a standing wave. A
lesser part is converted, giving the tidal wave a small progressive
component.

    The conversion of tidal energy to turbulence depends upon three
principal types of mechanism: boundary mixing, tidal jets and
plumes, and breaking of internal waves. The most universal of these
processes is boundary mixing, resulting from boundary form-drag and
friction in water of a stability usually low enough to ensure turbulence. The thickness of the affected boundary layer is a strong
function of the amplitude of the tidal current velocity, and therefore is expected to vary greatly within a fjord basin. Boundary
mixing is assumed to make itself felt in the interior of the basin
by lateral density currents acting within thin sheets. The process
may be greatly enhanced by baroclinic currents set up by internal
waves associated with the tidal motion. In fjords with constrictions, such as narrow sounds or shallow sills, the tidal current may
become strong enough to separate from the walls and form jets and
plumes penetrating the water masses of the fjord. According to
Drinkwater & Osborn (1975) such conditions are observed in Rupert
& Holberg Inlets, Vancouver Island. *Editor's note: see paper by*

D. J. Stucchi following this one. Stigebrandt (1979) has given a unified theory of the energetics of tidal flow through such constrictions, summarizing previous contributions by Glenne & Simensen (1963) and McClimans (1978). Tidal plumes of excess density may contribute to the generation of turbulence in the basin water.

Tides generate internal waves in the vicinity of sills and other topographic features interacting with the tidal flow in stratified water. In fjords internal waves of tidal period are often outstanding features, such as those observed in the Herdlafjord (Fjeldstad, 1964). The phenomenon may become particularly well developed where the sill reaches up to the pycnocline. Such is the case with the Oslofjord where the internal waves have heights up to 15 m (Johannessen, 1968). It was demonstrated that the internal tide diminished strongly away from the sill, indicating that the energy of the internal waves is subject to appreciable dissipation. The internal waves were therefore considered vital for transfer of energy to turbulence and consequent vertical exchange (Gade, 1970).

Similarly, a study of the vertical exchange processes in the deep water of Cambridge Bay, an arctic basin on the southern shore of the Candian Arctic Archipelago, pointed strongly towards internal waves of tidal origin as the main source of energy for generation of turbulence (Gade *et al.*, 1974). Analysis of the Richardson numbers suggested that the mixing took place along the boundaries (bottom) of the basin. A thorough discussion with reference to other more recent observations has been given by Perkin and Lewis (1978), based upon Stigebrandt's (1976) formulation of the problem.

Stigebrandt (1976) has studied the dynamics of internal waves generated near the sill in a two-layered fjord. He points out that oscillating barotropic currents - including meteorologically induced seiches - may cause generation of internal waves if the densimetric Froude number near the sill is less than 1. Supported by simple laboratory experiments, Stigebrandt's conclusion was that the internal waves would normally not be reflected from the sloping side walls, but would break and thus dissipate the energy. An interesting observation in the tank experiments was that the breaking would cause turbulence at the sloping of the wall to appreciable depths below the pycnocline whereas there was little evidence, if any, of turbulence in the upper layer. The breaking of internal waves would therefore have an effect similar to that of localized boundary mixing. Later observations have generally confirmed the principal behaviour of the theoretical internal tide in the Oslofjord (Stigebrandt, 1979). By relating the energy dissipation of the internal tides to the increase in potential energy, Stigebrandt found the overall flux Richardson number to be 0.05.

Also, wind generated currents are obviously of central importance in the generation of turbulence. However, such currents generally diminish with depth and reduce strongly below the pycnocline. Only disturbances of the density field in the form of internal seiches can be expected to contribute significantly to the field of turbulence below the pycnocline. Disturbances related to surface seiches are under such circumstances likely to be strongly, if not critically, damped (Stigebrandt, 1976).

A very special form of deep-water stirring takes place when water heavier than the resident water enters the basin from the sill. The potential energy released by the density current will then be almost completely converted to turbulence within the basin. From laboratory experiments of this kind Stigebrandt (personal communication), has determined the corresponding flux Richardson number to be less than 0.1, implying that more than 90% of the released energy is converted to heat. Determination of the corresponding process in nature is not reported in the literature, presumably because of the difficulties in establishing exact figures for the energy release, depending upon the initial velocity, the exact amount and the density of the intruding water. For a complete determination of the problem it is also necessary to know the density and the levels at which the resident water leaves the fjord. Further comments on the process are given in section 4.2.

The role of turbulence in diffusion is that of exchange and subsequent mixing of water particles. In addition, molecular diffusion will be present, but the effect of this on large scale eddy transports has generally been assumed to be negligible. It is possible that with the layered structure, molecular processes may be important across discontinuity surfaces. This would explain the higher eddy diffusivities observed for heat than for salt in the Oslofjord (Gade, 1970), in Loch Etive (Edwards & Edelsten, 1977) and in some North American fjords (Pickard, 1961).

Although virtually every scalar property of the water can be used in budgets to determine overall vertical eddy diffusivities within a fjord basin, coherent results have mostly been obtained from budgets of heat and salt. Apart from the fact that these can be considered conservative and therefore more easily employed, both variables are determining for the density and tend to be evenly distributed laterally. Inhomogeneities developing from localized turbulence will thus tend to be counteracted by the lateral currents set up by the associated pressure gradients.

From heat and salt budgets, overall vertical eddy diffusion coefficients have been established for selected fjord basins during stagnant periods. Thus, Gilmartin (1962) reported from Indian Arm, a British Columbia fjord, values ranging from 0.2×10^{-4} to 10×10^{-4} $m^2 s^{-1}$, mainly between 10^{-4} and $3 \times 10^{-4} m^2 s^{-1}$. Corresponding values computed from Norwegian fjords cover much the same range with a tendency towards lower values. The lowest known vertical diffusivities in fjords are of the order $10^{-6} m^2 s^{-1}$, and have been computed from observations from Framvaren (southern Norway), a shallow-silled fjord practically free of tides.

Examples of application of budget methods to fjord basins have also been given by Gade (1970), Goransson & Svensson (1975) and others. A penetrating discussion of the validity of the method has been given by Stigebrandt (1976).

Independent determinations of the vertical eddy diffusivity based upon release of an artificial tracer indicate values of an order of magnitude less than those derived drom overall budget methods (Bjerkeng *et al.*, 1978). Stigebrandt (1976) has interpreted the discrepancy as resulting from the assumed localized turbulence caused by internal wave breaking.

The computed eddy diffusivities are found to vary with the static stability. In both the innermost basins of the Oslofjord the dependency on the vertical stability is quite clear, K_z being related to the Väisälä frequency N by the following relation:

$$K_z = aN^{-2\alpha}$$

where α is a nondimensional constant close to 0.8, and a another constant. The latter differs widely from basin to basin (Gade, 1970). Examples of the computed diffusivities against N^2 are shown in Fig. 9 in a double logarithmic representation. Determinations of vertical diffusivities from other fjords support the relationship, but show different values of a and α. Thus, Goransson & Svensson (1975) found $\alpha = 0.6$ for the Gothenburg Byfjord, and values as low as 0.5 have been reported by Aure (1972).

4. DYNAMICS OF RENEWALS

4.1 Density currents and intrusions

Having entered the basin, renewing water is denser than its surroundings and sinks towards the basin bottom. It flows downslope from the sill as a turbulent density current. Such currents, or plumes, occur widely in the sea and their behaviour has been studied in laboratory experiments, for example, by Ellison & Turner (1959). The forces acting on an element of the plume are the force of gravity modified by the density contrast, a pressure gradient due to the changing plume thickness, the bottom drag, and a force due to the acceleration of entrained ambient fluid from its low velocity in the relatively quiet resident water to the velocity of the plume. The mean inflow across the plume boundary is commonly assumed to be proportional to the local plume mean velocity: the

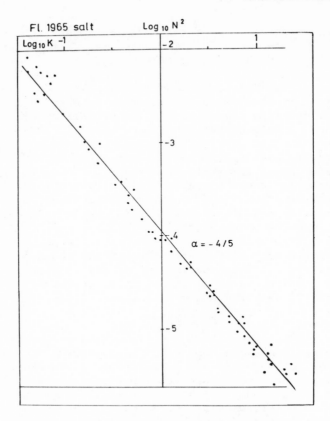

Figure 9.   Overall vertical diffusivity of salt in the Vestfjord
            (Oslofjord) versus the static stability in double log-
            arithmic representation.

constant of proportionality is the entrainment constant. If a
plume starts slowly, gravity dominates the force balance and the
flow accelerates; if the velocity is high, increased entrainment
slows it. The smaller the slope, the greater the stabilizing effect of gravity, which acts to suppress mixing. The entrainment
constant thus depends upon both Richardson number - suitably defined
in terms of density contrast, velocity and plume thickness - and
slope. It is greatest at large slopes and at small Richardson
number. Its functional dependence on slope and Richardson number
has been described by Turner (1973). The theory is scale independent and has been successfully applied to several basins. Smith
(1975) examined inflows to the Atlantic as isolated phenomena.
Manins (1973) modelled the deep water of the Red Sea with a combination of an advective turbulent plume and a diffusive flux up-

wards in the ambient fluid. An interesting aspect of his model, in which inflow extends to the bottom, is that it emphasises the dependence of the steady state density distribution in the interior upon the ratio of the total entrained buoyancy flux to the diffusive buoyancy flux out of the basin top. In the model, if this ratio is large there is effectively considerable recirculation within the resident water, which tends towards homogeneity. If the ratio is small, vertical density gradients in the ambient fluid are relatively greater. The observed shape of the density profile may thus be used to calibrate such models, which should find application in fjords. In a related formulation, Killworth & Carmack (1979) used the plume equations to describe the fate of relatively dense river water entering Kamloops Lake. In a Scottish fjord, Edwards & Edelsten (1977) showed that inflow formed a turbulent plume which behaved according to Turner's predictions. In similar work in the Saguenay Fjord, Seibert, Trites & Reid (1979), found acceptable values for bottom drag and entrainment, but were unable to account for potential energy changes in the basin solely as a result of energy injected by the plume. A more thorough study of the energy balance in the Saguenay Fjord is given by Reid *et al.* (unpubl. man.).

An important difference between the laboratory experiments and the natural basins is that plume density decreases downwards because of mixing, whereas resident density increases downwards in the ambient stratification. It is possible therefore that the plume will separate from the slope and spread into the interior (Killworth & Carmack, 1979; Smith, 1975; Seibert *et al.*, 1979) when the density contrast vanishes. These modifications to the theory clearly have application to the proper description of partial renewals in which the inflow does not attain to the bottom.

4.2 Partial renewal of the basin water

If an intrusion of ocean water does not occupy the entire volume below the sill septh of a fjord or a fjord basin, the renewal is said to be partial. Principally there are two ways renewals can be limited to being partial.

Intrusions of sea water not sufficiently dense to replace the deepest (bottom) resident water of a fjord basin will sink to their appropriate levels, spread out and begin to fill up the available space from that level and up. During this process the deeper water may remain relatively undisturbed although often giving evidence of being affected by mixing from above. Certainly, the release of potential energy by the falling of a blob or the sliding of a plume down the slopes of a fjord basin is bound to be converted to turbulence. In the less pronounced cases such as a thin plume reaching its appropriate level after a long journey from the sill, most of the energy may have already been spent (entraining ambient water

and overcoming bottom friction), and the new water may come to rest
without stirring up much of the resident basin water.

In more vigorous cases of renewals the intruding water may
reach its density level with considerable speed and entrain a significant portion of resident water both above and below. In the most
extreme cases the plume may "overshoot" the appropriate density
level so much as to tear up the pool of resident water below. When
finally coming to rest the new bottom water will have properties
lying between those of the resident and the intruding water. Thus,
a somewhat paradoxical situation, of sudden increase of the oxygen
content of a bottom water while the density appears to have decreased,
may occur. Such an event was seen in Bonnefjord in 1965 (Fig. 7).

The limitation of renewals to being partial in the sense described above is primarily a function of sill water density and the
density stratification of the basin water. As a rule of thumb, large
intrusions may cause bottom water renewal if the intruding water is
about as dense as the resident bottom water. For sea water entering
the fjord basin at a comparably low rate, it is a necessary condition
that the intruding water has a higher density than the resident
water for replacement of the bottom water to take place.

Deep water renewals may also be partial in the sense that although sill water is of sufficient density to replace the deepest
water in the basin, the duration of the event may not be sufficient
to allow the basin to be filled. Usually a renewal is considered
complete if comprising all the water of a basin up to the depth of
the sill. Complete deep water renewals are often seen as parts of
more or less complete renewals of the water masses of the entire
fjord.

4.3 Uplift of the resident water by intrusions

A conspicuous feature of intrusions of new water into a fjord
basin is the uplift of the resident water. The uplift is often
clearly indicated by the vertical displacement of the field of mass
or any other property having a discernible vertical gradient. A
universal feature of such uplifts is the vertical compression of
the layers as they expand laterally to fill the available horizontal
corss section of the basin. With complete renewals the resident water
is lifted above the level of the appropriate sill, although not necessarily above the main entrance sill of the fjord. In both cases
the resident water disappears from the basin but may be found at
higher levels throughout a major part of the fjord, although most
commonly least modified in the innermost parts. The latter feature
can possibly be attributed to outflow from the fjord at higher levels,
but appears to be the case also for partial renewals. In the latter
case the gradual fading of the typical properties of the remnants

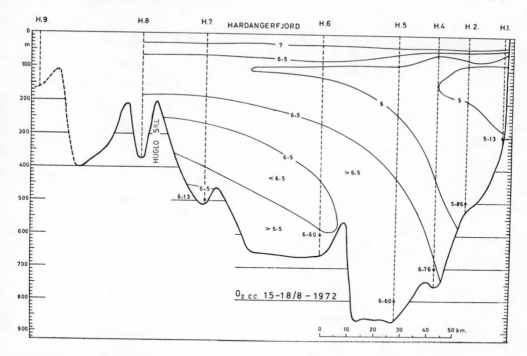

Figure 10. Isolines of oxygen concentration observed in the Hardangerfjord during period of deep water renewal. Rest of old resident water seen as intermediate layer in the inner parts of the fjord (upper right). Reproduced with permission from H. Svendsen and N. Utne, 1973.

of the basin water must be taken as indication of a higher degree of mixing in the outer portions of the fjord (Fig. 10).

It is often desirable to know the volume of a partial renewal. This can at times be obtained by direct integration of the newly advected water masses. The task may, however, not be so straight-forward as it seems because it is often virtually impossible to know how much of the resident water has been entrained into the advected water, or because the position of the (upper) interface between the new and the resident water may not be precisely defined.

The problem is most easily dealt with by determining the up-lift of the resident water. For this purpose it is convenient to make use of the elevation diagram of Fig. 11. The diagram is so constructed as to show the theoretical elevation H of the density surface originally situated at the level A, when all resident water below the level B, and not more than this, has been replaced by water advected from outside. In this case (bottom-water renewal)

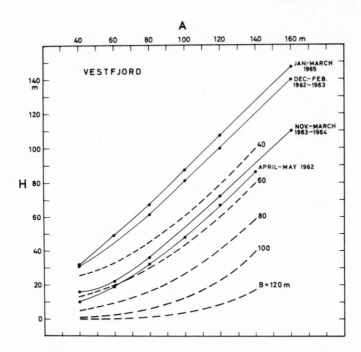

Figure 11.   Example of elevation diagram showing theoretical uplift
              (H) for density surfaces originally situated at the
              level A, for renewal under the selected level surfaces
              (B) (dashed lines). Also four events of deep water re-
              newals are indicated.

the mathematical formulation of the problem is simply

$$\int_{A}^{A+H} F dz = \int_{O}^{B} F dz$$

Here, F is the area of the horizontal cross section in the basin,
B the (equivalent) thickness of the new bottom water and z the vertical coordinate originating at the deepest point in the basin.

    The actually observed elevations of the isopycnal surfaces,
measured as the vertical distances between homologous points on
station curves taken before and after inflow, can now be entered.
Plotted directly into the diagram the actual elevation curve immediately reveals the level to which the influx has reached. It is
then a fairly simple matter to deduce how much water has entered

during the inflow period. Furthermore, comparison with the theoretical curves will also show whether the inflow really penetrated
to the bottom or an intrusion at higher levels took place.

Four separate events are recorded in Fig. 11. The two lower-
most (1962, 1964) indicate renewals up to the level of the sill to
the adjoining basin (~ 50 m.) and the upper curves represent complete renewals where the old resident water practically disappeared
from the fjord.

Volumetric determinations have also been attempted on the basis
of budgets of conservative or nearly conservative properties. Thus,
Anderson & Devol (1973) considered the sudden appearance of nitrates
in the deep water of Saanich Inlet after a period of inflow of new
water. Based upon knowledge of the concentration of nitrates in
the flushing water and the fair assumption that the natural conversion of nitrates was negligible during the relatively short time
of flushing, the change of concentration could be used for determination of the relative amount of new water in the final state.
For the reported cases the changes were from zero to about 10 µg-
atoms ℓ^{-1}, and could thus be fairly accurately determined. Anderson
& Devol (1973) also make use of observed changes in the TS-diagram
to relate the effect of a partial renewal to the fraction of admixed
new water of given characteristics. The actual volume of the renewal
is then obtained by integration over the affected depths of the basin.

Similar calculations have also been attempted for non-conservative substances in Lake Nitinat, a B. C. fjord, by Ozretich (1975).
In this case the volume of the intruding water could be established
on the basis of budgets of hydrogen sulphide being oxidized by the
dissolved oxygen introduced with the renewal. Because of the interval between the observations (about two years) the measurements were
adjusted for changes due to vertical diffusion.

4.4 Overmixed systems

The term overmixed was introduced by Stommel & Farmer (1952)
to characterize steady state estuaries in which the two way flow
through the mouth or any other constriction is maximized. The overmixed state depends on the dynamic control exerted by the mouth, or
a constriction, on the flow through the control section and it is
reached if sufficient energy is available for vertical mixing inside
the estuary (to cause maximum density gradients across the control
section). Steady state estuaries having a homogeneous, two-layered
structure fulfil these requirements if the interface lies at or
below the sill depth. In this case the two way flow through the
control section will be separated approximately half way down to the
bottom of the sill.

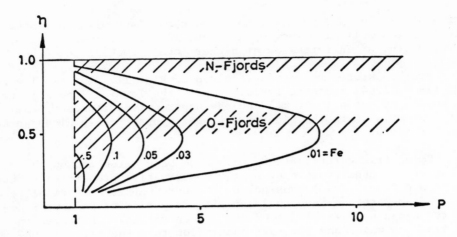

Figure 12.   Relative thickness η of the upper layer in a constriction
             with critical flow as function of the admixture ratio
             P in the estuary for selected values of the estuarine
             Froude number Fe. Position of N-fjords and O-fjords
             indicated. Reproduced with premission from A. Stigebrandt
             (unpublished manuscript), 1979.

        Although steady state conditions hardly exist, a number of
basins around the world appear to be overmixed in the sense that
the conditions above are otherwise fulfilled. Such is the case with
the Black Sea, the Baltic Sea and, in periods, Loch Etive, among
other fjords. Fjords are, however, rarely overmixed. This is due
to the rather small amount of energy normally available per unit
volume for mixing. The tidal energy tends to be reflected and the
intermediate water masses above sill depth are often too deep to be
significantly affected by wind mixing from above. Stigebrandt (1979)
has accordingly denoted such fjords N-fjords (normal fjords) in contrast to O-fjords (overmixed fjords). Overmixed fjords are therefore expected to be found among those which are relatively shallow,
strongly exposed, have appreciable tides, or have relatively little
fresh water discharge. Stigebrandt (unpubl. man.) has indicated how
N-fjords and O-fjords occupy different regions in the diagram of
Fig 12, where the relative thickness of the upper layer is shown as
function of the admixture ratio for fjords satisfying the Stommel
& Farmer (1952) criterion. In the case of tidally-driven overmixing
the simple theory of overmixed basins cannot be expected to apply
because the tidal currents at the sill interfere with the hydraulic
control (Stigebrandt, 1977). The problem of deep water circulation
in O-fjords, with particular reference to the Baltic, has been treated
by Walin (1977).

4.5 Multiple basins

Sills are common features of most fjords. Some fjords even have
a series of sills, dividing the fjord into a set of multiple basins.
In such a system deep water renewal is never a single process, but
a series of events which may or may not affect all the basins in
the fjord.

Consider for simplicity a single fjord arm divided into a
series of basins by a set of sills. We shall also assume the fjord
to be connected to the sea at one end only. In a system like this
it is often observed that the rate of vertical diffusion decreases
from the outermost to the innermost basin, apparently as a consequence of the decreasing velocity of the tidal currents. This observation is, however, not generally valid, owing to the varying depth
conditions of the sills and the respective basins. Nevertheless,
with a higher rate of vertical exchange in the outer basins, the
deep water there is more rapidly prepared for renewal and may, whenever that happens, be subject to displacement by intrusion from
outside.

Unless limited by time or density, the intrusion will cause
successive renewals of the basins, beginning with the outer one
and proceeding to the following whenever the first basin is completed.
As an intermediate stage the displaced resident water can flow into
the other basins if the appropriate density conditions are satisfied.
The process ends with all basins experiencing complete deep water
renewals.

The renewal of multiple basins is frequently limited by time
or density.   In the first case renewals may only occur in one or
several basins, beginning with the outer one. With appropriate density conditions the displaced resident water may then cause intrusions in neighbouring basins and there remain stagnant for some time.

If the renewal is density limited, intrusion to multiple basins
may likewise lead to partial (or complete) renewal in one or several
of the basins, although in this case not necessarily beginning with
the outer one. Displaced resident water can then only temporarily
occupy other basins, if at all. Also, entirely different circulatory features from those related above may be found. Particularly
where the vertical exchange in one part of the system is significantly
stronger than in other parts, basin water may eventually become of
sufficiently low density to allow water from more sheltered sections
to intrude at subsill levels, in extreme cases to descend to the
deepest levels in the basin.

5. MODELLING OF DEEP WATER RENEWALS

5.1 Deterministic models of renewal

    Time-dependent models which depend on fjord geometry and the
various forcings already discussed, are needed to examine the effects
of imagined engineering works or of changes in the weather or climate.
Although it is possible to build multidimensional numerical models
of a basin (Wilmot, 1974) and to impose some predictable forcing
at the sill (Dickson, 1973), a much simpler approach has so far
prevailed in fjords.

    Because of the usually small horizontal variation in the basin
the problem becomes one-dimensional, even merely two-layered, sufficiently specified by a comparison of deep and sill waters. The
period of active renewal and density currents cannot be described
in this way and is assumed instantaneous. If sill density exceeds
deep density, there is renewal: otherwise deep density decreases
steadily or at a rate proportional to the density difference. There
is assumed to be no limit to the supply of sill water, so that partial
renewals are not considered. This deceptively simple model and its
multi-layer derivatives was discussed by Welander (1974). The time-
mean properties of the deep water depend on details of the size and
timing of sill density maxima. Fig. 13 shows the sensitivity of
the model to an apparently small change in the forcing. A special
but widely relevant case of regular maxima with differing sizes is
examined in section 5.2. Welander's model has been applied to Loch
Etive by Edwards & Edelsten (1977), who used it to predict the consequences of hydroelectric interference with the natural run-off
{Edwards & Edelsten (1976)}. Inflow salinity of Loch Etive inner
basin is largely determined by the amount of run-off which has been
mixed down from the upper layers and has not yet been flushed out.
Suitably weighted run-off records can thus be used to predict sill
salinity: temperature has a known regular annual cycle: sill
density follows from temperature and salinity: bottom water property decay rates are known from observation; Welander's equations
were then used to hindcast bottom conditions during twelve years
when hydroelectric schemes operated and during the same period,
imagined without the schemes. The influence of the schemes proved
negligible because of their rather small storage capacity compared
to typical run-off during the flushing time of the intermediate
layers of the fjord. Figure 3 shows the irregularity of renewal
in this fjord, which hinders a statistical analysis on the lines of
section 5.2.

    The two layer model is plainly useful to investigate the time-
mean properties of deep water, either statistically or by particular
application, but it has at least two disadvantages. It does not
specify the sill water supply and it obscures the dependence of the

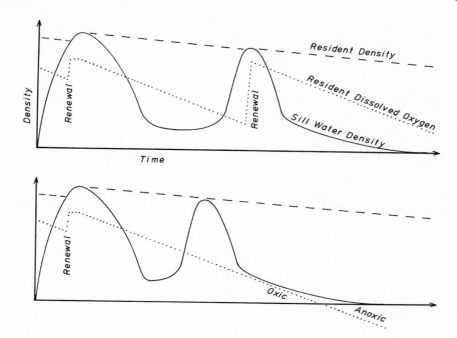

Figure 13.   The effect of a small alteration to sill density history.
             The slightly earlier arrival of the second peak in the
             lower record fails to satisfy the density condition for
             renewal.   Whereas in the upper case there is renewal,
             in the lower case there is no renewal and a non-conser-
             vative property such as oxygen continues to decline,
             even to the stage of anoxia.

deep decay rate upon any of the determining mechanisms.   Two papers
by Stigebrandt are therefôre particularly interesting.   In the
first, (Stigebrandt, 1977) the sill water supply is the sum of baroclinic flow, which for a given density difference increases with
sill width, and barotropic flow whose relative size decreases with
increasing sill width.   The sill transport is determined by sill
geometry and densimetric Froude number in the sill region and consequently the effect of sill engineering on the renewal rate may be
discussed.   In the second (Stigebrandt, 1976), the sill and its
associated barotropic flows are considered as a generator of internal waves which transfer energy to the basin.   The breaking of the
waves on the sides causes mixing and helps to dirve vertical diffusion.   The theory describes the rate of turbulent energy transfer
as a function of basin geometry, density structure and water level
fluctuation.   Although only about 5% of the transferred energy appears as an increase in the potential energy of the water column
in the Oslofjord, it is reasonable to assume that energy transfer

and diffusion rate generally increase together. It is thus possible
to discuss qualitatively the way in which sill alterations would
affect deep water decay rate.

5.2   Stochastic modelling of low frequency renewals

Insofar as the inflows to a fjord basin depend on the density
of a naturally-fluctuating stratification of the adjacent coastal
water, the renewals can be considered stochastically determined
and can be modelled accordingly. Gade (1973) considered basins
where the intrusions were seasonally triggered; the actual event of
basin water renewal depended upon the available water at sill depth
(at the most favourable moment) having a density equal to or greater
than that of the stagnant resident water. In that case, complete
(or bottom water) renewal would take place, otherwise the basin
water would remain stagnant for another year.

The model requires the annual density decrement of the stagnant
resident water to be specified or, preferably, determined from observations. Furthermore, the probability density distribution of the
density of the water available at sill depth must be known. For
simplicity it is assumed to be a normal distribution with standard
deviation σ. It is then possible to establish the conditional
probability density function $q(\rho/\rho_0)$, ρ_0 being the density of the
previous inflow to the basin. It follows that the marginal probability density function $f(\rho)$, *i.e.*, the probability density for inflow (renewal) regardless of the density of, or the time lapse since
the previous inflow, is given as

$$f(\rho) = \int_0^\infty f(\rho_0) \cdot q(\rho/\rho_0) \, d\rho_0$$

and can thus be determined.

An example of a marginal probability density function $f(\rho)$
for the Bonnefjord, the innermost basin of the Oslofjord, is given
in Fig. 14.

From the probability density functions the actual probabilities
for renewal after any number of years of stagnant conditions can be
computed, depending upon the density of the previous inflow. Likewise, the probability distribution for inflows separated by any number of years regardless of the density of the water introduced at
the previous inflow, is readily obtained by the marginal distribution:

$$Q(n) = \int_0^\infty f(\rho_0) \, Q(n/\rho_0) \, d\rho_0$$

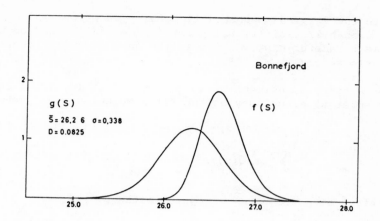

Figure 14. Adapted normal frequency distribution g(S) to sill level densities in the Vestfj. (Oslofjord), and computed marginal probability function f(S) of the density for influx of water to the adjoining basin Bonnefjord.

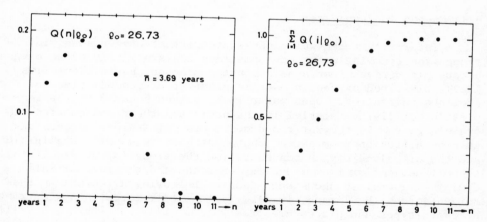

Figure 15. Conditional probability distribution $Q(n/\rho_0)$ of influx interval to the Bonnefjord (left), and conditional cumulative distribution of influx interval to the Bonnefjord (right).

where $Q(n/\rho_0)$ is the conditional probability distribution for inflow n years after the previous one. An example of a conditional probability distribution and the corresponding cumulative distribution for the Bonnefjord basin are given in Fig. 15.

Both the marginal probability density function $f(\rho)$ and the marginal probability distribution for inflow $Q(n)$ depend on only one single parameter σ/D, where D is the annual density decrement of the basin water.

For a quick assessment of a fjord basin it is useful to compute the interval expectancy as the first moment of the probability distribution

$$\bar{n} - \sum_{1}^{\infty} n.Q(n)$$

For practical purposes $\bar{n}$ can be approximated by

$$\bar{n} \approx 1 + 0.729 \; (\sigma/D)^{\frac{\sqrt{3}}{2}}$$

with sufficient accuracy, knowing the long term statistical behaviour of the adjacent (coastal) water.

6.   OTHER ASPECTS

6.1   Oxygen in deep water

Intermittent renewal causes intermittent oxygen supply. Although the renewing water may sometimes contain little or no oxygen because it derives from a neighbouring stagnant basin (Fonselius, 1969), dissolved oxygen is usually greatest at renewal time and declines thereafter. Deep water loses oxygen because of the respiration of living material in the basin and the decomposition of organic matter in the water and sediments. This loss exceeds any gain by diffusion from above, photosynthesis is usually negligible and the dissolved oxygen concentration therefore falls. As it falls, the diffusive flux from above may increase and the rate of fall diminish.   Fig. 16 shows such changes following a renewal in the Borgenfjorden (McClimans, 1973). Typical fall rates in a wide range of fjord types and places are from 2 to 8 ml/ℓ/year (Barnes & Collias, 1958; Bell, 1973, 1976; Cannon, 1975; Christensen & Packard, 1976; Coote, 1964; Edwards & Edelsten, 1977; Fonselius, 1969; Gade, 1963; Mackay & Halcrow, 1976; Matthews & Sands, 1973; Muench & Heggie, 1978; Strom, 1936). Christensen & Packard (1976) found that the majority of oxygen consumption in Dabob Bay was accounted for by the water column. However, in thinner bodies of stagnant water, where the flux of oxygen into the sediments depletes a shorter water column, the rate of fall might be expected to be greater than in thick bodies. This may explain a tendency for stagnant bodies only ten or twenty metres thick to fall at rates of 10 ml/ℓ/year or more (Craig, 1954; Edwards & Truesdale, 1980; McClimans, 1973).

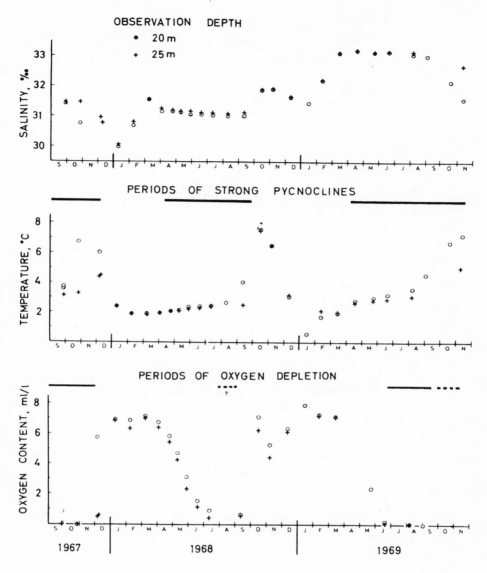

Figure 16. Bottom water properties in the Borgenfjorden. During
periods of strong pycnoclines bottom water stagnates:
its conservative properties change slowly: oxygen content falls more slowly as stagnation progresses, perhaps
because of increasing diffusive flux driven by the increasing oxygen gradient. Reproduced with permission
of T. A. McClimans, Physical Oceanography of Borgenfjorden (1973). Det Kongelige Norske Videnskabers Selskab
Skrifter No. 2. Universitetsforlaget, Oslo.

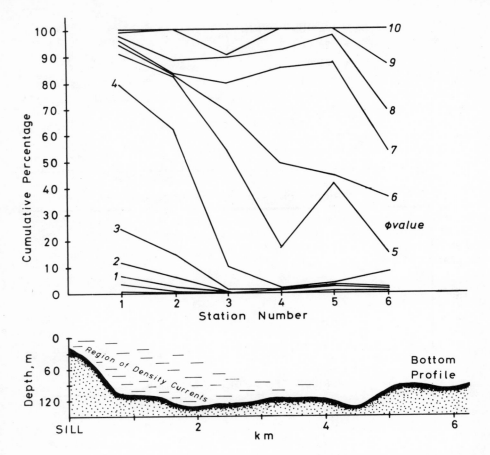

Figure 17.   The size analysis of surface sediment at the bottom of
             Loch Etive. Density currents which flow into the basin
             during renewals may be responsible for the increasing
             fineness of the sediment away from the sill. (ϕ:-1 to
             4: sand. ϕ: 4 to 8: silt. ϕ: more than 8: clay).

        If the fall continues, hydrogen sulphide may be formed.   The
basin then suffers from intermittent anoxia (Anderson & Devol, 1973;
Beyer, 1976; McClimans, 1973; Richards, 1965; Strom, 1936).   When
the basin is renewed, uplift of oxygen depleted water towards the
surface may go so far as to bring it to the very surface (Strom, 1936;
Ozretich, 1975), with associated noxious odours and fish kills.

6.2 Sediment distribution

The distribution of sediment on fjord bottoms depends upon the pattern of sources around the basin and upon the current regime. The coarse fraction of the sediment tends to decrease distally (Hoskin, *et al.*, 1978; Holtedahl, 1965), the source may be coastal or submarine. Apart from persistent tide and wind currents, another influence is that of turbidity currents (Hoskin & Burrell, 1972), which sort sediment to be coarsest where the currents are greatest. Such currents have been traced in the Hardangerfjord (Holtedahl, 1965) and must be recognized as generally important. However, it has been realized that density currents associated with renewal (section 4.1) may resuspend bottom sediment (Edwards *et al.*, 1980) and redistribute it such that in basin bottoms it is coarsest near the sill and becomes finer towards the head of the fjord. Gilbert (1978) suggested that this idea explains the grain size distribution in Pangnirtung Fjord. Fig. 17 illustrates a similar distribution in Loch Etive.

REFERENCES

Anderson, J. J. and A. H. Devol 1973: Deep water renewal in Saanich Inlet, an intermittently anoxic basin. Est. Coast Mar. Sci. 1, 1-10.

Aure, J. A. 1972: Hydrografien i Lindaspollene. Thesis, 120 pp. plus figures. (in Norwegian).

Barnes, C. A. and E. E. Collias 1958: Some considerations of oxygen utilization rates in Puget Sound. J. Mar. Res. 17, 68-90.

Bell, W. H. 1973: The exchange of deep water in Howe Sound Basin Pacific Marine Science Report 73-13.

Bell, W. H. 1976: The exchange of deep water in Alberni Inlet. Pacific Marine Science Report 76-22. Institute of Ocean Sciences, Patricia Bay, Victoria, B. C.

Beyer, F. 1954: Studies of a threshold fjord – Dramsfjord – in Southern Norway. I. Hydrography. Masters Thesis Part I. University Oslo Fac. Sci. Mim. 165 pp.

Beyer, F. 1976: Influence of freshwater outflow on the hydrography of the Dramsfjord in Southern Norway. In Freshwater on the Sea, 75-88. (Eds Skreslet, S., R. Leinebo, J. B. L. Matthews and E. Sakshaug. Association of Norwegian Oceanographers, Oslo.

Bjerkeng, B., C. G. Göransson, and J. Magnusson 1978: Investigations of different alternatives for waste water disposal from Sentralanlegg Vest. NIVA Rep. 013276 (in Norwegian).

Braarud, T. and J. T. Ruud 1937: The hydrographic conditions and aeration of the Oslofjord 1933-34. Hvalradets skrifter, 15, 56 pp.

Cannon, G. A. 1975: Observations of bottom-water flushing in a fjord-like estuary. Est. Coast Mar. Sci. 3, 95-102.

Cannon, G. A. and C. C. Ebbesmeyer 1978: Winter replacement of bottom water in Puget Sound. Estuarine Transport Processes. edited by B. Kjerfve, Univ. South Carolina Press, Columbia, S. C., 229-238.

Cannon, G. A. and N. P. Laird 1978: Variability of currents and water properties from year-long observations in a fjord estuary. In Hydrodynamics of Estuaries and Fjords (ed J. C. J. Nihoul), Elsevier, 515-536.

Christensen, J. P. and T. T. Packard 1976: Oxygen utilization and plankton metabolism in a Washington fjord. Est. Coast Mar. Sci. 4, 339-347.

Collias, E. E. and C. A. Barnes 1966: Physical and chemical data for Puget Sound and Approaches. January 1964-December 1965. University of Washington Technical Report 161, 231 pp.

Coote, A. R. 1964: Physical and chemical study of Tofino Inlet, Vancouver, B. C. M.S. Thesis, University of British Columbia, 74 pp.

Craig, R. E. 1954: A first study of the detailed hydrography of some Scottish West Highland sea lochs (Lochs Inchard, Kanaird, and the Cairnbawn Group). Ann. Biol. Copenh. 10, 16-19.

Dickson, R. R. 1973: The prediction of major Baltic inflows. Dt. Hydrogr. Z. 26, 97-105.

Drinkwater, K. F. and T. R. Osborn 1975: The role of tidal mixing in Rupert and Holberg Inlets, Vancouver Island. Limnol. Oceanogr. 20(4), 518-529.

Edwards, A. and D. J. Edelsten 1976: Control of fjordic deep water renewal by run-off modification. Hydrological Sciences Bulletin, 21, 3 pp. 445-450.

Edwards, A. and D. J. Edelsten 1977: Deep water renewal of Loch Etive: a three basin Scottish fjord. Estuar. Coast. Mar. Sci. 5, 575-595.

Edwards, A. D. J. Edelsten, M. A. Saunders and S. O. Stanley 1980: Renewal and entrainment in Loch Eil: a periodically ventilated fjord. NATO Fjord Workshop Proceedings.

Edwards, A. and V. W. Truesdale 1980: Speciation of iodine in Loch Etive, a Scottish fjord. NATO Fjord Workshop Proceedings.

Eggvin, J. 1943: The great exchange of water masses along the Norwegian coast 1940. Rapp. et Procès-Verbaux des Réunions, ICES, 112, 49-61.

Ellison, T. H. and J. S. Turner 1959: Turbulent entrainment in stratified flows. J. Fluid Mech. 6, 423-448.

Fjeldstad, J. E. 1964: Internal waves of tidal origin. Geofysiske publikasjoner, 25(5), 73 pp. plus tables.

Fonselius, S. H. 1969: Hydrography of the Baltic deep basins III. Fishery Board of Sweden, Series Hydrography Rept No. 23, 97 pp.

Gade, H. G. 1968: Horizontal and vertical exchanges and diffusion in the water masses of the Oslofjord. Helgolander wiss. Meeresunters, 17, 462-475.

Gade, H. G. 1963: Some hydrographic observations of the inner Oslofjord during 1959. Hvalradets skrifter. Nr. 46, 62 pp.

Gade, H. G. 1970: Hydrographic investigations in the Oslofjord, a study of water circulation and exchange processes. Rep. 24, Geophysical Institute, Bergen, 193 pp and figures.

Gade, H. G. 1976: Transport mechanisms in fjords. Fresh Water On The Sea, edited by Skreslet et al., Ass. Norwegian Oceanographers, Oslo, 51-56.

Gade, H. G. 1973: Deep water exchanges in a sill fjord. A stochastic process. J. Phys. Oceanogr., 3, 213-219.

Gade, H. G., R. A. Lake, E. L. Lewis and E. R. Walker 1974: Oceanography of an Arctic bay. Deep-Sea Res., 21, 547-571.

Gilbert, R. 1978: Observations on oceanography and sedimentation at Pangnirtung Fjord, Baffin Island. Maritime Sediments, 14, No. 1, 1-9.

Gilmartin, M. 1962: Annual cyclic changes in the physical oceanography of a British Columbia fjord. J. Fish. Res. Bd., Canada, 19, 922-974.

Glenne, B. and T. Simensen 1963: Tidal current choking in the landlocked fjord of Nordasvatnet. Sarsia, 11, 43-73.

Göransson, C. G. and T. Svensson 1975: The Byfjord: Studies of
     water exchange and mixing processes. Statens Naturvardverk,
     Sweden, 152 pp.

Gran, H. H. and T. Gaarder 1918: Über den Einfluss der atmosphär-
     ischen Veränderungen Nordeuropas auf die hydrographischen
     Verhältnisse des Kristiania-fjords bei Drobak im März 1916.
     Pub. de Circ., 71, 29 pp.

Helland-Hansen, B. and F. Nansen 1909: The Norwegian Sea. Rep.
     Norwegian Fishery and Marine Invest., 2, 390 pp.

Helle, H. B. 1978: Summer replacement of deep water in Byfjord,
     Western Norway: Mass exchange across the sill induced by
     coastal upwelling. Hydrodynamics of Estuaries and Fjords,
     edited by J. C. J. Nihoul, Elsevier, Amsterdam, 441-464.

Holtedahl, H. 1965: Recent turbidities in the Hardangerfjord,
     Norway. Colston Research Soc. Proc. 17, 107-140.

Hoskin, C. M. and D. C. Burrell 1972: Sediment transport and
     accumulation in a fjord basin, Glacier Bay, Alaska. Journal
     of Geology, 80, No. 5, 539-551.

Hoskin, C. M., D. C. Burrell and G. R. Freitag 1978: Suspended
     sediment dynamics in Blue Fjord, Western Prince William Sound,
     Alaska. Est. Coast. Mar. Sci., 7, 1-16.

Johannessen, O. M. 1968: Some current measurements in the Drobak
     Sound, the narrow entrace to the Oslofjord. Hvalradets
     skrifter, 50, 38 pp.

Killworth, P. D. and E. C. Carmack 1979: A filling-box model of
     river-dominated lakes. Limnol. Oceanogr. 24, 201-217.

Kullenberg, B. 1955: Restriction of the underflow in a transition.
     Tellus 7, 215-217.

Linde, E. 1970: Hydrography of the Byfjord. Univ. Bergen Geo-
     physical Institute, Rep. 20, 39 pp and figures.

Mackay, D. W. and W. Halcrow 1976: The distribution of nutrients
     in relation to water movements in the Firth of Clyde. In
     Freshwater on the Sea (eds S. Skreslet et al.). Association
     of Norwegian Oceanographers.

Manins, P. C. 1973: A filling box model of the deep circulation of
     the Red Sea. Mem. Soc. Royale de Sciences de Leige 6, 153-166.

Matthews, J. B. and A. V. Quinlan 1975: Seasonal characteristics of water masses in Muir Inlet, a fjord with tidewater glaciers. J. Fish. Res. Bd. Canada. 32(10), 1693-1703.

Matthews, J. B. L. and N. J. Sands 1973: Ecological studies on the deep-water pelagic community of Korsfjorden, Western Norway. Sarsia. 52, 29-52.

McClimans, T. A. 1973: Physical oceanography of Borgenfjorden Det Kongelige Norske Videnskabers Selskab Skrifter. No. 2, Universitetsforlaget, Oslo.

McClimans, T. A. 1978: On the energetics of tidal inlets to land-locked fjords. Marine Science Communications, 4(2), 121-137.

Muench, R. D. and D. T. Heggie 1978: Deep water exchange in Alaskan subarctic fjords. Estuarine Transport Processes, edited by B. Kjerfve. Univ. of So. Carolina Press, Columbia, S. C., 239-268.

Nutt, D. C. and L. K. Coachman 1956: The oceanography of Hebron Fjord, Labrador. J. Fish. Res. Bd. Canada, 13(5), 709-758.

Ozretich, R. J. 1975: Mechanisms for deep water renewal in Lake Nitinat, a permanently anoxic fjord. Estuarine Coastal Mar. Sci., 3, 189-200.

Pickard, G. L. 1961: Oceanographic features of inlets in the British Columbia Mainland Coast. J. Fish. Res. Bd. Canada, 18, 908-999.

Pickard, G. L. 1963: Oceanographic characteristics of inlets of Vancouver Island, British Columbia. J. Fish. Res. Bd. Canada, 20, 1109-1144.

Pickard, G. L. 1971: Some Physical Oceanographic Features of Inlets of Chile. J. Fish. Res. Bd. Canada, 28(8), 1077-1106.

Pickard, G. L. 1967: Some oceanographic characteristics of the large inlets of southeast Alaska. J. Fish. Res. Bd. Canada, 24, 1475-1506.

Pickard, G. L. and K. Rodgers 1959: Current measurements in Knight Inlet, British Columbia. J. Fish. Res. Bd. Canada, 16(5), 635-678.

Perkin, R. G. and E. L. Lewis 1978: Mixing in an Arctic fjord. J. Phys. Oceanogr., 8, 873-880.

Richards, F. A. 1965: Anoxic basins and fjords. In Chemical
    Oceanography Vol. I (Riley, J. P. and G. Skirrow, eds)
    Academic Press, London. 611-645.

Royer, T. C. 1975: Seasonal variations of waters in the northern
    Gulf of Alaska. Deep-Sea Res., 22, 403-416.

Royer, T. C. 1979: On the effect of precipitation and run-off on
    coastal circulation in the Gulf of Alaska. J. Phys. Oceanogr.,
    9, 555-563.

Saelen, O. H. 1950: The hydrography of some fjords in Northern
    Norway. Tromso Mus. Arshefter, 70, 1-102.

Saelen, O. H. 1967: Some features of the hydrography of Norwegian
    fjords. Estuaries, edited by G. H. Lauff, Am. Ass. Adv. Sc.
    Pub. 83, 63-71.

Seibert, G. H., R. W. Trites and S. J. Reid 1979: Deepwater
    exchange processes in the Saguenay Fjord. J. Fish. Res. Bd.
    Canada, 36, 42-53.

Skreslet, S. and H. Loeng 1977: Deep water renewal and associated
    processes in Skjommen, a Fjord in North Norway. Est. Coast
    Mar. Sci. 5, 383-398.

Skreslet, S. and B. Schei 1976: Hydrography of Skjommen, a fjord
    in North Norway. In Freshwater on the Sea, 101-108. (eds
    Skreslet, S., R. Leinebo, J. B. L. Matthews and E. Sakshaug).
    Assocation of Norwegian Oceanographers, Oslo.

Smith, P. C. 1975: A Streamtube Model for bottom boundary currents
    in the ocean. Deep Sea Res. 22, 853-873.

Stigebrandt, A. 1976: Vertical diffusion driven by internal waves
    in a sill fjord. J. Phys. Oceanogr. 6, 486-495.

Stigebrandt, A. 1977: On the effect of barotropic current fluct-
    uations on the two-layer transport capacity of a constriction.
    J. Phys. Oceanogr. 7, 118-122.

Stigebrandt, A. 1979: Observational evidence of vertical diffusion
    driven by internal waves of tidal origin in the Oslofjord.
    J. Phys. Oceanogr., 9

Stigebrandt, A. 1979: Some aspects of tidal interaction with fjord
    constrictions. Est. Coast. Mar. Sci. (in press).

Stommel, H. and H. G. Farmer 1952: Abrupt change in width in two
    layer open channel flow. J. Mar. Res., 11, 205-214.

Stommel, H. and H. G. Farmer 1953: Control of salinity in an estuary by a transition. J. Mar. Res. 12, 13-20.

Strom, K. M. 1936: Land-locked waters. Hydrography and bottom deposits in badly ventilated Norwegian fjords with remarks upon sedimentation under anaerobic conditions. Norske Vidensk. Akad. Skr., 1(7), 85 pp.

Svendsen, H. 1977: A study of the circulation in a sill fjord on the west coast of Norway. Marine Science Communications, 3, 151-209.

Svendsen, H. and R. O. R. Y. Thompson 1978: Wind driven circulation in a fjord. J. Phys. Oceanogr., 8, 703-712.

Turner, J. S. 1973: Buoyancy Effects in Fluids. Cambridge University Press. 367 pp.

Walin, G. 1977: A theoretical framework for the description of estuaries. Tellus, 29, 128-136.

Welander, P. 1974: Theory of a two-layer fjord-type estuary, with special reference to the Baltic Sea. J. Phys. Oceanogr., 4(4), 542-556.

Wilmot, W. 1974: A numerical model of the gravitational circulation in the Baltic. ICES Special Meeting on Models of Water Circulation in the Baltic. Paper No. 11.

THE TIDAL JET IN RUPERT-HOLBERG INLET

D.J. Stucchi

Institute of Ocean Sciences
P.O. Box 6000
Sidney, B.C. V8L 4B2
Canada

INTRODUCTION

Located at the northern end of Vancouver Island, Rupert and Holberg Inlets are part of the Quatsino Sound Group (figure 1). Together these two inlets form one basin (44 km in length and 170 m maximum depth) which is connected to Quatsino Sound by Quatsino Narrows, a long (4 km), slender (0.4 km) and shallow (25 to 30 m) channel. It was evident from the investigations of Drinkwater and Osborn (1975), Stucchi and Farmer (1976), and from the earlier observations of Pickard (1963) that there was intense mixing occurring in this fjord. The remarkable uniformity of water properties, high dissolved oxygen levels in the deep waters and the large and rapid temporal changes in the deep water properties were evidence that strong mixing was occurring. Drinkwater and Osborn (1975) identified the strong tidal flow through Quatsino Narrows and into Rupert Holberg Inlet as the process responsible for the intense mixing. Based on observations obtained in the strong tidal flow - tidal jet - between Hankin Pt. and the northern end of Quatsino Narrows, Stucchi and Farmer (1976) proposed a qualitative model of the exchange process in this fjord. A more extensive series of observations of the tidal jet was obtained in August and September 1975 in order to better understand the nature and dynamics of the tidal jet. Selected portions of this latter set of observations will be presented and discussed in this brief note. A complete discussion of these observations as well as additional information is to be published in another paper.

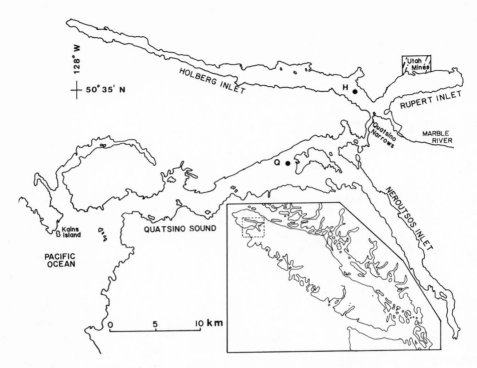

Figure 1.   Map of Quatsino Sound, Rupert-Holberg Inlet and
Neroutsos Inlet.

OBSERVATIONS

        Three current meter moorings (A,A-2 and B) were deployed on
the axis of the tidal jet, and an additional mooring (QN) was
deployed in Quatsino Narrows. The most striking feature in the
speed data (only speed is plotted since directions are predominantly
northerly during inflow) plotted in figure 2 is the change in the
vertical speed profile over the measurement period. From August
29th to September 3rd the tidal jet was concentrated in the upper
part of the water column with the highest speed near the surface.
The profile is almost completely reversed from September 6th to 9th
when the strongest currents are observed near the bottom. The
corresponding salinity data with the addition of the salinity data
from mooring QN is plotted in figure 3. Since density is primarily
determined by salinity in these waters it is evident that the
density of the tidal jet, as determined by measurements in Quatsino
Narrows, is less than that of the ambient waters of Rupert-Holberg

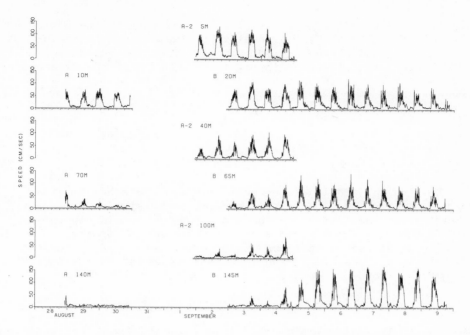

Figure 2. Time series of current speed obtained from moorings A, A-2 and B.

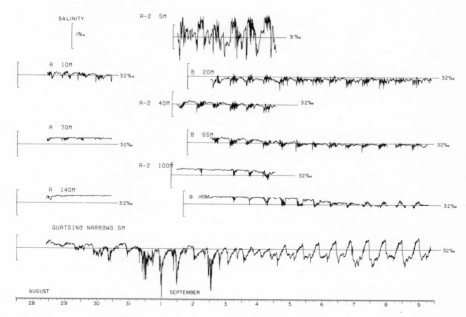

Figure 3. Salinity time series from moorings A, A-2, B and ON.

Inlet from August 29th to September 3rd, and that in the period
from September 6th to 9th the density of the inflowing jet is
greater than that of the ambient waters. Thus the current speed
and salinity data presented in figures 2 and 3 show the presence
of a buoyant jet which undergoes a transition to a negatively
buoyant jet.

BUOYANT JET

Comparison with Models

       The buoyant tidal jet and surface buoyant discharges from
power generating stations are in some respects similar turbulent
flows. Since the latter have been extensively modelled both in
the laboratory and mathematically the observations of the tidal
jet will be described in terms of the parameters used by these
models. In the near field the important parameter is the initial
densimetric Froude number, F_o, defined as

$$F_o = u_o / \sqrt{gh_o(\rho_a - \rho_o)/\rho_a}$$

where u_o is the initial discharge speed, ρ_a and ρ_o are the densities
of the ambient and inflowing fluid respectively, h_o is the depth
of the discharge channel and g is the gravitational acceleration.

       One aspect of the flow which is readily obtained from the
current meter data is the vertical penetration of tidal jet. Maximum
depths of the jet were determined for each hour of the flood tide
and plotted (figure 4) against the corresponding hourly averaged
Froude number computed for the flow. Despite the coarse description
of the vertical extent of the jet provided by the current meter
data, a simple relationship is evident, one in which the depth of
penetration is directly proportional to the Froude number. This
observed relationship also compares favourably with the predictions
of several mathematical models as shown in figure 4. It appears
that the buoyant tidal jet can be parametrized by the Froude number
in spite of the differences with the idealized flow considered in
models. It is significant to note that the buoyant jet can vertically mix to the bottom given a sufficient high Froude number for
the flow (i.e. $F_o > 7$), thus providing a more vigorous process for
the reduction of the deep water density than is normally the case
in other fjords.

Effect on Water Column

       The effect of the buoyant jet on the water column of the fjord
is evident in the salinity time series shown in figure 3. An overall

decline in the salinity is observed initially at 10 m, then at 40 m, 65 m, 100 m and finally at 145 m. Profiles of salinity obtained during this period also shows that the reduction in salinity occurs first near the surface and then progresses down to the bottom.

The observations described above are significant in that they demonstrate how the buoyant jet promotes the transition to the negatively buoyant inflow. A positive feedback loop is set up whereby the reduction of salinity initially near the surface facilitates the deeper penetration and mixing of the buoyant jet which in turn results in the reduction of salinity deeper down. The buoyant jet would eventually penetrate to the bottom and the net result would be a reduction of the density in the fjord water which in turn would make the transition to a negatively buoyant flow more probable.

NEGATIVELY BUOYANT JET

Dilution and Entrainment

During the period of the negatively buoyant inflow information on the lateral extent of the flow was obtained from observations off the axis of the flow. This information provided an upper limit for the lateral extent of the flow and enabled rough estimates of dilution and entrainment to be made. The volume flux, Q_B, through the cross-section of the flow at mooring B was calculated using the observed vertical velocity profile at mooring B, the observed lateral extent of the flow and an assumed lateral velocity profile. The dilution, D, was then calculated by taking the quotient Q_B/Q_O where Q_O is the volume flux at the northern end of Quatsino Narrows. Hourly values of dilution were calculated for the last 4 flood tides at mooring B resulting in an average dilution for the negatively buoyant flow of D = 5.3 ± 1.8. Although the average tidal inflow constitutes only 3% of the total basin volume, the tidal jet entrains and mixes several times the volume on each flood tide. At best the calculation of dilution provides an upper bound, nonetheless a significant fraction of the ambient waters is mixed on each flood tide.

The entrainment constant, E, as defined by Ellison and Turner (1959) is the constant of proportionality relating the entrainment velocity, to the mean velocity of the flow. The entrainment velocity u', of ambient fluid into the turbulent flow is given by

$$u' = (Q_B - Q_O)/A_E$$

where A_E is the entraining surface area. Calculated entrainment constants for the negatively buoyant flow yield an average entrain-

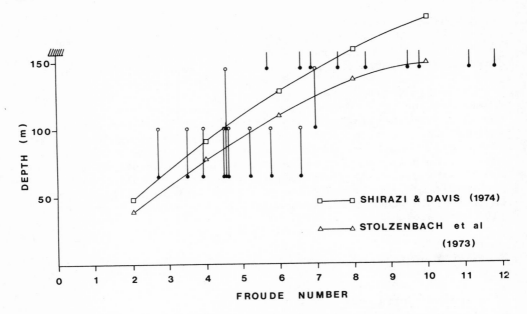

Figure 4.   Plot of vertical penetration of the buoyant jet versus
             hourly averaged Froude number, $\bar{F}_o$, for the flow.   Also
             plotted are the predictions of the mathematical models
             of Stolzenbach et al (1973) and Shirazi and Davis (1974).

ment constant, E = 0.05 ± 0.02. This value is about 2 to 3 times
larger than the experimentally determined entrainment constant for
a 2-dimensional plume on a similar slope of 5^0 to 10^0. One would
expect the entrainment constants to be larger for a 3-dimensional
flow since fluid can be entrained laterally and unlike vertical
entrainment lateral entrainment is not inhibited by buoyancy forces.

Exchange Rates

     The rate of exchange of the Rupert-Holberg Inlet waters
during a negatively buoyant inflow are estimated by considering
two simple exchange models. It is assumed in both models that
run-off is negligible, Quatsino Sound is an infinite source of
constant salinity water and that on each flood tide a fraction,
k ∿ 0.25, of the inflow water is water that was drawn out on the
preceding ebb tide.

     In the first model the inflowing water is assumed to mix
completely with all the ambient water. It can then be shown that

63% exchange of ambient waters occurs in about 3 weeks. The second
model is one of advective replacement in which that fraction, (1 - k),
of the following water which is Quatsino Sound water is assumed to
be injected at depth so that none of it is extracted on the following
ebb. In this extreme model 63% exchange occurs in about 2 weeks.
These two extreme models give 2-3 week time scale for 63% renewal
of the fjord waters during a negatively buoyant inflow. During
times of buoyant inflow the time scale for exchange will be longer
because the direct mixing of the jet will be concentrated in the
upper portions of the water column.

REFERENCES

Drinkwater, K.F. and T.R. Osborn. 1975. The role of tidal mixing
    in Rupert and Holberg Inlets, Vancouver Island. Limnology and
    Oceanography 20(4): 518-529.

Ellison, T.H. and J.S. Turner. 1959. Turbulent entrainment in
    stratified flows. J. Fluid Mech. 6: 423-448.

Pickard, G.L. 1963. Oceanographic characteristics of inlets of
    Vancouver Island, British Columbia. J. Fish. Res. Bd. Can.
    20(5): 1109-1144.

Shirazi, M.A. and L.R. Davis. 1974. Workbook of thermal plume
    prediction, volume 2, surface discharge. Pacific Northwest
    Environmental Research Laboratory Report EPA-R2-72-005b.
    Corvallis, Ore.

Stolzenbach, K.D., E.E. Adams and D.R.F. Harleman. 1973. A user's
    manual for three-dimensional heated surface discharge computa-
    tions. Ralph M. Parsons Lab. for Water Resources and Hydro-
    dynamics, Report No. 156. Dept. of Civil Eng., MIT, Cambridge,
    Mass.

Stucchi, D.J. and D.M. Farmer. 1976. Deep water exchange in
    Rupert-Holberg Inlet. Unpublished Manuscript, Inst. of Ocean
    Sciences, Patricia Bay, Victoria, B.C. Pac. Mar. Sci. Rept.
    76-10.

EFFECT OF OXYGEN DEFICIENCY ON BENTHIC MACROFAUNA

IN FJORDS

Rutger Rosenberg

Institute of Marine Research
Lysekil, Sweden

Benthic faunal reactions to oxygen deficiency in ten fjords and estuaries in northern Europe are reviewed. Chosen areas exhibit both permanent and temporary oxygen deficiency due to organic enrichment and/or geomorphological conditions. Number of species, abundance, biomass, and faunal composition are given in relation to oxygen concentration. Results show that the measured faunal parameters do not change gradually with oxygen concentration, but abruptly at approximately 2 mg O_2 litre $^{-1}$. Benthic faunal reactions to oxygen deficiency suggest the following grouping:

1. Non-tolerant species; most species
2. Tolerant species; mainly molluscs
3. Transitory opportunists; pioneer polychaete colonists appearing when conditions improve
4. Transitory emigrants; mobile species

The missing macrofaunal biomass in some waters around Scandinavia is roughly estimated and it is concluded that oxygen depletion due to geomorphological characteristics of water bodies wipes out significantly more animals than similar effects caused by organic pollution.

In this century of continuously increasing impoverishment of the ecosystem at many places in the sea, oxygen depletion is one of the most important symptoms, because of its obvious and sometimes visible effects. Oxygen depletion is commonly reported from fjords, where stagnation occurs in the basin inside the sill. The reduction in oxygen concentration is commonly promoted by organic enrichment (Fig. 1). Oxygen depletion begins at the bottom surface

499

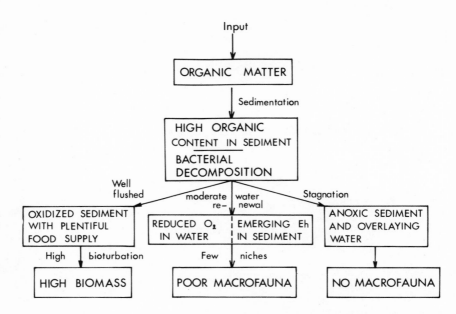

Figure 1. Simplified diagram showing the effects of organic input
          to the marine environment in relation to water renewal
          (from Pearson and Rosenberg 1978).

and affects the benthic organisms. In connection with oxygen reduction in the water the redox potential discontinuity in the
sediment rises. The benthic fauna in the sediments is reacting to
this by emerging and less living space is left. This induced stress
situation will successively kill most of the species, leaving those
adapted to reduced oxygen conditions as last survivors. The layer
of anoxic water occurring just above the bottom extends in some
cases only a few centimetres upwards. Naturally, such a situation
is difficult to record by traditional sampling methods. This
implies that oxygen depletion could be a much more frequent phenomenon close to the sediment surface than reported. Should stagnation
continue at the bottom the anoxic extension will probably increase
in thickness and be stratified by a vertical temperature and/or
salinity gradient.

    Anoxic conditions in marine environments could be either (1)
more or less temporary where macroorganisms never establish, or
(2) occurring periodically at irregular or rather regular intervals
interrupted by periods of faunal recolonization and establishment.
The development of the anoxia could be influenced by organic enrichment.

In the following I will give examples of the effects of oxygen
depletion on benthic macrofaunal structure and quantity, in stagnant
basins where oxygen deficiency is not mainly a consequence of pollution, and secondly, where stagnation in combination with organic
enrichment have resulted in oxygen depletion. The succession of
benthic species along increasing or decreasing environmental gradients has recently been described and discussed (Pearson and
Rosenberg 1978) and is not repeated here.

STAGNATION AND OXYGEN DEFICIENCY

The Byfjord, Sweden

The Byfjord is situated on the Swedish west coast. At the
entrance of the fjord is a sill at 12 m, which almost completely
prevents exchange of water below the halocline at 12 to 15 m. The
salinity above the halocline is 22 to 30%. The inner basin is,
apart from temporary water renewals, anoxic below an oscillating
depth level of 14 to 20 m (Fig. 2). The Byfjord receives treated

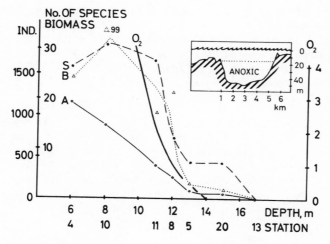

Figure 2. Number of species (S, per 0.5 m^2), abundance (A, per m^2)
and biomass (B, per m^2) in relation to decreasing oxygen
concentrations (mg/ℓ) by increasing depth in October 1971
in the Byfjord, Sweden. Oxygen concentration was at a
minimum at that time (see Rosenberg 1977). (*Hydrobia*
spp. excluded at station 8.)

waste water from a population of 45 000, but the anoxia is due to
the geomorphological characteristics of the fjord. The anoxic
conditions below the halocline structure the ecosystem in the fjord:
copepods and pelagic fish have low abundances compared to other areas
and a large quantity of the energy resources are utilized by epi-
faunal animals, mainly the blue mussel *Mytilus edulis* and the cionid
Ciona intestinalis (Rosenberg *et al*. 1977).

The distribution of macrofauna was limited to depths above
17 m in 1971 (Rosenberg 1977). The benthic communities were composed of rather short-lived species and displayed great seasonal
fluctuations. Among the most conspicuous species were *Corbula*
(= *Aloidis*) *gibba* and *Nephtys hombergi*, which characterized two
different communities separated by similarity analysis. The bottom
area without macrofauna in 1971 comprised an area of 3.41 km^2
(Rosenberg *et al*. 1977). The mean biomass in coastal areas below
15 m along the Swedish west coast has been estimated to 146 g m^{-2}
(Rosenberg and Möller 1979). Thus, the anoxic bottoms could have
accommodated approximately 500 tonnes wet weight of macrobenthic
animals, if oxygen concentrations were sufficient to support a
permanent fauna.

The oxygen concentration in the fjord decreased abruptly with
increasing depth over a few metres range. As an example a situation
prevailing for a few days in October 1971 is presented in Fig. 2.
Then, 4 mg ℓ^{-1} of oxygen was recorded at 10 m, whereas the concentration had dropped to zero at 14 m. A few days later the zero-
level was found at 17 m. This sharp decline in oxygen concentration
in the fjord creates a similarly sharp decline in number of species,
abundance and biomass. At an oxygen concentration of approximately
2 mg ℓ^{-1}, which at that time was found at 11 m, the diversity was
maintained fairly high (Shannon - Wiener H = 3.7).

In connection with water renewals in the Byfjord oxygen concentrations of below 1 mg ℓ^{-1} have been observed to rise towards
the surface over a few days duration, but without any notable effect
on the macrobenthos. Species repeatedly recorded at depths with
oxygen deficency were the polychaetes *Polydora ciliata, Ophiodromus
flexuosus;* the molluscs *Corbula gibba, Nassa reticulata;* and
Phoronis mulleri.

Lough Ine, Ireland

In Lough Ine anoxia develops annually as a consequence of
topography favouring temperature stratification, which normally
begins in March and increases until early summer. In 1970 the sharp
discontinuity of temperature in the summer at a depth of 20 to 30 m
resulted in oxygen depletion in the hypolimnion to a level of around
5 percent saturation. The salinity in the water column was stable,

between 34.4 to 34.9^0/oo. Faunal succession during the deoxygenation
process is described by Kitching *et al.* (1976). In July only a
few infaunal species were still present: *Pseudopolydora pulchra*,
Corbula gibba, *Scalibregma inflatum* and a few others. By mid-August
C. gibba was the only species recorded below 40 m and the distribution of *P. pulchra* was limited to 25 m level. The progressive
elimination of *P. pulchra* from below and upwards in 1970 is illustrated in Fig. 3. Later in winter or spring the species is re-
established in the loch.

The Limfjord, Denmark

The Limfjord ranges over an area of 1 500 km^2 and approximately
330 km^2 (22%) of the bottom area is influenced by annual oxygen
depletion in late summer resulting in mass mortality of benthic
animals. Recolonization seems to take place largely within a year.
The benthic fauna in the Limfjord has been investigated by Jörgensen
(in press). When diving during increasing oxygen depletion at the
bottom he observed motionless but still alive polychaetes laying on
the anoxic mud (*Nereis diversicolor, N. virens, Pectinaria koreni,
Heteromastus filiformis*). Molluscs (*Mya arenaria, Cardium edule,
Abra alba*) reacted at the same time by stretching their siphons up
into th e water. With continued oxygen depletion for another week
the polychaetes died. Now, the molluscs began to creep out of the
sediment and lay on the sediment surface. *Mya* bent their siphons
upwards and could stretch them 20-30 cm above the bottom. *Cardium,
Abra* and *Corbula gibba* lay scattered on the mud surface and could
perhaps survive for another week.

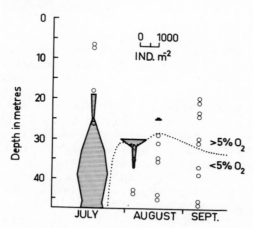

Figure 3.   Distribution and abundance of *Pseudopolydora pulchra*
            in relation to depth and oxygen saturation from July
            to September 1970 in Lough Ine, Ireland. Open circle
            indicates that no *P. pulchra* was found. (Modified
            after Kitching *et al.* 1976).

The Baltic Sea

     The Baltic Sea is like an enormous fjordic estuary, heavily
diluted in its northern parts with a salinity of around $3^0/oo$. The
salinity successively increases to approximately $15^0/oo$ in southern
parts. Renewal of the bottom water occurs through irregular inflows from the west (Kattegat). Absence of inflow of bottom water
results in stagnation and oxygen depletion in the deeper basins.
The extension of the anoxic bottoms increases with retardation of
water renewals and the processes have been described by Fonselius
(1969). In this century, bottoms completely devoid of macrozoo-
benthos or with only few animals show maximal extension in 1975.
According to Zmudzinski (1977) these bottom deserts extended over
84 000 km^2, of which 7 000 km^2 were located to the Bornholm basin
(Fig. 4). At the end of 1975 a large inflow of bottom water occurred and all the anoxic water was renewed.

     The vertical extension of the macro- and meiofauna in the
western central Baltic (see Fig. 4 point A) has been reported by
Elmgren (1975, 1979). He divided the faunal distribution into
three zones (see Fig. 5):

1. An upper zone down to 50 m with high oxygen content and rich
    fauna
2. An intermediate zone with decreasing oxygen concentrations with
    increasing depth. In this zone the macrofauna disappeared
    gradually, and at this lower border only the mobile polychaete
    Harmothoe sarsi was found along with abundant nematodes.
3. A lower zone with little or no oxygen virtually devoid of macro-
    fauna and extremely impoverished meiofauna communities of a few
    thousand nematodes per m^2.

     The biomass decreased from normally above 100 g m^{-2} in the
upper zone to less than 1 g m^{-2} in the intermediate zone and 0.01 g
m^{-2} in the lower zone.

     A joint sampling programme including several countries and
covering most of the Baltic was conducted in 1974 (Elmgren *et al*.,
in press). The results showed that in the northern parts, in the
Bothnian Bay, the meiofaunal biomass was higher than that of the
macrofauna. This was in agreement with Gerlach's (1971) ideas that
meiofauna is quantitatively more important relative to macrofauna in
brackish water than in marine areas. Similarly, the biomass ratio
between macro- and meiofauna was extremely low (2.2:1) in the oxygen-
stressed Byfjord (Rosenberg *et al*., 1977). Along a depth transect
in the Baltic the macrofauna dominated in the previously described
upper zone, whereas the importance of the meiofaunal biomass successively increased with depth and dominated below 60 m (Elmgren,
1979). Thus, there seems to be a general tendency for meiofauna
to increase its relative importance in relation to macrofauna along
environmental stress gradients.

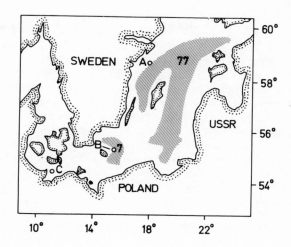

Figure 4.   Extension of the bottom deserts (shaded area) in the
Baltic in 1975 according to Zmudzinski (1977).   The
desert area in the central Baltic was 77 000 km^2.
Examples of benthic faunal composition given in the text
are from western central Baltic (A), the Bornholm basin
(B), and the Kiel Bay (C).

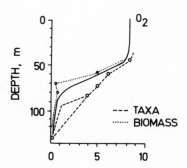

Figure 5.   Oxygen concentration (mℓ/ℓ), number of taxa and biomass
(g/m^2, wet wt), all on x-axis, in relation to depth in
the western central Baltic in 1972.   (Modified after
Elmgren 1975).

The Bornholm Basin in the southern part of the Baltic, with
a maximum depth exceeding 100 m, is also periodically anoxic in its
deeper parts i.e. below about 80 m (Fig. 4, point B). The transitory
return of benthic animals has been described by several authors
e.g. Leppäkoski (1969, 1971, 1975). Based on his data a schematic
presentation of the changes of number of species and abundance in
relation to the oxygen oscillations has been constructed (Fig. 6).
It is clear from the diagram that the improved oxygen conditions in
1969 was immediately followed by repopulation of previously defaunated bottoms. The colonists were atlantic-boreal and cosmopolitan
species of the opportunistic type. They were replacing the Arctic
relics inhabiting those bottoms before the severe stagnation in the
1950's. Thereby, the previous dominance of suspension feeders was
switched to predominantly deposit-swallowing species. The dominants
below 75 m in 1970 were the polychaetes *Scoloplos armiger*, *Harmothoe
sarsi*, *Trochochaeta multisetosa* and *Heteromastus filiformis*.
Leppäkoski concluded that *Scoloplos armiger* utilizes the accumulated
food after prolonged stagnation and that it is one of the most tolerant species to low oxygen concentrations in the southern Baltic.

Speculatively, the biomass wiped out in the deep basin can be
roughly estimated. Taking the mean biomass (wet wt) recorded at
six stations below 80 m in 1951-1952 from Leppäkoski (1969; Table 3)
approximately 20 g m^{-2} is obtained. The area without macrofauna in
the Bornholm basin in 1975 was 7 000 km^2 (Fig. 4) and this suggests
a lost biomass of 140 000 tonnes. If the same 20 g is applied to
the whole desert area in Fig. 4, which is double the size of Denmark,
one would arrive at a figure of close to 1.7 million tonnes of
macrofaunal biomass missing because of the anoxic conditions.

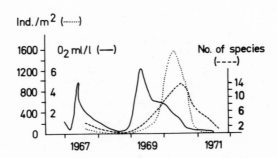

Figure 6. Number of species, mean abundance and oxygen concentration
           during 1967 to 1971 below 80 m in the Bornholm basin
           (southern Baltic). The diagram is based on data from
           Leppäkoski (1969: Table 2; 1971: Fig. 2; 1975: Fig. 1).

Kiel Bay, situated in the south-western Baltic (Fig. 4, point c), is relatively rich in marine species compared to the rest of the Baltic. The reason is the comparatively higher salinity, 13–20⁰/oo, and inflow of benthic larvae from the Kattegat in the north. An impoverished fauna is found in the deeper channels below 20 m of Kiel Bay, where oxygen depletion occurs annually in late summer. Faunal composition of these channels has been described by Arntz (1977), and Arntz and Brunswig (1976; in press). An *Abra alba* community dominates bottoms below 15 m where oxygen is sufficient. The mean biomass of this community was 93 g wet wt m^{-2} in 1968. The biomass decreased abruptly at bottoms affected by periodic oxygen depletion (Fig. 7), where the biomass is only a few grams m^{-2}. Periodic dominants here were the polychaetes *Capitella capitata*, *Harmothoe sarsi*, *Scoloplos armiger* and *Pectinaria koreni* along with the priapulid *Halicryptus spinulosus*, and when conditions were periodically improved the mobile cumacean *Diastylis rathkei*. Also other species showed a considerable recruitment in the spring fol- lowing these catastrophies and subsequent recovery of the sediments during the winter. The few species which permanently seem to have withstood the oxygen depletion were old specimens of the bivalves *Astarte elliptica*, *Mya truncata* and *Cyprina islandica*. Similarly, in Lubeck Bay east of Kiel Bay the last survivors in a region of low oxygen concentration were *C. islandica*, *Harmothoe spp.*, *H. spinulosus* and *D. rathkei* (Schultz 1968).

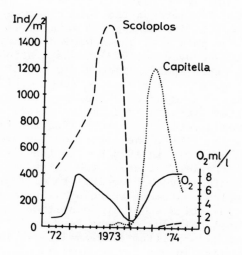

Figure 7.   Abundance of the polychaetes *Scoloplos armiger* and *Capitella capitata* in experimental boxes in relation to oxygen concentration during 1972 to 1974 at 20 m depth in Kiel Bay (south-west Baltic). (Modified after Arntz 1977).

Arntz (1977), in his report of recolonization experiments, clearly demonstrated the extremely rapid recolonization by *Capitella capitata* as soon as the oxygen concentration increased subsequent to almost anoxic conditions (Fig. 7). It seems that the first influx of oxygen rich water in September must have carried *Capitella* larvae ready to settle on this previously deserted bottom.

Organic enrichment in combination with poor water exchange is known to cause oxygen depletion in fjords and estuaries. The best reports concerning such effects on benthic faunal communities are from enrichment by sewage and pulp mill effluents. The well documented effects of the pulp mill in the Saltkällefjord in Sweden was estimated to reduce the macrobenthic biomass in an 1.5 km^2 area by about 60 tonnes. Similarly, a pulp mill effluent in the Gulf of Bothnia (northern Baltic) wiped out all macrofauna in a 1 km^2 area, also estimated to 60 tonnes of missing biomass (Rosenberg 1976). Subsequent to pollution abatement and improved oxygen conditions in the Saltkällefjord the first colonizers were *Capitella capitata* and *Scolelepis fuliginoa* (Rosenberg 1972). These two species were also among the dominants on the heavily enriched bottoms outside a pulp mill discharge point in Loch Eil, Scotland (Pearson, 1975).

The Idefjord, Sweden-Norway

Another example of the deterioration by pulp mill wastes in fjords is from the Idefjord on the border between Sweden and Norway. The whole water column inside the two main sills is anoxic most of the year due to the pulp mill and sewage discharges from Halden (Fig. 8). In 1967-68 Dybern (1972) found a strong reduction in the number of macrobenthic species from the mouth of the fjord and inwards. At about point A in Fig. 8 he recorded 130 species, at point B 80, and at point C 35 species. Inside point C no macrofauna was found on the bottoms (shaded area in the figure). The numbers given in Fig. 8 are from a recent report by Afzelius (1979) and summarize the years 1974-1978. Thus, the situation has been similar over the last ten years. However, in 1925 a rich fauna (24 species) was found by Agassiz-trawling close to the island west of Halden (Jäger-skjöld, 1971), indicating the possibility of faunal recovery following the pollution abatement which will be introduced this year. The shores were investigated at 10 localities and Afzelius reported no animals on the localities closest to Halden. However, further away in the inner branch littoral animals were present again, indicating oxic conditions in the surface water.

The Oslofjord has also changed due to pollution during the last century. In reports of investigations performed a hundred years ago neither pollution, decaying sediments, nor azoic areas were mentioned according to Beyer (1968). He listed three faunal groups, which were abundant in his epibenthic samples on polluted bottoms:

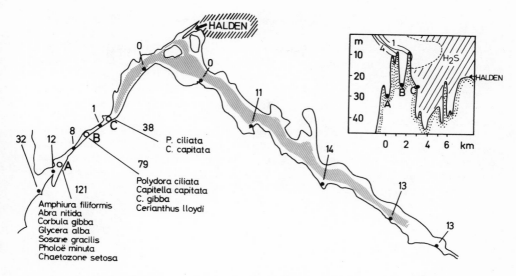

Fugure 8. The Idefjord between Sweden and Norway. Number of macrobenthic species and dominants found at the sites A, B and C during 1974 to 1978. No benthic fauna was recorded below 2 m inside site C (shaded area). Number of species qualitatively collected at ten sites along the shores are indicated by the filled circles. Faunal data from Afzelius (1979). Inserted is the extension of H_2S and oxygen concentration in August 1968 (from Dybern 1972). The discharge of pulp mill wastes occurred at Halden.

Spionidae, Hesionidae and Nudibranchia. Among the spionides *Polydora ciliata* seems to have been the dominant.

DISCUSSION

Oxygen depletion is a widespread phenomenon in stagnant marine basins. It occurs naturally and in combination with organic enrichment, which acts as an accelerator towards depletion. De-oxygenation of the bottom water affects the interstitial water and results in an emerging redoxcline concentrating the infauna at the top sediment layer. This process is accelerated by temperature, and periodic annual oxygen depletion is known to occur predominantly in later summer in the northern hemisphere (Pearson and Rosenberg, 1978). The examples given above of effects of oxygen depletion on benthic animals in fjords and estuaries are all from northern Europe. It is evident, however, that as a consequence of anoxia vast areas are missing macroscopic animals and, thereby, enormous potentials for fish food production.

In several of the examples vertical oxygen concentration gradients were prevailing. However, the benthic communities are not
responding to this gradual change in oxygen concentration with a
grandual decrease in richness and diversity. Rather, the notable
deterioration of the communities seems to begin rather abruptly at
a concentration of approximately 2 mg ℓ^{-1} of oxygen as indicated
by the generalized SAB-diagram (Species-Abundance-Biomass) in Fig. 9.
Below that concentration only a few species seem to be able to
survive prolonged periods and no species are directly favoured by
increased abundance during such conditions. Indirectly, however,
opportunists are favoured in that they take advantage when conditions
improve. They are the first colonists and utilize the accumulated
food until they are outcompeted by second invaders.

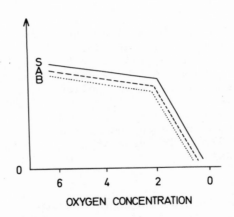

Figure 9.   Schematic presentation of macrobenthic SAB-curves
            (Species-Abundance-Biomass) in relation to approximate
            oxygen concentrations (mg/ℓ).

Based on the results presented in this review, it seems reasonable to categorize macrobenthic species in relation to their tolerance to low oxygen concentrations in nature as follows:

1. Non-tolerant species - species not connected with oxygen deplet-
    ion, i.e. most species.
2. Tolerant species - species surviving periodic oxygen depletion.
    Species belonging to this group are mainly molluscs (e.g.
    Astarte, Corbula, Mya, Cyprina). (Similar results have been
    reported based on laboratory experiments, e.g. Theede 1973).

3. Transitory opportunists – species colonizing the bottoms as soon
    as oxygen conditions improve. These species tolerate, even as
    larvae and juveniles, organically enriched sediments with low
    redox-potentials. Species belonging to this group are mainly
    polychaetes (e.g. *Capitella, Polydora, Pseudopolydora,
    Scolelepis*).

4. Transitory emigrants – mobile species inhabiting the vicinity
    of the oxygen depleted bottoms. When conditions improve for
    shorter or longer periods they emigrate into the area. Repre-
    sentative species are errant polychaetes (*Harmothoe, Ophio-
    dromus*) and cumaceans (*Diastylis*).

    It has repeatedly been argued that opportunistic species,
mainly capitellides and spionides, are exceptionally tolerant to
low oxygen concentrations. However, in all cases presented here,
opportunists are pioneers on deserted bottoms very soon after
oxygen is renewed. (See also Hiscock & Hoare 1975; Watling 1975).
Such is their adaptation, and they are not particularly tolerant
to oxygen depletion, as shown in experiments by e.g. Henriksson
(1969).

    It is impossible with our present knowledge to evaluate the
ecological consequences of the vast areas missing macrofauna due to
oxygen deficiency. The production of demersal fish-food is lacking
in these anoxic areas, but oxic waters in the vicinity could have
been stimulated to increased production, e.g. anoxia leads to increased release of phosphorus which might increase primary production and at the end more food is produced for the benthic animals.
Thus, oxygen depletion could have both negative and positive effects
if the total production of larger regions are considered.

    The direct effects on the benthic macrofauna in anoxic areas
can be estimated and compared by calculating the missing biomass,
i.e. the macro- benthic biomass that probably would have been present if the area was permanently oxic. The results given in Table
1 are to some extent based on rough approximations and should not
be used by administrators to make economic calculations, but are
rather put forward to stimulate ecologists to future studies about
the ecological consequences of such vast defaunated areas. References to the calculations of most of the figures are given in the
previous sections. The missing biomass in the Limfjord at the time
of most serious oxygen depletion in autumn could be approximately
71 g m^{-2}, which was presented as a mean value in shallow waters
above the halocline for adjacent coastal areas in Sweden (Rosenberg
amd Möller 1979). The area without macrofauna in the Idefjord
(Fig. 8) has been defined to depths below 10 m and was calculated
from a nautical chart, i.e. it is an underestimate. The mean
value of missing macrofaunal biomass in the Idefjord is taken from
the mean presented below the halocline in adjacent waters (Rosenberg
amd Möller 1979). It is evident, based on the figures in the table,

Table 1. Estimation of missing macrofaunal biomass (wet wt)
is some anoxic waters in Scandinavia. For references
see text.

	AREA in km^2	MEAN BIOMASS g m^{-2}	BIOMASS MISSING tonnes	YEAR
Byfjord	3.41	146	498	1971–74
Saltkällefjord	1.5	–	60	1968
Baltic	84000	20	1680000	1975
Limfjord	330	71	23420	1973–78*
Idefjord	13.9	146	2029	1967–78

that oxygen depletion creates enormous defaunated bottoms in
Scandinavia and that the most extensive effects are due to natural
reasons, i.e. geomorphological characteristics of the water bodies
leading to stagnation, and that only a minority is caused by organic
pollution.

REFERENCES

Afzelius, L. 1979: Nasjonalt program for overvåkning av vannres-
surser. Utvikling og status i Iddefjordens biologi. Norsk
institutt for vannforskning, rapport 0-75038, 50 pp.

Arntz, W. E. 1977: Biomasse und Production des Makrobenthos in den
tieferen Teilen der Kieler Bucht im Jahr 1968. Kieler
Meeresforsch. 27, 36–72.

Arntz, W. E. and D. Brunswig 1976: Studies on structure and
dynamics of macrobenthos in the western Baltic carried out by
the joint research programme "Interaction sea-sea bottom"
(SFB 95-Kiel). In: Proceedings of the 10th European Symposium on Marine Biology, 1975, Vol. 2, edited by G. Persoone &
E. Jaspers, University Press, Wetteren, Belgium, 17–42.

Arntz, W. E. and D. Brunswig 1980: Zonation of macrobenthos in the
Kiel Bay channel system and its implications for demersal fish.
Pr. morsk. Inst. ryb. Gdyni (in press).

* periodically

Beyer, F. 1968: Zooplankton, zoobenthos, and bottom sediments as
    related to pollution and water exchange in the Oslofjord.
    Helgolander wiss. Meeresunters. 17, 496-509.

Dybern, B. I. 1972: The Idefjord - a destroyed marine environment.
    (In Swedish with English summary). Fauna och Flora 67, 90-103.

Elmgren, R. 1975: Benthic meiofauna as indicator of oxygen condi-
    tions in the northern Baltic proper. Merentutkimuslait.
    julk., Skr. 239, 265-271.

Elmgren, R. 1979: Structure and dynamics of Baltic benthos com-
    munities, with particular reference to the relationship between
    macro- and meiofauna. Kieler Meeresforsch., Sonderheft 4,
    (in press).

Elmgren, R., R. Rosenberg, A. B. Andersin, S. Evans, P. Kangas,
    J. Lassig, E. Leppäkoski and E. Varmo: Benthic macro- and
    meiofauna in the Gulf of Bothnia (Northern Baltic). Pr. morsk.
    Inst. ryb. Gdyni (in press).

Fonselius, S. H. 1969: Hydrography of the Baltic deep basins III.
    Fishery Board of Sweden, Ser. Hydrography No. 23, 1-97.

Gerlach, S. A. 1971: On the importance of marine meiofauna for
    benthos xommunities. Oecologia (Berl.) 6, 176-190.

Henriksson, R. 1969: Influence of pollution on the bottom fauna of
    the Sound (Oresund). Oikos 20, 507-523.

Hiscock, K. and R. Hoare 1975: The ecology of sublittoral communities
    at Abereiddy Quarry, Pembrokeshire. J. Mar. Biol. Ass. U. K.
    55, 833-864.

Jägerskjöld, L. A. 1971: A survey of the marine benthonic macro
    fauna along the Swedish west coast 1921-38. Kungl. vetenskap-
    och vitterhetssamhället, Göteborg, pp. 1-146.

Jörgensen, B. B. 1980: Seasonal oxygen depletion in the bottom
    waters of a Danish fjord and its effect on the benthic com-
    munity. Oikos (in press).

Kitching, J. A., F. J. Ebling, J. C. Gamble, R. Hoare, A. A. Q. R.
    McLeod and T. A. Norton 1976: The ecology of Lough Ine. XIX.
    Seasonal changes in the western trough. J. Anim. Ecol. 45,
    731-757.

Leppäkoski, E. 1969: Transitory return of the benthic fauna of the
    Bornholm Basin, after extermination by oxygen insuficiency.
    Cah. Biol. mar. 10, 163-172.

Leppäkoski, E. 1971: Benthic recolonization of the Bornholm basin
    (Southern Baltic) in 1969-71. Thalassia jugosl. 7, 171-179.

Leppäkoski, E. 1975: Macrobenthic fauna as indicator of oceanization
    in the southern Baltic. Merentutkimuslait. julk. Skr. No. 239,
    280-288.

Pearson, T. H. 1975: The benthic ecology of Loch Linnhe and Loch
    Eil, a sea-loch system on the west coast of Scotland. IV.
    Changes in the fauna attributable to organic enrichment. J.
    Exp. Mar. Biol. Ecol. 20, 1-41.

Pearson, T. H. and R. Rosenberg 1978: Macrobenthic succession in
    relation to organic enrichment and pollution of the marine
    environment. Oceanogr. Mar. Biol. Ann. Rev. 16, 229-311.

Rosenberg, R. 1972: Benthic faunal recovery in Swedish fjord
    following the closure of a sulphite pulp mill. Oikos 23,
    92-108.

Rosenberg, R. 1976: The relation of treatment and ecological effects
    in brackish water regions. Pure Appl. Chem. 45, 199-203.

Rosenberg, R. 1977: Benthic macrofaunal dynamics, production, and
    dispersion in an oxygen - deficient estuary of west Sweden.
    J. Exp. Mar. Biol. Ecol. 26, 107-133.

Rosenberg, R., I. Olsson E. Ölundh 1977: Energy flow model of an
    oxygen - deficient estuary on the Swedish west coast. Mar.
    Biol. 42, 99-107.

Rosenberg, R. and P. Möller 1979: Salinity stratified benthic
    macro- faunal communities and long-term monitoring along the
    west coast of Sweden. J. Exp. Mar. Biol. Ecol. 37, 175-203.

Schultz, S. 1968: Rückgang des Benthos in der Lübecker Bucht.
    Monatsberichte Deutschen Akad. Wiss. Berl. 10, 748-754.

Theede, H. 1973: Comparative studies on the influence of oxygen
    deficiency and hydrogen sulphide on marine bottom invertebrates.
    Neth. J. Sea Res. 7, 244-252.

Watling, L. 1975: Analysis of structural variations in a shallow
    estuarine deposit-feeding community. J. Exp. Mar. Biol. Ecol.
    19, 275-313.

Zmudzinski, L. 1977: The Baltic deserts. Ann. Biol. 32, 50-51.

BENTHIC BIOLOGY OF A DISSOLVED OXYGEN DEFICIENCY EVENT IN HOWE SOUND, B. C.

C. D. Levings

Department of Fisheries and Oceans
4160 Marine Drive
West Vancouver, B. C. V7V 1N6

INTRODUCTION

This report presents benthic biological observations related to dissolved oxygen (D.O.) conditions before, during and after 2 renewals of bottom water in Howe Sound, a fjord in B. C. characterized by aperiodic renewals below sill depth (Bell, 1973). Trawl surveys began in August 1976, to provide baseline data for use in ocean dumping studies. A dramatic decline in catches behind the sill was observed in August 1977; this prompted additional surveys which were continued until a large exchange of deep water occurred in March 1979. The complete data set and further interpretations are presented elsewhere (McDaniel *et al.*, 1978; Levings 1980).

The outer portion of Howe Sound is actually an embayment of the Strait of Georgia, separated from its major basin by a relatively deep sill (110 to 200 m). A much shallower sill (30 to 70 m) separates this embayment from an inner basin, which has a maximum depth of 298 m (Fig. 1). Bell (1973) provided a detailed review of water characteristics and processes involved in bottom water renewal of the inner basin. Exchanges extend to the bottom waters possibly every 3 or 4 years. Behind the sill, at 200 m, salinity ranged from approximately 30.1 to 30.6 $^0/_{00}$ during the $2\frac{1}{2}$ years of observation involved in our work. At a similar location and depth, temperature varied from about 8.2^0 to 8.5^0C over the period 1971 to 1973 (Bell 1973).

Stations for trawling (15 cruises) and hydrography (22 cruises) were aligned along the longitudinal axis of the Sound as shown in Fig. 1. Biological samples were obtained using a small otter trawl (6 m throat) with 3.2 cm mesh body web and a 1.3 cm mesh cod-end liner. Usually 3 replicate hauls were obtained at each station.

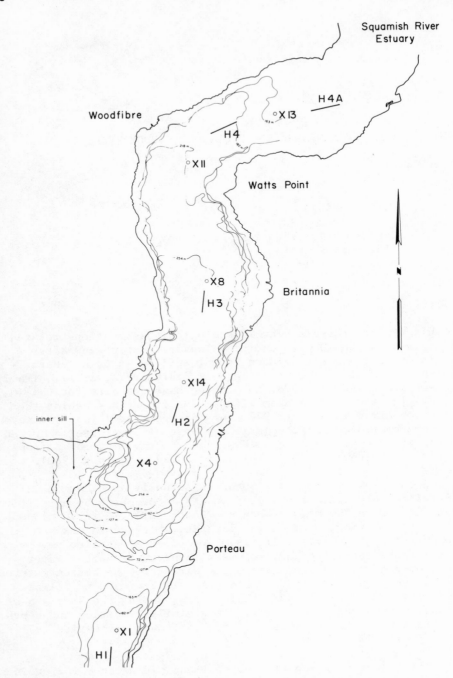

Figure 1.   Chart of Howe Sound showing bathymetry and sampling loc-
            ations in the inner basin.   Trawl stations have prefix H
            and bottle cast stations X.

Differences in community composition throughout time at individual stations were contrasted using numerical catch data (number
per trawl) for each taxa. In cluster analyses, the similarity index used was the euclidean distance measure and times were sorted
using centroid analysis. The computer programme used was the
"Cluster" routine documented in Fox and Guire (1976).

WATER CHARACTERISTICS

Dissolved oxygen features were analyzed by plotting D. O.
changes with time and on longitudinal sections of Howe Sound. Data
are most extensive from a station located just inside the inner
sill (station X4). At 250 m, D. O. values declined from approximately
3.0 mg ℓ^{-1} in August 1976 to about 0.5 mg ℓ^{-1} in August to October
1977 (Fig. 2). Renewal of deep water did not occur in winter 1976-
1977. An incursion did occur in November 1977, but D. O. values
again declined during 1978 so that by November of that year values
were below 0.5 mg ℓ^{-1}. Sometime during November 1978 to early
March 1979 a large incursion of oxygen-rich water occurred, so that
D. O. values at almost all depths increased to > 6.0 mg ℓ^{-1}.

Longitudinal sections of D. O. for each sampling time show
patterns (see Levings 1980) which complement the above results.
D. O. values outside the inner sill were always above 3.5 mg ℓ^{-1}.
Oxygen deficient bottom waters extended from behind the inner sill
to off Watts Point (Fig. 1) and were mainly confined to depths of
over 180 m. Approximately 24 km^2 of benthic habitats were affected
by the prolonged D. O. depletion in 1977.

BIOLOGICAL RESULTS

Table 1 lists the most frequent (occurring in 70% of catches)
and abundant taxa at the various stations. Several dominant taxa
at stations H1, outside the inner sill, were infrequent at other
stations, for example the holothurian *Chiridota* sp., the asteroids
Ctenodiscus crispatus and *Pseudarchaster parelli* and the poriferan
Iophon pattersoni. Most of the other frequently-occurring taxa
were obtained at all the other stations. Fish taxa were most frequent at the shallowest station (H4A, 125 m). *Lycodes diapterus*
was the only fish species which occurred frequently both inside
and outside the inner sill.

Changes in abundance and diversity (number of taxa) of invertebrates and fish were most striking at deep stations inside the
inner sill (H2, H3), and catches had decreased dramatically by
August 1977 (Fig. 3). Fish abundance decreased gradually over
February to August 1977. Catches of invertebrates, especially the
heart urchin *Brisaster latifrons* actually increased over the period

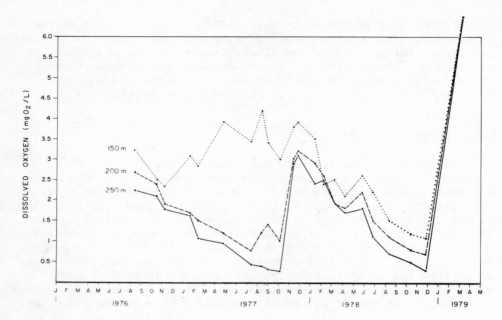

Figure 2.   Changes in dissolved oxygen concentration (mg O_2 ℓ^{-1})
at three depths at station X4, north of the inner sill
of Howe Sound.

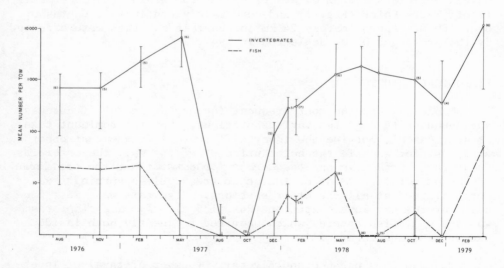

Figure 3.   Temporal changes in number of taxa, number of invertebrates
and number of fish per trawl inside the inner sill of
Howe Sound (combined data from stations H2 and H3).   Mean
values with ranges are shown, and numerals beside data
points indicate number of samples.

Table 1. Taxa occurring in 70% or greater of trawl catches at stations in Howe Sound. Data from H2 and H3 are from samples previous to August 1977.

H1
Iophon pattersoni[1]
Actinostola spp.[2]
Colus halli[3]
Neptunea phoeniceus[3]
Natica russa[3]
Fusitriton oregonensis[3]
Yoldia thraciaeformis[4]
Macoma nasuta[4]
Nucula tenuis[4]
Sternapsis fossor[5]
Pandalus borealis[6]
Pandalopsis dispar[6]
Spirontocaris spp.[6]
Munida quadrispina[7]
Brisaster latifrons[8]
Ctenodiscus crispatus[9]
Pseudarchaster parelli[9]
Chiridota sp.[10]
Molpadia intermedia[10]

Lycodes diapterus[11]
Dasycottus setiger[11]

H2
Colus halli[3]
Macoma nasuta[4]
Yoldia thraciaeformis[4]
Spirontocaris spp.[6]
Munida quadrispina[7]
Brisaster latifrons[8]

H3
Neptunea phoeniceus[3]
Spirontocaris spp.[6]
Munida quadrispina[7]
Brisaster latifrons[8]

Lycodes diapterus[11]
Lyopsetta exilis[11]
Parophrys vetulus[11]
Microstomus pacificus[11]

H4
Neptunea phoeniceus[3]
Colus halli[3]
Spirontocaris spp.[6]
Munida quadrispina[7]
Brisaster latifrons[8]
Molpadia intermedia[10]

Lycodes diapterus[11]
Lumpenella longirostris[11]
Parophrys vetulus[11]
Hydrolagus colliei[11]
Lyopsetta exilis[11]

H4A
Neptunea phoeniceus[3]
Colus halli[3]
Pandalus borealis[6]
Spirontocaris spp.[6]
Crangon communis[6]

Lycodes diapterus[11]
Lycodopsis pacifica[11]
Bathyagonus nigripinnis[11]
Microstomus pacificus[11]
Parophrys vetulus[11]
Theragra chalcogramma[11]
Merluccius productus[11]

1 = Porifera; 2 = Cnidaria; 3 = Gastropoda; 4 = Bivalvia; 5 = Polychaeta; 6 = Decapoda Natantia; 7 = Decapoda Reptantia; 8 = Echinoidea; 9 = Asteroidea; 10 = Holothuroidea; 11 = Pisces.

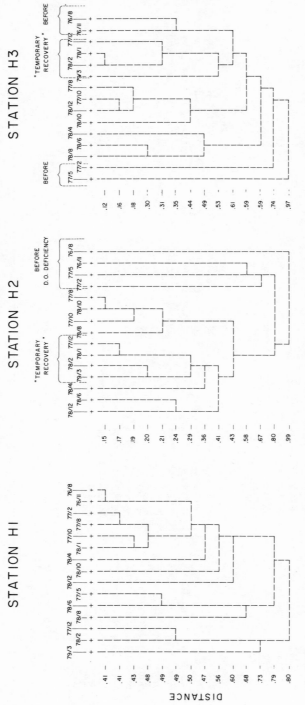

Figure 4.   Dendrograms showing similarity levels of trawl catches through time at stations south (H1) and north of the inner sill (H2, H3) of Howe Sound. The similarity measure used was euclidean distance with centroid sorting.

February to May before declining in August. At station H2 *B. latifrons* usually accounted for > 50% of catches but no living specimens were observed after August 1977. Catches at and after this date were characterized by urchin tests without tissue. In August other invertebrates such as the gastropods *Neptunea phoenicius* and *Colus halli* were still decomposing so samples were very odorous.

Cluster analyses for sampling times showed similar patterns to those observed with abundance and taxa data. Dendrograms generated with data from station H1, outside the inner sill, show that community structure at most of the sampling times (9 out of 14) was similar at above the 0.5 level (Fig. 4) so that the data set was relatively homogeneous. Analyses for station H2 show clearly that communities changed dramatically after August 1977 (Fig. 4), as did computations using data from station H3. In the latter case, August and November 1976 linked together at a high similarity level, but February and May 1977 were apparently quite dissimilar. This may have reflected the withdrawal of most of the fish species from this habitat. The dendrograms also reflect the recovery of the communities from D. O. deficiency, as data from December 1977, January and February 1978, and March 1979 were linked at both stations H2 and H3 (Fig. 4).

BIOLOGICAL RESPONSES

Patterns of migration associated with D. O. depletion and renewal reflected behaviour and reproductive patterns. Certain slow moving fish species, for example the brown cat shark *Apristurus brunneus*, apparently suffered direct mortality as carcasses of this species were obtained in trawls in August 1977. Other fish such as the lemon sole *Parophrys vetulus* and the eel pout *Lycodes diapterus* presumably escaped from the low D. O. areas by moving into shallow waters. *P. vetulus* catches gradually declined over the period August 1976 to May 1977, when D. O. values decreased to < 1.0 mg ℓ^{-1} and this species was never taken again at the deeper stations. In contrast, *L. diapterus* catches shifted directly in response to D. O. values.

Most of the sedentary invertebrates were eventually killed by the prolonged D. O. deficiency, but their behaviour before death can only be inferred from differences in catch patterns. For example catches of the urchin *B. latifrons* were actually increasing over the period February to May 1977. It is possible the urchins were migrating to shallower depths in the sediments, where they became more available to our trawl. The time scale for restoration of the urchin population and other invertebrates dependent on pelagic larvae is very difficult to predict. Recruitment depends on water exchange and biological factors such as the proximity

of reproducing organisms and suvival of larvae. At the last survey
there was no evidence of recruitment and growth of *B. latifrons*,
and the infaunal communities behind the inner sill remain bizarre.
The recovery of the epibenthic communities was attributable to the
rapid invasion of the deep basin habitats by mobile decapods,
primarily juvenile *Munida Quadrispina* and Spirontocaris sp.

REFERENCES

Bell, W. H. 1973: The exchange of deep water in Howe Sound basin.
     Pacific Marine Science Rep. 73-13. Institute of Ocean
     Sciences, Sidney, B. C., 35 pp.

Fox, D. J. and K. J. Guire 1976: Documentation of MIDAS (3rd. Ed.).
     Statistical Research Laboratory, University of Michigan,
     Ann Arbor, Michigan. 203 pp.

Levings, C. D. 1980: Demersal and benthic communities in Howe
     Sound basin and their responses to dissolved oxygen deficiency.
     (manuscript in preparation.)

McDaniel, N. G., C. D. Levings, D. Goyette and D. Brothers 1978:
     Otter trawl catches at disrupted and intact habitats in Howe
     Sound, Jervis Inlet, and Bute Inlet, B.C., August 1976 to
     December 1977. Fisheries and Marine Service Data Rep. No. 92.
     77 pp., West Vancouver, B.C.

RENEWAL AND ENTRAINMENT IN LOCH EIL; A PERIODICALLY VENTILATED SCOTTISH FJORD

A. Edwards, D. J. Edelsten, M. A. Saunders, S. O. Stanley

Scottish Marine Biological Association
P.O. Box 3
Oban, Argyll, Scotland

INTRODUCTION

The entrance sill of Loch Eil, a Scottish fjord, is so re-stricted that two layer flow is impossible during most of the tidal cycle, and a tidally pulsed circulation develops in the fjord. However, bottom water is well connected to outside water and on a time scale of a year there is no stagnation. Current metering has shown that this bottom water stagnates during neap tides, when oscillatory currents occur, and that renewals happen regularly, during the period of spring tides. The renewal sequence is surface ebb: low water: inflow to intermediate depth: density current inflow to the bottom: and lastly, high water. Entrainment of ambient fluid into the density current is estimated and compared with other fjords and laboratory experiments. Sedimentation rates are shown to be particularly high during renewals.

LOCH EIL

Loch Eil is a fjord at the head of Loch Linnhe on the West Coast of Scotland (Pearson, 1970). A narrow sill at Annat (Fig. 1) separates it from Upper Loch Linnhe, which is itself separated from the open sea by another sill at Corran (Fig. 2). Loch Eil has been much studied by biologists because of the discharge of cell-ulose waste from a pulp mill on the north shore of Annat narrows. Because the fjord is in an area of high freshwater runoff, with consequent high stability of the water column, it is possible that the bottom water stagnates {Edwards and Edelsten (1977)} and that it could thus be particularly sensitive to an increased organic load. Some support for this idea comes from Pearson (1975), who

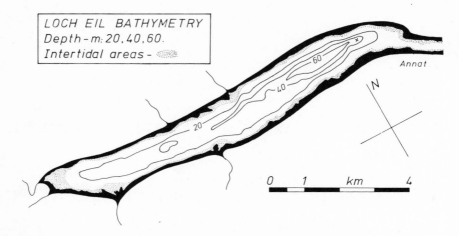

Figure 1.   Bathymetry of Loch Eil and the position of the current
            meter station X.

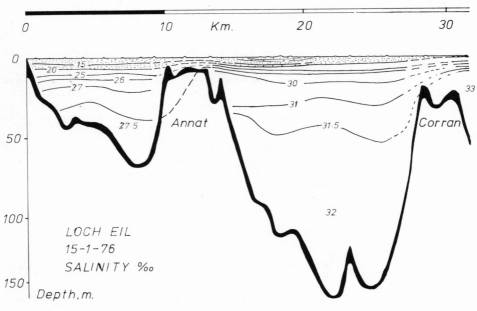

Figure 2.   A winter salinity distribution in Loch Eil and Upper Loch
            Linnhe.   Deep water is of a type found outside the basin at
            sill depth.   Isolines in the sill region (dashed) are only
            drawn for connection of the basins: water over Annat sill
            is usually well mixed vertically and its salinity varies
            rapidly in time.

has described many of the important benthic faunal changes which
followed the start of the pulp mill, and the present work is therefore intended to investigate the flushing and ventilation of the
bottom water in contact with the deep benthos.

TEMPERATURE AND SALINITY

The longitudinal temperature and salinity distributions of
the Upper Linnhe-Eil system were surveyed monthly during 1975 and
1976. Figure 2 shows a representative salinity section which dis-
plays some of the persistent features of the hydrography. Maximum
stratification is always found above sill depths where there is a
brackish surface layer fed by the runoff whose principal source is
at the seaward end of Annat narrows. The salinity in this layer is
inversely related to runoff, so that it is lowest in winter and
highest in summer. There is a similar seasonal pattern with a
smaller range of about 27-32 $^0/_{00}$ in Loch Eil Deep Water, which
has the same salinity as Upper Linnhe Water at the depth of Annat
sill (27.5 $^0/_{00}$ in Fig. 2.). A similar connection is also always
true of temperature of these two waters, for at no time was there
a difference of more than a few tenths of a degree C. between them.
Because this difference is small compared to the upper seasonal
range of about 8^0C, there is no measurable time lag between the
temperatures of the two waters, and so we conclude that Eil Deep Water
is not isolated from Upper Linnhe Water at Annat sill depth for
periods exceeding a month and is well connected to it by a regular
advection over the sill, occurring at least monthly. Such advection
has been demonstrated elsewhere when sill water density exceeds
bottom water density after a period of bottom water density reduction by upwards diffusion of salt or downwards diffusion of heat
(Gade, 1973).

BOTTOM CURRENTS

To test the diffusion-advection model, recording current meters,
Plessey type MO21, were moored near the bottom at station X (Fig. 1)
in 1976. Figure 3 shows the daily net displacement during the early
part of the year, and tidal range. There is a clear association of
spring tides with net inflow. During neaps, currents were usually
small (of the order of a few cm/sec), were sometimes oscillatory
with a period that was a harmonic of the locally dominant semidiurnal
tide, and gave no significant net displacement either landwards or
seawards. Conditions were quite different during springs: towards
high tide, a strong landward current of as much as 70 cm/sec could
develop: total displacement during a 12½ hour tidal cycle became
positive, corresponding to a mean inflow as much as 10 cm/sec, and
typically about 5 cm/sec.

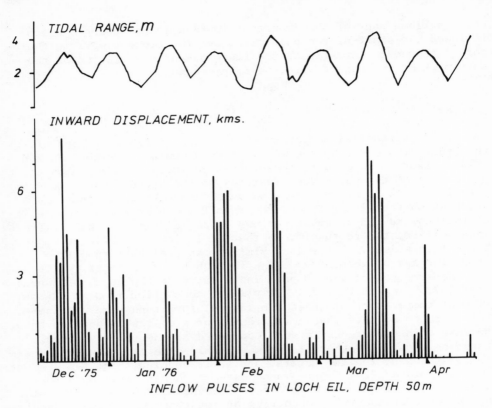

Figure 3.   Tidal Range and the daily net landwards displacement at
            X measured by Plessey MO21 current meter at a height of
            2 metres above bottom. Only measurements greater than
            9 cm/sec over 10 minutes were used to obtain the dis-
            placement.

        The main spring – neap pattern in Fig. 3 is modified by run-
off and by the memory of the system. January rainfall was twice
that of the other months. The response time of near surface salinity to runoff change in Scottish fjords is about a week (Landless
and Edwards, 1976) and thus the surface layers of the system probably
responded to increased January runoff by lowering their salinity
within the month. This lowering reduced the January supply of high
density water at the sill so that inflow was inhibited during spring
tides in the middle of the month. Conversely, reduced rainfall in
February enhanced the inflow at the beginning of that month. That
the system has a memory similar to that discussed by Welander (1974)
is suggested by the inhibition of inflows during the last springs
of February and March: these followed exceptionally high springs

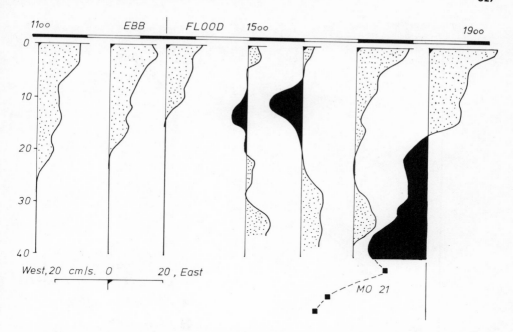

Figure 4.   Spring tide. Inward and outward current speeds at
            depth intervals of 2 metres, east of X. Flooding water
            enters Loch Eil as an intermediate intrusion or as a
            bottom density current.

in the middle of the month which may have injected sufficient dense
water that diffusion was inadequate to prime the system for strong
inflows in the last springs. Despite the inhibitions, net inflow
occurs monthly or fortnightly throughout the current record, supporting our interpretation of the hydrographic surveys.

AN INFLOW EVENT

     Figure 4 shows current profiles measured through the water
column 200 m east of station X in a spring tide. In the ebb tide,
water flows out of the fjord from the upper layers. This flow reduces and in the first part of the flood tide there appears an intermediate inflow at about 10 m. This inflow is not seen in the pen-
ultimate profile, perhaps because the flows are not uniform across
the fjord, but reappears at the bottom of the last profile towards
the end of the flood. The figure also shows simultaneous currents
measured by three meters at X: these data, and others on the same
day, show that at X the main inflow current was vertically limited
to a region about 5 metres off the bottom. We conclude that during

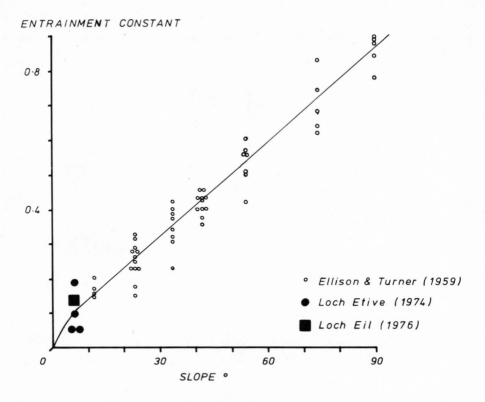

Figure 5.   The entrainment constant as a function of bottom slope
             for various situations.   The Loch Etive values are from
             Edwards and Edelsten (1977).

the late flood of spring tides, inflowing water from the Annat sill
becomes dense enough to switch from intermediate intrusion to downslope travel as a density current of limited height.   It is interesting to estimate the entrainment into this current by assuming
that the upper seawards flow in the last profile is caused by entrainment of the upper water into the turbulent density current.

If a typical velocity of the density current is V, and the
velocity of entrained fluid normal to the upper surface of the density current is v, the entrainment constant E is given by

$$v = EV$$

and we obtain E = 0.13; this and values from Loch Etive and Ellison
and Turner's (1959) experiments are shown in Fig. 5.   The error in
E with this estimate is about +0.1,-0.05, but the general agreement
suggests to us that density current theory is an adequate framework
in which the dynamics of bottom water renewal can be modelled.

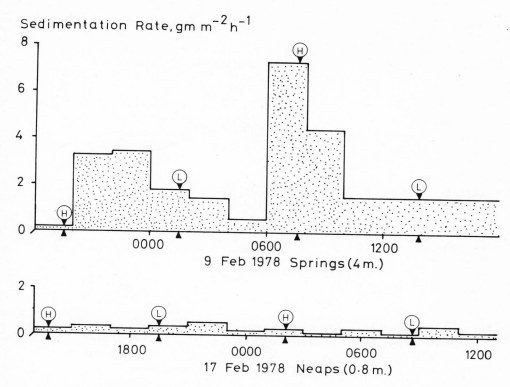

Figure 6.   Sedimentation rates one metre above the bottom 2 kilo-
            metres west of X during neap and spring tides in 1978.
            H and L are the predicted times of high and low water.

SEDIMENTATION

        Some of the consequences of the above pattern of currents in
Loch Eil were found in 1978 by mooring sedimentation traps one metre
above the bottom 2 km west of X during spring and neap tides. The
traps were changed every two hours so as to examine short-term sedimentation rates. Figure 6 shows the results of this experiment.
During neap tides the rate varies little and is not related to the
tide. However, during springs the rate peaks near high water and
its mean value is ten times the neap average. Either the spring
density currents cause considerable resuspension, or they bring in
high sediment load from the sill direction, or, most likely, they
do both. Because bottom stress increases non-linearly with current
speed, we expect resuspension to become more important with greater
bottom currents and we thus suspect that physical aspects of the
history of chemical exchange between the sediments and the bottom
water will be dominated by a few major inflow events rather than by
any of the long-term means of bottom conditions.

ACKNOWLEDGEMENT

    We are grateful to R. Bowers for his temperature salinity pro-
filer and to our ships masters and crews for their help.

REFERENCES

Edwards, A. and Edelsten, D. J. 1977: Deep Water Renewal of Loch
    Etive: a Three Basin Scottish Fjord. Estuar. Coast. Mar. Sci.
    5, 575-595.

Ellison, T. H. and Turner, J. S. 1959: Turbulent Entrainment in
    Stratified Flows. J. Fluid Mech. 6, 423-448.

Gade, H. G. 1973: Deep Water Exchange in a Sill Fjord: a
    Stochastic Process. Journal of Physical Oceanography, 3,
    213-219.

Landless, P. J. and Edwards, A. 1976: Economical Ways of Assessing
    Hydrography for Fish Farms. Aquaculture 8, 29-43.

Pearson, T. H. 1970: The benthic ecology of Loch Linnhe and Loch
    Eil, a sea loch system on the West Coast of Scotland. I. The
    physical environment and distribution of the macrobenthic
    fauna. J. Exp. Mar. Biol. Ecol. 4, 261-286.

Pearson, T. H. 1975: The benthic ecology of Loch Linnhe and Loch
    Eil, a sea-loch system on the West Coast of Scotland. IV.
    Changes in the benthic fauna attributable to organic enrich-
    ment. J. Exp. Mar. Biol. Ecol. 20, 1-41.

Welander, P. 1974: Theory of a two-layer Fjord-Type Estuary, with
    special reference to the Baltic Sea. J. Phys. Oceanogr. 4,
    542-556.

DEEP WATER RENEWALS IN THE FRIERFJORD - AN INTERMITTENTLY ANOXIC BASIN

Jarle Molvær

Norwegian Institute for Water Research
P.O. Box 333, Blindern
Oslo 3, Norway

INTRODUCTION

The Frierfjord is located in the southern part of Norway (see Fig. 1). At the entrance it has a narrow sill, 23 m deep, with inside depths of 98 m. The length of the fjord is about 10 km, and the surface area 18 km^2. The volume is $810 \cdot 10^6 \ m^3$. Between the Frierfjord and the open sea (Skagerrak) another basin is situated - Langesundsfjord - with a sill of 50 m and a maximum depth of 123 m.

The freshwater runoff to the fjord is completely dominated by the Skien river, varying from 50 m^3/s to 700 m^3/s, with approximately 270 m^3/s as an annual mean. The tidal variation in the area is approximately 0.2 m. Due to effluents from pulp and chemical industry along with municipal wastewater from a population of approximately 90,000, the Frierfjord is seriously polluted by nitrogen and phosphorous compounds, organic matter, metals and halogenated hydrocarbons. Since 1974 comprehensive studies of the pollution- problems have been carried out by several institutions. Below we present some results from the studies of deep water exchange processes.

The nine hydrographic fjord-stations (see Fig. 1) were visited with 4-6 weeks intervals from March 1974 to December 1977. Water samples were collected with Nansen bottles; temperature was read from reversing thermometers; the salinity was calculated from conductivity measurements and the oxygen concentration was measured by Winkler titration. At the sill Aanderaa current meters at 18 m depth were used to monitor in- and outflows. Vertical current profiles were obtained with gelatine current meters.

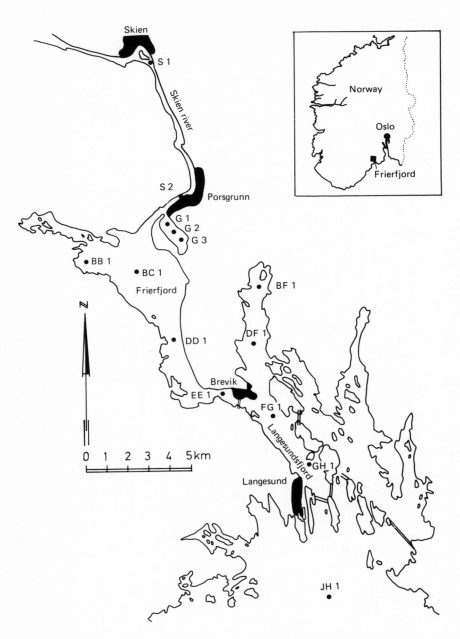

Figure 1.   The Frierfjord area with sills (=) and water sampling
            stations.

Wind observations were made daily at 06, 12 and 18 GMT at Langøytangen
lighthouse on the coast.

RESULTS

   From physical and chemical characteristics the watermass in
the Frierfjord can conveniently be divided into three layers:
a)   A surface layer: thickness usually 3-6 m, salinity < $10^0/_{00}$.
b)   An intermediate layer: usually between 8 m and 25-35 m depth.
c)   Deep water: below 25-35 m, salinity 33-34.5 $^0/_{00}$.
Due to the large freshwater supply there is a strong estuarine
circulation in the fjord. In winter the prevailing wind direction
on the coast is north to north-east. In the summer the prevailing
wind direction is south to south-west.

   The hydrographic conditions on the coast are illustrated in
Fig. 2. The formation of dense water on the coast in January-April
each year can in part be attributed to upwelling caused by northerly
winds as shown by Aure (1978). However, it is also likely that the
general hydrographic situation (low runoff, cooling from the surface)
plays an important part for the formation of dense water in winter.
It is interesting that the upwelling is mainly correlated to offshore
wind, and not to wind blowing along the shore as for the Norwegian
west-coast (Helle 1978).

   The density above sill depth in the Langesundsfjord shows a
strong co-variation with the density-variation at corresponding
depth on the coast. Increasing density in the coastal water results
in an ordinary inflow across the sill, with an opposite directed
current higher up in the water column. Decreasing density in the

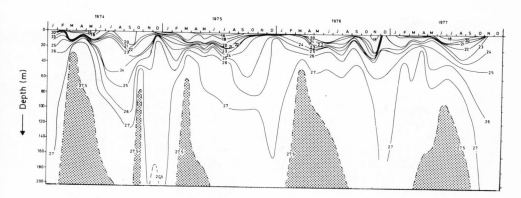

Figure 2.   Variations in Density (σ_t) in the coastal water (st.
            JH-1) from February 1974 to December 1977.

coastal water creates a "reverse inflow" with dense fjord water
flowing out an a return current of coastal water higher up in the
water column. These density-variations in the intermediate water
of the Langesundsfjord in their turn create similar inflows –
outflows across the Brevik sill. The typical timelag between the
onset of strong offshore wind and inflow over the Frierfjord sill
is 3 days. Deep water renewals in the Frierfjord and the Langesundsfjord regularly take place in January – May each year. Fig. 3 shows
an oxygen isopleth (ml O_2/l) at st. BC-1, Frierfjord, for 1974-77.
Large deep water renewals are indicated with arrows.

The extent of the renewals in the Frierfjord may be calculated
from total phosphorous (TOTP) budgets, assuming TOTP is a quasi-
conservative parameter for 4-6 weeks after a renewal of the deep
water. Considering that TOTP-concentrations may have been reduced
from 200 µg P/l to 80 µg P/l, this is a reasonable assumption.

The deep water is divided into horizontal segments (see table
1) and the renewal percentage for each segment is calculated from
the following budget:

$$Q_{new} \ (P_b - P_{new}) = Q \ (P_b - P_a)$$

where Q_{new} = volume of new water in the segment

Q = total volume of the segment

P_b = mean TOTP-concentrations in the segment before the
renewal

P_a = mean TOTP-concentration in the segment after the
renewal

P_{new} = mean TOTP-concentrations in the inflowing water.

The extent of the deep water renewal in March-April 1974
calculated by this method is shown in table 1, P_{new} = 35 µg P/l.

The negative percentage for 20-30 m illustrates the lifting
of former deep water into the intermediate layer (see fig. 4).
From TS-diagram the corresponding renewal percentage between 60 m
and bottom have been caluclated to 75-80%. Over a period of 39
months (March 1974-May 1977) inflows have occured each year between
February and May. The deep water renewal for each year have been
(from TOTP-budgets):

March – April 1974	:	75 – 80%
March – April 1975	:	15 – 20%
January – February 1976	:	Approx. 10% (estimate)
February – May 1977	:	80 – 85%.

Table 1

Deep water renewal in the Frierfjord, March–April 1974.

Depth interval m	Q $10^6 \ m^3$	P_b $\mu g \ P/1$	P_a $\mu g \ P/1$	Renewal %
20–30	102.5×10^6	40	65	14
30–40	86.0×10^6	90	60	55
40–50	69.5×10^6	150	65	74
50–60	54.0×10^6	195	64	82
60–80	82.0×10^6	205	76	76
80–bottom	30.5×10^6	230	85	74

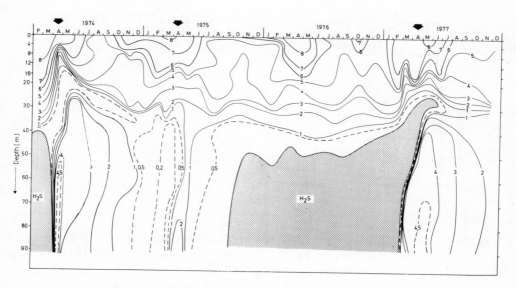

Figure 3.   Variations in oxygen concentrations ($m\ell \ O_2/\ell$) at st. BC-1, Frierfjord, from Feb. 1974 to Dec. 1977.

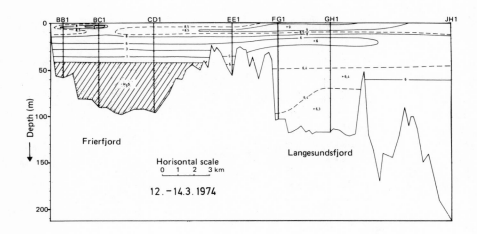

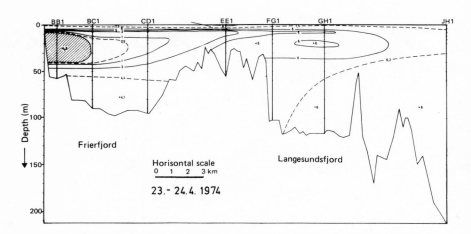

Figure 4. Sectional diagrams of oxygen content (ml O_2/l) shortly
before (12. - 14.3.1974) and during (23.-24.41974) a deep
water renewal in the Frierfjord.

CONCLUSIONS

1. The mainly wind-induced density variations in the coastal water
    are very important for renewal of the intermediate and deep
    water in the Frierfjord and the Langesundsfjord.

2. TOTP-budgets may give valuable information on the extent of
    deep water renewals.

3. Long data series are needed when deep water exchange processes
    in a sill fjord are studied.

REFERENCES

Aure, J. 1978. Den norske kyststrøm utenfor Langesund i juni og
    november 1974. Report 1/78 from the Norwegian Coastal Current
    project. (In Norwegian).

Helle, H.B. 1978. Summer replacement of deep water in Byfjord,
    western Norway: Mass exchange across the sill induced by
    coastal upwelling. In: Hydrodynamics of Estuaries and Fjords,
    ed. by J.C.J. Nihoul. Amsterdam. pp. 441-464.

A RADIOTRACER STUDY OF WATER RENEWAL AND SEDIMENTATION IN A SCOTTISH FJORD

I. G. McKinley and M. S. Baxter

Institute of Geological Sciences
A. E. R. E. Harwell
Oxon. OX11 ORA

INTRODUCTION

The British Nuclear Fuels Limited reprocessing plant at Windscale in Cumbria provides a unique and very useful source of radionuclides for tracer studies in British coastal waters e.g. Jefferies, Preston and Steele, 1973; Baxter and McKinley 1978. Of the range of species present, the radiocaesium nuclides ^{134}Cs ($t_{\frac{1}{2}} \sim 2.1$ y) and ^{137}Cs ($t_{\frac{1}{2}} \sim 30.2$ y) are particularly applicable due to (a) their suitable half-lives and water residence times; (b) their well specified, relatively large ($\sim 10^4$ Ci/month) and variable inputs and (c) their convenience of measurement via ion-exchange extraction and γ-spectrometry, MacKenzie et al, 1979. In this paper the application of these tracers to the study of water mixing and sedimentation in Loch Goil, a fjordic sea loch in the northern part of the Clyde Sea area, will be discussed. Loch Goil itself is ~ 10 km long by 1 km wide and runs $\sim$ NW from its mouth at the junction with Loch Long. This loch reaches depths of ~ 80 m in its deepest basin but is separated from Loch Long by a sill at ~ 20 m.

Vertical profiles of water samples (10 ℓ) were obtained by pumping from selected depths at stations in the deepest part of Loch Goil and in Loch Long just outside the Loch Goil sill. These profiles were obtained regularly (approximately every month) over a period of 2 years at the Loch Goil station and 1 year in Loch Long. Samples were analysed for radiocaesium by quantitative ion-exchange onto Postassium Hexacyanocobalt (II) ferrate (II) (KCFC) followed by high resolution γ-spectrometry, MacKenzie et al 1979, McKinley 1979.

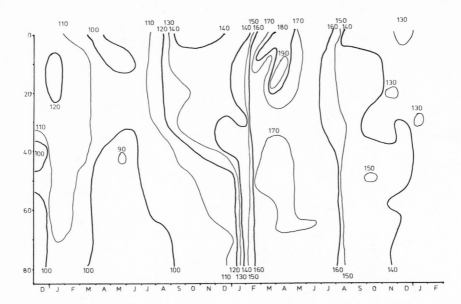

Figure 1. ^{137}Cs isopleths in L. Goil (Dec. 1975–Feb. 1978).
Vertical Axis – depth in metres, concentrations in
disintegrations per minute per litre.

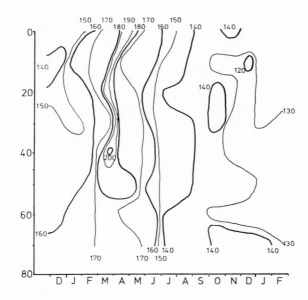

Figure 2. ^{137}Cs isopleths in L. Long (Dec. 1976–Feb. 1978).
Vertical axis – depth (m), concentrations in dpm ℓ^{-1}.

The results of ^{137}Cs analyses of Loch Goil samples may be summarised as ^{137}Cs isopleths on a depth/time plot (Fig. 1) and may thus be readily compared with the observed ^{137}Cs concentrations in the Loch Long source water (Fig. 2). From these diagrams it may be seen that Loch Goil is characterised by fast surface (<30m) exchange at all times of the year. A limited barrier to vertical mixing develops in late spring although slow exchange of deep water does occur over this period. During late winter and early spring there is apparent fast mixing and exchange at all depths possibly due to an external flushing process. We see, however, a behavioural difference between the autumns of the two years studied; in 1976 development of extreme stratification was observed over the period August-December with a marked barrier to mixing at ∿40m. In 1977, however, fast vertical mixing was evident from July-September followed by a more limited degree of stratification. These variations in mixing behaviour may readily be correlated with density profile changes at these stations – fast deep water replacement in Loch Goil occurring only when the density of Loch Long water at sill depth is greater than that of the Loch Goil deeps.

In a quantitative analysis, described in McKinley 1979, a simple box model may be used to describe mixing at times when flushing processes are slow, $e.g.$ during the period January-March 1977, it may be calculated that about 50% of the deep water (>50m) body was exchanged each month. This model does, however, break down during periods of very fast flushing, $i.e.$ when deep water residence times are less than the sampling interval, and also during periods of strong vertical stratification. During the latter, however, the presence of significant water transport is demonstrated by increases in the ^{134}Cs/^{137}Cs ratio (August-November 1977). By applying a one-dimensional diffusion model to this increase, McKinley 1979, a vertical eddy diffusion coefficient in the range $7 < K_z < 25$ cm^2 sec^{-1} may be obtained, the exact value depending on whether or not steady state assumptions are made.

Although caesium is generally regarded as effectively conservative in sea water, the high particulate flux found in coastal regions sequesters a small percentage into sediment (≈0.1% of that passing through the Clyde Sea area) corresponding to a caesium residence time of 660 years. While this removal is insignificant in terms of the total water inventory, it is sufficient to be used effectively as a tracer of sedimentation.

By comparing the radiocaesium distribution in three sediment cores, obtained by a soft-landing corer from the deep basin of Loch Goil over a period of 2 years, it is apparent that the data do not reflect a simple accumulation process (Fig. 3). The profiles may, in fact, be explained by postulating a mixing process operating in the upper 4 cm of the sediment column coupled to a mean sedimentation rate of around 200 mg/cm^2/y. From the distri-

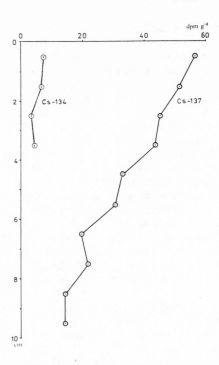

Figure 3.   a) L. Goil sediment radiocaesium profile LGC2.
              Vertical Axis - depth (cm) Horizontal Axis - activity
              (dpm g^{-1}).
              b) L. Goil sediment radiocaesium profile LGC9.
              Broken Line - ^{134}Cs activity   Depth in cm (left scale)
              Light Line - ^{137}Cs activity
              Heavey Line - ^{137}Cs activity (salt corrected) depth in
              mg cm^{-2} (right scale)
              c) L. Goil sediment radiocaesium profile LGC12.
              (lines As (b) above)

bution of ^{134}Cs is the 'mixed' upper sediment a mixing coefficient
in this region of $2 \leqslant K \leqslant 24$ cm^2/y may be estimated.   Both the sedimentation rate and mixing coefficient obtained by this method are
consistent with ^{210}Pb data for these cores (see Swan, 1978).

     An anomaly exists, however in that a similar interpretation
of the radiocaesium profile in a core from a shallower part of
the loch yields a sedimentation rate of around 530 mg/cm^2/y
(below a mixed zone about 6 cm deep), in marked contrast with the
value of 123 mg/cm^2/y obtained by ^{210}Pb.   Although this disparity
is, as yet, unexplained, it is noticeable that ^{226}Ra profiles in
the sediments of this loch are also extremely unusual, see Swan,
1978, and MacKenzie, 1977 for further discussion.

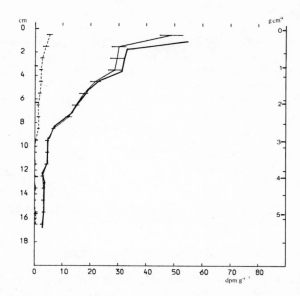

Figure 3b.   See caption on facing page.

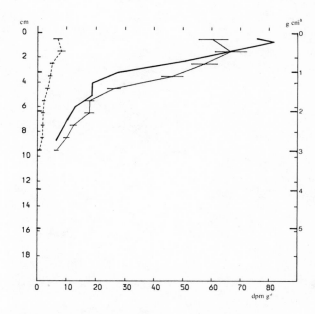

Figure 3c.   See caption on facing page.

ACKNOWLEDGEMENTS

    Thanks are extended to the Clyde River Purification Board for
Provision of ship time for sample collection and to A. B. MacKenzie
and D. S. Swan for natural series data. This work was supported by
a Natural Environment Research Council grant, while the first author
was at the Department of Chemistry, University of Glasgow.

REFERENCES

Baxter, M. S. and McKinley, I. G. 1978: Proc. Roy. Soc. Edin.
    76B, 17–35.

Jefferies, D. F., Preston, A. and Steele, A. K. 1973: Mar. Poll.
    Bull. 4, 118–122.

MacKenzie, A. B. 1977: Ph.D. thesis, Glasgow University.

MacKenzie, A. B., McKinley, I. G., Swan, D. S., Baxter, M. S. and
    Jack, W. 1979: J. Radioanal. Chem. 48, 29–47.

McKinley, I. G. 1979: Ph.D. thesis, Glasgow University.

Swan, D. S. 1978: Ph.D. thesis, Glasgow University.

DESCRIPTION OF WATER RENEWAL PROCESSES IN THE GROS MECATINA ESTUARY

J. Bobbitt

Nordco Limited
P.O. Box 8833
St. John's, Newfoundland, A1B 3T2
Canada

The estuary of the Gros Mecatina River is located on La Basse Côte Nord, Québec and exchanges water with the Gulf of St. Lawrence. It is a small fjord-type estuary of approximately six kilometers in length, and representative of many estuaries along this shore and the coast of Labrador. Three-quarters of the distance from the head of the estuary there is an island in the middle with narrow passages on both sides containing sills, restricting the exchange of water with the sea.

This study was undertaken to determine the exchange processes between sea water and the estuarine water below sill depth and to estimate the flushing time of the surface water. Salinity and temperature profiles carried out at four stations along the estuary during summer showed that there were three layers of distinct water masses. The upper layer was very stratified and ranged from slightly brackish water at the surface to saltier tidal water underneath. The middle layer below a depth of approximately 10 meters was less stratified and originated from the sea water between depths of 4.5 and 6 meters brought in over the sill by the tide and sinking to its equilibrium level in the vicinity of the sill. The bottom layer contained water remaining from the preceding winter and still maintaining a temperature of approximately $-1^{\circ}C$, even though the maximum depth of the estuary is only about 46 meters.

The exchange rate between the three water masses depends on the stability of the water column and the amount of turbulence present. The layer Richardson numbers were calculated for depth intervals of 4 meters, and found to have very high values ranging from 0.8 to 7.0 × 10^4, indicating suppression or extinction of turbulence. This was in sharp contrast to winter when the estuary was ice covered and only the first few meters of water were stratified. This stratification was maintained by the ice cover which caused a boundary layer to be formed with the result that most of the fresh water remained in this boundary layer near the ice surface. The remaining water at greater depths was fully turbulent with layer Richardson numbers ranging from 0.022 to 0.185.

Measurements from tide gauges showed that the tides were standing waves of the mixed, mainly semi-diurnal type. The tidal curves inside the estuary were much more asymmetrical than those outside, with the water level showing a quicker rise and a slower fall with the velocity of flood higher than the velocity of ebb. Both places reached high water at the same time but there was a phase lag during other stages of the tide. There was a linear relationship between the phase lag at low water and the tidal amplitude demonstrating that this lag depended upon the amplitude of the tide, which was associated with the volume of water transported inside the sill. The tidal choking coefficient had an average value of 1.06 due to the presence of the shallow sill and the narrow width of the estuary at the sill's location, tending to restrict the water from flowing out. The combined effect of having a river flowing in at the head of the estuary and prevailing winds from the opposite direction has been attributed to aiding this effect.

The rate of flushing of the surface water during summer and winter was determined by the distribution and concentration of the river water. Since most of the river water was concentrated in the upper layers, only this section had to be considered. Any river water mixed with seawater below sill depth remains in the estuary during the summer months due to the presence of the sill and the stratification of the water column. The flushing time given by $t = \dfrac{\text{volume of fresh water}}{\text{river discharge}}$ was found to be approximately one month for both summer and winter even though the amount of river discharge during winter was only about one-third of that during summer. The flushing time of one month represents the removal rate of only soluble pollutant entering the estuary by way of the river. However, there will not likely be any pollutants from the river because it drains an uninhabited area with no industries. Any soluble or dispersed pollutants would more likely be introduced to the estuary by the tidal water of the Labrador Current. The results of this study demonstrates that this water cannot be flushed from the estuary during the summer and will remain as it enters the estuary over the sill and the stability of the water column. The water of the intermediate layer is not readily mixed upward due to the strong stratification at a depth of approximately

10 meters and the lack of any significant amount of turbulence throughout the water column.

The water of the intermediate and bottom layer can be renewed only during the winter season when an extreme amount of turbulence takes place below the top highly stratified layer and the incoming tidal water is denser than during the summer months. Since it takes approximately one month for the top few meters of water above sill depth to be flushed, it would take a long time for that of the whole estuary to be renewed. However, due to the long duration of winter, the ice cover usually lasts for at least six months of each year, and this is probably long enough for the exchange to take place. This study indicates that the long time required for the water of fjord-type estuaries to become exchanged makes tham very sensitive areas in the event of an oil spill along the Labrador coast where hydrocarbon exploration is presently taking place. Any dispersed pollutants in the Labrador Current will be introduced to the estuaries and fjords with sills at an intermediate level and become trapped until the following winter when the renewal processes can commence.

CHARACTERISTICS OF FLOW OVER A SILL DURING DEEP WATER RENEWAL

G.A. Cannon and N.P. Laird

Pacific Marine Environmental Laboratory
ERL, NOAA, 3711-15th Avenue, N.E.
Seattle, WA 98105, U.S.A.

INTRODUCTION

Puget Sound is a fjord-like estuary connecting through Admiralty Inlet and Deception Pass to the Strait of Juan de Fuca and then to the Pacific Ocean (Figure 1). About 98% of the tidal prism flows through Admiralty Inlet over a sill of about 64 m depth near Port Townsend. The main basin has depths exceeding 200 m and extends south about 60 km from the major junction with Admiralty Inlet near Possession Point to The Narrows, a constriction of about 44 m sill depth separating a southern basin (Figure 2). Within the main basin, the 183-m contour deliniates a deeper section approximately 50 km long and 3-5 km wide. Numerous rivers enter the Puget Sound system, but the Skagit entering in the north supplies more than 60% of the freshwater. Intrusions of new deep water have been shown by hydraulic model studies to occur when flood-tide ranges at Seattle exceed 3.5 m. At this range, Strait of Juan de Fuca water transits Admiralty Inlet sill in one flood period and thus is least mixed (Farmer and Rattray, 1963).

During 1975-76 a year-long mooring to record currents, temperature, and salinity at several levels was maintained midway along the main basin just off Seattle as an outgrowth of month-long observations made at the same location during the winters of 1972 and 1973 (Cannon and Laird, 1978). The addition of deep water during these observations, as in the earlier observations, showed winter intrusions at this mid location to occur 8 to 10 days following periods when the flooding tides will be greatest in Admiralty Inlet over the sill. These observations, however, indicated some major differences from the earlier winter-replacement observations. First there were occurrences of large tidal ranges during November-January

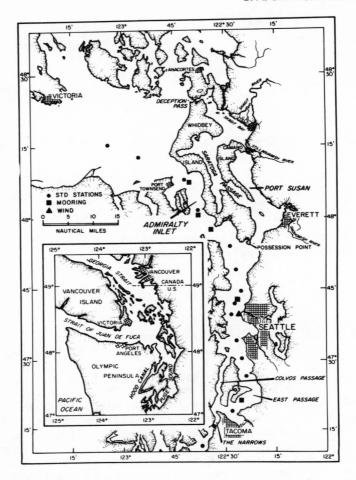

Figure 1. Puget Sound region showing mooring locations.

with almost no or very small inflow. Second, there were occurrences
of inflow during low tidal ranges during August-October and during
March-April. These latter two periods also were different in that
the summer-fall interval was when the most dense water could enter
the Sound, and the winter-spring interval was about when the least
dense water was in the Sound.

One apparent factor in determining when there will be intrusions
of deep water is the degree of mixing which takes place when crossing
the finite-length entrance sill. During summer-fall the water
entering Admiralty Inlet from seaward apparently is sufficiently salt
(dense) that it does not matter that more than one flood tide is
required to transport it across the sill, which also causes
additional mixing. At this time of year, there also is minimum

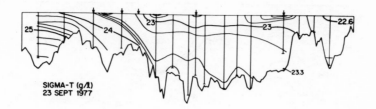

Figure 2. Sigma-t through Admiralty Inlet and along Puget Sound, north on the left. Arrows indicate moorings and vertical lines indicate STD stations shown in Figure 1. Bottom water inflow is indicated by the 23.4 contour.

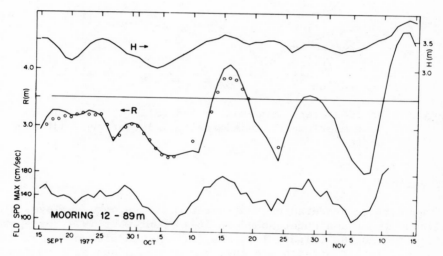

Figure 3. Range (R) and height (H) from tide tables of greatest daily Seattle flood tides (circles from pressure gauge); and maximum flood tide near bottom at the north end of Admiralty Inlet.

fresh water entering Admiralty Inlet in the surface waters from Puget Sound. The additional mixing still appears to result in new bottom water for the Sound. These flows of new bottom water, however, still occur at about fortnightly intervals. Thus, in late summer-early fall 1977 observations were made to better describe the flow characteristics over the Admiralty Inlet sill during bottom water replacement in Puget Sound. These observations also included the smaller tidal ranges noted from the year-long study.

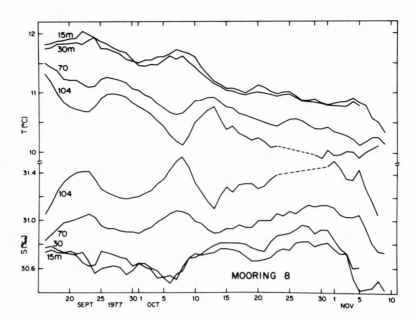

Figure 4. Daily average temperature and salinity from the southern
           most mooring (8) in Admiralty Inlet at the indicated
           depths.

OBSERVATIONS

     Subsurface Aanderaa current meter moorings were deployed along
Admiralty Inlet and Puget Sound on 14-15 September 1977 for approximately two months (Figure 1). The current measurements were made
at various levels at each mooring, but near-surface at 15 m and 5 m
above bottom were common at each location. Temperature and salinity
measurements were made twice, during low flood tidal ranges and
during higher tidal ranges when the flood tide range exceeded 3.5 m.
As it turned out, inflow occurred during both sets of STD's. The
low tidal range case was analogous to the 1975-76 year-long observations when inflow also occurred in late summer during low tidal
ranges (Figures 2 and 3).

     Details of the variations in water properties throughout the
approximate 2-month interval can be seen in daily averages from the
Aanderaa meters. Three occurrences of inflow at approximately fortnightly intervals represented by increasing salinity (density) and
decreasing temperature were observed in about the bottom 30 m of
water at the moorings in Admiralty Inlet (Figure 4). These changes
occurred during increases in bottom flow, which always was directed
into the estuary. There was a general decrease in temperature

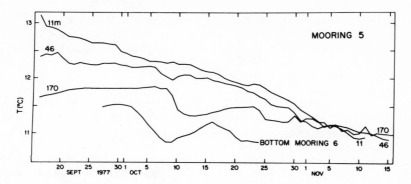

Figure 5. Daily average temperature from the mooring at the south
end of Puget Sound at the indicated depths and from bottom
at mooring (6) midway along the main basin off Seattle.

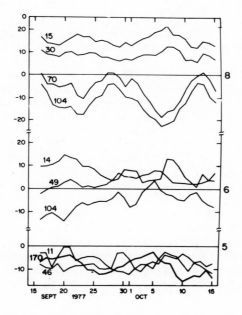

Figure 6. Daily average along-channel currents (cm/s) at the south
ends of Admiralty Inlet (8) and the main basin (5) at
the indicated depths (m).

during the two months, and salinity returned to about the same values
following the first two increases. The apparent steadily decreasing
salinity following the third increase may be the commencement of the
relatively large autumn decrease shown in the data from 1975-76.
Instruments in the top 50 m of water, however, indicated a reverse

effect, temperature increased and salinity decreased during the
increased bottom-water inflows, particularly during the first two.
The first set of STD observations was made at the end of the first
inflow (Figure 2). Hence, new bottom water was observed midway
along the main basin of Puget Sound.

Observations at the south end of Puget Sound near Tacoma also
showed a general decrease in temperature and fairly uniform salinity
except near surface (Figure 5). Temperature became uniform throughout the water column about the end of October. Of special significance were the step decreases in bottom-water temperatures. These
were identical to those observed farther north near Seattle during
winter 1973 and autumn 1975, and they are the first observations
of step decreases at any other location in the main basin. Although
the bottom meter failed at the mid mooring (6), a temperature sensor
on a pressure gauge was used to obtain daily average near-bottom
temperatures. About 7 days were required for the step decreases to
transit the approximately 45 km between the two moorings, implying
an average speed of about 8 cm/sec. The minimum temperatures
during these step decreases were $0.4-0.5^0$C warmer at the southern
end of the main basin than at the more northern mooring.

The records were low-pass filtered for periods greater than
30 hr, and then daily averages were computed for conceptual convenience (Figure 6). Of significance is the relatively large fortnightly variation in the currents in Admiralty Inlet. It is most
obvious at the south end of Admiralty Inlet (mooring 8) where the
upper two current meters showed increased outflow when the lower
two meters showed increased inflow. The increased inflow in the
deeper water corresponded directly with the intervals of increasing
salinity and decreasing temperatures. The maximum inflow occurred
simultaneously with the property extremes. At the southern end of
the main basin (mooring 5), flow at the bottom had a large fortnightly variation of about 5 cm/sec. The largest bottom flow
occurred on 10 October and was coincident with the arrival of the
temperature step decrease. The average speed at the bottom at
mooring 5 from 5-10 October was about 8 cm/sec. which was what was
required to transport the temperature front between the two moorings
in about 7 days.

REPLACEMENT PROCESSES

Still unanswered is when do some of the flows of new bottom
water into Puget Sound commence. The increased inflow commencing
about mid-October occurred about when the Puget Sound flood tide
range exceeded 3.5 m. However, the first two inflows were observed
during low tidal ranges. The first peaked about 23 September and
the second, about 8 October. These two inflows occurred during
minimum amplitudes in tidal currents at the bottom in Admiralty

AI 8-104m

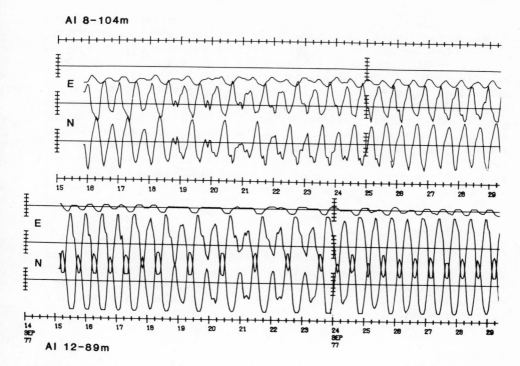

Figure 7.   Near-bottom east-west (upper) and north-south (lower)
             currents from moorings at the south (top) and north
             (bottom) ends of Admiralty Inlet. Scale ticks are
             10 cm/sec. (Ignore other graphs.)

Inlet (Figure 7). However, the duration of the flooding currents
appeared inversely related to their maximum amplitude. At the north
end on 20 and 29 September and on 7 October, the durations and
amplitudes were about 8½, 7, and 8½ hr, and 134, 158, and 93 cm/sec,
respectively. No increased bottom inflow occurred during 29
September. The durations and amplitudes at the south end of
Admiralty Inlet were about 10½ and 6 hr, and 100 and 90 cm/sec on
20 and 29 September, respectively. Finally, both duration and
maximum amplitude of the flooding current were large when the flood
tide range exceeded 3.5 m. On 16 October they were about 8 hr and
170 cm/sec at the north end.

     Another possibility speculated by Cannon and Laird (1978) was
that the water was sufficiently dense that it did not matter that
more than one tidal cycle was required to transit the sill causing
further mixing and reduction in density. The currents in Figure 7
can be used to show this. During neap tides on 22 September when
inflow occurred, the excursions for a successive flood-ebb-flood
were 26, -4, and 12 km at the north end and 14, -3, and 12 km at

the south end of Admiralty Inlet. The net excursions were 34 and 23 km, respectively. The relatively short ebb excursions provided little time for mixing, and the resultant salinity at the south end was considerably more (about 0.5%) than at the beginning of the first flood. During spring tides on 28 September, the excursions for a successive flood-ebb were 23 and -14 km at the north end and 11 and -9 km at the south end and were more nearly equal. More mixing could occur, and the resultant salinities at the south end were less than on 22 September. During the inflow around 7 October, the net excursions were again large and the resultant salinities higher. We still are not sure how to predict these inflows occurring during the flood tidal ranges less than 3.5 m.

ACKNOWLEDGEMENTS

This work was sponsored by the MESA Puget Sound Project of NOAA. Their support was greatly appreciated. Pat Laird was lost at sea off Hawaii in December 1978. He was a major contributor to this work and will be sorely missed. Contribution number 413 from PMEL/NOAA.

REFERENCES

Cannon, G.A. and N.P. Laird. 1978. Variability of currents and water properties from year-long observations in a fjord estuary. In: J.C.J. Nihoul (editor), Hydrodynamics of Estuaries and Fjords. Elsevier, Amsterdam, E. Oceanogr. Series, 23, pp. 515-535.

Farmer, H.G. and M. Rattray. 1963. A model of the steady-state salinity distribution in Puget Sound. Dept. of Oceanography, Univ. of Washington, Seattle, Tech. Rept. No. 85, 33 pp.

THE MEIOBENTHOS OF FJORDS: A REVIEW AND PROSPECTUS

Brian Michael Marcotte

Department of Biology
University of Victoria
Victoria, B.C. V8W 2Y2   Canada

INTRODUCTION

The marine meiobenthos is a diverse assemblage of small (0.5 – 0.1 mm) metazoans living on and in the bottom of the sea. It includes most invertebrate phyla, especially nematodes and copepod crustaceans. Although meiofauna are ubiquitous, abundant and productive members of benthic ecosystems, their ecology in deep-water fjords is little studied. The present report will review the ecology of meiofauna of fjords in the context of meiobenthic ecology in general. A prospectus for meiofauna research in fjords will complete this report.

REVIEW

Energy-trophic dynamics

The abundance of hard bodied meiofauna (taxa preserved with formalin fixation) in sundry marine habitats is given in Table I. In general, meiofauna are more abundant in eutrophic and shallow-water habitats (saltmarsh, Baltic Sea) than in oligotrophic seas (Mediterranean and Sargasso Seas) and deep-water environments. The abundance of meiofauna in fjords is known to this author in only two, unpublished reports. Harrison (Dept. of Oceanography, University of British Columbia) has found 400 – 1300 individuals/10 cm^2 (in a depth range 63 to 217 m) living in northern Howe Sound. Ellis and Ashwood-Smith (University of Victoria) report 292–1770 individuals/10 cm^2 in Trevor Channel at a depth of 175 m. Both of these sites are in British Columbia. These data indicate that meiofauna are about as abundant in fjords as in typical open shelf environments.

Table 1. The abundance of meiofauna in sundry environments.

	PLACE	ENVIRONMENT	MEIOFAUNA Ind – 10 cm^{-2}	REFERENCE
Estuarine	Georgia (USA)	Salt marsh	12,400	Teal and Wieser,1966
	Denmark	Sands	500 – 2000	Fenchel,1969
	France (Med. Sea)	35 m	150	Guille and Soyer,1968
	Baltic Sea	Shallow mud	2200	Elmgren,1976
		Deep mud	1000	Elmgren,1976
Non-estuarine	Bermuda (Sargasso)	3 – 35 m	560	Coull,1968
	North Sea	146 m	904 – 3363	McIntyre,1964
Slope	North Sea (Spitzbergen)	1250–1750 m	1170	Thiel,1972
Deep Sea	Atlantic (Josephine Bank)	1250–1750 m	470	Thiel,1972
	Atlantic (Iberian Deep Sea)	5250–5750	210	Thiel,1972
	South Africa (Walvis Ridge)	5170	312	Thiel,1972

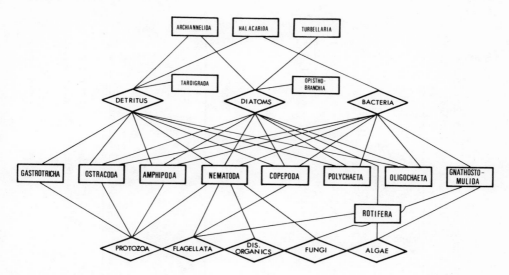

Figure 1.   Pathways for energy flow into the meiofauna. Energy
            proceeds from diamonds to rectangles in this diagram.
            Data upon which this figure is based were reviewed by
            Coull (1973).

     To appreciate the energetic role and relative importance of
meiofauna, abundances must be translated into biomass. Table II
compares the biomass (biomass in g 10 cm^{-2} and numbers in individual - 10 cm^{-2}) of meiofauna and macrofauna in a variety of environments of increasing water depth. Meiofauna outnumber macrofauna
in all environments. Macrofauna biomass is greater than that of
meiofauna in all shallow-water habitats. In deep water, however,
meiofauna biomass exceeds that of macrofauna. Elmgren (1976) has
shown this same trend in data confined to the Baltic Sea. In so
far as the Baltic is a large fjord, this trend might be expected
for fjords in general. The reason for this switch in biomass dominance to small animal sizes in deep (oligotrophic) waters is unclear (Thiel, 1975). (See also, below).

     Turnover and respiration rates are also essential to calculat-
ions of productivity. In Table III, estimates of the respiration
and turnover rates of meio- and macrofauna age given. Meiofauna
are energetically 3-5 times more active than an equivalent biomass
of macrofauna. Thus, whenever macro-/meiofauna biomass estimates
fall below 5:1, meiofauna can be expected to dominate the productivity of the metazoan benthos.

Table II. Comparison of meio- and macrofauna abundances and biomass.

PLACE	DEPTH	MEIOFAUNA		MACROFAUNA		MEIO-/MACRO-	
		NUMBERS	BIOMASS	NUMBERS	BIOMASS	NUMBERS	BIOMASS
Massachusetts (USA) (Wigley & McIntyre,1964)	366 m	127	–	1695	–	75:1	–
France (Mediterranean) (Guille & Soyer,1968)	15–91 m	38–354	0.00012– –.000751	1.38–7.08	0.00846– 0.06704	210:1– 1095:1	1:2– 1:12
Atlantic Ocean	250–750 m	–	0.017	–	0.03	–	1:2
(Island Faroer Rucken	750–1250 m	–	0.001	–	0.014	–	1:14
Josephine Bank)	1750–2250 m	–	0.005	–	0.023	–	1:4
(Thiel,1972)	2250–2750 m	–	0.011	–	0.008	–	1.5:1

Table III. Respiration and turnover rates of meio- and macrofauna.

TAXON	RESPIRATION RATE mm^3O_2 h^{-1} G^{-1} (20^0C)	GENERATIONS – $YEAR^{-1}$	TURNOVER RATE (YIELD) $Generation^{-1}$ $Year^{-1}$
MEIOFAUNA	200 – 2000	3	9–10
MACROFAUNA	Small: 200–500 Large: 10–100	1	2
MEIO/MACRO (Gerlach, 1970)	5:1	3:1	3:2 3–5:1

	TOTAL COMMUNITY RESPIRATION $m\ell$ O_2 m^{-2} h^{-1}*	BACTERIAL RESPIRATION	MEIOFAUNA & MICRO-FLORA-FAUNA RESPIRATION	MACROFAUNA RESPIRATION
Smith,1973	76.02	48.0%	39.1%	12.9%

*Average of one year

The energy relationships of meiofauna must include a study of paths through which energy flows into, within, and out of the meio-benthos. Figure 1 shows how energy enters the meiobenthos. There are three principle routes reported in the literature: bacteria, diatoms, and detritus (Coull, 1973). The term "detritus" usually designates unidentifiable particles in the guts of fixed individuals. The trophic significance of these particles is unknown. If "detritus" means only structural carbohydrates and ash derived from macrophytes and metaphytes, its energetic value may be low since meiofauna are not known to digest cellulose. If "detritus" means the microflora and fauna on these structural elements, then the trophic reality of this term is reduced to "bacteria", "diatoms" etc.

Energy circulates within the meiobenthos through predation (Figure 2). Important carnivores are copepods, nematodes and cnidarian polyps. This aspect of meiobenthic ecology is little studied.

Energy leaves the meiobenthos (Figure 3) either directly as food for higher trophic levels or indirectly through nutrient re-cycling. Nutrient recycling by meiofauna is very little studied and is not considered further. Meiofauna contribute to production of benthic macrofauna and natant nekton. That meiofauna are eaten by fish has been long known (*e.g.* Bruun, 1949). That meiofauna – especially harpacticoid copepods – are crucial for the juvenile ecology of Pacific salmon (*e.g.* pink and chum salmon: Kaczynski, Feller and Clayton, 1973) is a recent discovery. The effect of pre-dation on meiobenthic community structure (*e.g.* interactions among competing prey species) is unknown. In so far as salmon and other fish often occur in fjords, knowledge of the ecology of meiofauna may be very important.

Anoxia

A major oceanographic feature of many fjords are periods of low O_2 tension often occuring seasonally. Many macrofauna species are lost with anoxia (Leppakoski, 1969). Recolonization may be very slow. Many meiobenthic taxa are also lost with anoxia, but some taxa can survive – even thrive – in anoxic environments (Fenchel and Riedl, 1970). These taxa are generally soft bodied (Turbellaria, Gnathostomulida) and are lost with standard field fixation techniques. The energetic importance of these animals is unknown.

Meiofauna rapidly recolonize benthic habitats after anoxia has passed. Coull (1968) studied the recovery of meiofauna populations in Devil's Hole, Bermuda after a period of anoxia. He found that of the principle hard-bodied meiofauna, nematodes recovered first, then copepods.

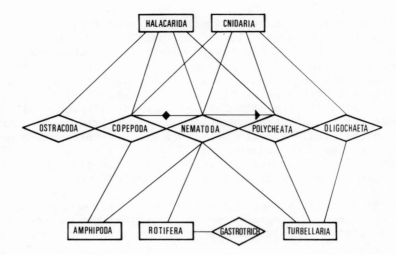

Figure 2.   Pathways for energy flow among meiofauna taxa.   Energy
            proceeds from diamonds to rectangles in this diagram.
            Data upon which this figure is based were reviewed by
            Coull (1973).

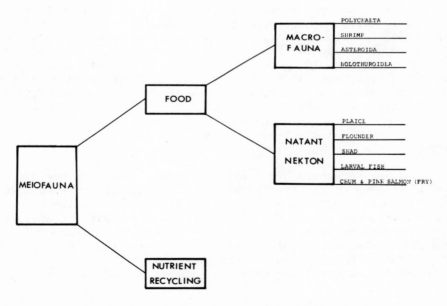

Figure 3.   Pathways of energy flow out of the meiobenthos.   Data upon
            which this figure is based were reviewed by Coull (1973)
            and in the present text.

PROSPECTUS

Allometric Continua

Many copepod crustaceans have been described from Nordic fjords (Sars 1903-1919). In Figure 4 some of these copepods have been re-drawn. The allometric transformations required to describe a natant and epiphytic copepod based on the morphometrics of a pelagic cop-epod are continuous. They involve a broadening of the prosome re-lative to the abdomen. These taxa inhabit environments with prop-erties which appear discontinuous to us. We recognize the sediment/sea water interface as a major discontinuity. The morphometrics of these copepod do not substantiate our perception. There *is* a discontinuity in the allemetry of copepods living inbenthically, however. The ecological significance of these continuous trans-formations has not been studied but may indicate that for copepods at least, *there is no bottom to a fjord, it just gets thicker.* Further evidence is needed to evaluate this hypothesis. Chemists, geologists and physicists are not copepods but they might seek in-sight into fjordic processes from this trend in the evolution of copepod morphology.

Further analysis of Figure 4 indicates that the genital somite lies at an allometric nexus. For non-inbenthic copepods, the dis-tance from this nexus to the end of the abdomen is always the same. This conservative feature may be due to conservative copulatory or egg carrying functions.

This allometric analysis leads to a further possible general-ization. Figure 5 lists possible combinations in Cephalothorax/Abdomen width and abdomen length. In this contingency table all copepod morphologies can be simply resolved. The evolutionary significance of this reduction is unclear but may provide important insight into copepod serial homologies and limits to morphological change.

Body Size and Environmental Frequency

Organisms adapt to at least 4 properties of environmental variables. 1) Mean: Organisms adapt to the mean value of variables *e.g.* temperature. Thus polar (cold) and tropical (hot) meiofauna can be characterized. 2) Variance: The value of some environmental properties varies widely seasonally or on a diel basis. Thus in temperate habitats, temperatures may vary from freezing in winter to $+30^0C$ in summer. In the tropics, temperature variance is less. 3) Predictability: Environmental variables in some habitats vary in a predictable manner; in other habitats (even different variables in the same habitat) do not. Survival "strategies" involving adapt-

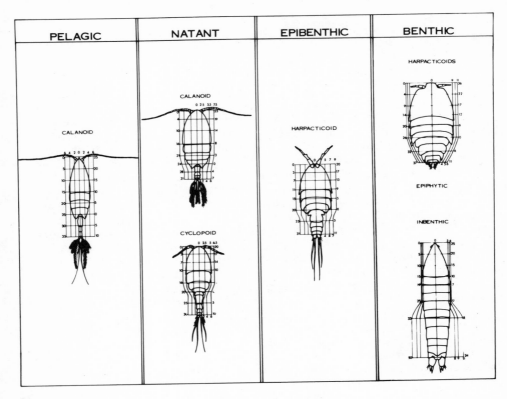

Figure 4.   Allometry of copepod taxa in fjords. Copepods redrawn
            at similar scale from Sars (1904-1919). Coordinates
            on the pelagic calanoid are orthogonal. All other co-
            ordinate systems are transformed by the morphologies
            illustrated.

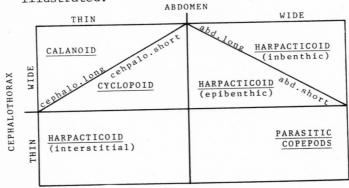

Figure 5.   Copepod taxa resolved into a contingency table illustrat-
            ing the relationship of cephalothorax and abdomen width
            and length.

ations imbedded in the genes of an organism are practical only in un-
predictable environments (Slobodkin and Sanders 1968). In impred-
ictable environments, non-genetic strategies (broad within pheno-
type niche breadth) may be the rule *e.g.* great aclimatization potential. 4) Frequency: Environmental properties have a frequency
of variation. Organisms may adapt to environmental frequencies by
altering their generation times, reproductive potential and body
size (Bohrer, unpublished manuscript). Fenchel (1974) has shown
that the generation times of animals is directly proportional to
their size. Their reproduction rate is inversely related to body
size. Bohrer has pointed out that similar relationships may exist
between the "generation time", "reproductive rate" and size of diffusion eddies in the ocean. He implies that animal body sizes and
its dependent variables, are adapted to the size/frequency of
physical events in the sea. If so, the dominance of small animal
sizes in oligotrophic seas and in the deep sea implies high frequency
events dominate the ecology of these environments. These events
may be physical or biological (competition, predation). If so,
a re-examination of the frequency of chemical, physical and geological events in deep water environments is appropriate. It
should be noted however, that animal size is an over-determined
variable. The disproportionate presence of small size animals could
also be due to the physics of energy transfer between trophic levels
in an oligotrophic sea (Odum, 1975).

In any event, it seems clear that, *meiofauna do not have rhythms,
they are rhythms.* Their body size/frequency may be adapted to the
rhythms, the frequencies of variations of physical and biological
factors in their environment.

Anaerobiosis and Metazoan Evolution

The physiology of anaerobic meiofauna is little studied. Some
meiofauna may survive short periods of anoxia by carrying O_2 bound
to respiratory pigments. These taxa would be lost during periods
of prolonged anoxia in a fjord.

Other meiobenthic taxa may exhibit true anaerobic metabolism.
For them O_2 may be a poison. For them respiratory pigments, if
present, would serve to keep O_2 away from metabolic pathways, not
transport it to them.

If respiratory pigments protecting metabolic pathways from O_2
can be shown to be the rule in ancestral meiobenthic forms, then
some meiobenthic taxa might have evolved in an ancient anoxic sea.
Thus some meiofauna might have evolved early in the history of life
on earth. Their simple morphologies and life histories are evidence
of antiquity not secondary specialization to a meiobenthic world.

If so, metazoan evolution might have progressed farther toward complex invertebrate taxa (pseudocoels etc.) well in advance of high
O_2 tension (The Pasteur Effect) at the start of the Cambrian. Such
a possibility would force a wholesale reconstruction of phylogenetic
trees.

Boaden (1975) has shown that for one group of flat worms, O_2
carrying pigments are primitively protective devices. The biology
of meiofauna in periodically anoxic fjords might provide an important opportunity to further test these hypotheses.

SUMMARY

Meiofauna in deep-water fjords are little studied. Available
data indicate that 1) meiofauna are abundant, 2) meiofauna biomass
can exceed that of macrofauna especially in deep water, 3) meiobenthic productivity is an important link in fjordic food chains,
4) meiobenthic taxa can survive periods of low O_2 tension, and
5) there may be a continuum of morphometric change in copepod crustaceans from the plankton to the benthos of fjords. The following
speculative points are discussed:

1. The community dynamics of the plankton and meiobenthos may be
    strongly linked.
2. The frequency of environmental change is as significant as its
    mean value and variance.
3. Respiratory pigments can protect some anaerobic meiofauna from
    O_2 poisoning, *i.e. meiofauna 'breathe deep the gathering gloom'*.

It is clear from this review and prospectus that research on
the ecology of meiofauna in fjords is woefully inadequate at present. Important issues may be resolved by the study of these communities.

REFERENCES

Boaden, P. J. S. 1975: Anaerobiosis, meiofauna and early metazoan
    evolution. Zool Scripta. 4(1), 21-24.

Bruun, A. 1949: The use of nematodes as food for larval fish. J.
    du Conseil. 16, 96-99.

Coull, B. C. 1968: Shallow water meiobenthos of the Bermuda plat-
    form. Ph.D. Thesis, Lehigh University. 189 pp.

Coull, B. C. 1973: Estuarine meiofauna: a review: trophic relat-
    ionships and microbial interactions. In Stevenson, L. H. and
    R. R. Colwell (Ed.) Estuarine Microbial Ecology. Univ. South
    Carolina Press: 499-512.

Elmgren, R. 1976: Structure and dynamics of Baltic benthos communities, with particular reference to the relationship between macro- and meiofauna. 5th Symposium of the Baltic Marine Biologists, Kiel. 43 pp.

Fenchel, T. 1969: Ecology of the Marine microbenthos I. The
quantitative importance of ciliates as compared with metazoans
in various types of sediments. Ophelia 4, 121-137.

Fenchel, T. 1974: Intrinsic rate of natural increase: the relationship with body size. Oecologia 14, 317-326.

Fenchel, T. and R. J. Riedl 1970: The sulfide system: a new biotic community underneath the oxidized layer of marine sand
bottoms. Mar. Biol. 7(3), 255-268.

Gerlach, S. A. 1971: On the importance of marine meiofauna for
benthos communities. Oecologia 6, 176-190.

Guille, A. and J. Soyer 1968: La fauna benthique des substrates
meubles de Banyuls-sur-Mer. Premiers données qualitatives et
quantitatives. Vie Milieu 19, 323-360.

Kaczynski, V. W., R. J. Feller and J. Clayton 1973: Trophic
analysis of juvenile pink and chum salmon (Oncorhynchus gor-
buscha and O. keta) in Puget Sound. J. Fish. Res. Board.
Can. 30, 1003-1008.

Leppäkoski, E. 1969: Transitory return of the benthic fauna of
the Bornholm Basin, after extermination by oxygen insufficiency.
Cah. Biol. Mar. 10, 163-172.

McIntyre, A. D. 1968: Ecology of marine meiobenthos. Biol. Rev.
44, 245-290.

Odum, H. T. 1974: Combining energy laws and corollaries of the
maximum power principle with visual systems mathematics.
In. Levin, S. A. (Ed.) Ecosystem Analysis and Prediction.
SIAM-SIMS Alta, Utah. 239-263.

Sars, G. O. 1903-1919: An Account of the Crustacea of Norway.
Bergen.

Slobodkin, L. B. and H. L. Sanders 1969: On the contribution of
environmental predictability to species diversity. In Wood-
well, G. M. and H. H. Smith (Ed.) Diversity and Stability in
Ecological Systems. Brookhaven Nat. Lab., 82-95.

Smith, K. L. 1973: Respiration of a sublittoral Community. Ecology
54(5), 1065-1075.

Teal, J. and W. Wieser 1966: The distribution and ecology of nem-
    atodes in a Georgia salt marsh. Limnol. Oceanogr. 11, 217-222.

Thiel, H. 1972: Die Beduntung der Meiofauna in Küstenfernen ben-
    thischen Lebensgemeinschaften verschiedener geographischer
    Regionen. Verhand. Deutsch. Zool. Ges. 65, 37-42.

Thiel, H. 1975: The size structure of the deep sea benthos. Int.
    Revue ges. Hydrobiol. 60(5), 574-606.

Wigley, R. L. and A. D. McIntyre 1964: Some quantitative comparisons
    of offshore meiobenthos and macrobenthos south of Martha's
    Vineyard. Limnol. Oceanogr. 9, 485-493.

MACROBENTHOS OF FJORDS

T. H. Pearson

Scottish Marine Biological Association
P.O. Box No. 3
Oban, Argyll, Scotland

INTRODUCTION

The historical development of studies on macrobenthic community
ecology has been closely related to surveys of fjordic populations
stemming from the classical work of Petersen and his successors in
Danish areas. This relationship is not fortuitous since fjordic
populations are rich, accessible and economically important as the
basis of extensive inshore fisheries. Thus the original surveys
of Petersen and Boysen Jensen (1911) of the Limfjord constitute not
only the first truly quantitive surveys of benthic macrobiology but
provided the initial information on which Petersen was to base his
concept of benthic animal communities, a concept which was to provide
the stimulus and framework for fifty years of development in marine
benthic ecology. The original surveys in the shallow and highly
productive Limfjord complex were complemented by extensive surveys
in the Kattegat and Belt areas and by a survey of the Christianafjord
(Oslofjord) (Petersen, 1913, 1915 a b, 1918) which was to provide
the first quantitative analysis of macrofaunal distributions throughout a deep fjordic basin. Thus studies carried out in both shallow
and deep fjordic basins played a formative part in the development
of Petersen's ideas on benthic ecology. The environmental contrast
between the Limfjord and the Oslofjord is large but, for the purposes
of the following review and discussion, both are considered to be
fjords in more than their common nordic nomenclature. Both are enclosed marine areas whose connection with the open sea is restricted
by sills or shallow areas. These two requirements, enclosure and
restriction of water exchange, encompass the essential physical
restraints which are the framework for the biological characteristics
of fjords, as will become apparent from the ensuing discussion.

SUMMARY REVIEW OF INVESTIGATIONS OF FJORDIC BENTHIC BIOLOGY

As a part of this review a comprehensive list of studies of the macrobenthos in fjords has been compiled, this is included with text references in the bibliography following this paper. Geographical analysis of the published fjord macrobenthos studies shows that only three surveys have been carried out in the southern hemisphere fjords compared with 74 in the northern. Although this disproportion does reflect the fact that there are many more fjord complexes in arctic and northern boreal areas than in the south, it is also symptomatic of the dearth of benthic studies in the southern hemisphere. The only information on macrobenthos from the Chile/Patagonian fjord complexes is contained in the reports on the Lund University Chilean Expedition 1948-49 where general dredge samples were made of benthic fauna from a variety of stations throughout the Chilean coast, including a number of areas in southern fjord complexes (see Brattstrom & Dahl (1951) for a general account). Taxonomic studies of many different groups from these collections are available, but no general ecological studies have been published. Information from the fjords of the southern island of New Zealand is confined to two papers by Hurley (1964) and McKnight & Estcourt (1978) who report on the distribution of macrobenthic communities in relation to major physical variables in Milford, Caswell and Nancy Sounds. No actual details of abundance , species numbers or biomass figures are reported, however.

The published work from Japan concerns embayments rather than true fjords but, since circulation is restricted in these areas by sills, the benthic conditions are similar to those found in boreal fjords. Thus the work is included for comparison. Original surveys by Miyadi (1941a, b, c, d) in many such areas have been complemented recently by Kawabe (1975) in Ago Bay, Kikuchi and Tanaka (1978) and in Kobe Bay with studies of faunal distributions in relation to deoxygenation.

The macrobenthos of fjords in the Strait of Georgia, Puget Sound and adjacent areas in British Columbia and Washington was first surveyed by Shelford *et al* (1925, 1935) and recently by a number of workers most notably Lie and Ellis. A brief summary of work in the area with an overall assessment of community distribution and standing crop in this area is given by Ellis (1971). The only published study of more northern Pacific fjords is of Port Valdez, in Prince William Sound, S. Alaska, where extensive quantitative surveys have been carried out prompted by the establishment of an oil terminal in the area Feder *et al* (1973). Benthic fjord studies in the eastern Canadian Arctic (see Ellis, 1959) include comparisons of species composition and biomass of the macrofauna of both high and low arctic fjords in Ellesmere Island and Baffin Island. The most extensive general surveys of boreal/arctic and arctic fjords have been carried out by Thorson and Spark in the fjord complexes of East Greenland.

The utility of these studies has been greatly enhanced by the expert attention given to the polychaete and lammellibranchiate Mollusca in the collections made in these areas by Ocklemann and Wesenberg-Lund (1953) who provide detailed comparison of faunal distributions of these groups in the East Greenland fjords with distributions in other high arctic, arctic and boreal areas. These studies together provide the most comprehensive faunal lists available for any fjordic area. In the northern European area the most intensively studied fjord has been the Gullmarfjord in western Sweden which was initially sampled by Petersen (1914) as part of his comprehensive surveys in the marine areas bordering the Danish coasts. Later comprehensive surveys by Molander and Gilsen in 1928-30 were complemented by Lindroth (1935) and more recently by Leppakoski, Bagge and most notably Rosenberg who have carried out extensive analyses of benthic macrofaunal changes in relation to the onset and abatement of organic pollution from the pulp and paper industry in the area. Molander's sampling surveys in 1927-28 were conducted in a number of Swedish west coast fjords, the results of which in the form of detailed analyses of comparative community structure in the various areas, were finally published in 1962. Rosenberg and Moller (1979) have recently complemented Molander's surveys by detailed studies in the same areas in addition to some other west coast bays and fjords in which faunal distributions are related to salinity boundaries and changes over fifty years separating the two sets of surveys are described. A similarly extensive series of surveys in bays and fjords on the Baltic coast of Sweden by Landner *et al* (1977) relates faunal distributions primarily to the incidence of pollutants in the various areas surveyed. These surveys together make the Swedish coast the most extensively surveyed fjordic area in so far as benthic macro-fauna is concerned. By comparison, the much more extensive fjordic coast of Norway has been relatively neglected in benthic population studies with the exception of the Oslofjord (Christianafjord) where Petersen's original quantitative sampling in 1915 has been followed up by Brock, Christiansen, Beyer and very recently by Mirza (pers. comm.) although only in the latter study has quantitative sampling of the infauna comparable with Petersen's original data been under-taken. In recent years comprehensive and extensive studies have been initiated in the Lindespollene, a small shallow fjord near Bergen {see Taasen & Evans (1977) for descriptive account} and further north in the Trondheim fjord {see Gulliksen (1977)} which both include extensive quantitative sampling of the benthic macro-fauna. Surveys of fjordic benthos in the British Isles has been confined to only a few areas. In the south of Ireland extensive studies of littoral and shallow water communities over a number of years (see Kitching and Ebling 1957 for summary) have recently been extended to include studies of changes in deep water benthic pop-ulations in response to seasonal deoxygenation (Kitching *et al.*, 1976). On the Scottish west coast detailed studies of the benthos of Lochs Etive and Creran have been published by Gage (1972, 1974) and long term changes in the benthic populations of Lochs Linnhe and Eil have been described by Pearson (1970, 1971, 1975) and related in particular to organic enrichment (Pearson and Rosenberg, 1978). Detailed studies of energy flow in sediments and benthic production have been carried out in Loch Ewe and Loch Thurnaig by, among others, McIntyre *et al*. (1970) Trevallion *et al*. (1970) and Davis (1978).

Fjords in the Shetland Isles to the north of the Scottish mainland have been extensively surveyed recently following large scale oil-related industrial development in the area. The benthic populations of the area are described in detail by Pearson & Stanley (1977), Pearson & Eleftheriou (in press) and Addy (in press).

The benthos of the more northerly islands of the North Atlantic has been the subject of investigations by, principally, Spark in the 1930's who briefly reported the distribution of major benthic communities from fjords and inlets in the Faroes and Iceland. Brotzky (1930) described the benthos of a semi-fjordic area in Spitzbergen.

The majority of the studies commented on above have been made in what may be termed true fjordic areas *i.e*. elongated deep glaciated basins connected to the open sea by narrow, shallow entrances. However, it has been noted above that the Danish fjords in which much of the formative work in benthic ecology was undertaken are relatively shallow and have many lagoon like characteristics. The early work by Petersen, Blegvad and co-workers in the Danish fjords has more recently been intensively added to by Muus (1957) and Rasmussen (1973). Muus has provided a comparative study of the ecology of the shallow areas of the Danish fjords which is an outstanding attempt at a comprehensive review of the stability, productivity and interrelationships of the major faunal elements of these areas. A similarly detailed account of the general biology of a single fjord, the Isefjord, is given by Rasmussen, who collects together and analyses data on faunal change in the fjord over a period of thirty years.

ENVIRONMENTAL FACTORS AFFECTING THE DISTRIBUTION AND ABUNDANCE OF FJORDIC MACROBENTHOS

What conditions, if any, affect the macrobenthic fauna of fjords in a unique manner or take on a greater emphasis in fjordic areas when compared with shelf or oceanic benthos? A brief consideration of this question will lead to an analysis of the effect of those factors considered to have a significant effect on the distributions and abundance of fjordic benthos based on published information.

Fjords are, by definition, enclosed marine basins with a restricted and shallow connection to the open sea. Thus the two factors
of over-riding significance on their sedimentary biology when compared with open sea areas is their proximity to land and the restriction of water exchange and circulation within the basin. The
former affects, in particular, the input of organic and inorganic
nutrients to the sediments, the sedimentary composition and the
temperature and salinity regimes. The latter also affects temperature
and salinity and may bring about either permanent or intermittent
stagnation in the bottom waters resulting in oxygen deficiencies
and the accumulation of ammonia and sulphides in the sediments and
bottom waters. In addition biological interactions in the macrofauna may be enhanced by enclosure effects, either by concentrating
predators or by restricting the free exchange of planktonic larvae
within the basin. Fig. 1. presents the interrelationship of these
various factors diagramatically. For convenience and simplicity
the impact of each factor on the benthos will be discussed in turn
before attempting a synthesis of their net effect on benthic productivity and fjordic ecology in general. It is important to bear
in mind, however, that all the factors are interdependent, and
changes attributed to any one particular environmental effect may
more properly be considered as the resultant of a complex of inter-
related factors.

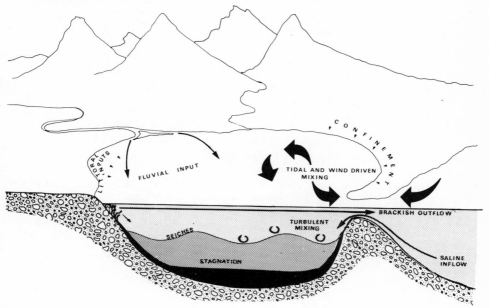

Figure 1.   Schematic representation of the various environmental
            factors affecting the distribution and abundance of
            fjordic macrobenthos.

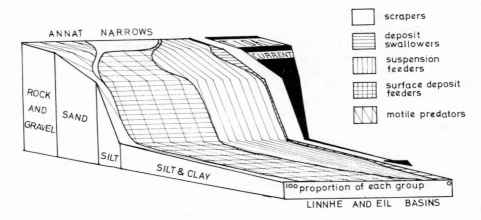

Figure 2.   Distribution of various trophic groups in the benthos
           of Loch Eil, Scotland, in relation to depth, sediment
           type and current speed (from Pearson, 1971).

CURRENT SPEED AND SEDIMENTARY COMPOSITION

     The majority of authors who have reported on the distribution
of benthic species from a particular area have related their occurrence to the grain size of the sediments in which they occur. Thus
early workers such as Petersen (1918), Molander (1928) and Spark
(1929, 1933) accounted for the occurrence of the various communities
they described by relating them primarily to the distribution of
particular types of sediment *e.g.* Petersen associated the *Venus*
community with 'pure sand' and the *Brissopsis-chiajei* community with
'pure clay', while Molander combined sedimentary composition with
salinity and temperature variations to define the limits of the
major faunal associations described in the Gullmarfjord. Such obvious environmental/faunal correspondances have formed the basis
of most subsequent descriptive benthic faunal survey work both within and without fjordic areas. Within a fjordic basin, however, the
distribution of sedimentary particles and the close relationship
of this factor to current speeds and water movement within the basin
assume a special significance. In areas of high tidal amplitude the
flows into and out of such basins create significant mixing of the
vertical water column within the basin and cause bottom currents of
considerable velocity in the entrance and sill areas (*e.g.* Edwards
et al., in press). Such currents have a considerable influence on
the composition and distribution of the benthic fauna. Pearson (1970,
1971) showed that the distribution of different trophic groups in
the benthos was related to the relative current speeds in various
parts of the basin of Loch Eil (Fig. 2). Similarly Gage (1972)
showed that the ratio of suspension feeders to deposit feeders

diminished as tidal currents decreased with increasing depth of the
basins and distance from the sill areas in two Scottish sea lochs.
Highly specialised and productive faunas dominated by large filter
feeding and herbivorous epifaunistic species are particularly
associated with the shallow sill areas of fjordic systems with considerable tidal exchange. Detailed ecological studies of such populations have been carried out over a prolonged period in Lough Ine
on the south-west coast of Ireland {see Kitching and Ebling (1967)
for a detailed summary of the range of studies undertaken on single
species and various communities in the area}. Erwin (1977) has
recently reported on the extent of similar communities in the tidal
narrows and shallow areas of Stratford Lough in Northern Ireland,
and Lande and Gulliksen (1973) have described the benthic fauna of
the tidal rapids of the Borgenfjord in Norway. Lateral circulation
patterns within a basin may have a significant influence on the
distribution of benthos. Miyadi (1941) suggested that the lower
biomass and abundance of macrobenthos and the absence of certain
species on the east side of Ise Bay in Southern Japan could be
attributed to the inflow of saline oceanic water on the western
side of the bay and the compensatory outflow of brackish water on
the east. This type of lateral circulation is not common in deep
fjordic situations however.

VERTICAL CIRCULATION AND NUTRIENT SUPPLY

     In deep basins the pattern of circulation within the basin has
an important bearing on the distribution and productivity of the
macrobenthos. An interesting comparative example of this was provided by MacIntyre (1964) who compared the fauna found on the Fladen
ground in the North Sea with that of an area of similar sediment
in Loch Nevis, a typical Scottish west coast fjord. Faunal composition on the two grounds was similar but the standing crop of
macrobenthos on the Nevis ground was some five times larger than
that in the North Sea area. This was attributed to the greater
exchange between the bottom and surface waters during the summer
in Nevis, although winter nitrate levels in the bottom waters of
both areas were similar, leading to a more rapid circulation of
nutrients and hence to a greater primary production (Fig. 3) which
could favour a richer bottom fauna. Conversely, the poor vertical
mixing on Fladen during the summer restricted the nutrient supply
to the euphotic zone thus reducing primary production. Davis (1975)
presents detailed evidence of the importance of sedimentary nutrient
recycling in maintaining production in both the plankton and benthos
of another Scottish sea-loch. In Loch Thurnaig he showed that
ammonia release from the sediment was sufficient to provide most of
the inorganic nitrogen required for phytoplanktonic production, and
that such release was significantly increased by wind-driven turbulence in the loch (Fig. 4). Rasmussen (1973) suggested that the
high productivity of the Isefjord in Denmark {detailed by Nielsen

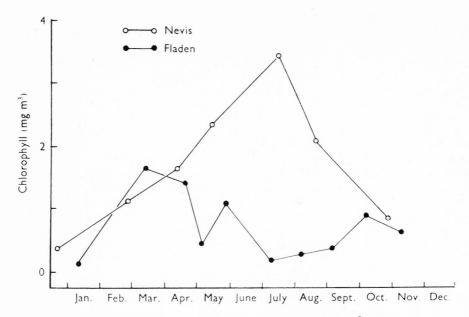

Figure 3.   Annual fluctuation in chlorophyll (mg/m³) in Fladen
            (North Sea) and Loch Nevis at 10 m depth (from McIntyre,
            1961).

(1951)} could be attributed to the continuous supply of nutrients
from the bottom muds brought to the surface layers by wind-driven
mixing of the water column. Nielsen (1958) demonstrated the occurrence of enhanced phytoplankton productivity based on nutrient
regeneration from the bottom waters of the Godthaab Fjord in W.
Greenland brought about, he suggested, by mixing engendered by strong
tidal currents inside the fjord. In contrast, the offshore area
outside the fjord showed vertical stratification of the water column
and low productivity. (Fig. 5). The exact reverse of this effect
may occur in deeper fjord basins where temperature or salinity discontinuities bring about stratification of the water column resulting
in the isolation of the bottom water from the euphotic zone during
the productive summer period. Such conditions are commonly found
in deep arctic and subarctic fjords where the presence of brackish
ice-melt water at the surface generally creates stable density
gradients in the enclosed basins, inhibiting vertical circulation
of nutrients. Thorson (1934) suggests that these circumstances
may be partially responsible for the low biomass of benthos in the
deep basins of the east Greenland fjord complexes. Similarly Curtis
(1977) showed that productivity of polychaetes in the Disko area
of east Greenland was over 100 times greater on the exposed coastal
areas subject to wave induced currents than in the vertically strat-

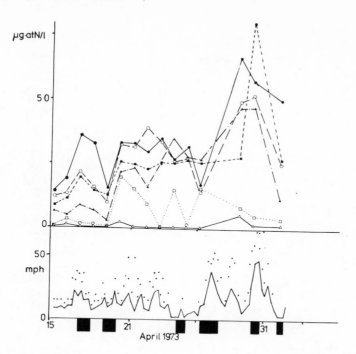

Figure 4. Nitrate concentrations in the water column of Loch
Thurnaig compared with wind speed, and periods of heavy
swell. Nitrate concentrations at 25 m ●—●, 20 m○—○,
15 m ■—■, 10 m ▲—▲, 5 m □—□, 1 m △—△. Periods of heavy
swell marked by solid bars. Average wind speed mph
(solid line) and strong gusts (dots) are shown in lower
graph, (from Davies, 1975).

ified waters of the sheltered fjord areas. A further factor inhibiting vertical circulation in the basins of arctic fjords is
the presence of shore bound sea-ice cover throughout the larger
part of the year. This significantly reduces the wind energy input
to these basins thus reducing mixing in the water column (Laker &
Walker, 1976).

GLACIAL AND MELT-WATER NUTRIENT INPUTS

    Spark (1933) and Thorson (1934) in their descriptions of the
benthic communities of the east Greenland fjord complexes suggested
that the low biomass figures recorded from the deep inner basins
of these areas might be attributable to their isolation from the
warmer and more productive water of the outer coastal areas. These
inner basins are fed by glacial melt-water however and this is a

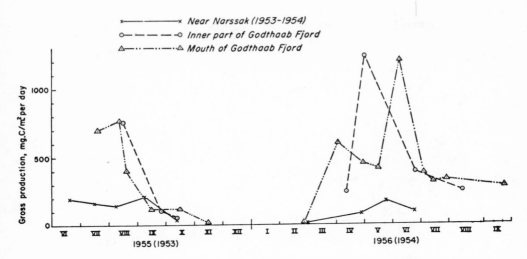

Figure 5.   Gross planktonic production in three areas of W. Green-
            land in 1955-56. X——X Offshore area, Δ——Δ Mouth of
            fjord, 0----0 inner fjord (from Nielsen, 1957).

probable source of nutrient input to the fjordic system.   Thorson
himself noted the presence of a rich and varied benthic fauna some
100 m from the mouth of rivers discharging to the Isfjord, part of
the Franz Joseph fjord system of east Greenland.   Stott (1936) and
Hartley and Fisher (1936) recorded the presence of rich concentrations of zooplankton species in the vicinity of melt-water inputs
from glacier faces in the Isfjord of Spitzbergen.   These attracted
large flocks of feeding seabirds which, in turn, increased the
nutrient levels of the area with their droppings.   Here also the
bottom fauna in the vicinity of the melt-water inflows was found
to be relatively rich.   Apollonio (1973) compared nutrient levels
in a glaciated and an unglaciated fjord in Ellesmere Island and
showed there to be significantly higher levels of silicate and
nitrate in the glaciated fjord.   Phosphate levels, however, were
similar in both areas.   He concluded that this supported the contention that nutrient levels were enhanced by glacial inputs since
the surrounding rocks contained soluble nitrate salts but only insoluble phosphates.   One further factor which may increase the
productivity of arctic fjords is the disturbance of bottom sediments
by the ploughing effect of large icebergs which ground on the bottom
sediments.   Vibe (1939 and 1950) noted that the bottom sediments
of sounds and fjords in the Thule and Upernavik areas of west
Greenland were frequently disturbed by ploughing icebergs and suggested that this might be a factor causing the recycling of nutrients
to the water column.   Golikov and Averincev (1977) also noted the
considerable sedimentary disturbance caused by icebergs in various
parts of the arctic ocean shelf.   Despite such possible sources of

nutrient enhancement and recirculation evidence from population studies of the benthos of arctic fjords conclusively suggests that they are considerably less productive than equivalent fjords in boreal areas (see below).

Mention should also be made of the possibility of nutrient enrichment from the presence of sea bird colonies on the edges of certain fjordic areas. However, Vibe (1950) noted that in the Thule district of north-west Greenland there were no sea-bird cliffs in the interior of fjords where icing conditions persisted well into the breeding season. The large Little Auk and Guillemot cliffs in the area were all situated near the outer coasts.

EUTROPHICATION AND STAGNATION

The phenomenon of eutrophication as it is known to occur in fresh water basins may and does occur in fjordic basins also, but in the marine situation the lacustrine sequence of high nutrient input leading to superabundant phytoplankton growth followed by excessive deposition of carbon at the sediment surface may be modified by a number of other effects. Indeed the final result of the eutrophic process, the development of an anoxic regime in the bottom waters and sediments trapping large amounts of carbon and nutrients in the lower levels of a stratified and stagnant water column, may be brought about either totally or in part by other events than those involved in the usual lacustrine eutrophic sequence. The existence of water exchange across the sill/basin entrance in marine basins and thus nutrient loss to the system and the input of hydrodynamic energy via tides reduces the chances of both superabundant planktonic growth and stagnation occurring. However, anaerobic conditions may be brought about if stagnation occurs due to low hydrodynamic energy input levels or if carbon and/or nutrient inputs are increased to a sufficiently high level. The latter two factors may create anaerobic sedimentary conditions even beneath a fully aerated water column should the biological demand (B.O.D.) of the sediments reach a sufficiently high level. There is a complex relationship between the factors affecting the hydrodynamic energy inputs to a fjordic basin e.g. tidal range, sill depth, wind induced turbulence and fresh water input and the resultant oxygen, temperature and salinity regimes of the bottom waters of the basin. An outline scheme showing the major relationships between these various factors is given in Fig. 6. In so far as effects on the benthos of the basin are concerned, the interplay of hydrographic factors result in three generalised bottom water states: i) permanently oxygenated bottom water ii) intermittently anoxic bottom water, iii) permanently anoxic bottom water. Examples of all these various states have been described from fjords although not often in great detail. Permanently oxygenated bottom waters are found in the majority of Scottish (e.g. MacIntyre, 1961, Pearson,

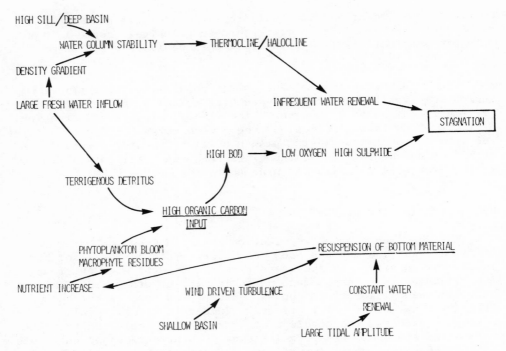

Figure 6.   Inter-relationships between hydrodynamic energy inputs
             to fjordic basins and the physical and chemical charact-
             eristics of the bottom waters.

1970, Gage, 1972) Icelandic (*e.g.* Spark, 1937, Einarsson, 1941)
and Greenland fjords (*e.g.* Thorson, 1933, 1934, Spark, 1933, Vibe,
1939) and in most British Columbian fjords (Ellis, 1971). Under
these conditions the benthic communities are stable in relation to
a seasonal time scale and are dominated by relatively large long-
lived mollusc and echinoderm species. Recruitment to such communities
is dependent on the availability of space and the predator/prey
balance during the post-larval settlement of potential community
members (Thorson, 1966, McCall, 1977) *i.e.* the communities are biologically accommodated or controlled (Sanders 1968) and the only
way in which they may differ in the fjordic situation from those
on similar grounds in open waters is through the agency of biological interactions peculiar to enclosed marine areas (see below).

     The benthos of fjords having intermittently anoxic bottom
water, being subjected to rapidly changing physical circumstances,
fluctuates between physically and biologically controlled communities
and thus covers a wide range of population and species variations.

Benthic population studies in fjords having naturally occurring
intermittently anoxic bottom waters have been described recently
by Kitching *et al*. (1977) from Lough Ine in Southern Ireland and by
Pearson and Stanley (1977), and Pearson and Eleftheriou (in press)
from Sullom Voe in Shetland. In these cases the establishment of
a summer thermocline results in the progressive deoxygenation of
the deep bottom water of the fjord leading to the eventual elimination of the benthic macrofauna beneath the thermocline. In the
periods of low oxygen levels prior to complete anoxia and following
the turnover of the water column in autumn when reoxygenation takes
place, the benthos is dominated by small rapidly breeding opportunistic species (for definition of opportunist see Grassle and
Grassle 1974, McCall, 1977) generally belonging to the Spionid and
Capitellid polychaetes. As conditions improve these species are
successively replaced by slower growing, longer lived species.
Macrobenthic successions associated with oxygen depletion and
organic enrichment in such areas have been reviewed by Pearson and
Rosenberg (1978) who summarised the progressive change in communities
pictorally (Fig. 7). In the absence of further deoxygenation the
benthic communities will eventually stabilise at a structure closely
approximating that described from similar areas not influenced by
oxygen depletion *i.e.* the communities will be structured by large,
relatively long lived species and have a high diversity. However,
if oxygen depletion is a seasonal phenomenon as described in Lough
Ine by Kitching *et al*. (loc. cit.), then the communities in those
areas subjected to depletion remain in the early stages of succession and are composed only of opportunistic species and post-larval
or juvenile stages of the later members of the succession, which
develop in the inter-depletion period and are eliminated with the
onset of anoxia which, in the case of Lough Ine occurs beneath a
summer thermocline (Fig. 8). In some fjordic basins stagnation
may be an aperiodic phenomenon brought about by the concurrence of
irregular hydrographic factors. This is the situation obtaining
in the inner basin of Sullom Voe in the Shetland Isles described
by Pearson and Eleftheriou (loc. cit.). Under these circumstances
benthic recolonisation and succession following an anoxic period
may proceed through several seasons (Fig. 9 a & b) and may even
approach final stability ('climax' in classical terminology) prior
to the onset of a further period of deoxygenation. Recolonisation
of the deeper anoxic areas of such basins is facilitated by the
existence of large populations on the sides of the basins above the
anoxic areas (Fig. 10 a & b). Although no other studies of aperiodic
intermittent anoxia appear to have been reported in detail Hurley
(1964) suggests that the fauna of the inner basin in certain southern
fjords in New Zealand is similarly controlled by intermittent anoxia
in the bottom waters and Dybern (1972) notes the recolonisation of
some deeper areas of the Idefjord in northern Sweden following the
brief reoxygenation of a normally anoxic area. In such areas a
knowledge of the hydrographic history of the basin is necessary in

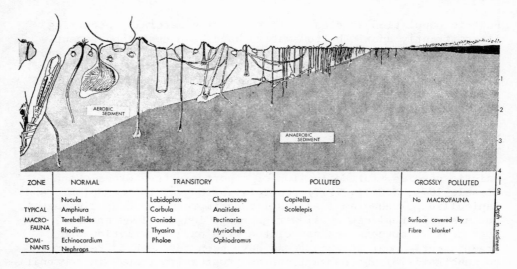

ZONE	NORMAL	TRANSITORY		POLLUTED	GROSSLY POLLUTED
TYPICAL MACRO-FAUNA DOMI-NANTS	Nucula Amphiura Terebellides Rhodine Echinocardium Nephrops	Labidoplax Corbula Goniada Thyasira Pholoe	Chaetozone Anaitides Pectinaria Myriochele Ophiodromus	Capitella Scolelepis	No MACROFAUNA Surface covered by Fibre `blanket`

Figure 7. Pictorial representation of the macro-benthic successions associated with oxygen depletion and organic enrichment (from Pearson and Rosenberg, 1978).

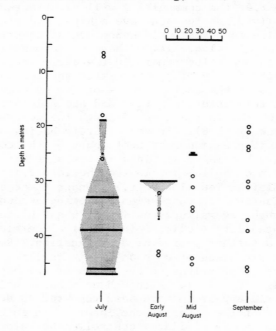

Figure 8. Progressive elimination of the small spionid worm *Pseudopolydora pulchra* from below upwards in the Western Trough of Lough Ine, Ireland in response to the establishment of anoxia beneath a summer thermocline, (from Kitching *et al.*, 1976).

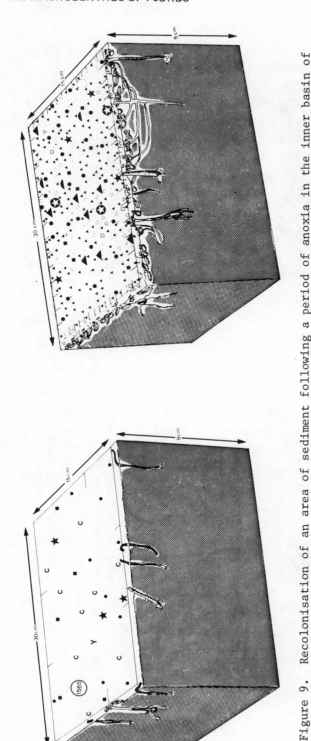

Figure 9.   Recolonisation of an area of sediment following a period of anoxia in the inner basin of Sullom Voe, Shetland. (a) Box diagram illustrating the density and distribution of animals in 0.005 m² of sediment immediately following a period of deoxygenation. Macrofaunal species are represented on the surface by symbols and on the sides of the diagram by scale drawings showing their relative positions in the sediment. Oxidised sediment is represented by light stippling, anoxic sediment by heavy stippling, (from Pearson and Eleftheriou, 1980). Key to the symbols shown in these two diagrams and in Fig. 11.

● *Notomastus latericeus*   ■ *Scalibregma inflatum*   c *Capitella capitata*   Y *Spio filicornis*
★ *Pectenaria koreni*   ▲ *Abra alba*   ∞ *Corbula gibba*   ▽ *Prionospio sp.*
□ *Thyasira flexuosa*   ✪ *Glycera alba*   ◯ *Ophiodromus flexuosus* ▲ *Tellina fabula*
○ *Sipunculoidea*   ◠ *Phaxas pellucidus*   ◣ *Spisula elliptica*   e *Other species*

(b) Animals present in the same area two years after the period of anoxia (original).

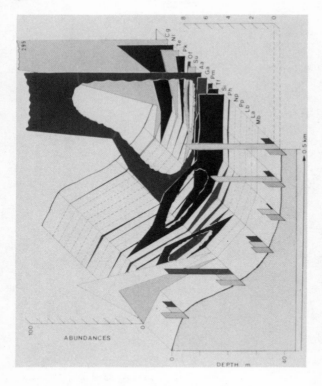

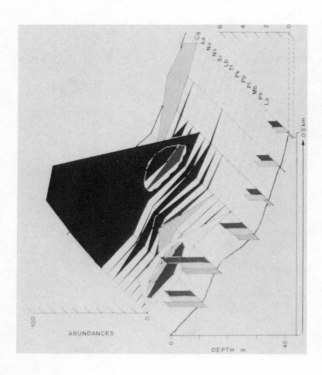

Figure 10. Changes in the abundance of the dominant macrofaunal species with increasing depth along a
transect across the inner basin of Sullom Voe during a period when the basin was aerated
(a) and following a period of anoxia (b) when the large populations of deposit feeding
species previously found in the deeper sediments now occupied the slopes above the level
of the thermocline (36 m). Histograms of the changes in total abundance, species numbers
and biomass with depth are also shown. {(a) original, (b) from Pearson and Eleftheriou,
1980}. Key to species shown:
Aa, *Abra alba*; Cg,*Corbula gibba*; Ga, *Glycera alba*; La, *Lepidopleurus asellus*;
Lb, *Lucinoma borealis*; Mb, *Mysella bidentata*; Nl, *Notomastus latericeus*; Np, *Nephtys* sp.;
Ns, *Nemertea* sp.; Nu, *Nucula* sp.; Of, *Ophiodromus flexuosus*; Ph, *Pholoe minuta*;
Pk, *Pectenaria koreni*; Pm, *Prionospio malmgreni*; Pp, *Pherusa plumosa*; Pt, *Pomatoceros
triqueter*; Si, *Scalibregma inflatum*; Te, *Tellina fabula*; Tf, *Thyasira flexuosa*;
Su, *Spisula elliptica*.

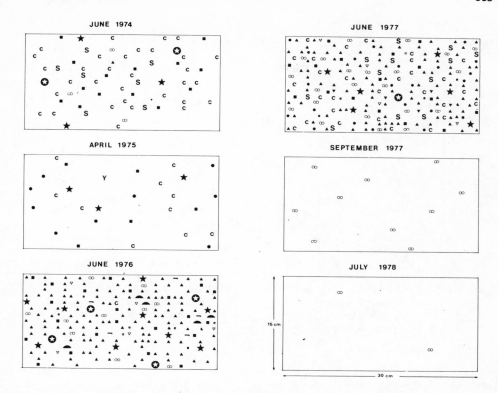

Figure 11. Changes in the density of macrofaunal species in 0.05
m^2 of sediment in the centre of the inner basin of
Sullom Voe over a four year period (for key to symbols
see fig. 9). Deoxygenation in late 1974 and again in
1977, resulted in elimination of the macrofauna. Recovery took place between these events, initially with
large juvenile populations of a few species in '75-'76
developing into a more diverse mature community by early
1977. The basin was again anoxic in the summer of 1978
thus precluding any redevelopment that summer, (from
Pearson and Eleftheriou, 1980).

order to interpret the status of the benthic communities at any one
time, although the general course of such a history may be inferred
from the composition of the communities (Fig. 11).

    Faunal studies on fjords containing permanently anoxic bottom
water have been reported from Sweden (Rosenberg, 1976, 1977,
Rosenberg and Moller, 1979, Dybern, 1972, Fonselius, 1976) and
Norway (Hjort and Dahl, 1900, Beyer, 1968). In such basins
macrofauna is absent from the deep areas overlaid by deoxygenated
water. Fauna is present only around the sides of the basin *i.e.*

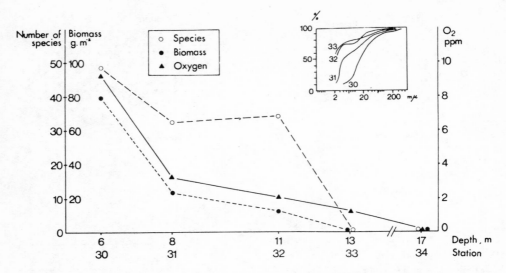

Figure 12.   Changes in number of species and biomass and oxygen
              content with increasing depth in the Byford. Sweden.
              Fauna is confined to the oxygenated area above 13 m,
              (from Rosenberg, 1977).

beneath the oxygenated upper water layers (Fig. 12).   In such basins
the level of the oxycline may vary seasonally in which case a tran-
sitional zone temporarily occupied by opportunistic species during
oxygenated periods may be present.   Eutrophication caused by organic
pollution has occurred intermittently or semi-permanently in many
boreal fjords in recent years.   Such conditions bring about benthic
ecosystem changes identical with those described for naturally
occurring deoxygenated basins, thus work describing such areas will
not be described here, but has been substantially reviewed recently
by Pearson and Rosenberg (1978) and Pearson (in press).

BIOLOGICAL INTERACTIONS AND ENCLOSURE EFFECTS

     The effect of enclosure imposes its own characteristic constraints
on the ecological interactions which modify the development and
distribution of fjordic macrobenthos.   There is little direct evidence
concerning the factors controlling recruitment in benthic communities, although recent work involving predator exclusion and controlled recolonisation experiments (Woodin 1974, Arntz *et al.*, 1977,
Rhoads *et al.*, 1978, McCall, 1978) have thrown some light on the
relative importance of invasion, larval settlement and predation in
controlling the development of communities.   Curtailment of the free
exchange of planktonic larvae between fjords and open coastal water

might materially affect the development of fjordic communities.
Pearson (1971) suggested that the blurring of community boundaries
frequently observed in fjordic populations might be attributable to
the concentration of larvae in such areas caused by the restricted
water exchanges resulting finally in their settlement on less than
optimal substrates. Rasmussen (1973) and Gage (1972) also suggested
that the lack of clear community boundaries in fjords might be
caused by larval confinement. Many authors have commented on the
lack of defined communities in fjordic benthos. Indeed Petersen
himself (1915) noted that the communities on the steeper slopes
of the Oslofjord were illdefined and difficult to categorise although he outlined definite communities from most parts of the fjord.
Although Molander (1928, 1960) described a detailed community
structure in the benthos of the Gullmarfjord in west Sweden, Lindroth (1935) working in the same fjord concluded that discrete communities did not occur. Recent workers in other areas e.g. Pearson
(1970), Stephenson et al. (1972), Gage (1972), Lie (1978) have
regarded benthic population structure in fjords as conforming more
closely to the continua concept as conceived, for instance, by
Mills (1968) rather than to the accepted framework of Petersonian
community structure. Weese in Shelford et al. (1935) attempted to
link a 'seral' succession of communities in a marine embayment in
Puget Sound to the terrestrial on shore sere in the manner of the
lacustrine succession of classical plant ecology. This attempt was
not particularly convincing because of the lack of any obviously
recognisable interdependent successional phases linking across the
littoral boundary. Although a parallel to the lacustrine succession
may be seen in the transition from salt-marsh communities to wet-
land meadows and subsequent terrestrial communities there appear
to be few fjordic areas evolving along such lines.

SALINITY AND TEMPERATURE FLUCTUATIONS

     Variations in salinity and temperature in fjords may be more
severe than in open sea areas and thus influence the composition
of the benthic communities to a greater extent. There are two
principal factors underlying this variability in fjords. Firstly,
the majority of fjords act as channels for fresh water, either
glacial melt water of catchment runoff, flowing to the sea. Such
freshwater flows are concentrated in the surface layers, usually
in the upper 10 metres but occasionally reaching as deep as 20
metres after storms e.g. Hurley (1964). Benthos in the shallow
areas subjected to intermittent low salinities is characteristically
dominated by the lamellibranch Macoma in northern areas. Thorson
(1933) noted that in the Franz Joseph Fjord complex shallow water
areas were inhabited by an impoverished Macoma calcaria community
where the number of species present were reduced both by the influence of ice abrasion during the winter months and by the low
salinities experienced during the spring melting of the ice.

Similarly in the Hurry Inlet of the Scoresby Sound complex in east
Greenland Thorson (1934) showed that those inshore areas of the
benthos affected by outflowing low salinity water were impoverished
when compared with areas below this influence. Recently Rosenberg
and Moller (1979) have shown that in many fjords on the west coast
of Sweden the benthic communities are divided by a halocline at
15 m depth maintained by the brackish Baltic current water combined
with local run off. Above this halocline the communities are relatively unstable and dominated by physical and chemical factors
whilst, below it, the communities are more diverse and tend to be
biologically structured. Gage (1972) found the fauna of the shallow
subtidal areas of Loch Etive in Scotland to be dominated by brackishwater species responding to the fluctuating salinity conditions in
the surface waters of the loch. However, in the near-by Loch Creran,
where the surface salinities were much higher, there was little
difference between the shallow and deeper water benthos. The major
factor of note in differentiating the temperature regime in fjords
from that of the open sea is the emponding effect of a silled basin.
Denser water retained in the basin behind the sill may differ considerably in temperature from the open waters beyond the fjord.
This is the case in the deep basins of the east Greenland fjord complexes reported by Spark (1933) and Thorson (1934) where the basin
water is warmer than the polar current water beyond the fjord. The
impoverished biomass of the benthos of these basins is attributed
by Thorson to their isolation from the more nutrient rich polar
current water.

PRODUCTION AND ENERGY FLOW THROUGH FJORDIC BENTHOS

     No detailed studies have been undertaken of the productivity
of fjordic benthos as distinct from open sea benthos, thus a de-
finitive discussion on the relative productivity of fjords is pre-
cluded. However, in recent years some general studies have been
made on the energy flow through fjordic basins, a number of gen-
eralising principles have been put forward to explain some of the
observed anomalies in the distribution of benthic biomass in inshore and embayed areas, and a number of apparently valid analogies
have been drawn between sediment productivity in fresh water lakes
and in marine basins. Thus a consideration of this evidence may
allow the presentation of some general conclusions on the comparative productivity of, and energy utilization in, fjords.

     Evidence presented in the forgoing sections suggests that the
two primary constraints in benthic production *i.e.* the availability
of carbon and of nutrients may be less restrictive in small basins
than in the open sea. Not only is there a considerable input of
allochthonous material to fjords but the input from macroalgae in
such areas may be of considerable importance (Hargrave, 1973,
Johnston *et al.*, 1977). Davies (1975) studied the energy flow

through the benthos of Loch Thurnaig in the west of Scotland and
showed that the carbon input of some 30 gC/m^2/day was approximately
balanced by community metabolism. In this shallow oxygenated system
much of the input was autochthonous and dependent·on the recycling
of nutrients, particularly nitrogen, from the benthos to the water
column, a process which was shown to be considerably enhanced by
wind-driven turbulence. An interesting contrast to the situation
in Loch Thurnaig was provided by Rosenberg *et al*. (1977) who
described the energy flow in the Byfjord on the west coast of Sweden.
This fjord has little tidal energy input and contains permanently
anoxic bottom water in a 50 m deep basin behind a 12 m sill. It
was estimated that the benthic macrofaunal production in those areas
of the fjord above the oxycline was about 30 gC/m^2/year (Rosenberg
1977b) (*i.e.* an identical figure to that given by Davis for the
Scottish loch, although in the former case the estimate was derived
from biomass comparisons between successive samples and in the latter
form community respiration data). Between 20 or 25% of the total
carbon input to the Byfjord was estimated to be lost to the deep
sediments in contrast to the net circulation of carbon in Loch
Thurnaig. Thus stagnation in marine fjordic systems with appreciable
carbon input levels leads to the creation of a carbon sink as is
the case in fresh water lakes. Two further factors which have been
noted as influencing the productivity of aquatic basins are temperature and the depth of the basin. Hargrave (1973) suggested that
temperature was not as important as carbon input in limiting the
carbon flow through benthic communities, basing his conclusions on
the comparison of a number of studies in marine and fresh water
areas. However, Johnston and Brinkhurst (1971) found that production
of benthic macroinvertebrates in Lake Ontario could be predicted by
the expression P = B T^2/10. (P production, B biomass, T temperature).
This latter study also showed that benthic production in the eutrophic area of the lake accounted for a much lower proportion of the
available carbon than in the oligotrophic areas of the lake, thus
leading to substantial carbon accumulation in the eutrophic areas.

Although few production studies have been undertaken in fjords
one feature of the distribution of fjordic benthos obviously related
to production, *i.e.* the diminution of total biomass with increasing
depth has aroused speculation and comment in previous studies.
Petersen (1915) first noted that the deep waters of the Oslofjord
were poor in biomass when compared with areas in the Kattegat. This
he attributed to the colder temperatures of the Oslofjord deeps.
Thorson (1934) and Ockleman (1958) both commented on the marked
decrease of biomass with depth in the fjord complexes of east
Greenland, and a similar trend can be seen in the biomass data published from amny fjords. Fig. 13 summarises this information from
a number of areas. Hargrave (1973) showed that in many aquatic
basins the oxygen uptake of the sediments was a function of the ratio
of primary production to mixed layer depth *i.e.* the deeper the basin
the lower the benthic production but suggested that this held true

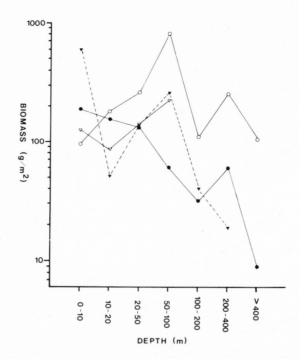

Figure 13.   Changes in biomass with depth in various fjordic areas.
              0——0 Data from fjords abutting the Strait of
              Georgia, Pacific N. W. Coast, given by Lie (1968) and
              Ellis (1968).   ▲——▲ Data from Petersen (1915) on the
              Skagerak and Oslofjord area.   ▽——▽ Data from various
              fjords on the west coast of Sweden, given by Rosenberg
              and Moller (1979).   ●——● Data from the Scoresby Sound
              complex, given by Thorson (1934).

only for basins receiving principally autochthonous carbon inputs.
However, the information in Fig. 13 suggests that the relationship
may have a general validity even in areas receiving large allochthonous inputs, *e.g.* Western Sweden. However, the rise in biomass
at the foot of the slope of the basins illustrated suggests that
this may be a perturbation caused by the accumulation of allochthonous material at the bottom of the slope. Thorson (1957) discussed the observation that, in addition to a depth gradient in
biomass, there was a topographical gradient in east Greenland fjords
with biomass diminishing towards the interior of the fjord. This
he attributed to the low carbon inputs in the interior of these
glaciated fjords.

     General observations of biomass level conceal a range of changes
taking place in the fjordic communities when compared over such
gradients. In general the animals concerned are smaller at greater

depths and belong to different trophic and taxonomic groups (*e.g.*
see Golikov and Averincev p. 359). However, diminishing biomass
does suggest a lower carbon input which itself may be a function
of both temperature and hydrodynamic energy input to the basin
concerned.

CONCLUSIONS

     Having considered to a greater or lesser degree the extent of
our knowledge of fjordic macrobenthos, the various environmental
and biological factors which distinguish fjordic sediments from
other benthic habitats and aspects of the productivity of such areas,
it remains to attempt a brief summary of those features which may
be of particular importance to this habitat.

     Fjords have frequently been considered as specialised types of
estuaries (*e.g.* Cameron and Pritchard, 1963). In so far as they are
usually the channels for large surface fresh water flows this is
true, however, their special characteristics of emponded bottom
waters with restricted water exchange results in a biological system
which differs widely from true estuarine systems. Their most common
general characteristic is to act as either carbon or nutrient sinks,
thus emulating on a small scale the infinitely greater ocean basins.
The study of such oceanographic microcosms allows for an easier
appreciation of the relative importance of the various environmental
factors which shape the ecosystems of such basins. The intricate
relationships between temperature, salinity, oxygen and hydrodynamic
energy input which control the diversity and productivity of the
benthos have been described in some detail and perhaps the most
justifiable conclusion to be drawn from the evidence presented is
that the ecology of each fjord is a result of a unique blend of
these factors. However, some (over) generalisations may be made
in the interests of simplicity and these are summarised in Table 1.
Differentiation into Boreal or Arctic fjords is primarily a distinction in temperature and associated icing effects. Most small
marine basins could be roughly assigned to one of the four general
categories listed, and thus a general guess as to the productive
range of the benthos could be made on the basics of their general
physical characteristics. Fjords with low hydrodynamic energy
inputs tend to act as sinks for either carbon or nutrients whereas
in high energy fjords the maximum energy flow through the benthos
is achieved and the systems are limited by either nutrient deficiency
in boreal areas or by carbon deficiency in the Arctic.

BIBLIOGRAPHY

Addy J. (in press): The macrobenthos of Sullom Voe. Proc. R. Soc.
     Edinbr. Section B.

Table 1.   General biological characteristics of fjordic basins
           related to their overall physical characteristics.

		BOREAL	ARCTIC
HYDRODYNAMIC ENERGY INPUT	HIGH	High biomass High diversity High productivity NUTRIENT LIMITED	High biomass Medium diversity Low productivity CARBON LIMITED
	LOW	Low biomass Low diversity Intermittently high productivity CARBON SINKS	Low biomass Low diversity Low productivity NUTRIENT SINKS

Apollonio, S. 1973:   Glaciers and nutrients in Arctic seas.   Science,
     180, 491–493.

Arntz, W. E. and H. Rumohr 1978:   The 'Benthosgarten':   Field
     experiments on benthic colonisation in the western Baltic.
     II.   Subsequent successional stages.   Kieler Meeresforsch.
     Sonderbd. 4, 94.

Bagge, P. 1969:   Effects of pollution on estuarine ecosystems.
     I.   Effects of effluents from wood–processing industries on
     the hydrography, bottom and fauna of Saltkallefjord (W. Sweden).
     Merentutkimuslait. julk. No. 228, 3–118.

Bertelsen, E. 1937:   Contributions to the animal ecology of the
     fjords of Angmagssalik and Kangerdlugssung in East Greenland.
     Medd. om Grönland, 108, 3; 1–58.

Beyer, F. 1968:   Zooplankton, zoobenthos and bottom sediments as
     related to pollution and water exchange in the Oslofjord.
     Helgolander wiss. Meeresunters. 17, 496–509.

Blegvad, H. 1914:   Food and conditions of nourishment among the
     communities of invertebrate animals found on or in the sea
     bottom in Danish waters. Rep. Dan. Biol. Stn. 22, 41–78.

Blegvad, H. 1951. Fluctuations in the amounts of food animals of
    the bottom of the Limfjord in 1928-1950. Rep. Dan. biol. Stn.,
    52, 3-16.

Brattegard, T. 1967. Pogonophora and associated fauna in the deep
    basin of Sognefjorden. Sarsia, 29, 299-306.

Brattstrom, H. and E. Dahl. 1951. Reports of the Lund University
    Chile expedition 1948-49. I. General account, lists of
    stations, hydrography. Lunds Universitets Arsskrift N.F.
    Avd. 2, 46, 8, 1-86.

Brotzky, V.A. 1930. Materials for the quantitative evaluation
    of the bottom fauna of the Storfjord, (E. Spitzbergen).
    Berichte wiss. Meeresinst. Moscow, 4,3, 47-61.

Cameron, W.M. and D.W. Pritchard. 1963. Estuaries. In The Sea.
    Vol. II, edited by M.N. Hill. Interscience Publishers, John
    Wiley and Sons, New York and London.

Christiansen, B. 1958. The foraminifer fauna in the Dröbak sound
    in the Oslo Fjord (Norway). Nytt magasin for Zoologi, 6, 5-92.

Curtis, M.A. 1977. Life cycles and population dynamics of marine
    benthic polychaetes from the Disko Bay area of West Greenland.
    Ophelia, 16, 9-58.

Davis, J.M. 1975. Energy flow through the benthos in a Scottish
    sea loch. Mar. Biol., 31, 353-362.

Dybern, B.I. 1972. Ideforden - en förstörd marin miljö. Fauna
    och Flora, 67, 90-103.

Edwards, A., D. Edelsten and S.O. Stanley (in press). Renewal and
    entrainment in Loch Eil: a periodically ventilated Scottish
    fjord. Proceedings of the 'Fjord Oceanographic Workshop'.,
    Victoria, Canada, June 1979.

Einarsson, H. 1941. Survey of the benthonic animal communities
    of Fax Bay (Iceland). Medd. Komm. Danm. Fisk.-og Havunders.,
    Ser. Fiskeri, 11, 1, 1-46.

Ellis, D.V. 1960. Marine infaunal benthos in arctic North America.
    Arctic Institute of North America Technical Paper No. 5, 1-53.

Ellis, D.V. 1967. a & b, and 1968 a,b. & c. Quantitative benthic
    investigations I-V. Fisheries Research Board of Canada Techni-
    cal Reports Nos. 26, 35, 59, 60 and 73. (Manuscript reports).

Ellis, D.V.   1970.   A review of marine infaunal community studies
     in the Strait of Georgia and adjacent inlets.   Syesis, 4, 3-9.

Erwin, D.G.   1977.   A diving survey of Strangford Lough: the benth-
     ic communities and their relation to substrate – a preliminary
     account.   In 'Biology of Benthic Organisms', edited by B.F.
     Keegan, P.O. Ceidigh and P.J.S. Boaden.   Pergamon Press, Oxford.

Evans, R.A.   1977.   The shallow-water soft-bottom benthos in
     Lindaspollene, Western Norway.   2.   Estimators of biomass
     for some macro-infauna species.   Sarsia, 63, 97-111.

Feder, H.M., G.J. Mueller, M.H. Dick and D.B. Hawkins.   1973.
     Preliminary benthos survey.   In 'Environmental studies of
     Port Valdez.   Edited by D.W. Hood, W.E. Sheils and E.J. Kelley,
     Institute of Marine Science, University of Alaska, Occasional
     Publication No. 3; 305-386.

Fonselius, S.H.   1976.   Eutrophication and other pollution effects
     in North European waters.   Meddn. Havsfiskelab. Lysekil, 202.

Gage, J.   1972.   A preliminary survey of the benthic macrofauna
     and sediments in Lochs Etive and Creran, sea-lochs along the
     west coast of Scotland.   J. mar. biol. Ass. U.K. 52; 237-246.

Gage, J.   1974.   Shallow water zonation of sea-loch benthos and its
     relation to hydrographic and other physical features.   J. mar.
     biol. Ass. U.K. 54; 223-249.

Gage, J. and P.B. Tett.   1973.   The use of log-normal statistics
     to describe the benthos of Lochs Etive and Creran.   J. Anim.
     Ecol. 42; 373-382.

George, J.D.   1977.   Ecology of the pogonophore *Siboglinum fiordicum*
     Webb in a shallow-water fjord community.   In Biology of benthic
     organisms edited by B.F. Keegan, P.O. Ceidigh and P.J.S. Boaden,
     Pergamon Press, Oxford.

Gilsen, T.   1930.   Epibioses of the Gullmeer Fjord.   1-11:   1877-
     1927.   No. 3 (123 pp.) No. 4 (380 pp.).

Golikov, A.N. and V.G. Averincev.   1977.   Distribution patterns of
     benthic and ice biocoenoses in the high latitudes of the
     polar basin and their part in the biological structure of the
     world ocean.   In 'Polar Oceans', edited by M.J. Dunbar.   Arctic
     Institute of North America, 682 pp.

Grassle, J.F. and J.P. Grassle.   1974.   Opportunistic life histories
     and genetic systems in marine benthic polychaetes.   J. Mar.
     Res. 32; 253-284.

Gulliksen, B. 1977. The Borgenfjord project 1967-76. Progress
    report (in Norwegian with English summary). Trondhjem
    Biologiske Stasjon.

Gulliksen, B. 1978. Rocky bottom fauna in a submarine gulley
    at Loppkalven, Finnmark, Northern Norway. Estuar. cstl mar.
    Sci. 7; 361-372.

Hargrave, B.T. 1973. Coupling caron flow through some pelagic
    and benthic communities. J. Fish. Res. Board Can. 30;
    1317-1326.

Hartley, C.H. and J. Fisher. 1936. The marine foods of birds in
    an inland fjords region in West Spitzbergen. Part 2. Birds.
    J. Anim. Ecol. 5; 370-389.

Hjort, J. and K. Dahl. 1900. Fishing experiments in Norwegian
    fjords. Report on Norwegian Fishery and marine investigations
    Vol. 1, No. 1. Kristiania.

Holthe, T. 1977. A quantitative investigation of the level-bottom
    macrofauna of Trondheimsfjorden, Norway. Gunneria, 28; 1-20.

Hoos, R. 1976. Environmental assessment of an ocean dumpsite in
    the Straight of Georgia, British Columbia, Canada. In 'Dredging:
    Environmental effects and technology. Proceedings of Wodcon
    VII, San Francisco, California'; 407-436. (Original not seen.
    Quoted in McDaniel et al., 1976).

Hopkins, C.C.E. and B. Gulliksen. 1978. Diurnal vertical migration
    and zooplankton - epibenthos relationships in a north Norwegian
    Fjord. In 'Physiology and Behaviour of Marine Organisms',
    edited by D.S. McLusky and A.J. Berry. Permagon Press, Oxford.

Hurley, D.E. 1964. Benthic ecology of Milford Sound. In 'Studies
    of a southern fjord', edited by T.M. Skerman. Memoir N.Z.
    Oceanographic Institute 17 (N.Z. Dept. of Scientific and
    Industrial Research Bulletin No. 157). pp. 79-89.

Johnson, M.G. and R.O. Brinkhurst. 1971. Production of benthic
    macroinvertebrates of Bay of Quinte and Lake Ontario. J. Fish.
    Board Can. 28; 1699-1714.

Johnston, C.S., R.G. Jones and R.D. Hunt. 1977. A seasonal carbon
    budget for a laminarian population in a Scottish sea-loch.
    Helgoländer wiss. Meeresunters. 30; 527-545.

Kawabe, R. 1975. Benthic animals of Ago Bay. Science Report
    of Shima Marineland. No. 3.

Kirboe, T. 1979. The distribution of benthic invertebrates
    in Holback fjord (Denmark) in relation to environmental
    factors. Ophelia, 18; 61-81.

Kikuchi, T. and M. Tanaka. 1978. Ecological studies on benthic
    macrofauna in Tomoe Cove, Amakusa. 1. Community structure
    and seasonal change of biomass. Publications from the Amakusa
    Marine Biological Laboratory, Kyashu University. 4; 189-213.

Kitching, J.A. and F.J. Ebling. 1967. Ecological studies at Lough
    Ine. Adv. Ecol. Res. 4; 197-291.

Kitching, J.A., F.J. Ebling, J.C. Gamble, R. Hoare, A.A.Q.R. McLeod
    and T.A. Norton. 1976. The ecology of Lough Ine. XIX.
    Seasonal changes in the Western Trough. J. Anim. Ecol. 45;
    731-758.

Lake, R.A. and E.R. Walker. 1976. A Canadian arctic fjord with
    some comparisons to fjords of the Western Americas. J. Fish.
    Res. Board Can. 33; 2272-2285.

Lande, E. and B. Gulliksen. 1973. The benthic fauna of the tidal
    rapids of Borgenfjorden estuary, North Tröndelag, Norway.
    K. norske Vidensk. Selsk. Skr. 1973, 1; 1-6.

Landner, L., J. Nilsson and R. Rosenberg. 1977. Assessment of
    industrial pollution by means of benthic macro-fauna surveys
    along the Swedish Baltic coast. Vatten 3; 324-379.

Larsen, K. 1935. The distribution of the invertebrates in the
    Dybsø fjord, their biology and their importance as fish food.
    Rep. Dan. biol. Stn., 39; 1-30.

Leppäkoski, E. 1968. Some effects of pollution on the benthic
    environment of the Gullmarsfjord. Helgolander wiss. Meeres-
    unters. 17; 291-301.

Leppäkoski, E. 1975. Assessment of degree of pollution on the basis
    of macrozoobenthos in marine and brackish-water environments.
    Acta Acad. abo., Ser. B. 35; 1-90.

Levings, C.D. 1976. River diversion and intertidal benthos at
    the Squamish river delta, British Columbia. In 'Fresh Water
    and the Sea', edited by S. Skreslet, R. Leinebø, J.B.L. Matthews
    and E. Sakshaug. The Association of Norwegian Oceanographers,
    Oslo.

Lie, U. 1968. A quantitative study of benthic infauna in Puget
    Sound. Fisk Dir. Skr. Ser. Havunders. 14 (5); 229-556.

Lie, U. 1969. Standing crop of benthic infauna in Puget Sound and off the coast of Washington. J. Fish. Res. Board. Can. 26; 55-62.

Lie, U. 1973. Distribution and structure of benthic assemblages in Puget Sound, Washington, U.S.A. Mar. Biol. 21; 203-223.

Lie, U. and R.A. Evans. 1973. Long term variability in the structure of subtidal benthic communities in Puget Sound, Washington, U.S.A. Mar. Biol. 21; 122-126.

Lindroth, A. 1935. Die associationen de marinen weichböden. Zool. Bidrag Uppsala, 15; 331-366.

McCall, P.L. 1977. Community patterns and adaptive stategics of the infaunal benthos of Long Island Sound. J. mar. Res. 35; 221-266.

McCall, P.L. 1978. Spatial-temporal distributions of Long Island Sound fauna: the role of bottom disturbance in a near shore marine habitat. In 'Estuarine interactions', edited by M.L. Wiley, pp. 191-219. Acedemic Press. New York.

McDaniel, N.G., R.D. MacDonald, J.J. Dobrocky and C.D. Levings. 1976. Biological surveys using in water photography at three ocean disposal sites in the Strait of Georgia, British Columbia. Tech. Rep. Fish. mar. Ser. No. 713; 1-41 pp. (Manuscript Report).

McIntyre, A.D. 1961. Ouantitative differences in the fauna of boreal mud associations. J. mar. biol. Ass. U.K., 41; 599-616.

McIntyre, A.D. 1964. Meiobenthos of sub-littoral muds. J. mar. biol. Ass. U.K., 44; 665-674.

McIntyre, A.D., A.L.S. Munro and J.H. Steele. 1970. Energy flow in a sand ecosystem. In 'Marine Food Chains', edited by J.F. Steele, pp. 19-31. Oliver and Boyd. Edinburgh.

McKnight D.G. and I. Estcourt. 1978. Benthic ecology of Caswell and Nancy Sounds. In 'Fjord Studies: Caswell and Nancy Sounds New Zealand, edited by G.P. Glasby, Mem. N.Z. oceanogr. Inst., 79; 85-94.

Mills, E.L. 1969. The community concept in marine zoology, with comments on continua and instability in some marine communities: a review. J. Fish. Res. Bd. Can., 26; 1415-1428.

Miyadi, D. 1941a. Indentation individuality in the Tanabe-wan. Mem. mar. Obs. Kobe, 7; 471-482.

Miyadi, D. 1941b. Marine benthic communities of the Beppu-wan. Mem. mar. Obs. Kobe, 7; 483-502.

Miyadi, D. 1941c. Marine benthic communities of the Ise-wan and the Mikawa-wan. Mem. mar. Obs. Kobe, 7; 503-524.

Miyadi, D. 1941d. Ecological survey of the benthos of the Ago-wan. Annotnes zool. Japan., 20; 169-180.

Molander, A.R. 1928. Animal communities on the soft bottom areas in the Gullmarfjord. Kristinebergs Zool. Station. 1877-1927, No. 2.

Molander, A.R. 1962. Studies on the fauna in the fjords of Bohuslän with reference to the distribution of different associations. Ark. Zool. Ser. 2, 15; 1-64.

Moore, H.B. 1930. The muds of the Clyde Sea area. 1. Nitrogen and phosphate content. J. mar. biol. Ass. U.K., 16; 595-607.

Moore, H.B. 1931. The muds of the Clyde Sea area. III. Chemical and physical conditions: rate and nature of sedimentation and fauna. J. mar. biol. Ass. U.K., 17; 325-358.

Muus, B.J. 1967. The fauna of Danish estuaries and lagoons. Distribution and ecology of dominating species in the shallow reaches of the mesohaline zone. Meddr Kommn Danm. Fisk.-og Havunders., N.S., 5; 316 pp.

Nielsen, E. Steeman. 1951. The marine vegetation of the Isefjord. Meddr Kommn Danm. Fisk.-og Havunders, Serie Plankton, 5; 1-114.

Nielsen, E. Steeman. 1958. A survey of recent Danish measurements of the organic productivity in the sea. Rapp. P.-v. Reun. Cons. perm. int. Explor. Mer., 144; 92-95.

Ockelmann, W.K. 1958. Marine lamellibranchiata. Meddr Grønland, 122; 1-256.

Pearson, T.H. 1970. The benthic ecology of Loch Linnhe and Loch Eil, a sea-loch system on the west coast of Scotland. 1. The physical environment and distribution of the macrobenthic fauna. J. exp. mar. Biol. Ecol. 5; 1-34.

Pearson, T.H. 1971. The benthic ecology of Loch Linnhe and Loch Eil, a sea-loch system on the west coast of Scotland. III. The effect on the benthic fauna of the introdution of pulp mill effluent. J. exp. mar. Biol. Ecol., 6; 211-233.

Pearson, T.E. 1975. The benthic ecology of Loch Linnhe and Loch Eil, a sea-loch system on the west coast of Scotland IV. Changes in the benthic fauna attributable to organic enrichment. J. exp. mar. Biol. Ecol. 20; 1-41.

Pearson, T.E. (in press). Marine pollution effects of pulp and paper industry wastes. Helgoländer wiss. Meeresunters.

Pearson, T.E. and A. Eleftherious (in press). The benthic ecology of Sullom Voe. Proc. R. Soc. Edinb. Section B.

Pearson, T.H. and R. Rosenberg. 1978. Macrobenthic succession in relation to organic enrichment and pollution of the marine environment. Oceanogr. Mar. Biol. Ann. Rev., 16; 229-311.

Pearson, T.E. and S.O. Stanley. 1977. The benthic ecology of some Shetland Voes. In 'The biology of benthic organisms', edited by P.F. Keegan, P.O. Ceidigh and P.J.S. Boaden, pp. 503-512, Pergamon Press, Oxford.

Pearson, T.H. and S.O. Stanley. 1979. Comparative measurement of the redox potential of marine sediments as a rapid means of assessing the effect of organic pollution. Mar. Biol., 53; 371-379.

Petersen, C.G.J. 1913. Valuation of the sea. II. The animal communities of the sea bottom and their importance for marine zoogeography. Rep. Dan. biol. Stn., 21; 1-44.

Petersen, C.G.J. 1914. Appendix to Report 21. Rep. Dan. biol. Stn., 22; 1-7.

Petersen, C.G.J. 1915a. On the animal communities of the sea bottom in the Skagerak, the Christiania Fjord and the Danish Waters. Rep. Dan. biol. Stn., 23; 3-28.

Petersen, C.G.J. 1915b. A preliminary result of the investigations on the valuation of the sea. Rep. Dan. biol. Stn., 23; 29-32.

Petersen, C.G.J. 1918. The sea-bottom and its production of fish food. A survey of the work done in connection with valuation of the Danish waters from 1883-1917. Rep. Dan. biol. Stn., 25; 1-62.

Petersen, C.G.J. and P. Boysen Jensen. 1911. Valuation of the sea. I. Animal life of the sea bottom, its food and quantity, Rep. Dan. biol. Stn., 20; 3-81.

Petersen, G.H. 1977. General report on marine benthic investigations near Godhaven, West Greenland. Ophelia, 16; 1-7.

Rasmussen, E. 1973. Systematics and ecology of the Isefjord marine
     fauna (Denmark). Ophelia, 11; 1-507.

Rhoads, D.C., P.L. McCall and J.Y. Yingst. 1978. Disturbance
     and production on the estuarine sea floor. Amer. Sci., 66;
     577-586.

Rosenberg, R. 1972. Benthic faunal recovery in a Swedish fjord
     following the closure of a sulphite pulp mill. Oikos, 23;
     92-108.

Rosenberg, R. 1974. Spatial dispersion of an estuarine benthic
     faunal community. J. exp. mar. Biol. Ecol., 15; 69-80.

Rosenberg, R. 1976. Benthic faunal dynamics during succession
     following pollution abatement in a Swedish estuary. Oikos,
     27; 414-427.

Rosenberg, R. 1977a. Benthic macrofaunal dynamics, production and
     dispersion in an oxygen deficient estuary of west Sweden.
     J. exp. mar. Biol. Ecol., 26; 107-133.

Rosenberg, R. 1977b. Effects of dredging operations on estuarine
     benthic macrofauna. Mar. Pollut. Bull., 4; 187-188.

Rosenberg, R. and P. Möller, 1979. Salinity stratified benthic
     macrofaunal communities and long term monitoring along the
     west coast of Sweden. J. exp. mar. Biol. Ecol., 37; 175-203.

Rosenberg, R., I. Olsson and E. Ölundh. 1977. Energy flow model
     of an oxygen-deficient estuary on the Swedish west coast.
     Mar. Biol., 42; 99-107.

Sanders, H.L. 1968. Marine benthic diversity: a comparative
     study. Am. Nat., 102; 243-282.

Shelford, V.E. and E.D. Towler. 1925. Animal communities of San
     Juan Channel and adjacent areas. Pub. Puget Sd. Biol.
     Stn. 5; 31-73.

Shelford, V.E., A.O. Weese, L.A. Rice, D.I. Rasmussen and A. MacLean,
     1935. Some marine biotic communities of the Pacific coast of
     North America. Part. 1. General survey of the communities.
     Ecol. Monogr. 5; 249-332.

Soot-Ryen, T. 1924. Faunistische untersuchungen im Ramfjorde.
     Tromsø Mus. Arsh., 45 (1922), 6; 1-106.

Soot-Ryen, T. 1959. Pelecypoda. Reports of the Lund University
     Chile Expedition 1948-49. No. 35. Lund Univ. Arsskr., N.F.,
     2, 55, 6; 1-86.

Spark, R.   1929.   Preliminary survey of the results of quantitative
    bottom investigations in Iceland and Faroe waters.   Rapp. P.
    -v. Reun. Cons. perm. inte. Explor. Me. 57; 1-28.

Spark, R.   1933.   Contributions to the animal ecology of the Franz
    Joseph Fjord and adjacent East Greenland waters.   1-11.
    Meddr. Grønland, 100, 1; 1-38.

Spark, R.   1937.   The benthonic animals of the coastal waters.
    The Zoology of Iceland, 1, 6; 1-45.

Spark, R. and S.L. Tuxen, 1928-71.   The Zoology of the Faeroes,
    1 part 2 and 3 part 1.   Andr. Fred. Høst & Søn, Copenhagen.

Stephenson, W., W.T. Williams and S.D. Cook.   1972.   Computer
    analyses of Petersen's original data on bottom communities.
    Ecol. monogr., 42; 387-415.

Stott, F.C.   1936.   The marine food of birds in an island Fjord
    region of West Spitzbergen.   Part 1.   Plankton.   J. Anim.
    Ecol., 5; 356-369.

Taasen, J.P. and R.A. Evans.   1977.   The shallow-water soft-bottom
    benthos in Lindaspollene, Western Norway.   1.   The study area
    and main sampling programme.   Sarsia, 63; 93-96.

Tanaka, M. and T. Kikuchi.   1978.   Ecological studies on benthic
    macrofauna in Tomoe Cove Amukusa.   II.   Production of
    Musculista senhousia (Bivalvia, Mytilidae).   Publ. Amakusa
    Mar. Biol. Lab. Kyusha Univ., 4; 215-233.

Thorsen, G.   1933.   Investigations of shallow water animal
    communities in the Franz Joseph Fjord (East Greenland) and
    adjacent waters.   Meddr. Gronland, 100, 2; 1-70.

Thorsen, G.   1934.   Contributions to the animal ecology of the
    Scoresby Sound Fjord complex (East Greenland).   Meddr. Grønland,
    100, 3; 1-68.

Thorsen, G.   1966.   Some factors influencing the recruitment and
    establishment of marine benthic communities.   Neth. J. Sea
    Res., 3; 267-293.

Trevallion, A., J.H. Steele and R.R. C. Edwards.   1970.   Dynamics
    of a benthic bivalve.   In 'Marine food chains', edited by
    J. Steele, pp. 285-295, Oliver and Boyd, Edinburgh.

Vibe, C.   1939.   Preliminary investigations on shallow water animal
    communities in the Upernavik- and Thule-districts (North west
    Greenland).   Meddr. Grønland, 124, 2; 1-42.

Vibe, C. 1950. The marine mammals and the marine fauna in the
    Thule district (North west Greenland) with observations on ice
    conditions in 1939–41. Meddr Grønland, 150, 6; 1–115.

Wesenberg-Lund, E. 1953. The Zoology of East Greenland. Polychaeta,
    Meddr Grønland, 122, 3; 1–169.

Wismer, N.M. and J.H. Swanson. 1935. A study of the animal commun-
    ities of a restricted area of the soft bottom in the San Juan
    Channel. Ecol. monogr., 5; 333–354.

Woodin, S.A. 1974. Polychaete abudance patterns in a marine soft-
    sediment environment: the importance of biological inter-
    actions. Ecol. Mongr., 44; 171–187.

THE EFFECT OF SEDIMENTATION ON ROCKY BOTTOM ORGANISMS   IN BALSFJORD, NORTHERN NORWAY

R. A. Evans, B. Gulliksen and O. K. Sandnes

Marine Biology Station
University of Tromsø
P. O. Box 2550
9001 Tromsø, Norway

INTRODUCTION

Sediment-animal relationships are many-sided: Sediment may
serve as food, shelter, or substrate for the benthic fauna; mechanical action of sediment may remove epifauna or an accumulation
of sediment may cover filter feeders; the animals themselves may
influence the amount of sediment accumulating on the bottom. Treatment of such interrelationships may be found in MOORE (1977) and
RHOADS (1974), among others.

In order to study the effect of sedimentation on rocky bottom
organisms, a near vertical rocky wall was selected for investigation.
The wall was located at Haugbergnes in Balsfjord, Northern Norway
(69^0 31'N, 19^0 1'E). It was about 500 m wide and extended downwards to a depth of c. 100 m, having some overhangs. The general
hydrography of the area has been described by SÆLEN (1950), and
the primary production investigated by EILERTSEN (1979).

Sedimentation was investigated by GULLIKSEN (in prep.) using
sediment traps. Traps suspended at 5, 10, 20 and 30 m depth from
October 1976 to November 1977 indicated an increasing amount of
sedimented material with depth (Fig. 1). The maximum value at
30 m depth was nearly 12 g dry weight/m^2/day, while the value at
5 m depth was 2-3 g dry weight/m^2/day. Seasonal fluctuations in
the rates of sedimentation were recorded with highest values during
spring and summer. At 5 and 10 m depth, maxima in the spring corresponded with phytoplankton blooms while maxima at 20 and 30 m
depth were more evenly distributed throughout summer.

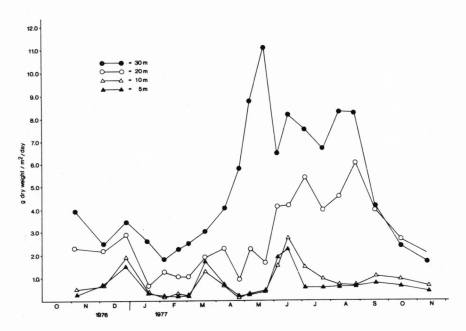

Figure 1.   Seasonal fluctuations in sedimentation at Haugbergnes
            at 5, 10, 20 and 30 m.

        The vertical distribution of the fauna from the surface to
50 m depth was studied using a Hasselblad SWC camera in an underwater housing. During the period March 2–8, 1977, 1/4 m^2 areas
were photographed at 2 m depth intervals. The number of solitary
organisms and per cent cover of colonial organisms were determined
from the photographs.

        A rock overhang at 12–15 m depth, and protruding 0.5 to 1.0 m,
was selected for a study of the effect of angle of inclination on
sedimentation and thus effect of sedimentation on organisms.   Samples were taken from above and below the overhang using a diver
operated suction sampler working on the "air-lift" principle.
(BARNETT & HARDY 1967, HISCOCK & HOARE 1973).

        The results of the vertical distribution study are presented
in Fig. 2.   There is a clear vertical gradient in the distribution
of the organisms. With increasing depth (and sedimentation, see
Fig. 1) the encrusting forms give way to species which by their
erect nature (bryozoans, solitary ascidians, actinians and pori-
ferans) or their motility (scallops) may rise above the sediment
covered surface.

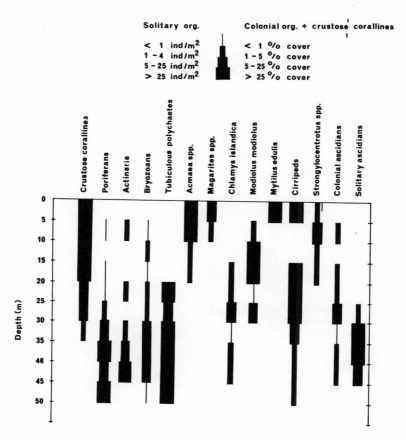

Figure 2.   Vertical distribution of organisms at Haugbergnes.
Per cent cover determined from stereophotographs taken
at 2 m depth intervals.

A summary of the results of the overhang study is shown in
Fig. 3. In addition, it should be mentioned that 100% cover by
barnacles on the underside of the overhang, was observed while
diving whereas there was almost a complete lack of barnacles on
the upper side. This seemed to be due to the barnacles inability
to cope with the sedimentation on the upper side.

Species affinities were noted -- for example, the amphipods
and pycnogonids appear to prefer bryozoans as a substrate for
attachment. *Modiolus modiolus* appears to have its own epifauna
and infauna, the latter consisting of the organisms which live in
the sediment trapped between the mussels.

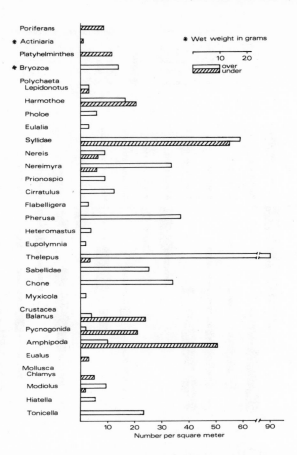

Figure 3.   Distribution of organisms on the upper and under sides
            of an overhang at Haugbergnes (number per square meter).

     Thus sedimentation is one factor of many affecting the species
composition of a given hard substrate area. It is also a factor
with feedback from the assemblage, in that it may be increased or
decreased by their presence and/or activity of organisms. Seasonal
fluctuations in rates of sedimentation and abundance/activity
of organisms suggest that no sharp boundaries exist between "hard"
and "soft" bottom areas.

ACKNOWLEDGEMENTS

     We would like to thank Gerd Wibeche Pettersen and Aagot Svendsen
for technical assistance, and Bjørnar Seim for diving assistance.
The work was supported by grants from the Norwegian Research Council

for Science and the Humanities and the Norwegian Marine Pollution
Research and Monitoring Program.

REFERENCES

Barnett, P. R. O. and B. L. S. Hardy. 1967: A diver-operated quant-
    itative bottom sampler for sand macrofauna. Helgoländer, Wiss.
    Meeresunters, 15, 390-398.

Eilertsen, H. C. 1979: Planteplankton, minimumsfaktorer og primær-
    produksjon i Balsfjorden, 1977. Cand. Real. Thesis, University
    of Tromsø. 130 p. (Phytoplankton, limiting factors and primary
    production in Balsfjorden 1977).

Gulliksen, B. (in prep.): Sedimentation at Haugbergnes 8 (69⁰ 31'N
    19⁰ 1'E), Balsfjord, Northern Norway.

Hiscock, K. and R. Hoare. 1973: A portable suction sampler for rock
    epibiota. Helgoländer. Wiss. Meeresunters, 25, 35-38.

Moore, P. G. 1977: Inorganic particulate suspension in the sea
    and their effect on marine animals. Oceanogr. Mar. Biol.
    Ann. Rev. 15, 225-363.

Rhoads, D. C. 1974: Organism-sediment relations on the muddy sea
    floor. Oceanogr. Mar. Biol. Ann. Rev. 12, 263-300.

Sælen , O. H. 1950: The hydrography of some fjords in Northern
    Norway. Tromsø Mus. Årsh. 70(1), 1-93.

SUPRABENTHIC GAMMARIDEAN AMPHIPODA (CRUSTACEA) IN THE PLANKTON OF

THE SAGUENAY FJORD, QUEBEC *

Pierre Brunel, Rejean de Ladurantaye[1,2] and Guy Lacroix[1]

Département des Sciences Biologiques
Université de Montreal
Box 6128, Montreal, H3C 3J7 Canada

INTRODUCTION

The Saguenay Fjord (100×2-3 km) is continuous, through a shallow sill (20 m) with the lower and middle St. Lawrence estuary. For the purposes of our biological work, the bottom of the fjord can be divided into two regions: the downstream trough (2 stations 238-256 m) and the downstream shallows (2 stations, 124-154 m) associated with a 70 m sill, north of the entrance to the estuary. Two muddy bottom areas, the deep central basin (8 stations, 233-274 m) and the upstream slopes (6 stations 90-162 m) near the head of the fjord are also considered. Cold (0.2-4.0°C), saline (26-31°/$_{oo}$), and well oxygenated water is pumped twice a day over the entrance sill from water upwelling at the upstream end of the estuary. A cold enclave located at 20-100 m depth, and pushed upstream after winter cooling, occurs over the central basin and slopes in spring and summer.

We want to show that the Saguenay Fjord plankton has more gammaridean amphipods than three adjacent ecosystems, and that these gammarideans probably originate from the near bottom suprabenthic layer. Ladurantaye (1979) and Rainville (1979) collected most of the data reported here using 0.5 m Bongo and standard closing nets (158 and 569 µm mesh). Over the period 1974 to 1976, a total of 1071 tows were completed during April to October, at the 18 stations

* Contribution to the program of GIROQ (Groupe Interuniversitaire de Recherches Océanographiques de Québec). A paper giving full data will be published elsewhere, Brunel *et al.*, 1980.
[1] Département de Biologie, Université Laval, Québec, P.Q., G1K 7P4, Canada.
[2] Now at Pêches et Océans Canada, Box 15500, Québec, P.Q., G1K 7X7, Canada.

referred to above. Data from eight other planktonic sampling series in the St. Lawrence estuary and Gaspé peninsula waters were also used. Because of the low densities and associated errors of estimates, mainly frequency data are used in this report.

RESULTS

The Saguenay Fjord plankton yielded 34 species of gammarideans. Mean density per region ranged from 0.2 to 6.2 individuals/100 m³. Total frequencies for the whole water column ranged from 19 to 25% (n=669) over the rocky stations. In contrast, gammaridean frequency outside the fjord was always <22%. In Gaspe waters, over bottom depths of 30 to 247 m, 13 species (8 found in the Saguenay plankton) had a total frequency of 1 to 6% (360 oblique Clarke-Bumpus hauls at three levels and 169 hauls with horizontal closing nets at 6 levels). Night samples with the closing nets (54 hauls) yielded catch frequencies of 0 to 21%. In the Lower St. Lawrence estuary, over 30 to 150 m depths, 4 planktonic surveys (closing nets, Bongo nets, or two types of midwater trawls) have only taken 4 species of gammarideans with frequencies of 9 to 13%.

The two rocky regions of the fjord have more species (7/20) in common, than do the two muddy regions (5/14). Only 3 out of 31 species occurred in both the muddy and rocky areas. Gammarids occurred more frequently over the rocky areas (50 to 60% of the catches) compared to muddy bottoms (19 to 25% of catches). In rocky regions, 4 species showed frequencies greater than 5%: *Halirages fulvocinctus* (M. Sars) (55%) in the downstream trough (24 times more frequent than any other species in the fjord); *Melphidippa goesi* Stebbing (27%); *Eusirus propinquus* (G.O.Sars) 11%; and *E. cuspidatus* Kroyer (6%). The latter 3 species were all from the rocky shallows. From the muddy regions, 3 species are more frequent than 5%: *Acanthostepheia malmgreni* (Goes) (16%) in the deep basin, occuring very frequently at its downstream end; *Rachotropis aculeata* (Lepechin) (9%); and *Paroediceros propinquus* (Goes) (8%) over the upstream slopes. Distribution with depth and distance from the bottom differed between rocky and muddy regions. Over the two downstream rocky regions, the diurnal samples of *H. fulvocinctus* from Rainville (1979) show a decrease of frequency with distance from the bottom, but frequency at 25 m is higher over the rocky shallows than over the trough. Over the muddy regions (downstream trough), 8 out of 15 species, including *H. fulvocinctus*, seem to be equally or more abundant over the bottom compared to near it.

Day and night samples (July 1978) at 4 depths over the rocky downstream trough suggest that density changes of *H. fulvocinctus* are more closely related to nocturnal vertical migrations than with the twice-daily tidal inflow of water from the St. Lawrence estuary outsude the entrance sill. Daytime abundance was high close to the bottom, but at night fewer gammarids were taken in bottom tows.

All the fjord species, except *Metopa invalida* G.O.Sars, *E.
longipes* Boeck and *Gammaracanthus loricatus* (Sabine) are known from
extensive benthic sampling in the estuary and Gulf of St. Lawrence.
Twenty-two species are shelf or upper-slope species, 8 are known
from the inshore algal belt or infralittoral sands, and 4 from the
downstream trough are of indeterminate origin. Of the 22 offshore
species, 6 belong to the family Oedicerotidae, 4 to Calliopiidae
and 4 to Eusiridae. Seven other families are represented by 1 or
2 species. Eleven of the 22 offshore forms have dorsal keels and
large eyes associated with active swimming. All but one (*P. propinquus*) of the 11 have strong gnathopods with prey-catching potential. Only one of the inshore species, *G. loricatus*, has dorsal
keels and large eyes. All 4 Eusiridae were taken in greater density
above the bottom and thus are considered the most pelagic, possibly
truly planktonic.

Extensive sampling over muddy bottom (110–120 m) with closing
nets mounted on a bottom sled (Brunel *et al*, 1978) in the Lower
St. Lawrence estuary (Besner, 1976) and the Gulf of St. Lawrence
(Brunel *et al*, unpublished) are available for comparison. Except
for *Schisturella pulchra* (Hansen), all 12 of the offshore species
from the two muddy regions of the Saguenay were much more frequent
in the suprabenthic boundary layer. For the rocky bottom species,
data are less conclusive as the bottom sled sampling did not focus
on rocky substrates. Among Saguenay species from both bottom types,
all but two, *Arrhys phyllonix* (M. Sars) and *P. Propinquus*, had
greater swimming indices (upper net (1.1 to 1.4 m from bottom) *vs*.
lower net (0.3 to 0.6 m from bottom)) than those of the whole
gammaridean fauna. Mean planktonic densities (approximately 4 to
6 ind./100 m^3) over the rocky regions in the Saguenay are very much
reduced compared to upper net catches in the lower estuary (15 ind./
m^3) or to lower net catches in both the lower estuary and Gaspé
waters (up to 1286 ± 714 ind./m^3).

DISCUSSION AND CONCLUSIONS

Gammaridean frequency and density in the Saguenay plankton are
significantly higher than in adjacent ecosystems, except in the
north channel of the middle estuary where midwater trawls have
caught them in 37–46% of tows (mostly *Eusirus cuspidatus* with some
H. fulvocinctus), as compared with 86% in the fjord and none in the
lower estuary. Known currents in the north channel flow away from
the fjord sill. Outside the latter, none of the significant downstream species have been caught in the plankton, and very few specimens in the deep suprabenthic tows. Scattered knowledge on the
dominant downstream species indicate that they are usually pelagic
(Eusiridae, *Pardalisca*) or associated with algae, hydroids, *etc*. of
hard offshore grounds (*Halirages, Melphidippa*). This and evidence
given above thus suggests that pelagic gammarideans in the downstream part of the fjord originate either from the middle depths

of the rocky shallows or of the steep rocky cliffs of the downstream trough, rather than from upwelled water inflowing over the rocky sill. Like the Mysid *Mysis littoralis*, also endemic to the fjord (Ladurantaye, 1979), some are washed out of the fjord with the surface outflowing current, to be carried into the upwelling area and then mostly to the north channel of the middle estuary. Gammaridean diversity in the plankton is further enhanced by several inshore species which reflect greater vertical mixing. Upstream species in the fjord are more obviously derived from the suprabenthos over mud and stay closer to the bottom, except *R. aculeata*, which seems to behave like the endemic *Boreomysis nobilis* studied by Ladurantaye (1979).

Surface primary production in the fjord is very low (Côté and Lacroix, 1979), yet the upper 100 m of the downstream rocky regions up to the downstream end of the deep basin are rich in physiologically healthy, potentially productive and long-living cells, mostly dinoflagellates, which are replenished twice daily in the fjord with water flowing over the sill from the lower part of the productive St. Lawrence upwelling region (Côté and Lacroix, 1978; Therriault *et al*, 1979). Planktonic gammarids in the fjord which have weak gnathopods and may mostly be suspension feeders like *Melphidippa* (Enequist, 1949) are presumably held in mid-water sectors by this stable and predictable planktonic food source. Species from the upstream muddy slope are predominantly predators, and may be attracted by the abundant and diverse assemblage of small copepods which develops there (Rainville, 1979). However, this region may also be rendered less attractive by the anoxic conditions in the sediments due to wood debris and petroleum pollution (Loring, 1978).

ACKNOWLEDGMENTS

Financial help was received from the National Research Council of Canada (development grants to GIROQ and operating grants to P. Brunel and G. Lacroix), the Quebec Ministry of Education (to GIROQ) and the Donner Canadian Foundation (to GIROQ). Lucie Huberdeau, Lucie McDuff and Luc Rainville contributed much to the identifications and data processing.

REFERENCES

Besner, M. 1976: Ecologie et échantillonage des populations hyperbenthiques d'Amphipodes Gammaridiens d'un écosystème circalittoral de l'estuaire maritime du Saint-Laurent. Univ. Montréal, M.Sc. thesis, 183pp.

Brunel, P., M. Besner, D. Messier, L. Poirier, D. Granger and M. Weinstein 1978: Le traîneau suprabenthique Macer-GIROQ: appareil amélioré pour l'échantillonnage quantitatif étagé

de la petite faune nageuse au voisinage du fond. Int. Rev.
ges. Hydrobiol. 63, 829-843.

Côté, R. and G. Lacroix 1978. Capacité photosynthétique du phyto-
plancton de la couche aphotique dans le fjord du Saguenay.
Int. Rev. ges. Hydrobiol. 63, 233-246.

Côté R. and G. Lacroix 1979. Influence des débits élevés et variables d'eau douce sur le régime saisonnier de production
primaire d'un fjord subarctique. Oceanol. Acta 2, 299-307.

Enequist, P. 1949. Studies on the soft bottom amphipods of the
Skaggerak. Zool. Bidr. Uppsala 28, 297-492.

Ladurantaye, R. de 1979. Cycle saisonnier et répartition spatiale
des Mycidacés dans le fjord du Saguenay. Univ. Laval
M.Sc. thesis.

Rainville, L. 1979. Etude comparative de la distribution verticale
et de la composition des populations de zooplancton du fjord
du Saguenay et de l'estuaire maritime du Saint-Laurent.
Univ. Laval M.Sc. thesis, 175pp.

Therriault, J.-C., R. de Ladurantaye and R. G. Ingram 1979.
Particulate matter exchange processes between the St. Lawrence
Estuary and the Saguenay Fjord. In this volume 363-366.

FLOCCULATION, AGGLOMERATION, AND ZOOPLANKTON PELLETIZATION OF SUSPENDED SEDIMENT IN A FJORD RECEIVING GLACIAL MELTWATER

James P. M. Syvitski

Department of Geology
University of Calgary, Alberta, Canada
T2N 1N4

INTRODUCTION

This study was designed to investigate the sedimentation of suspended sediments in Howe Sound, a fjord in British Columbia which receives copious glacial run-off. The following problems were addressed: 1) How does the suspended load behave upon entry into a fjord? 2) What is the relationship between the sediment in the water column and that collected by sediment traps? 3) By what mechanisms do the suspended particles settle out - as single particles, by inorganic flocculation, by biological interaction or by other processes? 4) What are the *in situ* settling velocities of these particles?

Howe Sound lies within the Pacific ranges of the Coast Mountains and is bound on the south by the Georgia Lowland of the Coastal Trough. In the 42 km length of Howe Sound, there are two sills, one halfway down the fjord and the other at the fjord mouth. Only the northern 17 km of the fjord (river mouth to first sill) were investigated. This section has a maximum depth of 325 m, an average width of 3 km with a 35 to 70 m deep sill at its southern end. Further description of this fjord, the study detail, and the data base can be found elsewhere Syvitski (1978).

SEDIMENT PLUME AND OCEANOGRAPHY

The suspended load enters the fjord as part of the cold glacial meltwater. During the summer months, the river water skims across the surface of the fjord with little mixing in the surface layer (i.e. a 4 ^{0}/oo salinity increase from river mouth to inner sill,

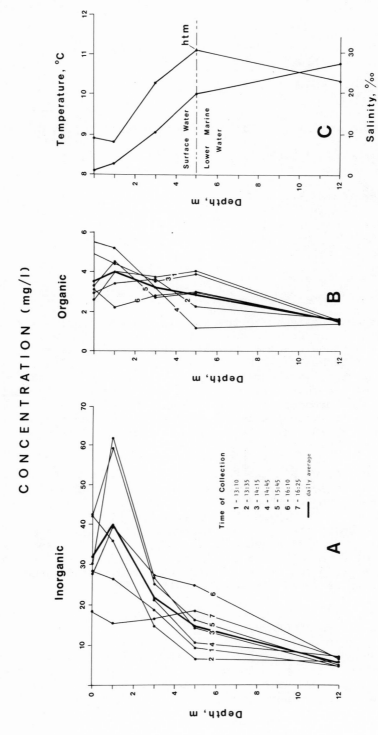

Figure 1. Typical vertical variation (April 26, 1977) during ebb tide of A) inorganic concentration and B) organic concentration. C) Daily average in temperature and salinity showing the boundary between the surface and marine waters.

a distance of 17 km). Substantial mixing does occur in the fall
and spring. During winter, when river discharge is lowest, wind
mixing can eliminate horizontal stratification, except at the river
mouth. Summer mixing, caused by wind generated turbulence and entrainment, produces an up-inlet current to replace the sea water
lost to the surface layer. This subsurface current which originates
from outside the sill, brings in warm saline water, causing the
halocline temperature maximum. The mixing between the currents
causes both salinity and temperature to increase in the surface
layer and the suspended load to decrease (Fig. 1). The even mixing
gives rise to the linear relationship of temperature, salinity and
inorganic concentration with distance from the river mouth (Fig. 2).
Any process that would increase the settling velocity of grains
does not significantly alter this state of simple dilution. The
dominant cause of temporal variation in the surface layer flow has
been related to daily wind forcing Buckley (1977). Surface current reversals, that occur during times of strong up-inlet winds,
would increase the chance of surface-layer sediment settling into
the lower marine water. These reversals would be responsible for
the 1 m maxima in suspended load at stations near the river mouth
as sketched in Figure 3.

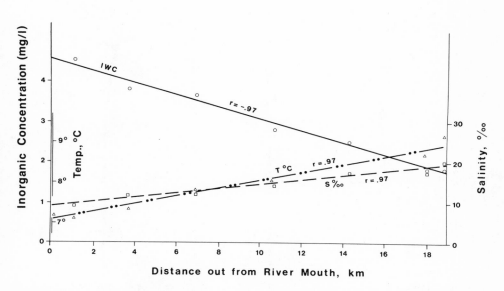

Figure 2.   Linear variation of inorganic concentration, temperature,
            and salinity of the surface water (0m) with distance
            from the river mouth.

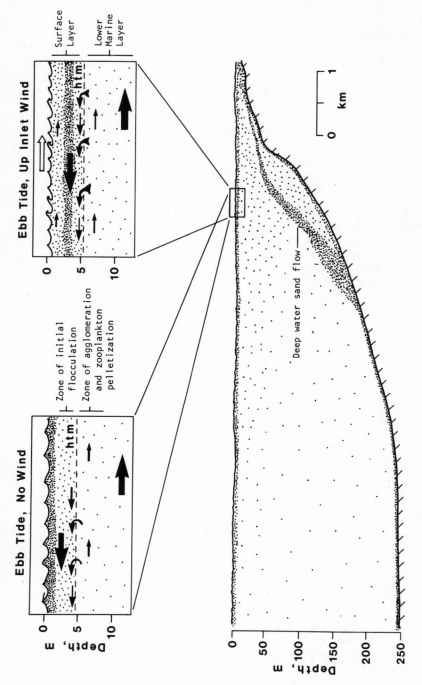

Figure 3. Sediment concentration in upper Howe Sound during a typical summer freshet day. Indicated are the deep water sand flow near the delta, and two conditions of water flow in the surface-layer.

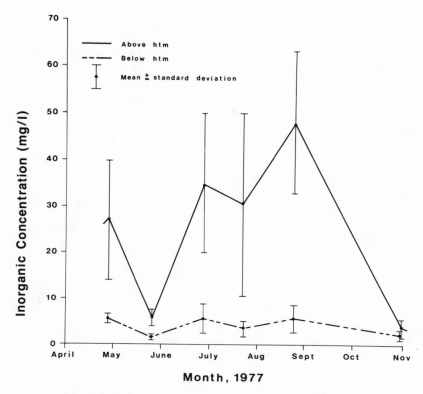

Figure 4.   Variations of inorganic concentration above and below
the halocline temperature maximum near the delta.   Note
the pre-freshet peak on April 26, 1977, which was
caused by heavy rain and subsequent run-off.

Most of the sediment in the lower marine water originates from
the surface layer, which accounts for the linear correlation between the mean daily concentrations of sediment at the surface with
that deeper in the water column (Fig. 4).   This suggest that the
response time between the two layers is fast (days), indicating
processes that increase the component particle settlement.

Linear correlation between mean organic and inorganic concentrations in the surface layer indicates the Squamish River as the
major organic input into the fjord.   Much of the river-borne seston
is flushed out of the bayhead delta during the river freshet, Levings
(1973), and much of the organic detritus falls close to the river
delta.

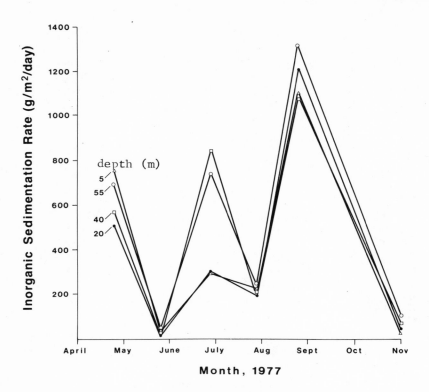

Figure 5.   Variation of inorganic sedimentation rates near the delta
from four levels of sediment traps. Note the collection
of a deep water sand flow on June 27 in the deeper traps.

SEDIMENTATION RATES

Howe Sound is a basin of largely inorganic deposition (97%)
in which the rate of sedimentation is in direct response to river
discharge. As the river discharge increases, its competence increases, so that both grain size and rate of sedimentation increase
proportionately. The small quantities of sedimentary organic detritus are directly related to the sedimentation of inorganic
grains. This suggests an intimate assocation during particle settlement. Stratification and presence of deep water currents can be
surmised from the variation of sedimentation rates between levels
at a given station and time (Fig. 5).

The lack of correlation between sedimentation rates and water
turbidity, surrounding the traps during the collection period, suggests that much of the suspended sediment in the water exists simply

as residual background through which larger particles settle. Therefore water turbidity cannot be used as a measure of the downward
flux of particles.

The discharge of sand in deep water from the delta is suggested
to explain the mid-depth maxima in sedimentation near the delta.
Sand too coarse to travel in the surface layer quickly settles out
only to become entrained by a current at depth flowing seaward, and
to settle out further down inlet (Fig. 3).

SIZE DISTRIBUTION CHARACTERISTICS

Individual laminae in bottom sediments have significant variation in mean size and in the shape of the sub-laminae size frequency
distributions of which they are composed (*e.g.* Grace, Grothaus and
Ehrlich, 1978) for example. One day's sediment collection in a
trap represents part of a sedimentary lamina on the bottom (i.e.
a sub-lamina). Summation of these distributions (biased by the
actual weight collected) was found to approximate the grain size
distribution of bottom sediment. This is unique to the fjord environment where bed load movement is minimal.

FJORD SUSPENSATES

Most particles settling in the fjord environment fall as:
1) Sand and silt grains containing attached clay particles. The
attached particles are usually less than 1 µm; 2) Clay clasts,
usually well-rounded, that range from 10 to 35 µm in diameter;
3) Mineral-bearing fecal pellets from pelagic zooplankton
(euphausiids, copepods). The average pellet size is 100 µm in
length and 38 µm in width. The pellets contain up to 98% clay
flakes; 4) Large-grain inorganic floccules composed mostly of clay
plates larger than 2 µm. The floc diameter ranges from 10 µm to
upwards of 70 µm with the size increasing with collection depth;
5) Colloidal flocs composed of less than 2 µm clay flakes intermixed with colloidal material. These flocs range from 15 µm up;
and 6) Inorganic-biogenic agglomerates (large grain flocs that contain over 30% biogenic components, such as broken diatom frustules).
These agglomerates tend to be large (>35 µm).

IN SITU SETTLING VELOCITY OF FJORD SUSPENSATES

Processes that cause enhancement of particle settling result
in a basic uniform settling velocity of 20 m/day for particles less
than 20 µm. A 20 µm particle also has a settling velocity of approximately 20 m/day in Howe Sound water. These two observations
support the Kranck Theory (see papers by K. Kranck, 1973 and 1975)

which states "...as progressively larger flocs are formed, the transport velocities become more uniform and particle collision less frequent. Eventually a {steady} state is reached in which the settling velocities of the largest flocs equal the velocities of the largest grains and further flocculation ceases....as flocs become larger they also become more unstable and the stage where all sediment particles have the same speed is approached, but not reached." In the case of Howe Sound, the "largest grains" are those over 20 µm. Settling velocities of particles less than 1 µm have been increased over 1400 times.

THEORY OF PARTICLE INTERACTION

     Particles enter the fjord as single entities except for a) silt size grains of quartz, feldspar and hornblende which already possess attached particles of clay, and b) clay clasts, the origin of which may be similar to river mudball aggregates. As the surface layer begins to mix with sea water, the suspended load is diluted downward, injecting some of the particles into the halocline. Inorganic salt flocculation begins to take place. Clay plates of equal size (below 20 µm) flocculate initially with edge-to-edge orientation. Non-platy minerals, however, continue to fall as single entities, although some get caught and become part of a floc. The initial floc-forming period does not cause an initial acceleration downwards, but increases the likelihood of further particle collision. The colloidal flocs being of low density are still moving downward faster by diffusion and mixing than by gravity settling. During the intial stage, stabilization of the flocs (in terms of electrostatic forces) is slow due to the dilute suspended concentrations, $i.e.$ 1 to 15 mg/ℓ. Organic cohesive forces start attaching part of the organic detritus that comes to contact with the flocs, forming agglomerates. Zooplankton, just under the halocline, can graze on the flocs and agglomerates, producing mineral-bearing fecal pellets. Those non-platy minerals, such as quartz and feldspar, that still settle as single entities, are not consumed by the zooplankton. Some large-grain flocs, due to size and settling velocity, also escape zooplankton grazing. These flocs continue to increase in size, developing face-to-face and face-to-edge particle orientations, until stable three-dimensional deep-water flocs are formed. Growth stops, since further particle attachment would create hydrodynamic instability. The colloidal flocs retain their two-dimensionality with depth and grow into snake-like flocs. These flocs are thought to produce the large vertically aligned mucous-like streamers observed in the water column during dives in northern Howe Sound (Levings & McDaniel, 1973). During descent, some of the agglomerates become colonized by bacteria. Since the agglomerates have both cohesive as well as electrostatic forces in operation, they develop into "formless" three-dimensional blobs (also visually observed, Levings & McDaniel). These blobs contain low density organic matter (i.e. dead phytoplankton) which would increase the porosity of the agglomerates, Sakamoto (1972).

CONCLUSION

Fjords receiving copious glacial run-off are depositional basins for immature clastic sediment intermixed with minor amounts of siliceous biogenic fragments. These fjords offer an excellent example of flocculation and agglomerative sedimentary processes in action.

REFERENCES

Buckley, J. R. 1977: The currents, winds and tides of northern Howe Sound. Unpubl. Ph.D. thesis U. of British Columbia, 238 pp.

Grace, J. T., Grothaus, B. T. and Ehrlich, R. 1978: Non-normal size frequency distributions of sand from divided laminae--- suggested reinterpretation of conventional size-frequency data. J. Sed. Petrology 48, 1193-1202.

Kranck, K. 1973: Flocculation of suspended sediment in the sea. Nature 246, 348-350.

Kranck, K. 1975: Sediment deposition from flocculated suspension. Sedimentology 22, 111-123.

Levings, C. D. and McDaniel, N. 1973: Biological observations from the submersible Pisces IV near Brittania Beach, Howe Sound, B. C. Fish. Res. Bd. Can. Tech. Rept. 409, 23 pp.

Levings, C. D. 1973: Intertidal benthos of the Squamish River Estuary. Fish. Res. Bd. Can. MS. Report, 1218, 60 pp.

Sakamoto, W. 1972: Study on the process of river suspension from flocculation to accumulation in an estuary. Tokyo Daigaku Jaiyo Ken Kyojo Bull. 5, 1-49.

Syvitski, J. P. M. 1978: Sedimentological advances concerning the flocculation and zooplankton pelletization of suspended sediment in Howe Sound, British Columbia: A fjord receiving glacial meltwater. Unpubl. Ph.D. thesis, U. of British Columbia, 291 pp.

DEPOSITIONAL PROCESSES IN AN ANOXIC, HIGH SEDIMENTATION REGIME

IN THE SAGUENAY FJORD

John N. Smith, Charles T. Schafer and Douglas H. Loring

Atlantic Oceanographic Laboratory
Bedford Institute of Oceanography
Dartmouth, Nova Scotia

INTRODUCTION

The sediment dynamics of north temperate estuaries are seasonally modulated by factors such as ice cover, wind speed, and by the
magnitude of the spring freshet. Anomalous events such as land-
slides, storms or consecutive years of high rainfall are also responsible for significant variations in annual sediment transport
and deposition rates in coastal marine environments. Bioturbation
of sediments deposited in these environments usually obscures much
of the sedimentary and paleontologic record, however large amounts
of organic detritus can lead to low O_2 concentrations in or above
sediments, causing reduced benthic activity, and unconformities may
be preserved. The head of Saguenay Fjord in Eastern Canada has
been identified as an area where detrital organic input to sediments
is comparatively high.

CORE DESCRIPTION

A 10 cm diameter Leheigh gravity core (Station 18-76, Fig. 1)
was collected in the upper reaches of the Saguenay Fjord on April
20, 1976 and was analyzed for various geochemical, textural, and
paleontological parameters to reconstruct the nature of depositional
processes in this part of the Fjord over the past decade.

The Saguenay Fjord occupies a long (93 km) narrow (1-6 km)
submerged valley that joins the St. Lawrence estuary. About 90%
of freshwater runoff into the fjord is derived from the Saguenay
River. The river has a mean monthly discharge rate (between 1960

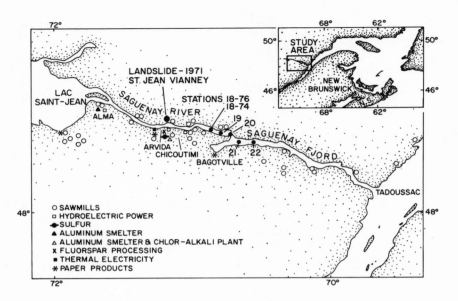

Figure 1.   Location of Saguenay Fjord and the coring stations
            (black dots) in relation to the St. Lawrence Estuary.

and 1975) of 1581 m^3s^{-1}, with monthly discharge levels as high as
4504 m^3s^{-1} leading to resuspension and transportation of bottom
sediments.

    A catastrophic landslide occurred on the west shore of Riviere
Petit Bras at Saint Jean Vianney near Arvida on May 4, 1971. This
event resulted in the displacement of an estimated 7.6 × 10^6 m^3
(25 × 10^6 tons) of older Mer de LaFlamme raised marine sediments
into the Riviere Petit Bras which drains into the Saguenay River.

    The sedimentological features of the core (Fig. 2) are high-
lighted by the occurrence of alternating black and grey layers of
sediment that were readily evident in the freshly split core. The
grey layers contain a comparatively large amount of find sand-size
quartz particles while the black layers are finer grained, and are
much richer in organic matter (Fig. 3). Much of the organic matter
in the core consists of woody fibers derived from upstream pulp and
paper and sawmill outfalls (Loring, 1975; Pocklington and Leonard,
1979).

    Calcareous foraminifera occur only in the grey sediment layers
where they are associated with fresh water thecamoebian specimens.
The calcareous forms represent a distinctive shallow marine association that is no longer common in the upper part of the fjord.

CORE 763III DESCRIPTION

COLOR	SEDIMENTARY STRUCTURES		MICROFOSSILS	LENGTH
BLACK	0-4cm - DISTORTED ORGANIC-RICH SEDIMENT GRADING INTO FINELY MOTTLED INORGANIC SEDIMENT.		THECAMOEBIANS 0-2cm ONLY	4 8
GREY	8-23cm - FINELY MOTTLED INORGANIC SEDIMENT GRADING INTO THIN, DISCONTINUOUS, WAVY LAMINATIONS CHARACTERIZED BY HIGHER ORGANIC COMPONENT.		12-13cm THECAMOEBIANS ONLY	12 16 20
BLACK	23-33cm - INCREASED AMOUNTS OF ORGANIC MATTER AND FINER DISCONTINUOUS LAMINAE ESPECIALLY BETWEEN 25cm and 30cm LEVELS.		27-28cm THECAMOEBIANS ONLY	24 28 32
GREY	32-37cm - INCREASED INORGANIC SEDIMENT COMPONENT. DISCONTINUOUS LAMINAE ARE COMPARATIVELY RARE BETWEEN 33cm AND 36cm LEVELS AND BETWEEN 39 AND 42cm LEVELS.		THECAMOEBIANS 39-40cm AND ONE SPECIMEN OF E. EXCAVATUM.	36 40 44
BLACK	44-47cm - DISCONTINUOUS WAVY LAMINAE. 47-52cm PRONOUNCED INCREASE IN CONTINUOUS RELATIVELY STRAIT LAMINAE WITH HIGHER CONCENTRATIONS OF ORGANIC MATTER BETWEEN 48-51cm.		49-50cm THECAMOEBIANS	48 52
DARK GREY	52-60cm - ALTERNATING LAYERS OF RELATIVELY STRUCTURELESS INORGANIC SEDIMENT AND DIS-CONTINUOUS, WAVY ORGANIC-RICH LAMINAE. 60-74cm - ORGANIC-RICH CONTINUOUS LAMINAE BECOME PRONOUNCED.		BARREN AT 56-57cm INTERVAL	56 60 64 68
LIGHT GREY	74-90cm - COMPARATIVELY INORGANIC SEDIMENT CHARACTERIZED BY FAINT RANDOMLY ORIENTED PARTICLES.		NUMEROUS FORAM-INFERA IN 78-79cm INTERVAL. E. SUBARCTICUM P. ORBICULARE T. ISLANDICA C. COMPLANATA THECAMOEBIANS	72 76 80 84 88
B L A C K	90cm - PRONOUNCED INCLINED UNCONFORMITY. SEDIMENTS ARE DISTINCTLY LAMINATED TO ABOUT 106cm. AT THIS DEPTH THE LAMINAE GIVE WAY TO ALTERNATING ORGANIC-RICH AND THINNER IN-ORGANIC LAYERS THAT EXTEND TO ABOUT THE 140cm LEVEL. ORGANIC LAYERS ARE EVIDENT AT THE 110, 114, 118, 123, 131 AND 138cm LEVELS AND SHOW INCREASED THICKENING STARTING AT THE 131cm LEVEL.		THECAMOEBIANS 90-100cm THECAMOEBIANS 90-100cm	92 96 100 104 108 112 116 120 124 128 132 136 140

Figure 2.   Sediment characteristics and occurrence of microfossil species in various intervals of core 76-3111.

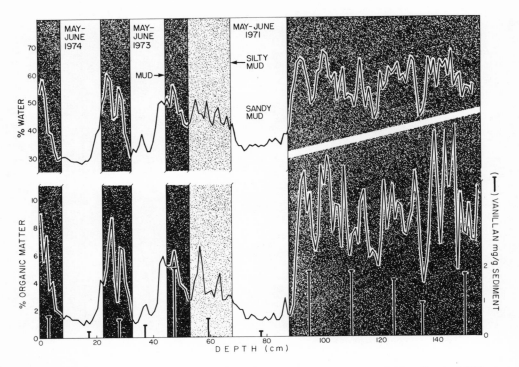

Figure 3.   Organic matter concentrations, % of water and vanillin
concentration in the sediment core 76-3111.

These species inhabited the Saguenay fjord area during the Mer de
LaFlamme marine transgression 8200 to 10,200 years before the present (Ochietti and others, 1977). Mer de LaFlamme sediments were
subsequently raised above the present river level as a result of
post glacial isostatic rebound.

DEPOSITIONAL PROCESSES AND TIME STRATIGRAPHY

High excess Pb-210 activities occur in the black layers of
sediment (Smith and Walton, 1979) while the activities in the grey
layers are an order of magnitude lower, only slightly greater than
the Ra-226 supported background levels. A sediment accumulation
rate of 4.0 ± 0.8 g cm^{-2} yr^{-1} (7 cm yr^{-1}) was estimated for the
core based on the Pb-210 results. For this rate of sediment accumulation, it is estimated that the upper boundary of the lowest
black layer of the core, which appears at a mass-depth (excluding

the mass of overlying grey sediment) of 23 gm cm^{-2}, was deposited during 1970 $\pm$ 2 years. Pb-210 dates for deposition of the upper two grey sediment layers correspond to 1972 $\pm$ 1 year and 1975 $\pm$ 1 year respectively.

The Pb-210 geochronology is consistent with the transport of a portion of Mer de LaFlamme sediment into the fjord during the St. Jean-Vianney landslide in May, 1971, resulting in the deposition of the lowest grey, sandy mud layer. The fossil and textural characteristics of the two upper, younger layers indicate that these layers represent landslide material that was initally deposited in river and tributary channels and subsequently transported from the channels to the head of the fjord during later periods of high river discharge. Exceptionally high river discharge events occurred in 1973 and 1974. These events are approximately synchronous with the estimated Pb-210 dates (1972 $\pm$ 1 yr, 1975 $\pm$ 1 yr) of deposition of the younger grey layers that comprise the $\overline{33}$ cm to 45 cm and the 8 cm to 23 cm intervals. The black sediment layers in the core represent organic-rich suspended matter deposited during alternating periods of low to mean river discharge levels.

One notable feature of the sediments deposited at Station 18 is that they include strata of older marine material which have resided in a freshwater geochemical environment for varying periods of time. The lowest grey layer was deposited in 1971 and has a distinctly lighter colour and a much richer marine fossil assemblage compared to the overlying grey layers. The paucity of foraminifera specimens in the 33 cm to 45 cm grey layer (deposited in 1973), and their absence in the uppermost grey layer (deposited in 1974), suggests reworking of residual Mer de LaFlamme sediment that was originally introduced into the river channel in May 1971.

High Hg levels occur in the lowest black layer of sediment while Hg levels in the upper black layers are substantially reduced (Fig. 4). In contrast, Hg levels in the grey layers are uniformly low. A direct positive correlation of Hg and organic matter of 0.73 (P > 0.001; n = 150) supports Loring's (1975) contention that terrigenous organic matter is the main scavenger of Hg in the fjord. The elevated Hg levels stem from outfalls associated with a chloralkali plant located at Arvida 17 km above Station 18. Despite rather large fluctuations in the Hg/O.M. ratio through the lowest black layer, the average value remains constant from the bottom of the core up to the 100 cm level (Fig. 4). Here the Hg/O.M. ratio decreases sharply from approximately 0.80 at 100 cm to 0.20 at 90 cm. Above the 90 cm level it decreases more gradually to a value of 0.04 near the top of the core.

By April 1971, one month prior to the landslide at Arvida, the chlor-alkali plant at this location had reduced their effluent releases of Hg from greater than 13 kg Hg day^{-1} to 0.2 kg Hg day^{-1}

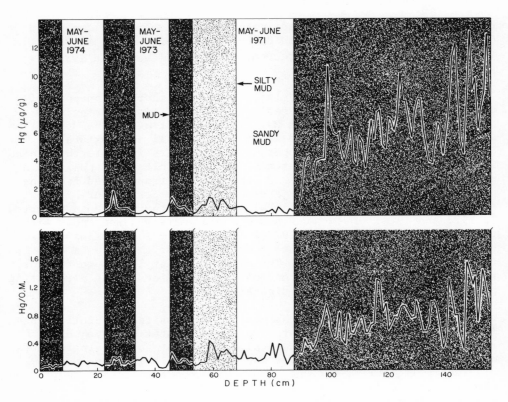

Figure 4.   Concentration (µg/g) of Hg and the ratio of Hg to organic
            matter (Hg/O.M.) in core 76-3111 (Station 18-76).

(Tessier, pers. comm.). Since the residence time of Hg in the water
column between Arvida and Station 18 is less than a month (Loring
and Bewers, 1978), the sharp decrease in the Hg/O.M. ratio measured
from the 100 cm level to the 90 cm level of the core is probably
related to this large decrease in Hg discharge. Although Hg levels
are lowest in the grey sediment layers, the Hg/O.M. ratio is
relatively constant in these layers compared to the adjacent black
layers. This suggests that even during the high river discharge
events that were responsible for the transport and deposition of
the grey sediment layers, the normal flux of Hg (associated with
organic matter) was maintained.

    The transport of landslide-derived material into the Saguenay
Fjord during periods of high river discharge since 1971 suggests
that this same fluvial mechanism may be responsible for the development of the closely spaced laminae observed in X-radiographs and
the pronounced variations in the percent of organic matter measured

through the pre-1971 portion of the core. Comparison of the organic matter profile (Fig. 3) with the fresh-water discharge record suggests that organic matter concentrations are increased in sediments deposited during periods of low river discharge. It appears that the three periods (1970, 1966 and 1964) of highest average monthly river discharge during the 10 year interval prior to 1971, are probably responsible for the deposition of the sediment strata at the 96-99 cm, 116-120 cm and 134-137 cm intervals, respectively.

REFERENCES

Loring, D. H. 1975: Mercury in the sediments of the Gulf of St. Lawrence. Can. J. Earth Sci. 12, p. 1219.

Loring, D. H. and J. M. Bewers 1978: Geochemical mass balances for mercury in a Canadian fjord. Chem. Geol. 22, 309-330.

Ochietti, S. and C. Hillaire-Marcel 1977: Chronologie 14C des Evenements Paleogeographiques du Québec Depuis 14,000 Ans. Geographie Physique et Quarternaire 31, 123-133.

Pocklington, R. and J. D. Leonard 1979: Terrigeneous organic matter in sediments of the St. Lawrence Estuary and the Saguenay Fjord. J. Fish Res. Board Can. 36, (in press).

Smith, J. N. and A. Walton 1979: Sediment accumulation rates and geochronologies measured in the Saguenay Fjord using the Pb-210 dating method. Geochim. Cosmochim. Acta (in press).

OBSERVATIONS OF THE SEDIMENTARY ENVIRONMENTS OF FJORDS ON

CUMBERLAND PENINSULA, BAFFIN ISLAND

R. Gilbert

Department of Geography
Queen's University
Kingston, Ontario
Canada

INTRODUCTION

Preliminary studies on the fjords which lead from Penny icecap on Baffin Island provide indication of the nature of sedimentation and its control in a region previously uninvestigated (figure 1). The head of Coronation Fiord is occupied by a major source of ice contact, marine sediment; Maktak Fiord is separated by 13 kilometres of active sandur from a similar large glacier; while in Pangnirtung Fiord a number of smaller glaciers are distributed farther up the valley from the fjord. Maktak and Coronation Fiords reach a maximum depth of approximately 550 metres seaward of their confluence whereas Pangnirtung Fiord has maximum depth of 160 metres behind a sill only 12 to 22 metres below mean sea level.

OCEANOGRAPHY

Morphology of the fjords, especially the sill in Pangnirtung Fiord, and the volumes of fresh and salt water entering the fjords are probably the major factors along with the strong winds which occur in these valleys and the rejection of brine with freezing in determining the pattern of circulation and the degree of exchange of fjord water with the sea beyond (Neilson and Hansen, this volume). Measurements of dissolved oxygen through the summer of 7 to 9 ml/l in the bottom water of the fjords indicates that circulation and exchange at depth is frequent especially in Pangnirtung Fiord where the tidal range of 6.7 metres occurs over the shallow sill. These conditions permit a rich and diverse bottom fauna to flourish (Gilbert, 1978). In Pangnirtung Fiord the brittle stars *Ophiocten*

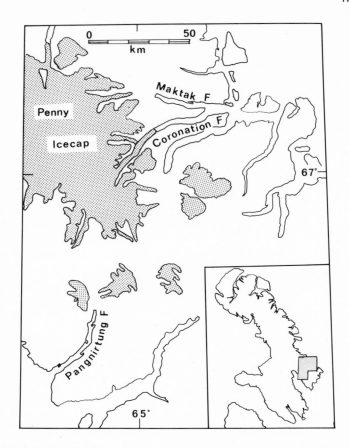

Figure 1.   Location of Maktak, Coronation and Pangnirtung Fiords.

sericeum and *Amphiophiura* sp., marine worms *Onuphis (Nothria)*
conchylega, and mollusc *Portlandia arctica* are especially important
in the bioturbation of the sediments.

SEDIMENTS

        The melting of glacial ice and snow provides water abundantly
charged with fine sediment. In this overflowing fresh water,
sediment concentrations are in excess of 20 mℓ/ ℓ near the deltas
at the heads of the fjords, but sediment is rapidly lost by settling,
so that by 20 to 30 kilometres down the fjord values are normally
2 to 3 mg/ℓ. In Pangnirtung Fiord interpretation of sedimentary
processes is not possible because bioturbation of the bottom deposits
is so severe. In addition, ice rafted drop-stones from the extensive

tidal flats around the shore further turbate the sediments, and
currents along the floor of the fjord near the mouth are apparently
sufficient to sweep the bottom clean of fine sediment. In Maktak
Fiord where the tidal range is 1.5 metres and the circulation is
probably less well developed, the density and diversity of benthic
fauna is less, even though the fjord is oxygen rich to the bottom.
Turbation is also less severe, and the sediments document three
distinct processes: settling of fine, highly flocculated rock
flour from suspension in overflowing fresh water, turbidity currents
generated by slumping of sediment deposited on the steep valley sides,
and the deposition of eolian silts transported from the sandur
surface and deposited on the ice during winter to be released by
melting in summer. The grainsize diagrams of figure 2 illustrate
the differences among these sediment types.

    In Maktak and Coronation Fiords, echo sounding indicates that
more sediment is being deposited along the right hand (south)sides
of the fjords, since their floors slope downward to the north by
several metres in most locations. Calculation of the Rossby Number
(Smith, 1978) indicates that for current velocity of less than about
0.3 m/s, Coriolis force will dominate. In Maktak and Coronation
Fiords surface currents are usually less than this, whereas in
Pangnirtung Fiord where tidal currents are strong, no cross-fjord
bottom slopes were detected. Observations of the muddy overflowing
water and examination of air photographs and satellite images
confirms these patterns of overflow. Similar patterns have been
observed in proglacial lakes (Gilbert, 1975; Smith, 1978).

    The fine sedimentary deposits are interrupted at intervals of
several centimetres by layers about 1 cm thick of coarser sediment
(figure 2) attributed to deposition from turbidity currents. In the
vicinity of riegels in Pangnirtung and Maktak Fiords, trenches up
to 200 metres wide and 15 to 20 metres deep occur in the otherwise
nearly flat floor. It may be that especially powerful turbidity
currents passing these constrictions at the riegels have sufficient
velocity to.erode the trenches. The presence of turbidities in
fjord sediments is well known (for example, Holtedahl, 1965) and
trenches have been observed in the floors of other fjords (Hoskin
and Burrell, 1972). The currents could not be generated directly
from inflowing fresh water since, to become more dense than the
fjord water which has salinity of 30 to 32 $^0/_{00}$, a concentration
of suspended sediment of more than 40 g/ℓ would be required. This
is 1 to 2 orders of magnitude greater than concentrations observed
in inflowing glacial streams. However, it is possible that the
currents may be generated from the slumping of sediment from the
steep valley sides. Mounds of slumped sediment and contorted
sediments are commonly observed in echograms. Figure 3 represents
a preliminary calculation of the possibility that such currents
would erode these trenches. Consideration of the flow velocity
equation (Middleton, 1966) with appropriate friction (Dick and

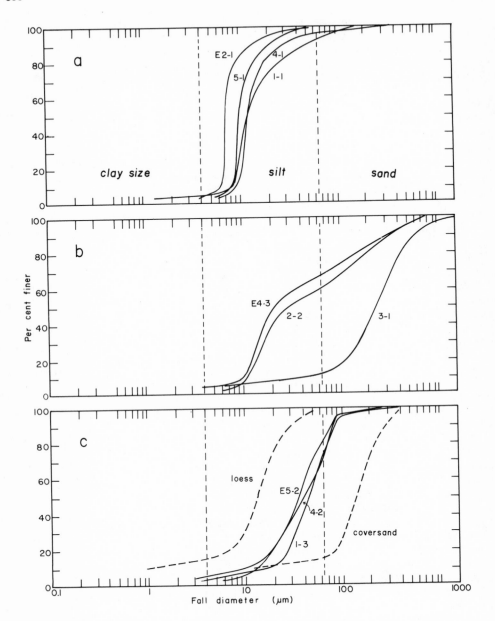

Figure 2.   Grainsize diagrams of flocculated sediment from Maktak
            Fiord.   Sediment deposited (a) from suspension in over-
            flowing water, (b) by turbidity currents, and (c) upon
            release by melting of eolian material.   Samples were taken
            from the fjord floor between 0.6 km (E3) and 13.7 km
            (E5) from the sandur.

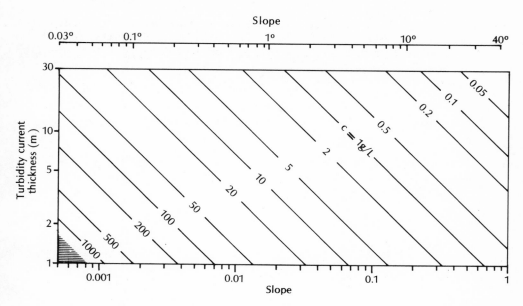

Figure 3. Calculated relation among variables governing the flow of a turbidity current (where c is the concentration of suspended sediment) of sufficient velocity to erode unconsolidated fine sediment on the fjord floor.

Marsalek, 1973) allows calculation of the concentration of sediment necessary to provide a density difference of sufficient magnitude to drive currents of different sizes down different slopes at a velocity sufficient to erode unconsolidated, fine sediment (about 0.3 m/s). Figure 3 shows that even on low slopes of the order of 1^0, currents of sufficient size to nearly fill the trenches observed require only moderate concentrations of sediment to drive them. Currents generated on the steeper valley sides might be considerably thinner and faster flowing.

REFERENCES

Dick, T.M. and J. Marsalek. 1973. Interfacial shear stress in density wedges. Canadian Hydraulics Conference, University of Alberta, Edmonton.

Gilbert, R. 1975. Sedimentation in Lillooet Lake, British Columbia. Canadian Journal of Earth Sciences, 12, pp. 1697-1711.

Gilbert, R. 1978. Observations on oceanography and sedimentation at Pangnirtung Fiord, Baffin Island. Maritime Sediments, 14, pp. 1-9.

Holtedahl, H.   1965.   Recent turbidities in the Hardangerfjord,
     Norway.   Submarine Geology and Geophysics, W.F. Whittard and
     R. Bradshaw (editors), Proceedings of the Seventeenth Symposium
     of the Colston Research Society, Butterworths Scientific Pub-
     lications, London, pp.   107-140.

Middleton, G.V.   1966.   Experiments on density and turbidity
     currents II.   Uniform flow of density currents.   Canadian
     Journal of Earth Sciences, 3, pp.   627-637.

Smith, N.D.   1978.   Sedimentation processes and patterns in a
     glacier-fed lake with low sediment input.   Canadian Journal
     of Earth Sciences, 15, pp.   741-756.

VOLUME OF WATER PUMPED AND PARTICULATE MATTER DEPOSITED BY THE

ICELAND SCALLOP (*CHLAMYS ISLANDICA*) IN BALSFJORD, NORTHERN NORWAY

Ola Vahl

Institute of Fisheries
University of Tromsø
P.O. Box 488
N-9001 Tromsø, Norway

INTRODUCTION

Suspension feeding bivalves obtain their food from organic particles in the surrounding water which is pumped through their mantle cavities. The particles are retained by the gill. All suspension feeding bivalves investigated retain, with 100% efficiency, particles down to a size of 9 µm, and many retain all particles larger than 2 to 3 µm (see Vahl 1979 a for review), *i.e.* suspension feeding bivalves retain almost all species of phytoplankton. Since the suspension feeding bivalves pump large volumes of water through their mantle cavities (see Jørgensen 1975 and Winter 1978 for review) large volumes are cleared of phytoplankton. Therefore, when suspension feeding bivalves occur in high densities, they harvest a sizeable portion of the phytoplankton production of the area in which they live. The present paper deals with this aspect of the biology of the Iceland scallop (*Chlamys islandica*).

MATERIAL AND RESULTS

The study was carried out on *Chlamys islandica* from a bed situated at 32 m below MLWS. The bed is found at depths of 20–60 m in Balsfjord, northern Norway (lat. $69^0 30'$N, long. $18^0 54'$E), and covers an area of 13 km^2. Since the surface area of Balsfjord is about 3.53×10^8 m^2 the scallop bed covers an area equivalent to 3.7% of the surface of the fjord.

The present study is based on an age-specific energy budget (Vahl 1979b) and a population energy budget (Vahl 1979c) for the

Iceland scallop, and on a knowledge of the hydrography (Schei 1977)
and phytoplankton production (Eilertsen 1979) of Balsjford.

The scallop population has not been quantitatively sampled
outside the study area but rough estimates, using a calibrated
scallop dredge, indicate that the density outside the study area is
lower. However, the scallops outside the area are larger, so that
the volume pumped by the scallops living on a given area of bottom
is probably much the same over the whole bed.

The water masses in Balsfjord are well mixed for most of the
year (Fig. 1), and a pycnocline is not established until June. The
spring phytoplankton bloom commences in late March or at the beginning of April. By the end of October when the pycnocline breaks
down, the phytoplankton production is almost nil (Eilertsen 1979,
Schie pers. comm.). The annual phytoplankton production in
Balsfjord is 109.7 gC m^{-2}, and 57% of this production has occurred
before the pycnocline is established (see Eilertsen 1979). The
phytoplankton production is probably underestimated by about 50%
(Eilertsen pers. comm.) and using a factor of 1.7 for converting
gC into dry weight (Cushing 1959), the total phytoplankton production in Balsfjord before the pycnocline is established (*i.e.*
production in the period 1st April to 30th June) is 5.63×10^4
tonnes dry weight. In the same period the scallops in the study
area have retained 1494 g organic matter m^{-2}. By extrapolating
this value, the whole bed would have retained 1.94×10^4 tonnes
dry weight of particulate organic matter, or 34% of the phytoplankton
production in Balsfjord during this period.

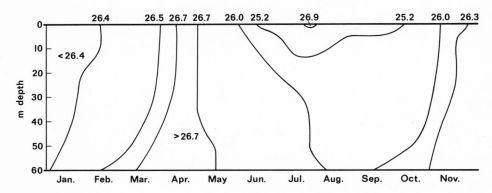

Figure 1. σ_t at Berg in Balsfjord (northern Norway) in 1976 (from
            Schei 1977).

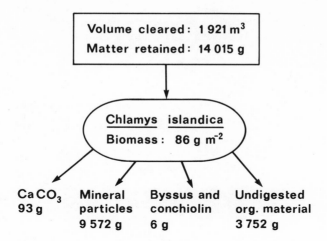

Figure 2.   Volume cleared of particles larger than (>4µm) and
            amounts deposited annually (m^{-2}year^{-1}) by *Chlamys
            islandica*.

Annually the scallops living in one square m of the study area
clear a volume of 1921 m^3 of water of 14,015 g of particulate material (Fig. 2). The annual metabolism and production amounts to
192 g m^{-2}, so that almost all of this material is deposited, either
as CaCO$_3$, mineral particles, byssus and organic material from the
shells, or is voided as undigested organic material.

The average particle load is 7.4 mg ℓ^{-1}. Since the currents
over the scallop bed reach about 70 cm sec^{-1} (Skreslet ·1973), the
major part of the particulate matter is probably involved in a con-
tinous process of deposition by the scallops and resuspension by
the water movement. From a knowledge (Vahl 1979b,c) of the amount
of matter metabolized (R), production of organic material by the
scallops (P) and the amount deposited as dead shells, between
April 1st and June 30th, an estimate can be made of how much of the
organic material deposited is available for resuspension (Table 1.).
This amounts to 1376 g m^{-2} which is 92% of the organic matter originally retained.

DISCUSSION

The total amount deposited (Fig. 2) is 13,423 g m^{-2} year. This
is well within the range reported in the literature. For example,
Haven and Morales-Alamo (1972) found that the annual biodeposition

Table 1.   Amount of particulate organic matter deposited by the
           Iceland scallop available for resuspension.

Part. org. matter retained   1494 g m^{-2} dry wt.
R + P   94 g m^{-2} dry wt.
CaCO$_3$ 24 g m^{-2} dry wt.   118 g m^{-2} dry wt.

= Available for resuspension   1376 g m^{-2} dry wt.

in an oyster bed was 7600 g and Verwey (1952) found that 200,000
g m^{-2} year was deposited by mussels and cockles.

The estimated 34% of the spring phytoplankton production retained by the scallops in Balsfjord is based on an extra-polation
from the study area to the whole scallop bed.   In Balsfjord, the
scallops are not the only animals which derive their food from the
phytoplankton production.   Since the fraction of phytoplankton production retained by the scallops is probably out of proportion to
their biomass in the fjord, the questions arise: How much of the
primary production retained by the scallops becomes available to
other organisms and by what processes is this achieved?

After the pycnocline is established, the scallops continue to
feed and grow throughout the summer (Vahl 1979b,c).   Since the
phytoplankton production is not directly available to them, because
of the pycnocline, their food at this time must be resuspended
organic material.   This may amount to the equivalent of 92% of the
organic matter originally retained.   This resuspended material has
a low C/N-ratio of 8-9 (Sundet 1979), which indicates a high protein
content.   Therefore, the resuspended material is an adequate food
source.   Since the high protein content cannot be attributed to
living phytoplankton, it must be due to microorganisms associated
with the particles.

The Iceland scallop, therefore, through its high population
density and pumping rates and low utilization of the retained material, deposits large amounts of organic material in spring when
phytoplankton is available.   Through microbial activity this material continues to have an adequate nutritive value and, when resuspended, becomes available to the scallops during the part of the
year when phytoplankton is not available.   On an annual basis, however, the metabolism and production of the scallops amounts to only
(192/1494) × 100 - 13% of the matter they retained before the pycnocline was established.   Therefore, about 87% of the organic matter
originally retained by the scallops is made available to other benthic organisms through the scallops' ability to concentrate particulate organic matter from dilute suspensions.

ACKNOWLEDGEMENT

Thanks are due to Dr. M. Jobling for correcting the English.

REFERENCES

Cushing, D. H. 1959: On the nature of production in the sea. Fish. Invest. Ser. 2,(6), 1-40.

Eilertsen, H. C. 1979: Planteplankton, minimumsfaktorer og primærproduksjon i Balsfjorden, 1977. Cand. Real. post-graduate thesis, University of Tromsø. 130 pp.

Haven, D. S. and Morales-Alamo, R. 1972: Biodeposition as a factor in sedimentation of fine suspended solids in estuaries. In Environmental framework of coastal plain estuaries. Ed. B. W. Nelson, Geology Society of America, Mem., 133.

Jørgensen, C. B. 1975: Comparative physiology of suspension feeding. Ann. Rev. Physiol. 37, 57-79.

Schei, B. 1977: Hydrografiske undersøkelser i Balsfjord 1976. Marinbiologisk Stasjon, University of Tromsø (mimeographed), 31 pp.

Skreslet, S. 1973: Water transport in the sill area of Balsfjord, North Norway, during vernal meltwater discharge. Astarte, 6, 1-6.

Sundet, J. H. 1979: Arsvariasjoner i C/N-forholet i partikulært materiale og i tørrvekt og biokjemisk sammensetning av haneskjell {Chlamys islandica (O.F. Müller)} far Balsfjorden, Nordnorge. Cand. Real postgraduate thesis, University of Tromsø. 64 pp.

Vahl, O. 1979a: ·Particle capture by filter feeding bivalves. CRC Handbook of nutrition and food. (in press).

Vahl, O. 1979b: Feeding, growth and an age-specific energy budget for the Iceland scallop {Chlamys islandica (O.F. Müller)} from 70°N. (sub judice).

Vahl, O. 1979c: A population energy budget for the Iceland scallop {Chlamys islandica (O. F. Müller)}. (sub judice).

Verwey, J. 1952: On the ecology of distribution of cockle and mussel in the Dutch Waddensea, their role in sedimentation and the source of their food supply, with a short review of the feeding behaviour of bivalve molluscs. Arch. Neerl. Zool. 10, 172-239.

Winter, J. E. 1978: A review on the knowledge of suspension feeding
    in lamellibranchiate bivalves, with special reference to art-
    ifical aquaculture systems. Aquaculture 13, 1-33.

SEDIMENT SURFACE REACTIONS IN FJORD BASINS

David Dyrssen

Chalmers University of Technology
University of Gothenburg
Sweden

INTRODUCTION

A phosphate molecule in the basin water of the By Fjord could very well have originated from a plankton formed in the upwelling areas off West Africa, carried across the Atlantic by the equatorial currents, brought up around Scotland into the North Sea by the Gulf Stream and from the Skagerack part of the North Sea into the surface water of the By Fjord on the west coast of Sweden. There the phosphate molecule could be built into a new plankton species in a spring bloom, which later on was eaten by zooplankton. Finally it reached the bottom of the By Fjord in the form of a fecal pellet. During the decomposition of the organic matter of these fecal pellets the organic phosphate eaters were hydrolyzed and the phosphate molecule was released into the basin water.

The By Fjord investigation was, however, not carried out to study the phosphate circulation of the Atlantic Ocean. The problem was just to find out how the installation of a phosphate removal unit in the sewage treatment plant of the city of Uddevalla would change the phosphate content of the surface water of the By Fjord. Fig. 1 shows the location of the fjord at $58^0 20.00'$ N and $11^0 52.65'$ E inside the islands of Orust and Tjörn. The cross section showing the sill (11 m deep) and the deep (50 m) is also included in Fig. 1. Further details are given in the paper of Svensson at this fjord oceanographic workshop.

AREA–DEPTH RELATIONS

The fjord area at different depths for the By Fjord basin are given in Table 1.

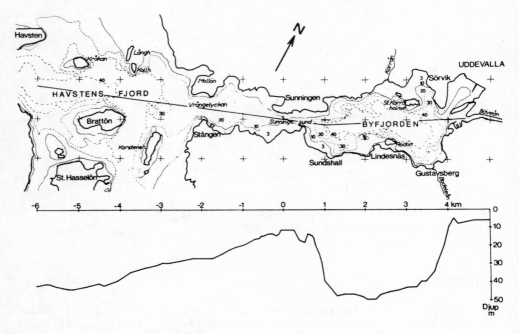

Figure 1.   Map and vertical cross section of the By Fjord.

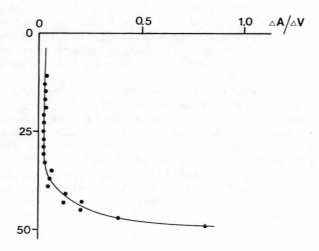

Figure 2.   Area per layer volume (m^{-1}) vs. depth (m).

Table 1

Area in km^2 (at depth in m) for the By Fjord basin

4.16 (10), 3.86 (12), 3.64 (14), 3.41 (16), 3.20 (18),
2.99 (20), 2.82 (22), 2.66 (24), 2.53 (26), 2.40 (28),
2.29 (30), 2.17 (32), 2.03 (34), 1.79 (36), 1.59 (38),
1.44 (40), 1.09 (42), 0.70 (44), 0.46 (46), 0.20 (48),
0.02 (50).

The tracer studies described in the paper of Svensson showed
that the horizontal mixing is completed in two weeks while the
vertical spreading takes many years. All results of the investigation are consistent with the fact that the release of inorganic
matter mainly takes place in the sediments and not in the water
column. Thus the bottom area per layer volume, $\Delta A/\Delta V$, has an
important influence on the depth profiles of the concentrations of
hydrogen carbonate, ammonium, phosphate and hydrogen sulfide. ΔA
is calculated from $A_n - A_{n+2}$ and ΔV from $A_n + A_{n+2}$ in Table 1. $\Delta A/\Delta V$ (in m^{-1}) is shown in Fig. 2. From this figure it may be seen
that $\Delta A/\Delta V$ is approximately 0.03 m^{-1} down to 33 m, but increases
considerably (up to 1 m^{-1}) below 40 m. If the release rates are
R mmol $m^{-2}d^{-1}$ then the influence per layer volume is $R\Delta A/\Delta V$ mmol
$m^{-3}d^{-1}$. Thus for the anoxic-sulfidic basin water below the pycnocline at 15-20 m the concentration vs. depth profiles should have
the shape of the curve in Fig. 2. However, as may be seen from
Figs. 3, 5, 6, and 7 the depth profiles for the alkalinity (A_t),
total carbonate (C_t), hydrogen sulfide, ammonium and phosphate do
not have the shape of the curve in Fig. 2.

This can be accounted for by vertical mixing that will spread
the released matter from the bottom water of the basin to the
upper parts. Furthermore, one should mention the observation from
the dumping of dredge spoils: The soft muds flow towards the deep
parts of the basin. This means that more material could be released
per unit area of the deeper parts.

THE VERTICAL DIFFUSION COEFFICIENT

In Fig. 7 the depth profiles have been normalized by $C_t:H_2S:
NH_4:PO_4 = 80:36\ 2/3:6\ 2/3:1$. For the depth region 40-50 m
the slope of the curve (dashed line in Fig. 7) has the value of
0.0024 μM PO_4 per cm. The rhodamine tracer experiments according
to Svensson imply that a linear diffusion equation can be applied
to the By Fjord basin. Thus the number of micromoles generated
per second from a bottom area of 1 cm^2 is

$$F = -\frac{D}{1000} \cdot \frac{dc}{dz}$$

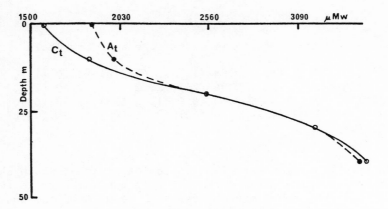

Figure 3.   Alkalinity (dashed line A_t) and total carbonate (C_t) depth profiles for the By Fjord.

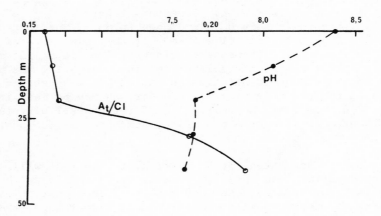

Figure 4.   Specific alkalinity in mM_w per o/oo Cl and in situ pH as a function of the fjord depth.

where $D = 0.04$ cm^2s^{-1} is taken from Svensson (Fig. 5) and $dc/dz = 0.0024$ μM PO_4 per cm. The different values of F for each species are obtained by scaling up dc/dz for each one. The result is

Total carbonate   6.6   mmol per m^2 and day
Hydrogen sulfide   3.0   - " -
Ammonium   0.55   - " -
Phospate   0.083   - " -

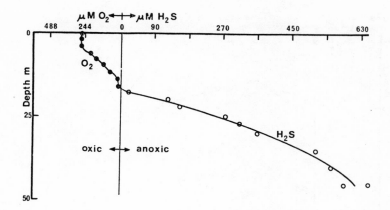

Figure 5. Depth profiles for oxygen and hydrogen sulfide for the
By Fjord.

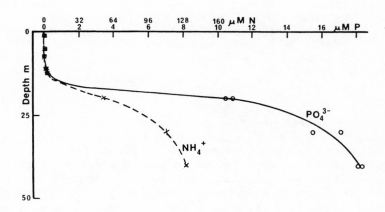

Figure 6. Depth profiles for dissolved inorganic nitrogen and
phosphate.

In order to vizualize these rates of bacterial decay it would
take approximately 20 years for a slice of white bread (23 g. 0.0081
m^2, 60% carbohydrate) to vanish from the bottom of the By Fjord.

CHEMICAL REACTIONS

There should be a rather steady, but seasonally varying, supply
of fresh material to the sediment of the fjord basin. Since the

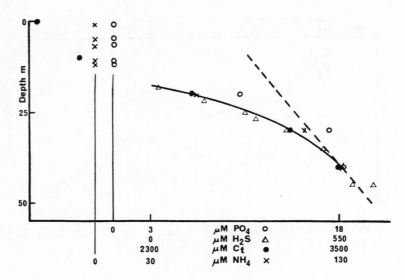

Figure 7.   Normalized values for total carbonate (●), sulphide (Δ),
            ammonium (×) and phosphate (O) for the stagnant bottom
            water of Byfjorden below 17 m. The concentration gradient
            - dc/dz is calculated from the dashed line.

salinity of the basin water is 30-31 $^0/_{00}$ the sulfate concentration
is approximately 25000 µM. Thus 630 µM hydrogen sulfide (see Fig. 5)
only represents the reduction of 2.5% of the sulfide. It is therefore reasonable to assume that the release of hydrogen carbonate,
hydrogen sulfide, ammonium and phosphate are controlled by pseudo
zero-order reactions. All organic material will of course not be
oxidized or hydrolyzed at the sediment surface. Some will be
embedded and subjected to slower decomposition (diagenesis) which
sets it mark on the interstitial water (cf. Murray et al., 1978,
Bågander, 1977, Engvall, 1978 and Holm, 1978). Gas formation
will disturb the soft sediments and bring some of the pore water
to the basin bottom water. The bulk diffusion coefficient of
molecules in interstitial waters is four orders of magnitude lower
($2 \cdot 10^{-6}$ cm^{2}s^{-1}) than those measured for the stagnant basin water
of the By Fjord (see Svensson). In general the pore water sulfate
is exhausted at a sediment depth of 20 cm.

The main carbon source available for bacteria in surface
sediments is most likely carbohydrate polymers in the form of mono-
and disaccharides (sugars) and polysaccharides (starch, dextrane,
cellulose). The di- and polysaccharides can be hydrolyzed, probably

by enzymatic processes, to glucose and other monosaccharides. By
fermentation and respiration reactions the monosaccharides are
oxidized to carbon dioxide and water. Before the anoxic condition
is reached in the fjord basin, the overall reaction can be generalized as:

$$(CH_2O)_6 + 6\ O_2 \rightarrow 6\ CO_2 + 6\ H_2O \qquad\qquad (1)$$

Hereby we have neglected the intermediate steps (cf. Lehninger,
1970). The production of carbon dioxide leads to a shift in the
disproportionation equilibria

$$2HCO_3^- \rightleftharpoons CO_2 + CO_3^{2-} + H_2O$$

and a decrease in $A_t - C_t$ (cf. Dyrssen, 1977). Thus the pH of an
anoxic basin will be low (see Fig. 4).

The oxidation of monosaccharides will, however, also take place
in sulfate-containing anoxic water. The decay reaction with sulfate-
reducing bacteria is

$$(CH_2O)_6 + 3\ SO_4^{2-} \rightarrow 6\ HCO_3^- + 3\ H_2S \qquad\qquad (2)$$

Since $A_t - C_t$ is not altered by this reaction pH will not
change. Fig. 3 shows that A_t and C_t increased by almost equal
amounts. Fig. 7 demonstrates an increase of H_2S by 550 μM for an
increase of A_t and C_t by 1200 μM instead of 1100 μM. This shows
that the stoichiometry of reaction (2) almost holds within the
limits of error. Considering that the basin water is almost certainly saturated with iron (II) sulfide (Dyrssen and Hallberg, 1979)
a loss of 10% hydrogen sulfide could be accounted for by the formation of FeS

$$Fe^{2+} + H_2S \rightarrow FeS(s) + 2\ H^+ \qquad\qquad (3)$$

By this reaction A_t could be lowered but not C_t.

The ammonium originates from the proteins settled on the
sediment surface. The peptide bonds are split by peptidases
(Lehninger, 1970).

$$R_1NHCOR_2 + H_2O \rightarrow R_1NHCOOH + R_2H \qquad\qquad (4a)$$

or

$$R_1NHCOR_2 + H_2O \rightarrow NH_2COR_2 + R_1OH \qquad\qquad (4b)$$

Amino acids may free the ammoniogroup by oxidative degradation,
(Lehninger, 1970):

$$RCH(NH_2)COOH + H_2O + FP \rightarrow RCOCOO^- + NH_4^+ + FPH_2$$

where FPH_2 represents the reduced form of the catalyzing flavoprotein enzyme. There is no record of the amino acid concentrations in the By Fjord basin water. Now we have a method that is also suitable for interstitial water (Lindroth and Mopper, 1979). In one run both ammonia and most amino acids may be determined. Ammonium is most likely the main decomposition product of proteins and we may therefore represent the overall reaction approximately by

$$NHCO + 2 H_2O \rightarrow NH_4^+ + HCO_3^- \qquad\qquad (5)$$

This reaction also leads to an equal increase in A_t and C_t by 100 μM according to Fig. 7.

The decay of organic matter in seawater and sediments leads to the formation of inorganic phosphate. Part of the phosphate on particles may be released under anaerobic conditions (Balzer, 1978, Holm, 1978). Transformation into soluble organic form through a microbial process has been proposed by Fleischer (1978). The organic phosphate may subsequently be hydrolyzed to inorganic phosphate. In the respiration process ATP is the end-product of the electron transport and oxidative phosphorylation (Lehninger, 1970) together with carbon dioxide and water. In ATP the adenine-ribose as well as the ribose-triphosphate groups can be hydrolyzed to adenine, orthophosphate and ribose. The latter, which is a monosaccharide, is oxidized according to reaction (2). The end-products of the decay of organic phosphate are therefore adenine together with carbon dioxide, water and inorganic phosphate

$$A(CH_2O)_6(PO_3)_3Mg^{2-} + 3 H_2O \rightarrow AH + (CH_2O)_6 + MgHPO_4 + 2 H_2PO_4^-$$
$$(6)$$

AH = adenine and HPO_4^{2-} are titrated with hydrochloric acid in an alkalinity titration. In the By Fjord basin the main part of the phosphate was inorganic phosphate (see Fig. 6). The ratio C_t:NH_4^+ :HPO_4^{2-}=80:6 2/3:1 seems to indicate that part of the phosphate is not of planktonic origin (C:N:P = 106:16:1), but comes from phosphate adsorbed on sediment particles. Furthermore Fig. 6 indicates a loss of nitrogen (e.g. by formation of N_2) or incomplete decay of proteins.

EXPERIMENTS WITH RESPIROMETERS, BOXES AND BELL JARS

The results of the release of hydrogen carbonate, ammonium and phosphate and the production of hydrogen sulfide in the By Fjord may be compared with experiments where the consumption of oxygen or sulfate and production of ammonium and phophate are measured. Some recent studies will be considered.

Oxygen Consumption and Sulfate Reduction

An oxygen consumption of 6 mmol m^2d^{-1} is equivalent with the production value of 3 mmol hydrogen sulfide per square meter and day obtained in the By Fjord investigation. Although the oxygen consumption by no means was of zero order Balzer (1978) obtained values between 4.5 and 7.9 mmol $m^{-2}d^{-1}$ in his three bell jar (3.1 m^2) experiments. In two experiments where the enclosed seawater (salinity approx. 17 to 21 $^0/_{00}$) became anoxic the production of hydrogen sulfide was 0.9 and 2.3 mmol $m^{-2}d^{-1}$. Since iron (II) was released from the sediments as the enclosed water became anoxic the low values could be explained by reaction (3), i.e. the formation of iron (II) sulfide. Jansson and Hagström (1978) obtained a sea-bed respiration of 10.3 mmol $m^{-2}d^{-1}$ in core samples from the Baltic Proper during BOSEX 77. Granéli (1977) measured the oxygen uptake of undisturbed lake sediment cores. He found that the uptake varied between 2 and 22 mmol $m^{-2}d^{-1}$ depending on the oxygen saturation. The uptake was bacteria-dominated with minor macrobenthic respiration. Only part of it was chemical. Granéli used both mercury (II) chloride and formalin to measure the chemical oxygen uptake. With sludge sediments Smith Jr. (1973) reached high values of 30 to 50 mmol m^{-2} d^{-1}, 50 to 70% of which was community respiration. In the San Diego trough at 1230 m he found 2.6 mmol $m^{-2}d^{-1}$(Smith Jr., 1974) while Mäkelä and Niemistö (1978) found a mean oxygen consumption of 5.3 mmol $m^{-2}d^{-1}$ in sediments from the Bothnian Sea. In box experiments (see Bågander, 1977) the sulfate reduction (and sulfide production) starts as a zero-order reaction with 5.3 to 6.4 mmol m^{-2} d^{-1}, but turns into a first-order reaction as the sulfate becomes exhausted. The experimental results for normal coastal and lake sediments are thus in the same range as the rates obtained in the By Fjord investigation (i.e. 3 mmol $m^{-2}d^{-1}$ for sulfate reduction corresponding stoichiometrically to an oxygen consumption of 6 mmol $m^{-2}d^{-1}$).

Release of Ammonium

In his bell jar experiments Balzer (1978) showed that the release of ammonium was rather small before the enclosed water became anoxic. Thereafter he obtained 1.6 and 2.6 mmol $m^{-2}d^{-1}$ as compared with 0.55 in the By Fjord investigation. He connected the large rates with the death of benthic organisms. Engvall (1978) obtained 0.55 mmol $m^{-2}d^{-1}$ in one series of box experiments (cf. Dyrssen and Hallberg). From other experiments she found that the concentration of nitrogen in the organic matter was crucial for the release rate.

Release of Phosphate

Balzer (1978) and Holm (1978) both show that phosphate is
fixed together with iron (III) hydroxide in oxic waters. When
the enclosed water becomes anoxic more phosphate is released
than corresponds to the actual content of the organic debris.
For an anoxic milieu Balzer obtained 0.73 to 0.74 mmol $m^{-2}d^{-1}$
which is considerably more than 0.083 obtained in the By Fjord
investigation. Holm (1978) obtained variable pseudo zero-order
release rates of 0.58 to 1.84 mmol $m^{-2}d^{-1}$, but for one box the
rate was as low as 0.18 mmol $m^{-2}d^{-1}$ (cf. Dyrssen and Hallberg,
1979).

PHOSPHATE BUDGET

Within the By Fjord investigation many different determinations
were made. From the exchange current + the inflow of river water
together with phosphate concentrations inside and outside the sill
the yearly discharge of phosphate from the fjord could be calculated
to 15.6 - 17.3 tons of phosphorus. The sewage discharge was 17.2
ton/y and the rivers carried 6.0 ton/y into the By Fjord. The flow
from the anoxic basin was calculated to be 2.1 ton/y. The balance
was obtained by adding 8.0-9.7 tons of phosphorus per year to the
basin by sedimentation. The phytoplankton production varied between
0.1 and 1 g C $m^{-2}d^{-1}$ which gave an average production for the By
Fjord of 1107 ton C/y. For a C/P weight ratio of 41 one arrives
at a formation of 27 tons of planktonic phosphate per year. The
concentration of particulate phosphate varied between 4 and 40
mg P/m^3, which multiplied with the volume of the surface water
gives 0.2 - 2.4 tons of phosphorus in the standing crop (biomass).
Thus the phosphorus is recycled in the surface water before it
settles to the bottom of the basin, maybe as fecal pellets from
zooplankton and fish. Fig. 8:1 summarizes the budget calculations.

It was estimated that Uddevalla could cut down the discharge
of phosphate from 17.2 to 6 ton/y. This would diminish the outflow
of phosphorus from the fjord by 50%, according to Fig. 8:2, but
have very little influence on the standing crop of phytoplankton.
The sedimentation and upwelling rates would be slightly smaller
(see Fig. 8:2).

In practice the total concentration of phosphate in the surface
water of the By Fjord was 50 µg/l before aluminium sulfate was
used to precipitate phosphate in the sewage works. Afterwards
it decreased to 35 µg/l. In the fjord outside the sill (Havstens
fjord, see Fig. 1) the total phosphate concentration in the surface
water was close to 25 µg/l both before and after the reduction of
phosphate at Uddevalla. Outside the islands in Skagerack the

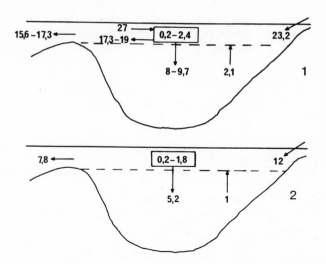

Figure 8.   Phosphate budget for the By Fjord before (1) and after
            (2) phosphate removal at the sewage works of Uddevalla.

concentration of total phosphate was 15 µg/l (0.5 µM), which is
the normal value for the eastern part of the North Sea. In this
part the North Atlantic Ocean water is diluted with Baltic Sea
water. Close to the Norwegian Sea the total phosphate concentration
of the North Sea is around 30 µg/l (1 µM). The efforts by the
city of Uddevalla to remove phosphate have therefore a very limited
range.

REFERENCES

Balzer, W. 1978. Untersuchungen über Abbau Organischer Materie
    und Nährstoff-Freisetzung am Boden der Kieler Bucht beim
    Übergang vom Oxischen zum Anoxischen Miliew. Report Nr. 36
    Sonderforschungsbereich 95. Universität Kiel.

Bågander, L.E. 1977. Sulfur Fluxes at the Sediment-Water Inter
    face - an in situ study of closed systems, Eh and pH. Thesis
    Dept. of Geology, Univ. of Stockholm, Sweden.

Dyrssen, D. 1977. The Chemistry of Plankton Production and
    Decomposition in Seawater. Oceanic Sound Scattering Prediction.
    E. by N.R. Andersen and B.Y. Zahuranec, Plenum, New York. N.Y.
    pp 65-84.

Dyrssen, D. and R. Hallberg. 1979. Anoxic Sediment Reactions -
    A comparison between Box Experiments and a Fjord Investigation.
    Chem. Geol. 24: 151-159.

Engvall, A.-G. 1978. The Fate of Nitrogen in Early Diagenesis
    of Baltic Sediments. A study of The Sediment-Water Interface.
    Thesis, Dept. of Geology, Univ. of Stockholm, Sweden.

Fleischer, S. 1978. Evidence for the Anaerobic Release of Phos-
    phorus from Lake Sediments as a Biological Process.
    Naturwissenschaften 65: 109-110.

Granéli, W. 1977. Measurement of Sediment Oxygen Uptake in the
    Laboratory using Undisturbed Sediment Cores. Vatten, No. 3:
    1-15.

Holm, N.G. 1978. Phosphorus Exchange through the Sediment-Water
    Interface. Mechanism Studies of Dynamic Processes in the
    Baltic Sea. Thesis, Dept. of Geology, University of Stockholm,
    Sweden.

Jansson, B.-O. and Å. Hagström. Sea-bed Respiration: Core Samples
    for Oxygen Consumption Measurements. Proc. XI. Conf. Baltic
    Oceanographers, Vol. 2: 806-820, Rostock, April 24-27, 1978.

Lehninger, A.L. 1970. Biochemistry. Worth, New York, N.Y., 833 pp.

Lindroth. P. and K. Mopper. 1979. Subpicomole, High Performance
    Liquid Chromatographic Determination of Amino Acids by
    Precolumn Fluorescence Derivatization with o-Phthaldialdehyde.
    Anal. Chem, to be published.

Murray, J.W., V. Grundmanis and W.M. Smethie, Jr. 1978. Inter-
    stitial Water Chemistry in the Sediments of Saanich Inlet.
    Geochim. Cosmochim. Acta 42: 1011-1026.

Mäkelä, K. and L. Niemistö. 1978. Notes on the Decomposition of
    Organic Matter in Sediments. Proc. XI Conf. Baltic Oceanogra-
    phers. Vol. 2: 643-652, Rostock, April 24-27, 1978.

Smith, Jr., K.L. 1973. New York Bight Sludge Dumping Area: In
    situ measurements of benthic oxygen demand. Science.

Smith, Jr., K.L. 1974. Oxygen demand of San Diego Trough Sediments:
    an in situ study. Limnol. Oceanogr. 19: 939-944.

Svensson, T. 1979. Tracer Measurements of Mixing in the Deep Water of a small stratified sill fjord. Proc. Fjord Oceanographic Workshop, Patricia Bay, June 4-8, 1979.

REDOX DEPENDANT PROCESSES IN THE TRANSITION FROM OXIC TO ANOXIC
CONDITIONS: AN IN SITU STUDY CONCERNING REMINERALIZATION, NUTRIENT
RELEASE AND HEAVY METAL SOLUBILIZATION

W. Balzer

University of Kiel
Olshausenstrasse
Kiel, F.R.G.

INTRODUCTION

Whenever the rate of supply of oxygen in natural waters cannot
cope with its consumption rate in organic matter oxidation, redox
gradients develop and reducing conditions occur. In the marine
environment the most widespread systems exhibiting redox gradients
are the upper layers of shallow water and continental shelf sediments. Under certain conditions of morphology, density stratification and organic matter supply the redoxcline may cross over into
the overlying water column leading there to a more or less transient
anoxic milieu (see for a review: Deuser, 1975, Grashoff, 1975).
Fjords being connected to the open ocean by a shallow sill, often
meet these preconditions.

Despite the structuring effect of redox gradients on the distribution of the biota reducing environments lead to some important
consequences: e.g. organic matter of reducing sediments or sediments
below anoxic waters seem to be preserved to a greater extent than
under oxic conditions possibly depending on the reduced energy
yield for microbial decomposition of organic matter under lowered
redox conditions; of great importance is the ability of sediment
surfaces becoming reduced to release huge amounts of inorganic
nutrients being normally trapped in the bottom; by its connection
to surface layers in the Baltic this process may become self-
accelerating; as a third consequence of developing redox gradients
the migration and redistribution of heavy metals may be mentioned.

Therefore, it should be of interest to study redox dependant
processes at the sediment/water-interface under in situ conditions

659

simulating redox turnovers by trapping a water mass in a closed
system.

AREA OF INVESTIGATION AND EXPERIMENTAL SYSTEM

    The investigations have been carried out at a water depth of
20 m in a restricted area of the Kiel Bight, Western Baltic (Fig. 1).
To simulate the transition from oxic to anoxic conditions an in situ
bell jar system enclosing 2094 ℓ seawater over 3.14 m^2 sediment
has been applied (Balzer, 1978). The processes were followed by
automatic probes for Eh, pH, temperature, and oxygen. The water in
the bell jar was homogenized by a remote controlled stirrer. Samples
were taken by SCUBA-divers and analyzed by convenient methods.

RESULTS AND DISCUSSION

    In the following results are summarized from several in situ
experiments having been conducted during different seasons of the
year, i.e. at different availabilities of freshly deposited organic
matter and different initial conditions of oxygen saturation.

Redox state during consumption of different oxidants

    The environmental changes accompanying the decomposition of
organic matter during consecutive steps of oxygen consumption,
denitrification and sulfate reduction are depicted schematically in
Fig. 2. Starting from saturation, oxygen is consumed at a slowly
decreasing rate in the range between 100 and 300 ml $m^{-2}d^{-1}$ depending
on the season and duration of the experiment (Balzer, 1978).
Even the initial consumption rates measured here all lie at the
lower end of the range reported in the literature leading thus to
the conclusion that the rates measured during short term experiments
over-estimate the biological activity probably as a result of
mechanical disturbance during initial set up and lack of time
allowed to the system for re-establishment of "normal environmental
conditions".

    The initial redox potential of about 400 mV, a value being
observed usually in well aerated clean waters (Cooper, 1937), rose
to potentials above 600 mV (Fig. 2) after a time delay of some days
that may be attributed to establishment of certain steady-state
conditions on the electrode surface. Despite the conceptual
limitations (Stumm, 1966) the measured potentials may be compared
with thermodynamic predictions. Even the high potential lies well
below the expected value for the O_2/H_2O systems but it corresponds
to the value calculated for the system O_2/H_2O_2 if certain assumptions

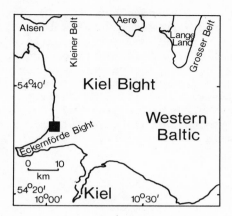

Figure 1. Area of investigation: Kiel Bight, Western Baltic.

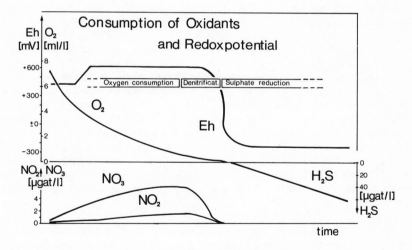

Figure 2. Redox potential (Eh) during consecutive steps of oxygen
          consumption, denitrification and sulfate reduction in
          the bell jar system.

are made about H_2O_2 concentration (Breck, 1974). From this view the
initial delay before achieving the long term oxic potential would
have been necessary to establish appropriate steady-state concentrations of H_2O_2 at the electrode surface. With the onset of
denitrification the redox potential starts to decline. Due to low
amounts of available nitrate denitrification is short intermediate
period leading finally to sulfate reduction and redox potentials
between -230 mV and -300 mV. These measurements are in good agreement with the value predicted for the systems SO_4^{2-}/HS^- or SO_4^{2-}/S^o
from thermodynamic considerations. Intermittent reaeration of the
system caused an instantaneous response of the redox electrode
and yielded reasonable potentials thus ruling out objections of
long lasting poisoning of electrodes.

Remineralization and nutrient release.

        The nutrients released from decomposing organic matter or
dissolved from the frustules of diatoms, respectively, exhibit quite
different behaviour during transition from oxic to anoxic conditions
(Fig. 3).

        Silicon is dissolved at a rate of 1250 $\mu gat \cdot m^{-2} \cdot d^{-1}$ averaged
over the year being independent of the redox transition; this is not
self-evident since thin organic coatings and iron compounds enriched
at the surface (Lewin, 1961) may impede dissolution under oxic
conditions.

        In contrast to ammonia which was released at a slowly increasing
rate during oxic conditions phosphate can be fixed to the sediments
under conditions of high oxygen saturation thus confirming the
results of Pomeroy et al. (1965), that sediments can act as a phosphate buffer in both ways. Under reduced oxygen saturation phosphate
is released at an annual mean rate of 55 $\mu gat \cdot m^{-2} \cdot d^{-1}$. Depending
on the abrupt redox decline phosphate afterwards is released at
a 12 times higher rate that tended to even increase. The accelerated release could be attributed to reduction of iron-hydroxo-
phosphates via correlation of iron and phosphate: a pH dependance,
which would suggest involvement of calcium phosphates could not be
found in this area.

        Among the different inorganic compounds of nitrogen ammonia
represents the most abundant released form at all environmental
conditions. This suggests ammonification to be a faster process
than nitrification. Increased ammonia release rates during anoxic
conditions are mainly due to higher availability of organic matter
arising from animals dying in the H_2S-milieu.

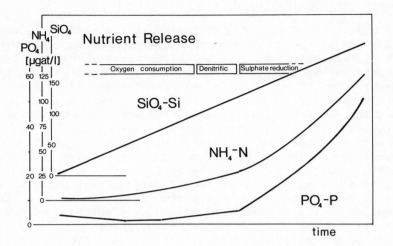

Figure 3.   Release of silicate, ammonia and phosphate from the sediment during the redox turnover.

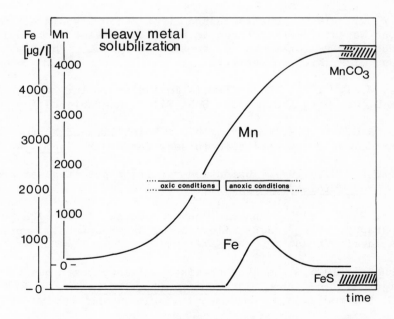

Figure 4.   Iron and manganese solubilization at the sediment/water interface during the redox turnover.

Heavy metal solubilization

The behaviour of iron and manganese at the sediment/water interface during the transition to anoxic conditions is depicted in Fig. 4. Solubility control by solid phases is exerted for iron during oxic and anoxic conditions. Oxic concentrations, until the abrupt redox change were in the range predicted by Byrne and Kester (1976); whereas anoxic concentrations were higher than expected for FeS solubility control, but seemed to be limited by a mineral phase. Manganese concentration is controlled by $MnCO_3$ in the anoxic milieu, but exhibited quite unusual behaviour during oxic conditions. The release of manganese at an increasing rate during high and constant Eh can be attributed to a metastable passing of divalent Mn from reduced layers through the oxic sedimentary microlayer into the oxic bottom water due to slow kinetics of oxidation to the thermodynamically stable tetravelent manganese (Balzer, in prep.). This is an important process in the diagenetic formation of manganese concretions because conditions of oxide formation are not necessarily adjacent to reducing environments thus preserving those oxides from redissolution to some extent.

REFERENCES

Balzer, W., 1978. Untersuchungen uber Abbau organischer Materie und Nahrstoff-Freisetzung am Boden der Kieler Bucht beim Ubergang vom oxischen zum anoxischen Milieu. PhD-Thesis, pp. 129, Kiel University.

Breck, W.G., 1974. Redox levels in the sea. In: The Sea, Vol. 5 E.D. Goldberg (ed.), pp. 895, Wiley-Interscience.

Byrne, R.H. and D.R. Kester, 1976. Solubility of hydrous ferric oxide and iron speciation in seawater. Mar. Chem., 4 255-274.

Cooper, L.H.N., 1937. Oxidation-reduction potential of sea-water. J. Mar. Biol. Ass. U.K., 22, 167-176.

Deuser, W.G., 1975. Reducing environments. In Chemical Oceanography, Vol. 3, J.P. Riley and G. Skirrow (eds.), pp. 564, Academic Press.

Grasshoff, 1975. The hydrochemistry of landlocked basins and fjords. In: Chemical Oceanography, Vol. 2, J.P. Riley and G. Skirrow (eds.), pp. 646, Academic Press.

Lewin, J.C., 1961. The dissolution of silica of diatom walls. Geochim. Cosmochim. Acta, 21, 182-198.

Pomeroy, L.R., E.E. Smith, C.M. Grant, 1965. The exchange of
    phosphate between estuarine water and sediments. Limnol.
    Oceanogr., 2, 167-172.

Stumm, W., 1966. Redox potential as an environmental parameter:
    conceptual significance and operational limitation. Proc.
    Third Intern. Water Poll. Res. Conf., 3.

REACTION AND FLUX OF MANGANESE WITHIN THE OXIC SEDIMENT AND BASIN WATER OF AN ALASKAN FJORD[1]

T.L. Owens, D.C. Burrell, and H.V. Weiss[2]

Institute of Marine Science
University of Alaska
Fairbanks, Alaska 99701, U.S.A.

Manganese occurs in two oxidation states – one insoluble, one mobile – over the redox potential range found in near surface estuarine sediments. This geochemical character leads to partial remobilization of deposited manganese in the sediments below the redox boundary and a flux into the oxic marine environment where oxidation and reprecipitation occur. Such closed sub-cycle behaviour is of considerable interest; not least because of potential control on the distribution of many other trace elements.

Using neutron activation analysis, we have determined the distribution of soluble and particulate manganese within the basin and near surface sediments of an Alaskan fjord (Table I, Figs. 1 and 2). Resurrection Bay has a basin (station depth 287 m) sep-arated from the Gulf of Alaska shelf by a sill at approximately 185 m. This basin is flushed annually during the summer-fall but, because of seasonal circulation patterns developed in the Gulf, thereafter remains effectively isolated (Heggie et al., 1977). The data of Table I was collected near the end of the oceanographic winter period (March). At this time, Heggie and Burrell (1979) have shown that the distribution (C) of a non-conservative species within the basin may be closely approximated by a vertical diffusion-reation model:

$$- \frac{dC}{dt} = \frac{\delta}{\delta z}\left(K_z \frac{\delta C}{\delta z}\right) + J \tag{1}$$

---

[1] Institute of Marine Science, Contribution No. 375
[2] Naval Ocean Systems Center, San Diego, California 92152, U.S.A.

Table 1

Manganese concentrations in basin water column and sediment,
Resurrection Bay, Station RES 2.5, March

| Depth[a] | Water Column | | Sediment | | | |
	Dissolved $\mu M \times 10^{-2}$	Particulate $\mu M \times 10^{-2}$	Pore Water Depth[b]	μM	Sediment Depth[c]	mg/g
3	5.04	10.60	2	65.9	0	2.92
8	2.62	9.37	0	288.0	7.5	0.92
18	–	5.66	– 2	922.4	27.5	0.94
28	1.64	3.82	– 4	194.0	52.5	0.86
38	1.57	3.39	– 6	121.6	97.5	0.95
58	1.79	2.29	– 8	107.0		
78	1.20	1.78	–10	82.6		
98	1.04	1.62	–12	96.3		
			–14	63.0		

a = Positive upwards from interface in metres
b = Positive upwards from reaction zone in centimetres
c = Positive down from interface in centimetres

From profiles of a number of chemical species in the sediment
pore waters at this locality, it appears that advective mixing of
the surface sediments is undetectable and that equation (1) may
generally apply to this environment also.

In the case of dissolved manganese, the major reaction (J) term
is oxidation. Following Hemm (1964) and Morgan (1967), we may define
an apparent first order oxidation rate constant k:

$$- \frac{d[Mn(II)]}{dt} = k [Mn(II)] \qquad (2)$$

We assume that oxidized manganese sorbs onto indigenous particles
in the basin column (*via* reactions that are rapid compared with the
oxidation kinetics) and that subsequent transport is largely due to
gravitational settling (W_s). Then the distribution of particulate
manganese (P) shown in Figure 1 is given by:

$$\frac{dP}{dt} = -W_s \frac{\delta P}{\delta Z} + k_w C_w \qquad (3)$$

and the concentrations of dissolved water column and pore water Mn
(II) (C_w and C_s; Figs. 1 and 2) may be modeled by:

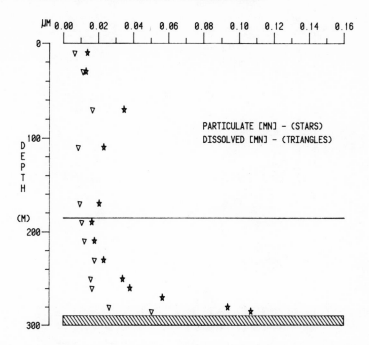

Figure 1.   Data point profiles showing exponential increases with
            depth in both particulate and dissolved manganese in the
            water column (Resurrection Bay, March).

$$\frac{dC_w}{dt} = \frac{\delta}{\delta Z}\left(K_z\frac{\delta C_w}{\delta Z}\right) - k_w C_w \qquad (4)$$

$$\frac{dC_s}{dt} = \frac{\delta}{\delta Z}\left(\phi D\frac{\delta C_s}{\delta Z}\right) - k_s C_s \qquad (5)$$

The distribution of manganese at the time of year sampled is
expected to approximate steady state. Assuming also constant turbulent (K_z) and effective ionic (D; constant porosity ϕ) ionic
diffusion coefficients to the interface. The solution to equations
(3) – (5) for particulate, dissolved and pore-water manganese
respectively are:

$$P-P_\infty = (k_w K_z)^{\frac{1}{2}}\left((C_o-C_\infty)/W_s\right)\ \exp\{-(k_w/K_z)^{\frac{1}{2}}\cdot Z\} \qquad (6)$$

$$C_w-C_\infty = (C_o-C_\infty)\ \exp\{-(k_w/k_z)^{\frac{1}{2}}\cdot Z\} \qquad (7)$$

$$C_s-C_{s\infty} = (C_{so}-C_{s\infty})\ \exp\{-(k_s/D)^{\frac{1}{2}}\cdot Z\} \qquad (8)$$

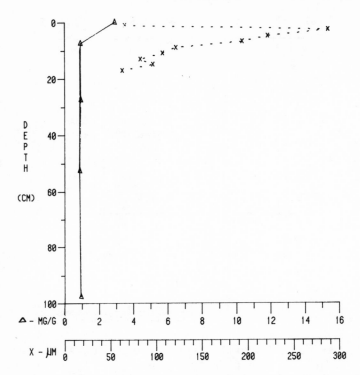

Figure 2.   Data point profiles showing surface enrichment of solid phase manganese and negative gradient of interstitial water manganese in the sediment (Resurrection Bay, March).

The interfacial concentration (C_o) of dissolved Mn(II) may be determined *via* a reiterative best fit (r = 0.895) of equation (7) to the data of Table I, setting C_∞ equal to the mean soluble concentration of manganese above sill height (1.04×10^{-2} µM) and using $K_Z = 3.5$ cm^{2}sec^{-1} (Heggie et al. 1977). We obtain $C_o = 5.71 \times 10^{-2}$ µM and an oxidation rate constant for the column (k_w) of 1.12×10^{-6} sec^{-1}. Applying the same reiterative fit technique (r = 0.995), and substituting computed values for C_o and k_w, allows solution of equation (6): $P_\infty = 1.60$ µM and $W_s = 1.03 \times 10^{-3}$ cm sec^{-1}. This latter corresponds to particulate manganese settling *via* particles of 20 µM mean diameter.

Similar treatment of the distribution of manganese in the oxic sediment zone using equation (8) is a considerably less satisfactory approach. Table I shows two relevant data points only, each of which is the mean of an approximately 2 cm sediment slice. Assumption of

Table II

Modeling of Pore Water manganese in the Oxic Sediment Zone
(Z positive up from reaction zone; see Table I)

Equation (see text)	$C_{s\infty} \times 10^{-2}$ (μM)	C_S (μM)			$k_S \times 10^{-6}$ (sec^{-1})	$D \times 10^{-6}$ ($cm^2 sec^{-1}$)	$F \times 10^{-7}$ (μ mole cm^{-2} sec^{-1})
		z = 0	z = 2cm	z = 3cm			
8	1.04	288	65.9	31.5*	0.82*	1.5[a]	2.84*
8	1.04	288	65.9	31.5*	1.1 *	2.1[b]	4.33*
8	1.04	288	65.9	31.5*	1.6 *	3.1[b]	5.71*
8	1.04	288	65.9	31.5*	2.1 *	3.8[c]	7.25*
9&10	1.04	288	0.99*	0.057	1.12*	0.14*	1.14

* Values computed from model

[a] Bender (1971)

[b] Elderfield et al. (unpublished MS)

[c] Li and Gregory (1974)

the steady state mode permits a second approach: C_s must equal C_o (5.71 × 10^{-3} μM) at the interface and the benthic flux of Mn(II) equals the rate of oxidation in the column. The latter relationships may be expressed as:

$$F_w = k_w \int C_w \, dZ \tag{9}$$

$$F_s = -D \frac{dCs}{dt} = (C_{so} - C_{s\infty})(k_s D)^{\frac{1}{2}} \cdot \exp\{-(k_s/D)^{\frac{1}{2}} \cdot Z\} \tag{10}$$

Values computed for the constants and relevant boundary conditions using these two approaches are summarized in Table II.

Application of the combined fjord basin sediment-water column model has yielded an apparent first order oxidation rate in both sediments and water in the order of 1×10^{-6} sec^{-1}: this gives a mean lifetime of Mn(II) under these conditions of around 10.3 days. The mean settling velocity of the particulate manganese phase is approximately 1×10^{-3} cm sec^{-1}. If the flux of remobilized manganese into the basin is taken as 1.14×10^{-7} μ moles cm^{-2} sec^{-1}, then the standing stock of "excess" manganese (ie., manganese transported from the sediment) may be computed at 0.27 μ moles cm^{-2}. Compared with a "natural" standing stock estimated at 0.76 μ moles cm^{-2}, the recycled fraction constitutes some 25% of the total water column manganese. These are lower limit estimates resulting from use of constant diffusion coefficients to the interface. Applying literature values of D (Table II) to the pore water data of Table I results in 50-90% of the benthic flux being oxidized in the zone between the interface and some 10 cm above. The major need now is for precise, very close interval sampling across the benthic boundary.

REFERENCES

Bender, M.L. 1971. Does upward diffusion supply the excess manganese in pelagic sediment? J. Geophys. Res. 76: 4212.

Heggie, D.T., D.W. Boisseau and D.C. Burrell. 1977. Hydrography nutrient chemistry and primary productivity of Resurrection Bay, Alaska 1972-75. IMS Report R77-2, Inst. Mar. Sci., Univ. of Alaska. 111 pp.

Heggie, D.T. and D.C. Burrell. 1979. Depth distributions of copper in the water column and interstitial waters of an Alaskan Fjord. In K.A. Fanning and F.T. Manhein (eds.), The Dynamic Environment of the Ocean Floor. D.C. Heath and Co., Lexington. In Press.

Hemm, J.D. 1964. Deposition and solution of manganese oxides. Geol. Surv. Water Supply Paper 1667-B.

Li, Y.H. and S. Gregory. 1974. Diffusion of ions in sea-water and in deep-sea sediments. Geochim. Cosmochim. Acta 38: 703-714.

Morgan, J.J. 1967. Chemical equilibria and kinetic properties of manganese in natural waters. In: Principles and Applications of Water Chemistry. Forest and Hunter eds.

SEDIMENT-SEAWATER EXCHANGES OF NUTRIENTS AND TRANSITION METALS

IN AN ALASKAN FJORD

David T. Heggie* and David C. Burrell

Institute of Marine Science
University of Alaska
Fairbanks, Alaska 99701

INTRODUCTION

Resurrection Bay (southcentral Alaska) is a single silled fjord-estuary having a basin depth of 290 m (at Station RES 2.5) and a sill at approximately 185 m. The inner basin waters are renewed annually during May–October, but during the oceanographic winter period the basin waters remain relatively isolated from a downwelling flushing of bottom waters that occurs in the outer fjord as a result of a coastal convergence. The seasonal hydrography and dissolved oxygen distributions within this basin have been described by Heggie *et al.* (1977).

Figures 1 and 2 show pore water profiles for NO_3-N, NH_3-N, SiO_4-Si, PO_4-P, titration alkalinity and solid phase and soluble concentrations of the transition metals Fe, Mn, and Cu within the surface sediments of the fjord basin (Station RES 2.5; May 1975). There is no evidence for significant advective transport in these sediments so that the profile inflexions indicate zones of chemical reaction and transport of species in the pore water is predominantly *via* ionic diffusion. Table 1 gives species concentrations at the base of the water column and for interstitial water extracted from approximately the upper 5.5 cm of the sediment. The computed fluxes across the sediment-seawater interface are for the ionic diffusion coefficients (D_a) assuming that the bottom water concentrations listed approximate those at the interface. In view of the large concentration differences involved this assumption is not believed to be a major source of error.

*Present address: Graduate School of Oceanography
University of Rhode Island, Narrangansett, R.I. 02882, U.S.A.

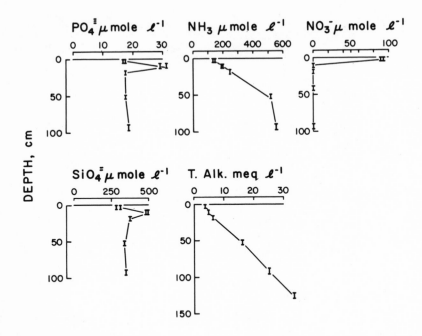

Figure 1.   Resurrection Bay pore water nutrient data May 1975.

Table 1.   Benthic boundary species concentrations and fluxes

Species	C (μ mole ℓ⁻¹) B.W.	I.W.	*D_a × 10⁻⁶ (cm² sec⁻¹)	F (μ mole cm⁻² yr⁻¹)
NO_3-N	36.5	91.7	11.6	7.0
NH_3-N	< 1.0	137.5	11.6	17.3
PO_4 -P	2.6	17.2	8.0	1.3
SiO_4-Si	54.8	295.5	3.0	7.9
Mn	0.06	304.9	3.8	12.6
Fe	0.01	1.7	4.2	0.08
Cu	0.03	0.16	4.2	0.006

B.W. = bottom water
I. W. = surface (∿ 0 - 5 cm) interstitial water
* Diffusion coefficients from Li and Gregory (1974) interpolated
  to 5°C.

Maximum nitrate (plus nitrite) concentrations over this period
within the oxygenated sediment layer were in excess of 90 μ mole ℓ⁻¹

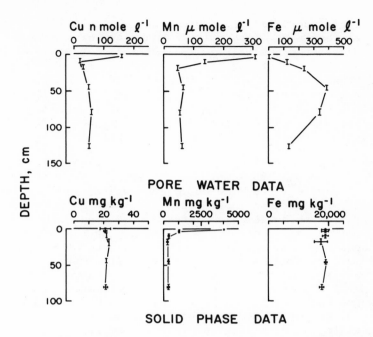

PORE WATER DATA

SOLID PHASE DATA

Figure 2.   Resurrection Bay pore water and solid phase transition
metal data, May 1975.

see Figure 1. In the absence of regeneration, surface porewater
concentrations should approximate those measured in bottom waters
following renewal periods ($\sim$ 30 μ mole ℓ^{-1}). Nitrate produced by
aerobic decomposition of biogenic material (C:N::106:15) can raise
surface sediment porewaters only up to approximately 50 μ mole NO_3-N
ℓ^{-1}. Since the measured nitrate concentration of the surface interstitial water cannot be accounted for by standard regeneration
stoichiometry it is suggested that the "excess" nitrate is due to
nitrification at the boundary. Added evidence for such an ammonia
oxidation hypothesis is supplied by the indicated gradients and
computed benthic flux of this species. The ammonia concentration
difference across the benthic boundary is in excess of two orders
of magnitude (Table 1; Fig. 1) and during 1973-75 ammonia concentrations at the bottom of the fjord basin at no time exceeded 1 μ
mole ℓ^{-1} (although values > 2 μ mole ℓ^{-1} were found 25 m above the
boundary in January and April 1975; Heggie *et al.*, 1977). The
computed benthic flux of 17.7 μ mole cm^{-2} yr^{-1} is capable of supporting a minimum gradient in ammonia of 1.6×10^{-7} μ mole cm^{-4}
(from a mean turbulent diffusion coefficient of 3.5 cm^2 sec^{-1} computed for sill height). This would be equivalent to a concentration
difference between the sediment interface and the top of the basin

at sill height (approximately 100 m) of 1.6 μ mole ℓ^{-1}; this latter
has not been observed. Interstitial water ammonia increases linearly
with depth to around 50 cm (Fig. 1). Diffusive transport along this
gradient was computed at not less than 2.7 μ mole cm^{-2} yr^{-1}; a flux
which, at steady state, would be insufficient to maintain the est-
'imated benthic boundary flux (17.7 μ mole cm^{-2} yr^{-1}) discussed
above. It is concluded that these several observations require
additional production of ammonia at the interface, part of which
is oxidized to nitrate.

Dissolved oxygen concentrations in basin waters during 1973-75
decreased from around 180 μ mole ℓ^{-1} afterreplacement (October) to
less than 50 μ mole ℓ^{-1} by the end of the winter (May). Mean annual
consumption rates for the inner basin have been estimated at 165 μ
mole ℓ^{-1} yr^{-1}, with consumption rates some 5 m above the sediment
at approximately 250 μ mole ℓ^{-1} yr^{-1}. Assuming that interfacial
oxygen concentrations approximate those measured at the base of the
water column, the minimum (molecular) diffusive flux of oxygen into
the sediment is calculated to vary seasonally between 22 and 5 μ
mole cm^{-2} yr^{-1}. A mass balance for NO_3-N estimates that 50 percent
of the oxygen consumed in the bottom water (5 m above the sediments)
during the winter of 1974-75 may be due to ammonia oxidation. Annual
net productivity over this period in Resurrection Bay has been
estimated at 19 mole C m^{-2} yr^{-1} (Heggie et al.,1977) some 30 percent of which ($\sim$ 6 mole C m^{-2} yr^{-1}) may be accounted for using conventional organic matter oxidation stoichiometry by oxidation in
deep and bottom waters, with less than an additional 1 percent
oxidized in the surface oxic sediments.

Maximum pore water concentrations of manganese (305 μ mole kg^{-1})
were found in the top 5.5 cm of sediments, decreasing to 50 μ mole
kg^{-1} at 14-22 cm depth, and thereafter remain relatively constant
to the maximum depth sampled (130 cm; Fig. 2). Maximum solid phase
concentrations (Fig. 2) occurred in the top 0.5 cm of sediment.
These distributions are attributable to diagenetic reduction and
mobilization below the redox boundary, with upward diffusion and
reprecipitation in the oxic zone. Pore water contents of iron increase with depth to a maximum of around 390 μ mole kg^{-1} over the
42-47 cm depth zone (Fig. 2). From this surface sediment gradient,
an ionic diffusion flux of approximately 1.8 μ mole cm^{-2} yr^{-1} was
computed, considerably larger than the estimated flux of 0.08 μ
mole cm^{-2} yr^{-1} across the benthic boundary (Table 1). It would
appear that remobilized reduced iron is efficiently trapped in the
surface sediments and that, unlike manganese in this environment,
little escapes into the water column.

The depth distributions of the redox species O_2, NO_3-N, Mn and
Fe are consistent with a model of organic matter degradation in the
sediments by oxidants in order of free energies made available to
the organism. These oxidants are consumed in the top 5.5 cm of this

Table II.   Sediment - seawater flux of Mn, Fe, Cu normalized to
            apparent equilibrium contents of solid phases

Element	Flux (F) $mg\ cm^{-2}\ yr^{-1}$	Equil. Solid phase conc. (C) $mg\ cm^{-2}$	C/F yrs
Mn	0.69	3.3	4.7
Cu	4×10^{-4}	0.2	500
Fe	4.5×10^{-3}	190	42,222

* Sediment density $2 \times 10^{-3}\ kg\ cm^{-3}$; 5 cm sediment depth.

Table III. Benthic flux, standing crop estimates and doubling
            time. Resurrection Bay 1974-75.

Species	Annual Mean Water column concs. $\mu\ mole\ \ell^{-1}$	Annual Mean Standing crop (C) $\mu\ mole\ cm^{-2}$	Flux (F) $\mu\ mole\ cm^{-2}\ yr^{-1}$	Doubling time (C/F) yrs
$NO_3 + NH_3$	20.	600	24.	25
$SiO_4^=$	36.7	1100	7.9	139
$PO_4^=$	1.9	56	1.3	43
Mn	0.013	0.37	12.6	0.03
Cu	0.008	0.23	0.006	38

fjord basin. The depth distribution of copper in Resurrection Bay
sediments is shown in Figure 2. Heggie and Burrell (1979) have
presented a model which incorporates remobilization in the surface
sediments with a flux into the water column and a removal reaction
in sub-surface sediments.

    A relative measure of the remobility of these transition metals
is given by normalizing the computed sediment-seawater flux to the
apparent equilibrium solid phase concentration of each metal as
shown in Table II. The relative remobilities are Mn > Cu > Fe.
Table III summarizes the computed interface fluxes together with
approximate standing stocks in the overlying water. The sub-sill
basin water of Resurrection Bay is renewed annually (Heggie et al.,
1977): approximately 60 percent of the fjord water volume. If the
time required to double the concentration of a species in the water
column via input across the benthic boundary were of the same order
as the residence time of the water then this transport could be an
important component of the estuarine geochemical balance: this
appears to be the case only for manganese in this environment. It
is inferred that, for the other species considered here, riverine
influx and regeneration in the water column are quantitatively more
important processes.

This paper is Contribution No. 374 of the Institute of Marine Sciences, University of Alaska.

REFERENCES

Heggie, D. T. and D. C. Burrell 1979: Depth distributions of copper in the water column and interstitial waters of an Alaskan fjord. In K. A. Fanning and F. T. Manhein (eds.), The Dynamic Environment of the Ocean Floor. D. C. Heath and Co., Lexington. In press.

Heggie, D. T., D. W. Boisseau and D. C. Burrell 1977: Hydrography, nutrient chemistry and primary productivity of Resurrection Bay, Alaska 1972-75.   IMS Report R77-2, Inst. Mar. Sci., Univ. of Alaska. 111 pp.

Li, Y. H. and S. Gregory 1974: Diffusion of ions in sea-water and in deep-sea sediments. Geochim. Cosmochim. Acta 38, 703-714.

REDOX SPECIES IN A REDUCING FJORD: THE OXIDATION RATE OF

MANGANESE II

Steven Emerson

Department of Oceanography
University of Washington
Seattle, Wa. 98195, U.S.A.

INTRODUCTION

In June 1977 ten redox sensitive chemical species were analyzed across the O_2/H_2S interface of the stagnant basin in Saanich Inlet. In this report I present the results for O_2, H_2S, Mn II and Fe II and calculate the oxidation rate for Mn II. Further details of the study are published elsewhere (Emerson et $al.$, 1979).

Saanich Inlet is an intermittently anoxic fjord located on Vancouver Island, British Columbia. The seasonal anoxia cycle has been described by Richards (1965) and is shown in Figure 1. Although the characteristics of this cycle differ from year to year, the general trend is for the water behind the sill to be isolated for about 6 months during winter and summer, causing oxygen depletion and the build up of H_2S, and for flushing to occur in late summer or early fall. The formation of a manganese-rich particulate layer just above the O_2/H_2S boundary during periods of anoxia has been described by Kiefer (1973). The purpose of this study is to invest-igate the distribution of dissolved manganese across this boundary and to estimate the oxidation rate based on the manganese distri-bution and previous knowledge of the mixing characteristics of the basin.

METHODS

The data presented here are from R/V T. G. $Thompson$ cruise TT-119 in July of 1977. The station from which the data were col-lected is at the maximum depth of the inlet, 220 m (48^0 33.1'N, 123^0 32.7'W). Samples were collected using 5-liter P.V.C. bottles

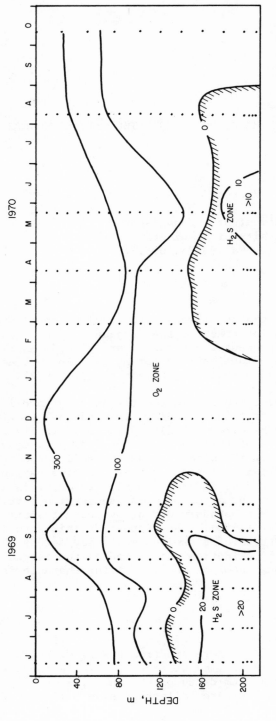

Figure 1.   The oxygen and hydrogen sulfide distribution in Saanich Inlet in previous years (Taken from Kieffer, 1973).

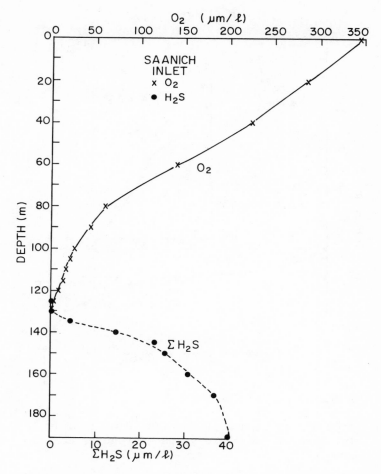

Figure 2.   Oxygen and hydrogen sulfide ($\Sigma H_2S = H_2S + HS^- + S^2$)
distribution in Saanich Inlet waters in July 1977.
The oxygen/hydrogen sulfide boundary is at 130 meters.

on a stainless steel hydrowire and by a pumping system (Anderson
and Richards, 1977).   The latter was used to collect closely spaced
samples at the O_2/H_2S interface.   Oxygen and hydrogen sulfide were
measured by the methods of Carpenter (1965) and Broenkow and Cline
(1969), for O_2, and Cline (1969) for H_2S.   Manganese and iron were
measured spectrophotometrically by the methods of Brewer and Spencer
(1971) and Stookey (1970).

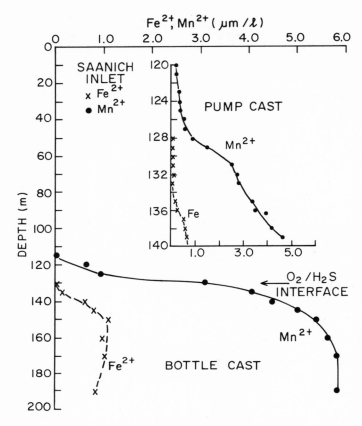

Figure 3.   Iron and manganese in Saanich Inlet in July 1977.   The
inset shows the closely spaced results derived from the
pump cast.

RESULTS

    The results are shown in Figures 2 and 3.   The O_2/H_2S inter-
face is clearly represented at a depth of about 130 m.   There is
virtually no measurable iron II (detectability = 50 μm/ℓ) in waters
which also have oxygen.   This is not the case, however, for manganese II which coexists at levels of nearly 2 μm/ℓ with oxygen.   The
difference in the behaviour of these species is a result of differing
oxidation rates.   In the following I pursue this difference by estimating the oxidation rate of Mn II based on the distribution in
Figure 3 and a steady-state model.

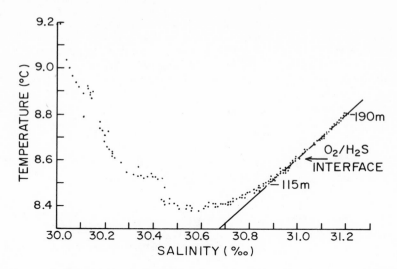

Figure 4.   Temperature-salinity relationship for Saanich Inlet in
             July 1977. The points are pump cast results.

DISCUSSION

        The temperature-salinity characteristics of the water at our
station in Saanich Inlet (Figure 4) shows that the depth range from
∿115 m to the bottom is a linear T-S region. Since a linear T-S
plot implies end member mixing and no advection, we will interpret
the manganese profile in this way. The distribution of manganese
in the oxic portion of the water column below 115 m is described by

$$\frac{\partial Mn}{\partial t} \; = \; K \, \frac{\partial^2 Mn}{\partial z^2} \; - \; R \qquad\qquad (1)$$

where K is a vertical "eddy diffusion coefficient" and R is an
oxidation rate term. Since Saanich Inlet never achieves a true
steady-state with respect to the redox sensitive chemical constituents, it is valid to assume a steady-state only if the residence
time of the species being considered is much shorter than the anoxic
cycle of the basin (Figure 1). We shall see shortly that this is
the case for Mn II, justifying the approximation that $\partial Mn/\partial t = 0$
in equation 1. We further assume that the oxidation rate of Mn II
is first order with respect to Mn II (R = kMn). Morgan (1964) has
reviewed the oxidation kinetics of manganese II and shows that the
rate is autocatalytic and depends on the particulate manganese concentration as well as oxygen and OH^- . The oxidation rate constant
which we derive is thus a "pseudo" first order constant in that we

assume the particulate manganese, OH^- and oxygen to be constant over the oxidation depth interval. This assumption deviates from reality only with respect to oxygen which varies from zero to about 10 $\mu m/\ell$ in this region. In the following treatment we derive an average oxidation rate for the 130-120 m depth region. Equation (1) can now be written as

$$0 = K \frac{d^2 Mn}{dz^2} - k\ Mn \qquad (2)$$

With the boundary conditions: $z = 0$, $Mn = Mn_0$ ($z = 0$ is at 130 m and z is positive upwards), and $z = \infty$, $Mn = 0$; equation (2) has the solution

$$Mn(z) = Mn_0 exp\{-z\sqrt{k/K}\} \qquad (3)$$

By fitting this equation to the data we derive a value for $\sqrt{k/K}$ of $0.0024 - .0032$ cm^{-1}.

The eddy diffusion coefficient at the oxygen-hydrogen sulfide interface in Saanich Inlet has been derived in previous studies by salinity and nutrient balances (Broenkow, 1969; Devol, 1975). The range of values for K from these investigations is $0.4 - 0.7$ cm^2/sec. In addition, Emerson *et al.* (1979) has shown that these values are consistent with the iodide distribution during the period of our sampling. We, thus, use 0.5 ± 0.2 as the value of the vertical eddy diffusion coefficient K. Using this value and the result of the solution to equation 3, we derive a rate constant for oxidation of $0.15 - 0.62$ day^{-1} or a residence time with respect to oxidation of $1 - 7$ days.

This value is many orders of magnitude greater than proposed for laboratory inorganic Mn II oxidation (Brewer, 1975) and Mn II oxidation at the Black Sea O_2-H_2S interface (Murray and Brewer, 1977). The reason for these differences is probably the different influence of catalysis by manganese oxidizing bacteria in these three examples. A chemical analysis of the particulate layer in Saanich Inlet during this study revealed that it is 30% by weight organic carbon. We take this as preliminary evidence for a rich bacterial population at the interface, some of which may be capable of oxidizing manganese. The final word on this hypothesis, however, awaits a thorough bacterial investigation of the particulate layer.

REFERENCES

Anderson, J. J. and F. A. Richards 1977: Continuous profiles of chemical properties in the euphotic zone of the DOMES study area in the Equatorial North Pacific. Department of Oceanography Special Report No. 78, University of Washington, Seattle, Washington, 97 pp.

Brewer, P. G. 1975: Minor elements in seawater. In: Chemical
    Oceanography, J. P. Riley and G. Skirrow, editors, Academic
    Press, 1, pp. 415-496.

Brewer, P. G. and D. W. Spencer 1971: Colorimetric determination
    of manganese in anoxic waters. Limnology and Oceanography,
    16, 107-112.

Broenkow, W. W. 1969: The distribution of nonconservative solutes
    related to the decomposition of organic material in anoxic
    basins. Ph.D. dissertation, University of Washington, 207 pp.

Broenkow, W. W. and J. D. Cline 1969: Colorimetric determination
    of dissolved oxygen at low concentrations. Limnology and Ocean-
    ography, 14, 450-454.

Carpenter, J. H. 1965: The Chesapeake Bay Institute technique for
    the Winkler dissolved oxygen method. Limnology and Ocean-
    ography, 10, 141-143.

Cline, J. D. 1969: Spectrophotometric determination of hydrogen
    sulfide in natural waters. Limnology and Oceanography, 14,
    454-458.

Devol, A. H. 1975: Biological oxidations in oxic and anoxic marine
    environments: rates and processes. Ph.D. dissertation,
    University of Washington, 208 pp.

Emerson, S., R. Cranston and P. S. Liss 1979: Redox species in a
    reducing fjord: equilibrium and kinetic considerations.
    Deep-Sea Research (in press).

Kieffer, G. J. 1973: On the nature of manganese in Sannich Inlet.
    M. S. thesis, University of Washington, 65 pp.

Morgan, J. J. 1964: Chemistry of aqueous manganese II and IV.
    Ph.D. dissertation, Harvard University, Cambridge, Massachusetts,
    244 pp.

Murray, J. W. and P. G. Brewer 1977: Mechanisms of removal of
    manganese, iron, and other trace metals from seawater. In:
    Marine Manganese Deposits, G. P. Glasby, editor, Elsevier,
    pp. 291-325.

Richards, F. A. 1965: Anoxic basins and fjords. In: Chemical
    Oceanography, J. P. Riley and G. Skirrow, editors, Academic
    Press, 1, pp. 611-645.

Cr SPECIES IN SAANICH AND JERVIS INLETS

R.E. Cranston

Geological Survey of Canada
Atlantic Geoscience Centre
Bedford Institute of Oceanography
Dartmouth, Nova Scotia B2Y 4A2

INTRODUCTION

Two stations in Saanich Inlet and one station in Jervis Inlet were sampled in July 1977 in order to determine the distribution of redox sensitive chemical species in the water column. Saanich Inlet becomes reducing in the bottom water during the summer while Jervis Inlet water remains oxygenated throughout the year.

Dissolved aqueous chromium exists in two oxidation states, Cr (III) and Cr(VI). In both fresh and salt water, it is predicted that Cr(VI) will be present in oxidized water as CrO_4^{2-}, while Cr(III) will be present in reduced water as $Cr(OH)_2^+$. Cr(III) complexes are known to have a rather stable nature displaying a residence time of about 30 days in oxidized water (Fukai and Vas, 1969; Schroeder and Lee, 1975). Solid phase reactions are also important in understanding the fate of Cr in natural waters. Cr(III) has a strong tendency to adsorb onto surfaces while Cr(VI) in very soluble.

METHOD

An accurate and precise method for the analysis of Cr species was developed prior to the cruise (Cranston and Murray, 1978). Cr (III) was co-precipitated with ferric hydroxide and collected on 0.4 μm Nuclepore filters. Cr(VI) was reduced to Cr(III) by ferrous hydroxide. The Cr(III) co-precipitated with the solid phase and was also collected by filtration. Particulate Cr was collected by filtering water only through 0.4 μm Nuclepore filters. On returning to the land based lab, 1 ml of 6 MHCl was added to each filter to

689

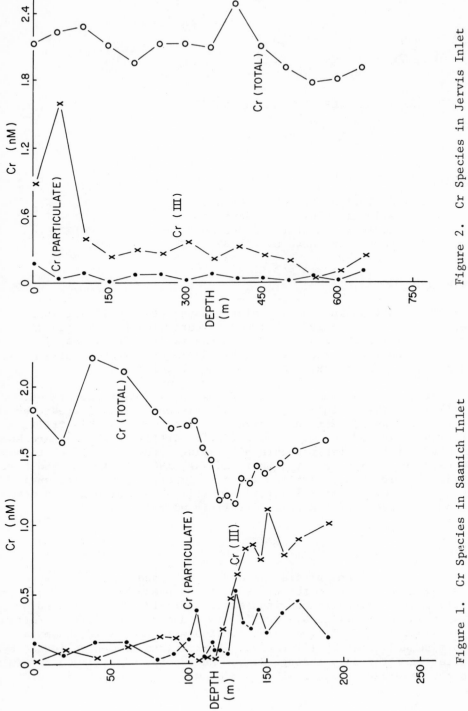

Figure 2. Cr Species in Jervis Inlet

Figure 1. Cr Species in Saanich Inlet

dissolve the bound Cr. The Cr was then determined by flameless atomic absorption analyses.

RESULTS AND DISCUSSION

In Saanich Inlet, oxygen and hydrogen sulfide profiles reach background concentrations at 130 m (Table 1). Phosphate values increased steadily with depth, while high ammonia values in the deep water were related to degradation of organic matter in the anoxic zone. In the case of Jervis Inlet, dissolved oxygen analyses indicated that the entire water column was oxygenated (oxygen concentrations >65 μM).

Thermodynamic calculations predict that Cr(III) should not be present in oxidized water. however it should be the dominant oxidation state in reduced waters. Cr(III) concentrations were found to be minimal in the oxidized waters to a depth of 115 m, then the concentrations steadily increased to 140 m, after which they were rather constant (Fig. 1). The particulate Cr profile shows low concentrations to about 125 m (except for possible contamination at 105 m). The maximum particulate concentrations occurred at 130 m and accounted for nearly 50% of the total Cr. Manganese particles are known to form at the redox discontinuity and it is suggested that Cr(III) was adsorbing onto the particles (Keifer, 1973). Cr(total) values decreased with depth, displaying a marked mid-depth minimum which coincided with the redox interface and resulted from the removal of Cr(III) by the particulate matter.

In Jervis Inlet, Cr(particulate) concentrations were minimal, while some Cr(III) was found in the upper 50 m (Fig. 2). Biological production may account for the Cr(III) as has been suggested for stations in the Pacific (Cranston, 1978). The Cr(total) profile is rather constant with depth; the average concentration being significantly greater than that observed for Saanich Inlet. Mathematical modelling can be applied to the Saanich data in order to study the kinetics of Cr(III) oxidation. If Cr(III) removal was immediate, Cr(III) would not be observed in the oxidized water. However, it was found in the zone immediately above the redox boundary. Applying a one dimensional diffusion - first order removal model, residence times for Cr(III) in the oxygenated water can be calculated (Tsunogai, 1971). In the layer which includes the redox interface (130-135 m), a residence time of 81 days was calculated for the Cr(III). In the oxygenated layer immediately above the redox boundary (125-130 m), a residence time of 56 days was obtained. In the next two 5 metre layers, the Cr(III) residence times were 13 days (120-125 m) and 1 day (115-120 m). It appears that as the oxygen concentration increases, the residence time for Cr(III) decreases, suggesting that oxidation is more important in removing the species than is adsorption.

Table 1

Saanich Inlet Analyses

Selected Depth (m)	O_2 (μM)	H_2S (μM)	PO_4^{3-} (μM)	NH_4^+ (μM)
1	344	<0.01	0.4	<0.01
80	60	<0.01	3.6	<0.01
130	0.6	0.21	5.6	4.4
190	<0.1	14.1	7.0	4.1

REFERENCES

Cranston, R.E. 1978. Dissolved chromium species in coastal waters.
    Current Research, Geol. Surv. Can. Paper 78-1A, 337-339.

Cranston, R.E. and J.W. Murray. 1978. The determination of dis-
    solved chromium species in natural waters. Anal. Chim. Acta
    99. 275-282.

Fukai, R. and D. Vas. 1969. Changes in the chemical forms of
    chromium on the standing of sea water samples. J. Ocean. Soc.
    Japan 25. 47-49.

Keifer, G.T. 1973. On the nature of manganese in Saanich Inlet.
    M.S. Thesis, University of Washington, Seattle, Wa., 65 pp.

Schroeder, D.C. and G.F. Lee. 1975. Potential transformations of
    chromium in natural waters. Water, Air and Soil Poll. 4,
    355-365.

Tsunogai, S. 1971. Iodine in the deep water of the ocean. Deep-
    sea Res. 18. 913-919.

THE CHEMISTRY OF SUSPENDED PARTICULATE MATTER FROM TWO OXYGEN

DEFICIENT NORWEGIAN FJORDS - WITH SPECIAL REFERENCE TO MANGANESE

Jens Skei

Norwegian Institute for Water Research
P.O. Box 333, Blindern, Oslo 3
Norway

SUMMARY

The study of the elemental composition of suspended particulate matter in seawater is essential to understand the interaction between sediments and water, physio-chemical reactions related to redox conditions, breakdown of organic matter and sedimentation processes in general. In fjords the amount and composition of micro particles are influenced by river inputs (terrestrial inorganic matter), biological production (plankton) and by authigenic phases precipitating near redox boundaries. The transport of these particles is governed by estuarine circulation and deep water exchange.

To demonstrate the usefulness of particulate matter studies in fjords, results from two intermittently anoxic fjords in South Norway are presented. Particulate matter was collected by pressure filtering $\sim$ 2 ℓ of water through 0.4 μm Nuclepore membrane filters. These filters have been analysed by X-rayfluoresence for several major elements, among them Al, P and Mn. Selected filters have been looked at in SEM and single particles have been analysed by EDAX. The composition of the bottom sediments at the localities of sampling is also known, as well as the rate of deposition measured by lead-210.

Bunnefjord (Fig. 1), a branch of Oslofjord, is a typical fjord basin with a 50 m deep sill at the entrance and a maximum basin depth of 154 m. The freshwater input is generally less than 10 m^3/s. Due to a restrictive deep water circulation, the bottom water occasionally turns anoxic. Normally the bottom water is flushed

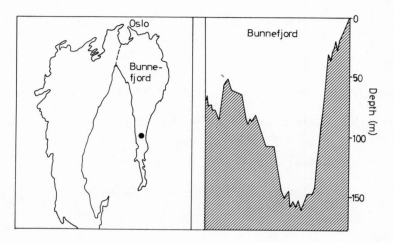

Figure 1.   Bunnefjord – area of investigation and bottom topography.
            (● = station).

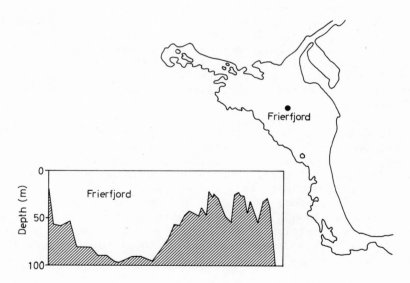

Figure 2.   Frierfjord – area of investigation and bottom topography
            (● = station).

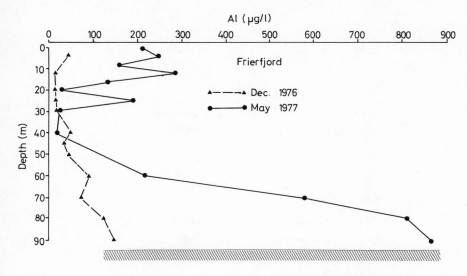

Figure 3.   Vertical distribution of particulate Al prior to (Dec.76) and after (May 77) deep water renewal in Frierfjord.

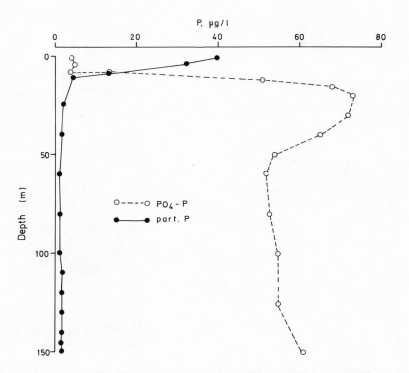

Figure 4.   Vertical distribution of particulate and dissolved P in Bunnefjord (June 77).

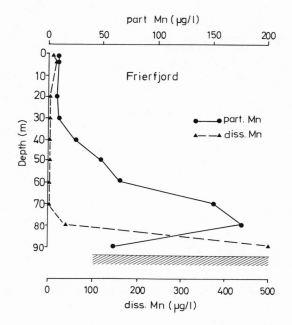

Figure 5. Vertical distribution of particulate and dissolved Mn
in Frierfjord (0.3 ml/1 O_2 at 90 m depth).

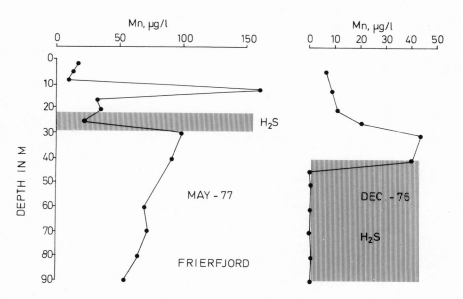

Figure 6. Depth profiles of particulate Mn in Frierfjord during
two stages of anoxic development.

every second or third year. The concentrations of oxygen drop rapidly following a deep water renewal. This implies a high rate of oxygen consumption in the bottom water.

Particulate matter was collected 8 times during 12 months from 15 depths in the middle of the deep basin of Bunnefjord. The aim of this study was to look at the seasonal variations of the particulate matter chemistry.

Frierfjord is another typical fjord basin (Fig. 2) with a 25 m deep sill and a maximum depth of 98 m. The freshwater input to Frierfjord is more than 20 times that of Bunnefjord. As a result the fjord has a very well developed halocline between 5 and 10 m and a brackish surface layer. The basin water is normally exchanged every year or every second year.

Particulate matter has been collected at a station located in the deep basin of Frierfjord during 1975-79. It was sampled during oxygenated periods, during anoxic conditions and immediately following a bottom water renewal.

The results from the analysis of suspended particulate matter in Bunnefjord and Frierfjord suggest:

1. The vertical and seasonal distribution of Al is controlled by terrestrial inputs to the surface water and to a lesser degree dependent on bottom sediment resuspension. Exceptions are periods of deep bottom water renewal which tend to resuspend the sediments significantly (Fig. 3). Horizontal transport of particulate Al mainly takes place above the halocline.
2. Particulate P residing in organic matter is abundant in the euphotic layer but decreases rapidly in the underlying water due to remineralization (Fig. 4).
3. The distribution of particulate Mn in the water column is controlled by diagenetic fluxes from the bottom sediments. Dissolved Mn diffuses out of the sediments and reprecipitates in the bottom water or at intermediate depths depending on the situation of the redox boundary (Fig. 5 and 6).

It appears that some Mn precipitates as mixed Mn-Ca-carbonates in the oxygenated water. This suggests that Mn-carbonate formation is not strictly dependent on anoxicity.

FLUORESCENCE MEASUREMENTS OF HUMIC SUBSTANCES AND LIGNIN SULFONATES IN A FJORD SYSTEM

Gunnar Nyquist

Department of Analytical Chemistry
University of Göteborg
S-412 96 Göteborg, Sweden

INTRODUCTION

Fluorescence measurements of humic substances and lignin sulfonates have been performed in the Idefjord and its vicinity since 1974. The method used is described by Almgren, Josefson and Nyquist (1979).

Humic substances and lignin sulfonates are fluorescent compounds containing phenolic structural elements. Lignin sulfonates are exclusively waste products from sulfite pulp mills, during the sulfite pulping lignins are released from the wood into the sulfite liquor. The high polarity of the sulfonic acid groups makes the lignin macromolecule water-soluble. Humic substances are naturally-occurring compounds, partly originating from lignins degraded by white-rot fungi. The low-molecular weight fractions of the humic substances (fulvic acids) are water-soluble and are transported to the sea by rivers.

Lignin sulfonates can be used as a fluorescent tracing substance for determinations of the spread and dispersion of wastewater outlets from sulfite pulp mills. It should be stressed that the flow patterns of lignin sulfonates are also representative of all other compounds in the wastewater, e.g. toxic compounds and inorganic nutrient compounds of ecological significance. Furthermore, fluorometric measurements of lignin sulfonates can be useful for studies of water exchange processes. The collection of water samples and the measurements are simple and fast procedures, opening up the possibility of covering a large sea area in a short time, which may improve our understanding of rapid shifts in complex water systems. Also humic substances can be very useful when water mixing processes are studied, expecially in estuaries.

AREA INVESTIGATED

The Idefjord is located at the Swedish west coast on the border
between Sweden and Norway. The fjord is separated from the adjoining
Singlefjord by two sills. The sill depth is 8-9 metres and there are
two main basins in the fjord.

The tidal exchange of seawater from Singlefjord gives a salinity
of 10-20 $^0/_{00}$ in the top layer of Idefjord. The bottom water of
the outer basin is exchanged once or twice a year, however, the inner
basin is, as a rule, anoxic. The Idefjord is polluted at the outlet of River Tista by wastes from the town of Halden, the pollution
consists of both sewage and wastewater from a paper and pulp mill.

MEASUREMENTS INSIDE THE IDEFJORD

Oxygen

Since the water exchange in the fjord is restricted, an oxygen
deficit easily arises, especially as the fjord is supplied with a
lot of organic substances. Organic material is brought to the
fjord by two rivers and from the town of Halden. The main source
of organic substances is the pulp and paper mill at Halden. The
oxygen-consuming substances in wastewater from a pulp mill are
found primarily in the carbohydrate fraction. Lignin sulfonates
decompose slowly and will only to a minor extent cause oxygen
consumption in the water.

The top water layer is seldom saturated with oxygen, and sometimes, the hydrogen sulfide extended to the surface.

Humic Substances

The observed variations of the humic concentration were moderate, the highest humic concentrations were found in the surface
water.

Lignin Sulfonates

The concentration of lignin sulfonates in the water is dependant upon the output of wastewater from the sulfite mill at the time
of sampling. The input of river water is also of importance,
because the river water causes an outward directed surface current.
During the summer, when the freshwater input is low and the hydrographic conditions often are stable, very high surface concentrations of lignin sulfonates were sometimes measured. On occasions,
with restricted or no production in the pulp mill, a remarkably
strong and rapid decrease of the lignin sulfonate concentration
occurred. Below the halocline, the concentration of lignin sulf-

onates was relatively low. Measurements of surface water samples both along and across the Idefjord showed that the lignin sulfonates were very unevenly distributed in the fjord, expecially near the outlet point, and the concentration in water samples collected at stations close to each other often differed considerably. This gives an idea of the difficulties in collecting representative surface samples.

MEASUREMENTS OUTSIDE THE IDEFJORD

By collecting water samples along and off the coast in the vicinity of the Idefjord, the dilution of lignin sulfonates and humic substances can be followed. There is an inverse relationship between these substances and the salinity when only physical mixing occurs; that is, the concentration of humic substances and lignin sulfonates decreases when the salinity increases. Therefore, additional inputs of humic substances or lignin sulfonates can be detected if the concentrations are plotted versus the salinity.

Further data and results of the measurements in the Idefjord area are given in Josefsson and Nyquist (1976) and Nyquist (1979).

REFERENCES

Almgren, T., B. B. Josefsson and G. Nyquist 1975: A fluorescence method in studies of spent sulfite liquor and humic substances in sea water. Analytica Chimica Acta, 78, 411-422.

Josefsson, B. and G. Nyquist 1976: Fluorescence tracing the flow and dispersion of sulfite wastes in a fjord system. Ambio, 5 183-187.

Nyquist, G. 1976: Data of fluorescence measurements in the Idefjord and Koster area. Report No. XVII, Department of Analytical Chemistry, S-412 96 Göteborg, Sweden.

Nyquist, G. 1979: Investigation of some optical properties of seawater with special reference to lignin sulfonates and humic substances. Thesis, to be published.

INDEX OF GEOGRAPHIC NAMES

In the following Geographical Index detailed place names are located by general area according to the following key:

Al	Alaska	B.C.	British Columbia (Canada)
C.A.	Canadian Arctic	C.E.	Canadian East Coast
Chile	Chile	Dk.	Denmark
F.R.G.	West Germany	Gr.	Greenland
Ice.	Iceland	Nor.	Norway
N.Z.	New Zealand	P.Q.	Province de Québec (Canada)
Scot.	Scotland	Sw.	Sweden
U.S.	United States	Wa.	Washington State (U.S.A.)

INDEX OF TAXONOMIC NAMES

713